ANDERSON'S

Flora of ALASKA
and adjacent parts of Canada

J. P. ANDERSON

Jacob Peter Anderson, 1874-1953

Maxine Morgan Williams

Flora of Alaska
and Adjacent Parts of Canada

an Illustrated Descriptive Text of All Vascular Plants
Known To Occur Within the Region Covered

By J. P. ANDERSON

Integrated and Indexed at the
Anderson Herbarium, by
RICHARD W. POHL, Curator

The Iowa State University Press, Ames, Iowa, U.S.A.

PUBLISHER'S NOTE

First published in 1959, *Flora of Alaska* represented the life work of Jacob Peter Anderson. Long out-of-print and now somewhat dated in terms of what is now known of the Alaskan flora and species taxonomy, we believe, however, that the book remains a valuable resource for the study of Alaska's diverse plant life. Our goal in making this work available once again is to provide a fieldguide-sized book useful throughout Alaska for researchers, students, and others interested in the state's flora. As the original book did not provide keys and descriptions to the important willow (*Salix*) genus, we have provided an addendum (page 514) with summer and winter keys to *Salix* from Dominique M. Collett's *Willows of Interior Alaska* (U.S. Fish and Wildlife Service, 2004). Other important works to the Alaska flora include Stanley L. Welsh's *Anderson's Flora of Alaska and Adjacent Parts of Canada* (Brigham Young University Press, 1974), based on this book but with completely new keys, text, and illustrations; and Eric Hultén's *Flora of Alaska and Neighboring Territories*, published in 1968 by Stanford University Press.

STEVE CHADDE
October 2019

AN ORCHARD INNOVATIONS REPRINT EDITION
ISBN-13: 978-1-951682-10-1

Originally published in 1959 by Iowa State University Press; reprinted with additions in October 2019 as an Orchard Innovations Reprint Edition.

Printed in the United States of America

Preface

Throughout most of his adult life, Jacob Peter Anderson devoted his major energies to the study of the Alaskan flora. His ultimate objective was to provide a usable Flora of Alaska, The Yukon Territory and the Northwest Territories west of Hudson's Bay. During the years 1942–53, while he was resident at Iowa State, he published nine fascicles of a preliminary version of this Flora, covering only Alaska, the Yukon Basin, and northwestern British Columbia. At the time of his death, Dr. Anderson was working on a revision of his Flora, to include the territory between Alaska and Hudson's Bay.

Statehood, increasing settlement, and military and naval operations have resulted in increasing interest in Alaskan natural resources, including the flora. Since Dr. Anderson's death in 1953, the continued and insistent demand for copies of his articles on the Alaskan flora has demonstrated the need for reprinting his work, even in the preliminary form in which it existed at the time of his passing. We have therefore decided to reissue the nine fascicles in a single volume, with an index to the taxa included in the work.

With the exception of correction of a few obvious typographical errors, no attempt has been made to alter the original publication, and it is reprinted here by offset from the original. Integrated pagination has been provided. No treatment of the genus *Salix* was included in the original papers, but all other major taxa of vascular plants are covered.

Publication of this edition has been made possible by the financial assistance of The Iowa State Journal of Science. Grateful acknowledgement is tendered to Dr. R. E. Buchanan, Editor of the Iowa State Journal of Science for his assistance in the preparation of the index, and to Mr. Marshall Townsend of the Iowa State University Press for technical assistance.

JACOB PETER ANDERSON

Jacob Peter Anderson, was born April 7, 1874, in Glenwood, Utah, and lived as a child in Nebraska. He received his formal education at the University of Nebraska and at Iowa State College. After graduation from Iowa State in 1913, he moved to Sitka, Alaska, where he was horticulturist at the Experiment Station. In 1916, he received his M.S. degree from Iowa State. In 1917, he opened the first greenhouse in Alaska and operated as a commercial florist in Juneau until 1940. He was active in territorial affairs as a legislator and census supervisor.

During his period of residence in Alaska, Anderson traveled widely to study the regional flora. His first herbarium was destroyed by fire in 1924. Undaunted, he began again and by 1940 had amassed the largest Alaskan collection in existence. In recognition of his studies, the University of Alaska granted him the honorary degree of Doctor of Science in 1940. A devoted botanist, Dr. Anderson gathered around him an enthusiastic group of amateurs of the Alaskan flora. His botanical correspondence and acquaintanceship were worldwide. After leaving Alaska in 1940 to establish residence at Iowa State, he devoted his full time to the writing and illustration of his Flora. During these years he returned to Alaska on four summer field trips. Despite infirmities of age, he never lost his devotion to his work, which he continued till a few weeks before his death. His collections, known as the J. P. Anderson Herbarium of Arctic and Boreal Plants, were bequeathed according to the terms of his will to Iowa State University and are preserved as a part of the University Herbarium.

R. W. POHL

August, 1959

Table of Contents

Introduction	1
Pteridophyta	6
Spermatophyta	23
Gymnospermae	24
Angiospermae	41
Index	525
Addendum. Keys to Willows (Salix) of Interior Alaska	514

Alaska saxifrage, *Saxifraga ferruginea,* page 289.

PTERIDOPHYTA AND GYMNOSPERMAE (FERNS TO CONIFERS)

INTRODUCTION

Few people realize the importance or the size and extent of Alaska. The region was originally considered most valuable for its furs, but fisheries and mining have long surpassed furs in value. Approximately 60 per cent of all the salmon canned in the whole world are caught in Alaskan waters, in addition to several million dollars worth of other fish and fish products. About $25,000,000 in gold was produced in 1941 beside lesser values of many other mineral products. The value of Alaska in the defense of the United States is now beginning to be realized.

The actual area of Alaska is nearly 587,000 square miles which is more than twice that of Texas and ten and a half times that of Iowa. It extends from 130 degrees west longitude to 173 degrees east longitude or about the same as from Eastport, Maine to the Pacific coast of Oregon. Due to the actual length of degrees decreasing from the Equator to the Poles the actual distance in miles is about that from Savannah, Georgia, to Los Angeles, California. Point Barrow is about the same latitude as North Cape, Norway, and is about 20 degrees north of the middle Aleutian Islands. This distance is about the same as that from the Canadian border north of western Minnesota to the mouth of the Brazos River south of Houston, Texas. It may surprise many to learn that the middle Aleutian Islands extend about 3 degrees farther south than southeastern Alaska, and cut across the great circle route from Seattle or Vancouver to Yokohama.

Naturally in such an extended area climate varies. The Pacific Coast districts are characterized by mild winters and cool summers. Zero temperatures are rare and in some places unknown, while the average summer temperatures are between 50 and 60° F. Precipitation is heavy, and there is much cloudy weather. The eastern part of this region, from Cook Inlet eastward, is largely covered by a heavy forest growth, some trees reaching very large size. Along the Alaska Peninsula and the Aleutian Islands, summer conditions are unfavorable to tree growth, probably due to the excessively cool summers.

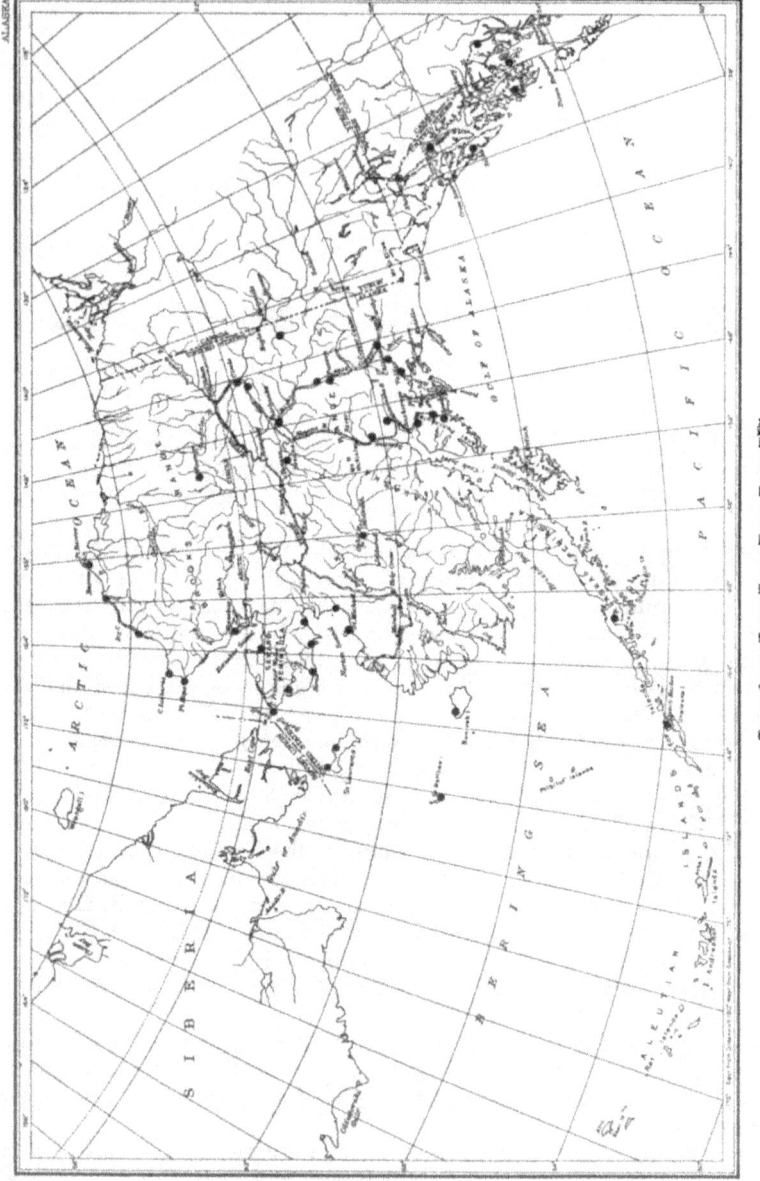

Stations where intensive collections were made by the author.

Following up the Bering Sea Coast we have a tundra-like formation merging into the true tundra farther north. These formations are treeless, covered in summer with a growth of low shrubs, grasses and sedges with intermixture of other herbaceous plants. In the true tundra the soil is permanently frozen, thawing a few feet at the surface each summer. The woody growth on these so-called barren lands consists mostly of dwarf birches, willows, and members of the Heather Family.

The interior districts are characterized by short, warm summers and long, cold winters. Summer temperatures are higher than in the coast districts, and the precipitation is comparatively light with long hours of sunshine. January temperatures in most places average below zero Fahrenheit. This region is largely covered by forest of moderate growth. Some limited districts, such as the head of Lynn Canal and Cook Inlet with the Matanuska Valley have climatic conditions intermediate between those of the coast and the interior.

From what has been said it can be perceived that there are three main types of vegetation in Alaska: 1. The heavily forested central and eastern Pacific Coast districts dominated by Sitka spruce and western hemlock. 2. The more lightly forested districts of the interior characterized by birches and white spruce. 3. The tundra and tundra-like districts of the Alaska peninsula, Aleutian Islands, Bering Sea littoral and Arctic slope. This latter type is also found in the other regions above timberline on the mountains, the alpine meadows occurring at successively lower altitudes until they meet the tundra. In types 1 and 2 are found muskeags or peat bogs. These are areas of a few square meters up, characterized by a surface covering of sphagnum moss underlaid with moss and other vegetation in various stages of decay, merging gradually into black muck, and the whole saturated with water. This mass of water-logged material may vary from less than 1 meter to many meters in depth. The surface usually is dotted with small ponds, and growing in the moss are very much stunted trees and characteristic shrubs and other plants, the whole having somewhat the appearance of tundra.

On February 1, 1914, the author started for Alaska. This was destined to be his home until October 1, 1941. The years 1914-16 were passed in Sitka as Horticulturist for the United States Agricultural Experiment Station. In 1917 he removed to Juneau and engaged in commercial floriculture under the title of Juneau Florists. In 1937 this business was sold with the idea of devoting more time to a study of the plant life of Alaska.

Having always been interested in plant life the author proceeded to collect the flora of the country as circumstances permitted. In November, 1924, the entire collection of more than 3,300 numbers was destroyed by fire. Some of these numbers were represented by specimens in the U. S. National Herbarium and elsewhere, but many were not. In 1925 a new collection was started which now numbers nearly 8,000; this, together with exchanges and specimens sent in by friends, brings the total of Alaska specimens in this herbarium up to about 10,000.

This collection is now permanently deposited at the Iowa State College of Agriculture and Mechanic Arts at Ames.

Besides places where residence was maintained, collections have been made at the following places: Ketchikan, Craig, Hyder, Skagway, Chitina, Valdez and along Richardson Highway to Fairbanks, Fairbanks and along Steese Highway to Circle, Franklin in the Fortymile district, Seward and other places on Kenai Peninsula, Matanuska Valley, Talkeetna, Healy, Manly Hot Springs, Wiseman, Takotna, Unalaska, St. Paul Island, Nunivak Island, St. Matthew Island, St. Lawrence Island, St. Michael, Stebbins, Unalakleet, Elim, Golovin, Nome, Teller, Cape Prince of Wales, Deering, Kotzebue, Kivelina, Point Hope, Cape Lisburne, Point Lay, Wainwright, and Barrow. Also fewer numbers were collected at other places and, by exchange and gift, specimens were obtained from places not visited by the author, especially from Mt. McKinley National Park and the Aleutian Islands.

Most of the literature dealing with the plants of Alaska is widely scattered in many publications, and at that is useful mostly to the professional botanist and not to the amateur. For this reason the author for some years has cherished the idea of writing a descriptive manual that, though scientifically accurate, would be so written that it would be of maximum usefulness to all persons interested in the flora, even though their botanical training was rather limited. The author has found many such persons, both among the permanent residents and among the tourists, who visit the territory every summer during normal times.

The author is not the only person working on the flora of Alaska. Dr. Eric Hultén of the Botanical Museum, Lund, Sweden, who is probably the world's foremost authority on arctic and boreal plants in general, in 1930 issued a "Flora of Kamtschatka" and in 1937 a "Flora of the Aleutian Islands." These are very good in listing the known species, presenting data on collections and giving useful comments, but contain no keys or descriptions, except of new species or varieties. Dr. Hultén is now at work on the "Flora of Alaska and Yukon," and the first part listing ninety-one species has been issued. This is on the same plan as his other publications listed above but does contain keys to the species but not to the families or genera. The author has consulted these works of Dr. Hultén and places much reliance on them. In addition, leading American manuals have been freely consulted, also papers on special groups in leading scientific publications, and lists of plants made by various collectors in Alaska.

Although the author's primary interest is in Alaska, species known to occur in adjacent parts of Canada are included. This additional area is Yukon Territory and the extreme northwestern part of British Columbia. These regions belong to the same floral provinces, and nearly all form a part of the Yukon River drainage system. The entire area covered is more than 800,000 square miles.

This manual aims to include all species of vascular plants known to occur within the geographical limits covered. In such a vast region, so sparsely settled, there remain large districts in which there has been no botanical collecting. Even in places where considerable collections have been made, other species may be found. The author collected in the vicinity of Juneau for many years, and every season when considerable scouting around was possible, species not before noted in the region were found. It is evident, therefore, that many more species will be found as further collections and studies are made. However, enough is known to give a good general idea of the flora and to include all the more common and widely distributed species.

The arrangement of the families in this manual is that usually followed in American descriptive manuals and the herbaria of the country, although it does not always reflect the true relationships.

The nomenclature followed is that known as the International Rules. The original rules as adopted in 1905 have been greatly modified and modernized by International Botanical Congresses since that time and are now generally used by American botanists. The synonyms as given are not complete, but include only those used in publications relating to Alaska or in American or Canadian descriptive manuals. Further, all names given are not necessarily synonymous, but the species may have been reported under that name due to error in determination, hence a synonym only in that sense.

Unless otherwise noted all illustrations were made by the author and are based on material collected in Alaska. The aim has been to make them as true to nature as possible and to bring out the differences between the species more clearly than short descriptions make possible.

In giving the range of a species the abbreviations of the states and Canadian provinces are those in general use. Circumboreal does not mean that the distribution is necessarily continuous around the northern regions, for many species have breaks in this continuity of occurrence. They do, however, occur in some part of North America, Europe, and Asia.

Measurements are given in the metric system. The following are the approximate equivalents in inches; 25 millimeters (mm.) equals 1 inch, 1 decimeter (dm.) equals 4 inches, 1 meter (m.) equals 39.37 inches.

As the completion of this manual will require several years, it has been thought best to publish it in parts as ready, so as to make such parts available to interested parties at once. This is all the more desirable as all published American manuals combined do not even mention some of the species occurring in Alaska, especially the western part of the territory.

Index and other accessory material will accompany the completed manual.

ACKNOWLEDGEMENTS

This work is being carried on at the Iowa State College of Agriculture and Mechanic Arts where the author moved his botanical collections in 1941 in order to have better facilities for this work. The facilities of the College were placed at his disposal, and the members of the Botanical staff have been helpful, especially the head of the Botany Department, Dr. I. E. Melhus, Dr. Geo. J. Goodman, Curator of the Herbarium, and Dr. J. C. Gilman, Editor of Iowa State College Journal of Science.

PHYLUM PTERIDOPHYTA

Plants containing woody and vascular tissues in the stem and producing spores which give rise to small, inconspicuous growths known as prothallia (gametophytes), on which the archegonia (female organs) and antheridia (male organs) are borne. The fertilization of an archegonium by a spermatozoid from an antheridium results in the large conspicuous plant we call a fern, a horsetail or a club-moss.

Known living species number about 7,000, three-fourths of them tropical. They appeared early in geological history, and were the predominant type of vegetation during the Carboniferous times, fossil forms often being found in connection with coal beds. This was probably about 250 million years ago.

1A. Spores produced in sporangia, which are borne on the back of the leaf, in spikes or panicles ....Order 1. *Filicales* P.6
    1B. Vernation erect or inclined, sporangia in spikes or panicles.
    .... Family 1. *Ophioglossaceae* P.6
    2B. Vernation coiled, sporangia reticulated ....
    .... Family 2. *Polypodiaceae* P.8

2A. Spores produced in sporangia, which are clustered underneath scales in a terminal conelike spike .... Order 2. *Equisetales*   P.18
    One family ....Family   *Equisetaceae*   P.18

3A. Spores produced in sporangia, which occur in the axils of scalelike or tubular leaves. .... Order 3. *Lycopodiales*   P.20
    1B. Spores all of one size. .... Family 1. *Lycopodiaceae*   P.20
    2B. Spores of two sizes:
        1C. Leaves scalelike. .... Family 2. *Selaginellaceae*   P.22
        2C. Leaves tubular. .... Family 3. *Isoetaceae*   P.23

OPHIOGLOSSACEAE (Adder's-tongue Family)

More-or-less succulent plants with fleshy rhizomes and stems bearing a leaf and one or more stalked spore-bearing spikes or panicles (sporophylls). Leaves simple or usually compound, not coiled in vernation, sporangia bivalvular, formed from the interior tissues of the sporophylls; prothallia subterranean, without chlorophyll.

Veins reticulate; sporangia in a spike. ....1. *Ophioglossum*
Veins free; leaves pinnatifid to tripinnate;
sporangia in panicles. ....2. *Botrychium*

1. OPHIOGLOSSUM (Tourn.) L.

Small herbaceous perennials with short, usually erect, fleshy subterranean rhizomes. Leaves erect, glabrous, fleshy, arising at the side of the apical bud; sterile blade simple, sessile or short-stalked, with reticulate venation; sporophyll a simple, slender, long-stalked spike; the large globose sporangia marginal in two ranks, transversely dehiscent. (Greek, tongue of a snake, alluding to the sporophyll.)

O. vulgatum L. Adder's-tongue

Fronds usually solitary, 1–4 dm. tall; sterile blade usually sessile, lanceolate to spatulate or ovate, 2½–12 cm. long × 1–5 cm. broad; spike 2–4 cm. long × 1½–3½ mm. wide, apiculate; sporangia 10–30 pairs. Our plant differs from the type in its large, thin, ovate, very distinctly veined sterile blade. It was described by E. G. Britton as *O. alaskanum* and may be regarded as a variety of the circumboreal *O. vulgatum*. It has been twice collected at Unalaska.

2. BOTRYCHIUM Sw.

Rootstock short, erect, with fleshy clustered roots, the bud for the succeeding season's frond embedded in the base of the stem; the blade pinnately or ternately compound; veins free, forking; the sporophylls pinnate to tripinnate with sessile, distinct sporangia on either side of its branches, forming large panicles in some species. (Name in allusion to the grapelike arrangement of the sporangia.)

1A. Sterile blade once to twice pinnate.
    1B. Segments reniform or fan-shaped. ....1. *B. lunaria*
    2B. Segments rounded. ....2. *B. boreale*
    3B. Segments acute. ....3. *B. lanceolatum*

2A. Sterile blade thrice pinnate.
    1B. Sterile blade long-petioled, arising from the base of the plant.
    ....4. *B. silaifolium*
    2B. Sterile blade sessile or nearly so, affixed to middle of the plant.
    ....5. *B. virginianum*

1. *B. lunaria* (L.) Sw. Moonwort

Fleshy, 4–30 cm. tall; sterile leaf-blade nearly sessile, borne about the middle of the plant, simply pinnatifid, the segments lunate or fan-shaped, entire or crenulate or even incised, often imbricate; sporophyll bent down in vernation, at maturity erect and surpassing the sterile blade; panicle 1–3 times pinnate. A circumboreal species occurring in Alaska from about the Arctic circle southward. (Fig. 1.)

2. *B. boreale* (Sw.) Milde. Northern Grape-fern

Fleshy, 4–25 cm. tall; sterile leaf nearly sessile, borne above the middle of the plant, triangular in outline, obtuse, pinnatifid, with the lower divisions crenately incised, all divisions crenate and often imbricate; sporophylls much as in *B. lunaria*. A form found at Unalaska has been described as var. *obtusilobium* Rupr. Circumboreal, in Alaska from Wiseman southward. (Fig. 2.)

3. **B. lanceolatum** (S. G. Gmel.) Angstr. Lance-leaved Grape-fern

Fronds 5–20 cm. tall; common stalk long; sterile leaf sessile, triangular, acute, 1–6 cm. long × 1–8 cm. wide, once or twice pinnately divided, the primary divisions ovate-lanceolate; sporophylls sessile or short-stalked, forming a diffuse panicle, the larger divisions ascending and often subequal. Arctic-alpine situations, Unalaska eastward. Circumboreal. (Fig. 3.)

4. **B. silaifolium** Presl. Leathery Grape-fern
    B. multifidum (Gmel.) Rupr. var. *robustum* (Rupr.) C. Chr.

Fronds 1–6 dm. tall, fleshy, coriaceus in drying; sterile leaf long-stalked, broadly triangular or pentagonal in outline, 10–30 cm. broad and about as long, subternately compound, the primary divisions 1–3-pinnately divided, the ultimate segments ovate or rhomboid, crenulate, obtuse; sporophyll long-stalked with diffuse panicle. Known from the Aleutian Islands and southeastern Alaska thence to Quebec.—Pa.—Wis. —N. Calif. (Fig. 4.)

5. **B. virginianum** (L.) Sw. Virginia Grape-fern

Fronds 1–7 dm. tall; stalk slender; sterile blade sessile or nearly so, spreading, thin and membranous, deltoid, 4–21 cm. long × 5–36 cm. wide, the ultimate divisions variously toothed or lobed; sporophyll long-stalked, 2–3-pinnate. Isolated and rare in southwestern Alaska. Occurs B.C.—Lab.—Fla. Also in Eurasia, Brazil, and Mexico.

POLYPODIACEAE (Fern Family)

Leafy plants with the rootstocks horizontal, often elongated, or shorter and oblique or erect, often stout; the leaves (fronds) coiled in the bud. Sterile fronds leaflike; fertile fronds (sporophylls) leaflike or more or less modified, bearing the sporangia on their lower surface or at their margins, usually in clusters (sori); sori naked or usually covered, especially when young, by a membrane (indusium); sporangia stalked, furnished with an incomplete ring of thickened cells (annulus), opening transversely; prothallia green, above ground.

1A. Sterile and fertile fronds different, pinnae of fertile fronds contracted.
    1B. Fertile fronds simply pinnate.
        1C. Sterile fronds simply pinnate. .... 1. *Blechnum*
        2C. Sterile fronds with pinnatifid pinnae. .... 2. *Struthiopteris*
    2B. Sterile and fertile fronds bipinnate. .... 3. *Cryptogramma*

2A. Sterile and fertile fronds similar.
    1B. Sori marginal, the indusia formed wholly or in part by the revolute leaf margins.

1C. Sori distinct, on underside of reflexed leaf-lobes. ....
.... 4. *Adiantum*
2C. Sori continuous or confluent. .... 5. *Pteridium*
2B. Sori dorsal on the veins.
1C. Sori roundish.
1D. Sori naked. .... 6. *Polypodium*
2D. Sori with wholly or partly inferior indusia.
1E. Indusia wholly inferior, the divisions stellate or hairlike. .... 7. *Woodsia*
2E. Indusia hood-shaped, attached at side, early deciduous. .... 8. *Cystopteris*
3D. Indusia superior.
1E. Indusia peltate, centrally attached. ....
.... 9. *Polystichum*
2E. Indusia orbicular-reniform, attached at the sinus.
.... 10. *Dryopteris*
2C. Sori oblong.
1D. Sori straight or slightly curved, fronds evergreen. ....
....11. *Asplenium*
2D. Sori usually curved, fronds herbaceous. ....
.... 12. *Athyrium*

1. BLECHNUM (L.) With.

Our species is a woodland fern with woody rootstock and fronds of two kinds, both with pinnate or pinnatifid blades. Sori in a continuous band next to the midrib, covered by a continuous membranous indusium arising under the margin of the pinna; indusium often lacerate, often reflexed at maturity. (Greek for some fern.)

B. *spicant* (L.) J. E. Smith.   Deer-fern
Lomaria spicant (L.) Desv.
Osmunda spicant L.
Struthiopteris spicant (L.) Weis.

Sterile fronds numerous, in a circular crown, evergreen; 2–10 dm. long; stipe rather short, brownish; blades linear-lanceolate, attenuate to both ends, cut to the rachis into linear, falcate segments, those near the base mere auricles, the segments entire or finely crenulate toward the apex; fertile fronds few, central, erect, 4–15 dm. long, with long reddish-brown stipes; blades pinnate, the pinnae narrowly linear.

Known from Atka and Kodiak Islands and common in the coast region from Cook Inlet eastward and extending to California. Also in Eurasia. (Fig. 5.)

2. STRUTHIOPTERIS Scop.

Coarse ferns with the fertile fronds rolled into necklace-like or berry-like segments and unlike the foliaceous sterile ones; sori round, borne on the back of the veins; indusium delicate, fixed at the inferior side of the sorus. (Name from struthio, ostrich; and pteris, fern.)

S. filicastrum All.   Ostrich-fern
  Matteucia struthiopteris (L.) Todoro.
  Onoclea struthiopteris (L.) Hoffm.
  Pteretis nodulosa (Michx.) Nieuland.

Rootstock stout, bearing a circle of sterile fronds with fertile ones in the center; sterile fronds up to 2 meters tall, narrowed at the base, pinnate, the pinnae once pinnatifid, 5–18 cm. long; fertile fronds shorter, with rigid upcurved, necklace-shaped pinnae; veins pinnate, free, simple; texture firm.

Woods; central and southern Alaska, especially abundant along the Alaska railroad from Talkeetna to north of Curry. Distribution circumboreal. (Fig. 6.)

3. CRYPTOGRAMMA R. Br.

Small ferns of rocky situations with dimorphous, tufted, 2–3-pinnate fronds. Sterile fronds foliaceous, with numerous, crowded, rather small, obtuse segments; sori in a continuous line at the free ends of the forked veins, confluent; indusia formed of the revolute, modified margins of the segments, which later open out. (Greek, in allusion to the hidden sori.)

Rootstock stout, short. .... 1. C. acrostichoides
Rootstock slender, creeping. .... 2. C. stelleri

1. C. acrostichoides R. Br.   Parsley-fern

Rhizome in massive tufts, chaffy; fronds numerous, the fertile 1–3 dm. long, erect, long-stalked, overtopping the short-stalked sterile ones; sterile blades ovate to ovate-lanceolate, the ultimate segments suboval, obtuse, serrulate; fertile segments elliptical or linear, 6–12 mm. long, about 2 mm. wide. Considered by some botanists as only a variety or subspecies of the Eurasian C. crispa (L.) R. Br.

Pacific coast and Bering Sea regions of Alaska—Baffin Bay—Colo.—Calif. (Fig. 7.)

The var. sitchense (Rupr.) C. Chr. is characterized by the broadly deltoid tripinnate sterile fronds with small, more deeply toothed segments. Same range.

2. C. stelleri (S. G. Gmel.) Prantl.   Slender Cliff-brake

Fronds scattered, arising singly from slender creeping rhizomes; pinnae few, the lower ones usually pinnatifid; segments of the sterile blades ovate or obovate, crenately lobed; those of the fertile ones linear to lanceolate.

East central Alaska—Lab.—Va.—Colo.—Wash. (Fig. 8.)

4. ADIANTUM (Tourn.) L.

Graceful delicate ferns of moist rocky woods and ravines with compound fronds having segments in the form of small leaflets and with dark-colored shining stipes. Sori marginal under the modified, sharply reflexed margins of the leaflets. (Greek, unwetted, in allusion to the leaflets repelling raindrops.)

A. *pedatum* L.   Maiden-hair Fern

Rhizome thickish, chaffy with shining, dark, chestnut-brown scales; fronds 2–8 dm. long, forking into 4–8-pinnate divisions, the longer ones 1–3 dm. long; segments very short-stalked, the lower margin formed by the midrib, the upper cut and toothed.

Coastal districts of Alaska—N. S.—Ga.—Ark.—Calif.—Asia. The form on the Pacific coast from Japan to Alaska to Calif. and on the Atlantic coast from Newf. to Mass. has pinnules with longer stalks and the upper margin more deeply cleft. It has been described as the var. *aleuticum* Rupr. (Fig. 9.)

5. PTERIDIUM Scop.

Coarse ferns of open or partly shaded situations with woody, branched, wide-creeping rhizomes. Sporangia borne in a continuous line under the margin of the frond, occupying a veinlike receptacle connecting the ends of the veins; indusium double, the outer one formed by the reflexed margin of the frond, the inner delicate and minute. (Diminutive of pteris, Greek name of ferns.)

P. *aquilinum* (L.) Kuhn var. *lanuginosum* Bong.   Western Bracken
    P. *aquilinum* (L.) Kuhn var. *pubescens* Underw.

Stipe erect, stout, 15–100 cm. long; blades triangular or deltoid-ovate, as long or longer than the stipe, subternately tripinnate, the lower divisions being bipinnate; segments variable, mostly oblong and entire, pubescent or strongly tomentose beneath, slightly hairy or glabrous above.

Southeastern Alaska—Mont.—Mex. Entire species quite cosmopolitan in distribution. (Fig. 10.)

6. POLYPODIUM (Tourn.) L.

Ferns of various habit, our species with creeping rootstock growing in moss. Fronds pinnately compound, usually articulated to the rhizome; sori round or elliptical, borne on the backs of the fronds, without indusia, veins free. (Greek, many and foot alluding to the knoblike prominences of the rhizome).

P. *vulgare* L. var. *occidentalis* Hook.   Licorice-fern
    P. *glycyrrhiza* D. C. Eat.
    P. *falcatum* Kell.

Rhizome hard, 3–5 mm. thick, covered with rusty-brown scales; fronds 1–6 dm. long, the stipe usually shorter than the blade, firm, naked; blades lanceolate, abruptly attenuate or caudate, pinnatisect; segments alternate, tapering from the middle or the base, serrulate; sori about midway between midrib and edge of the segments.

Common in the coastal districts and rare in the Yukon Valley, extending to California. Entire species circumboreal. (Fig. 11.)

7. WOODSIA R. Br.

Small ferns of rocky situations with densely tufted, pinnately compound fronds and round sori borne on the back of the free veins. Indusia placed under the sporangia, thin and often evanescent, roundish or stellate, small and open, or bursting at the top into irregular segments. (Joseph Woods (1776–1864) was an English architect and botanist.)

1A. stipe articulated near the base.
   1B. Fronds glabrous. ....1. *W. glabella*
   2B. Fronds with hairs or scales on the lower surface.
      1C. Primary segments about as broad as long. ....
      ....2. *W. alpina*
      2C. Primary segments longer than broad. ... 3. *W. ilvensis*
2A. Stipe not articulated near the base. ....4. *W. scopulina*

1. *W. glabella* R. Br.   Smooth Woodsia

Fronds tufted, pinnate, 3–16 cm. long; stipes smooth, usually straw-colored; pinnae deltoid to ovate, crenately lobed or parted, glabrous; indusia divided into narrow, hairlike, curving divisions.

Moist rocks, in most parts of our territory. Circumboreal. (Fig. 12.)

2. *W. alpina* (Bolton) S. F. Gray.   Alpine Woodsia

Rootstock short; fronds densely tufted, the blades narrowly lanceolate, 5–15 cm. long × 15–25 mm. wide; pinnae cordate-ovate to triangular-ovate, pinnately 5–7-lobed, sparingly hairy; sori near the margins, the indusia cleft into numerous hairlike filaments.

Moist rocks, of scattered distribution through most of Alaska and Yukon Ter. Circumboreal. (Fig. 13.)

3. *W. ilvensis* (L.) R. Br.   Rusty Woodsia

Fronds tufted, lanceolate, 8–20 cm. long; pinnae pinnately lobed, sparingly hairy above, hairy and with rusty chaff beneath; sori borne near the margins of the segments, somewhat confluent when old; indusia cleft into filiform segments.

In most of Alaska south of the Arctic Circle. Circumboreal. (Fig. 14.)

4. *W. scopulina* D. C. Eat.   Rocky Mountain Woodsia

Fronds numerous, borne close together, 6–35 cm. long, blades lanceolate, finely glandular-puberulent; pinnae oblong-ovate, deeply pinnatifid into 10–16-toothed segments; indusia delicate, cleft into narrow, spreading, flaccid segments.

Crevices of rocks, Kenai Penin. and southeastern Alaska—Calif.—Utah—S. Dak. Isolated station in eastern America.

8. CYSTOPTERIS Bernh.

Ferns of rather thin texture, on slender stipes with 2–4-pinnate blades. Sori roundish, borne on the backs of the veins; indusia delicate, hoodlike or flattish, attached at one side and partly underneath the sori, at first arched over them, later thrown back and withering, the sori then appearing naked. (Greek, bladder-fern.)

Blades lanceolate, 2–3-pinnate. .... 1. *C. fragilis*
Blades deltoid, 3–4-pinnate. .... 2. *C. montana*

1. *C. fragilis* (L.) Bernh.   Fragile-fern
   Filix fragilis (L.) Gilib.

Fronds somewhat clustered or slightly scattered; stipe slender, about as long as the blade, brittle, stramineous or brownish below; blades extremely variable, nearly or fully bipinnate; pinnae deltoid to lanceolate or ovate-lanceolate, acute to acuminate, narrowly decurrent on the rachis, the lower ones slightly reduced; the segments toothed or incised; veinlets excurrent to the marginal teeth; indusia convex, rounded or usually pointed, toothed or lacerate at apex.

Common and very variable. The most widely distributed and cosmopolitan of all ferns. (Fig. 15.)

2. *C. montana* (Lam.) Bernh.   Mountain Cystopteris
   Filix montana (Lam.) Underw.

Rhizomes slender, widely creeping; fronds scattered, stipe slender, blade often subternate, 3–4-pinnate, 5–15 cm. long and wide; lower pinnae much the largest; the pinnules pinnatifid to the winged rachis, the final segments oblong, deeply toothed or divided; indusia convex, acute, soon thrown back or withering.

Bering Str. to Yukon Ter. Circumboreal. (Fig. 16.)

9. POLYSTICHUM Roth.

Ferns of rather firm texture, with pinnate or pinnately decompound, tufted fronds from the crown of the rhizome, the divisions with sharply toothed or spinulose margins (except in *P. aleuticum*); sori round, indusia peltate, attached by the middle, persistent to caducous; veins free. (Greek, many rows.)

1A. Blades simply pinnate.
   1B. Low-grown, tissues thin. .... 1. *P. aleuticum*
   2B. Taller, coriaceous, pinnae with spinulose teeth.
      1C. Fronds short-stalked, lower pinnae reduced. 2. *P. lonchites*
      2C. Fronds longer stalked, lower pinnae about as long as those above. .... 3. *P. munitum*

2A. Blades bipinnate.
   1B. First upturned secondary segment longer than the others............
      .... 4. *P. andersoni*
   2B. First upturned secondary segment not conspicuously longer than the others .... 5. *P. braunii*

1. *P. aleuticum* C. Chr.   Aleutian Shield-fern

Fronds about 15 cm. tall, blades thin, pinnae not spinose or aristate and with the general appearance of *Woodsia alpina*.

Known only from a single collection made on Atka island.

2. *P. lonchites* (L.) Roth.   Holly-fern

Fronds rigidly ascending in a close crown, 1-6 dm. tall, bearing pinnae almost to the base, densely chaffy at base, lanceolate in outline, broadest near the middle; rachis more or less chaffy; pinnae numerous, close, densely spinulose-toothed, glabrous above, somewhat chaffy beneath; auricles on upper side, sori usually in two rows, indusia orbicular, nearly entire.

Woods, Pacific coast districts. Circumboreal. (Fig. 17.)

3. *P. munitum* (Kaulf.) Presl.   Dagger-fern

Fronds growing in a crown, 3-15 dm. tall; stipes 5-60 cm. long, together with the rachis decidedly chaffy; blades lanceolate, narrowed toward the base; pinnae numerous, spreading, 2-14 cm. long, sharply and often doubly serrate, the serrations with spinescent, often incurved teeth; indusia papillose-dentate to long ciliate.

In woods, southeastern Alaska—Mont.—Calif. (Fig. 18.)

4. *P. andersonii* Hopkins.   Anderson's Shield-fern

Similar in appearance to *P. braunii* but the rachis with proliferous buds, the first upturned pinnule conspicuously larger than the next, and the base of the pinnules decurrent and connecting, the blade scarcely bipinnate. Indusia ciliate-erose.

A rather rare woodland fern in southeastern Alaska and ranging to Mont. and Wash.

5. *P. braunii* (Spenner) Fée.   Prickly Shield-fern
P. alaskense Maxon.

Fronds in a crown, 2-6 dm. tall; stipe and rachis chaffy with both broad and narrow bright-brown scales; blades lanceolate, gradually narrowed toward the base; pinnae numerous, lanceolate; segments ovate, oblique, spinulose-toothed, beset with long, soft hairs and scales; indusia orbicular, small, nearly entire.

In woods, Pacific coast districts of Alaska. Circumboreal. (Fig. 19.)

The form described as *P. alaskense* may be regarded as a variety. It has pinnules which are more cuneate at the base, more ellipsoid in form, and have a broader attachment at the base.

10. DRYOPTERIS Adans.

Aspidium Sw. in part.

Mainly woodland ferns of upright growth; rhizomes various, fronds borne singly or in a crown, the fertile and sterile usually alike, 1-3-pinnate or decompound; sori roundish, dorsal; indusia when present roundish-reniform, fixed at its sinus. (Greek, meaning oak-fern.)

1A. Blades long-stalked, triangular, fronds scattered.
    1B. Blades longer than broad, pinnate-pinnatifid. 1. *D. phegopteris*
    2B. Blades as broad as long, 2-3-pinnate.
        1C. Fronds glandless. ....2. *D. linnaeana*
        2C. Rachis and lower surface glandular. ....3. *D. robertiana*
2A. Blades clustered from short stout rhizomes.
    1B. Blades 1-2-pinnate.
        1C. Blades small, thick. ....4. *D. fragrans*
        2C. Blades large, thin. ....5. *D. oreopteris*
    2B. Blades 2-3-pinnate. ....6. *D. austriaca*

1. *D. phegopteris* (L.) C. Chr.   Beech-fern
    Phegopteris phegopteris (L.) Underw.
    Thelypteris phegopteris (L.) Slosson.

Rhizome slender, wide-creeping; fronds scattered, 10-55 cm. long, the stipe usually longer than the blade, more or less scaly; blades triangular, long-acuminate, sparingly hairy on both surfaces, especially on the veins; pinnae mostly closely adnate, horizontal, linear-lanceolate, pinnatifid; segments oblong, obtuse, entire or slightly crenate; sori submarginal, naked.

In woods, most parts of Alaska—Greenl.—Newf.—Va.—Ohio—Wash. (Fig. 20.)

2. *D. linnaeana* C. Chr.   Oak-fern
    D. dryopteris (L.) Christ.
    Phegopteris dryopteris (L.) Fée.
    Thelypteris dryopteris (L.) Slosson.

Rhizomes slender, wide-creeping; fronds scattered, erect, 1-6 dm. long; stipe slender, much longer than the blade, from a chaffy blackish base; blades deltoid, 8-25 cm. long and wide, subternate, the 3 primary divisions stalked, 1-2-pinnate, the larger pinnules pinnately lobed or divided; lobes oblong, entire to serrate-crenate; sori small, without indusia; leaf tissue thin, glabrous.

In woods, common in most parts of Alaska. Circumboreal. (Fig. 21.)

3. *D. robertiana* (Hoffm.) C. Chr.   Scented Oak-fern
    Thelypteris robertiana (Hoffm.) Slosson.

Very similar to *D. linnaeana* in appearance but the stipe and blade bearing minute stalked glands. The lateral main divisions of the blade are also somewhat smaller in proportion.

In woods, central Yukon River district. Circumboreal.

4. *D. fragrans* (L.) Schott.   Fragrant Shield-fern
    D. aquilonaris Maxon.
    Aspidium fragrans (L.) Sw.

Rhizome chaffy with brown shining scales; fronds borne in a dense crown, 4-40 cm. long, aromatic; stipe short and with the rachis very chaffy; blade lanceolate, narrowed toward the base, bipinnate; pinnae triangular-lanceolate, the segments oblong, obtuse, adnate-decurrent, dentate or nearly entire; nearly covered by the large sori; indusia very large, persistent, ragged, somewhat glandular.

On rocks, Bering Sea—Wiseman—Matanuska east. Circumboreal. (Fig. 22.)

5. *D. oreopteris* (Ehrh.) Maxon.   Mountain Wood-fern
Thelypteris oreopteris (Ehrh.) Slosson.
Aspidium oreopteris (Ehrh.) Sw.

Fronds in a crown, ascending, glandular, 4–11 dm. long; stipes short, stipe and rachis somewhat scaly; blades lanceolate, tapering below; pinnae pinnatifid, broadest at base, glabrous or nearly so above, sometimes short-hairy on the veins and midrib below; segments oblong, obtuse, subentire, the margins finely hyaline-papillose; sori rather small, submarginal; indusia round-reniform, toothed, deciduous.

On mountain slopes, Pacific coast districts of Alaska. Circumboreal. (Fig. 23.)

6. *D. austriaca* (Jacq.) Woynar.   Spreading Wood-fern
D. dilatata (Hoffm.) A. Gray.
D. spinulosa (Muell.) Kuntze.
Aspidium spinulosum var. *dilatatum* Hook.

Rhizome chaffy; fronds in a crown, 3–12 dm. long, stipe stout, 15–50 cm. long, chaffy with brownish, often darker-centered scales; blades triangular to ovate, acuminate, nearly or fully tripinnate; pinnae unequally ovate or triangular; pinnules lanceolate to oblong, the larger ones not decurrent, pinnate or pinnately divided; the ultimate segments pinnatifid or toothed; indusia glabrous or sparsely glandular.

In woods, from Bering Sea east and south. Circumboreal. (Fig. 24.)

11. ASPLENIUM L.

Our species are small ferns of rocky ledges with simply pinnate leaves. Sori oblong or linear, oblique, borne on the veins; indusia straight or curved, attached by one edge, often nearly concealed by the sporangia at maturity. (Greek, alluding to the supposed medicinal properties.)

Rachis dark brown, shining. ....1. *A. trichomanes*
Rachis yellowish-green. ....2. *A. viride*

1. *A. trichomanes* L.   Maidenhair Spleenwort

Fronds tufted, 5–20 cm. long; stipes short; blades linear, somewhat narrowed toward base and apex; pinnae oval or oval-oblong, 3–8 mm. long, rigid, evergreen, sessile, the margins usually crenulate; indusia usually crenulate.

Southeastern Alaska, rare. Circumboreal—Africa—Australia—S. Am. (Fig. 25.)

2. A. viride Huds. Green Spleenwort

Fronds tufted, 4–20 cm. long, laxly ascending; stipes 1–7 cm. long, reddish-brown at base only; blades linear-lanceolate, pinnae up to 25 pairs, roundish-ovate to rhombic, obtuse, cuneate at base, the margins deeply crenate; sori at maturity becoming confluent, concealing the delicate indusia.

Pacific and Bering Sea districts. Circumboreal. (Fig. 26.)

12. ATHYRIUM Roth.

Medium to large ferns of upright habit growing in moist situations. Fronds usually large, long-stipitate, erect-spreading, ours 2–3-pinnate; veins free; sori dorsal, oblique to the midrib, oblong, or often crossing the vein and becoming horseshoe-shaped or roundish; indusia following the shape of the sori, attached along its length at the side next to the vein, delicate, sometimes minute or hidden. (Greek, shield-less, which seems hardly applicable.)

Fronds bipinnate ....1. A. filix-femina
Fronds tripinnate or nearly so. ....2. A. alpestre

1. A. filix-femina (L.) Roth. Lady-fern
A. filix-femina (L.) Roth. var. sitchense Rupr.
A. cyclosorum Rupr.
"Asplenium cyclosorum Rupr." (Henry's Flora of Southern B. C.)

Rhizomes erect or ascending, stout; fronds closely clustered, up to 2 m. long; stipes straw-colored, dark at base; blades lanceolate, attenuate toward both ends; pinnae linear to lanceolate, attenuate or acuminate, sessile; segments from crenate to incised or pinnatifid and dentate; sori oblong, linear or horseshoe-shaped; indusia subentire to toothed or ciliate. Our form is perhaps best classified as var. cyclosorum (Rupr.) C. Chr.

Central Alaska and Bering Sea—Wash.—Ida. The entire species is circumboreal. (Fig. 27.)

2. A. alpestre (Hoppe) Rylands. Alpine Lady-fern
A. americanum (Butters) Maxon.

Rhizomes short, stout; fronds in a crown, 2–9 dm. long; stipes short, sparsely scaly, straw-colored from a dark base; blades oblong-lanceolate, narrowed toward the base; pinnae triangular-lanceolate, their rachises very narrowly winged; pinnules stalked, somewhat obliquely incised, the lower pinnatifid or pinnate; segments sharply toothed; sori round, small; indusia minute and evanescent. The general appearance is quite lacelike. Our form differs somewhat from the old world form and is the var. americanum Butters.

Alpine-arctic situations, southeastern Alaska—Calif. The entire species is circumboreal. (Fig. 28.)

3. EQUISETACEAE (Horsetail Family)

Rushlike plants with perennial, blackish, creeping rhizomes and hollow, jointed, simple or often much-branched stems bearing toothed sheaths at the joints. Spores borne in a terminal cone formed of verticels of peltate bracts bearing on the under surface a few sporangia which open on the inner side; spores uniform, provided with 4 hygroscopic bands; prothallia minute, green, lobed.

EQUISETUM L.

The only genus. (Latin, Equus, horse; and setum, bristle.)

1A. Stem annual, spike rounded at top, stomata scattered.
   1B. Stems of 2 kinds, the fertile ones appearing earlier than the sterile ones.
      1C. Fertile stem simple, soon withering. ....1. *E. arvense*
      2C. Fertile stem later producing branches.
         1D. Branches simple. ....2. *E. pratense*
         2D. Branches compound. ....3. *E. sylvaticum*
   2B. Stems of one kind, branches simple or none.
      1C. Center cavity small. ....4. *E. palustre*
      2C. Center cavity large. ....5. *E. limosum*

2A. Stems perennial, evergreen; spike with a rigid tip; stomata in regular rows.
   1B. Central cavity wanting, stems filiform. ....6. *E. scirpiodes*
   2B. Central cavity present.
      1C. Stems slender, 5–10-grooved. ....7. *E. variegatum*
      2C. Stems medium, 8–12-grooved. ....8. *E. alaskanum*
      3C. Stems stout, 16–36-grooved. ....9. *E. hiemale*

1. *E. arvense* L.   Common Horsetail

Rhizome slightly angled, felted, tuber-bearing; fertile stems erect, light-colored, 5–25 cm. tall; sheaths pale, loose, with 8–12 brownish, lanceolate teeth; spike ovoid, peduncled; sterile stems erect to decumbent, 1–5 dm. long, 6–14-furrowed, the numerous branches in verticels, 3–4-angled, solid; teeth of sheaths lanceolate, sharp-pointed. An extremely variable species. Seven varieties have been recognized in Alaskan material, but these do not seem to be permanent.

Common throughout the territory. Circumboreal, N. and S. Africa, Canaries. (Fig. 29.)

2. *E. pratense* Ehrh.   Thicket or Meadow Horsetail

Stems 1½–4 dm. long with 8–12 ridges, the fertile developing a few branches, spreading in age, the sterile with numerous long, simple branches; sheaths green, loose, the teeth lanceolate with dark middle; branches 3-ridged; teeth of the sheaths deltoid; cone peduncled; rhizome solid, acutely angled.

In woods, Bering Str. east and south. Circumboreal. (Fig. 30.)

3. **E. sylvaticum** L.   Wood Horsetail, Bottle-brush

Stems 1–5 dm. tall, 8–14-ridged, both fertile and sterile developing copious verticillate compound branches; sheaths loose, cylindrical or campanulate, the upper portion brown with more or less cohering teeth; primary branches 4–5-angled, the branchlets 3-angled, the sheaths with 3 divergent teeth.

In woods, Bering Str. east and south. Circumboreal. (Fig. 31.)

4. **E. palustre** L.   Marsh Horsetail

Rhizomes without felt or tubers; stems 2–9 dm. long, slender, the 5–10 angles of the stem with deep, winglike but rounded ridges; branches long, ascending, hollow, 5–7-angled; sheaths loose, widened upward, and the apices acute-subulate; spikes short-peduncled, terminating the stem with smaller ones terminating some of the branches.

Wet places, Bering Str. east and south. Circumboreal. (Fig. 32.)

5. **E. limosum** L.   Swamp Horsetail
  E. fluviatile L.

Stems 5–15 dm. tall, 4–8 mm. thick, 10–30-grooved, with large central cavity, often simple, but more usually sparingly branched above with spreading or more often upcurved, 4–6-angled, simple branches; sheaths appressed with blackish, narrow, distinct teeth; cones short-peduncled.

In shallow water or swamps, Bering Sea east and south. Circumboreal. (Fig. 33.)

6. **E. scirpoides** Michx.   Little Horsetail

Stems tufted, simple or branched from the base, prostrate or weakly ascending, 5–15 cm. long, 6-ribbed by the deep grooving of the 3 angles, the ribs with a regular row of silica tubercles; sheaths loose, becoming black or dark brown, teeth 3, distinct, persistent, with a whitish border and a fragile subulate tip; cones 3–5 mm. long.

Damp situations, from above the Arctic Circle southward. Circumboreal. (Fig. 34.)

7. **E. variegatum** Schleich.   Northern Scouring-rush

Stems slender, tufted, 1–4 dm. long, 1–3 mm. thick, 5–10-grooved, the ridges bearing 2 lines of silica tubercles; sheaths loose, green below, dark above; teeth black with white border, persistent, with a filiform, deciduous tip; cones short-peduncled, 8–12 mm. long.

Throughout most of Alaska. Circumboreal. (Fig. 35.) A very small scirpoides-like form is var. *anceps* Milde.

8. **E. alaskanum** (A. A. Eaton), new comb.   Alaska Scouring-rush
  E. variegatum var. *alaskanum* A. A. Eaton ex Gilbert.[1]

Very similar to *E. variegatum* but much larger, growing up to at least 8 dm. long, the stems 2–4 mm. thick, 8–12-grooved.

Alaska Range and Bering Sea to Wash. Dr. Hultén has suggested

---
[1] Gilbert, B. D., List N. Amer. Pterid., p. 9, 1901.

that this type as well as similar forms elsewhere may have originated from the crossing of *E. hiemale* and *E. variegatum*. It is generally classified as a variety of *E. variegatum* but it appears to be quite distinct.

9. **E. hiemale** L. Scouring-rush

Stems stiff, 5–15 dm. long, 5–10 mm. thick, unbranched, or with a few slender branches near the top, 16–36-ridged, rough with 2 rows of tubercles on the ridges; central cavity large; sheath with dark base and teeth with a light band between, the teeth adhering in groups by their pale, membranous margins; spike pointed, sessile or nearly so, 1–3 cm. long. Our form is the var. *californicum* Milde. which is more robust than the type.

Aleutian islands and central Alaska—Calif.—N. Mex. The entire species is circumboreal. (Fig. 36.)

4. LYCOPODIACEAE (Club-moss Family)

Low, evergreen, often mosslike, usually trailing plants with erect or ascending fruiting branches. Leaves very numerous, usually stiff, imbricate, 1-nerved, lanceolate or subulate; sporangia in the axils of the ordinary leaves, or more often in spikes at the base of modified leaves (sporophylls); spores minute and all of one kind; prothallia fleshy, subterranean.

LYCOPODIUM L.

Sporangia flattened, usually reniform, 1-celled; spores copious, sulfur yellow, inflammable. (Greek, wolf's foot.)

1A. Sporangia in the axils of ordinary leaves, not in spikes. ....
.... 1. *L. selago*
2A. Sporangia borne in the axils of bracts arranged in spikes.
   1B. Spikes borne on bracteate pedicels more than 2 cm. long.
      1C. Branches flat, leaves in 4 rows. ....2. *L. complanatum*
      2C. Branches terete, leaves in many rows. ....3. *L. clavatum*
   2B. Spikes sessile or nearly so.
      1C. Aerial branches simple. ....4. *L. inundatum*
      2C. Aerial branches mostly branched.
         1D. Aerial branches treelike. ....5. *L. obscurum*
         2D. Aerial branches not treelike.
            1E. Leaves 4-ranked. ....6. *L. alpinum*
            2E. Leaves 5-ranked. ....7. *L. sitchense*
            3E. Leaves 8-ranked. ....8. *L. annotinum*

1. **L. selago** L. Fir Club-moss
   L. porophilum Lloyd and Underw.

Stem more or less curved below, erect or ascending above, 2–several times forked, forming tufts 3–20 cm. tall; leaves crowded, appressed or ascending, often spreading or reflexed near the base, narrowly triangular-lanceolate or subulate, acute, usually entire, those bearing sporangia slightly shorter; plant usually producing gemmae. Very variable, the shade forms being dark green; the alpine and arctic form (var. *adpressum* Desv.) has closely appressed leaves and is yellowish-green in color.

Moist, rocky situations, throughout our range. Circumboreal. (Fig. 37.)

2. *L. complanatum* L.   Ground Cedar

Main stem creeping on or slightly below the surface of the ground; aerial branches yellowish-green, 4–40 cm. tall, usually much branched, the branches flattened, glaucous, with minute, decurrent, 4-ranked leaves, the lateral broad, the upper narrow, incurved, the lower small; peduncle 1–10 cm. long, bearing 1–4 spikes, each 1½–4 cm. long; sporophylls broadly ovate, acuminate, erose.

Bering Sea and central Alaska east and south. Circumboreal. (Fig. 38.)

3. *L. clavatum* L.   Running Pine

Main stem creeping, often 1–3 m. long; ascending branches 4–35 cm. tall, pinnately branched; leaves crowded, about 1×4 mm., many-ranked, linear-subulate, incurved-spreading, entire or denticulate, mostly bristle-tipped; peduncles 4–10 cm. long, branched at apex and bearing 2–4 spikes, the whorled or scattered bracts mostly bristle-tipped; sporophylls deltoid-ovate, abruptly acuminate, usually bristle-tipped, the margins membranous and erose.

Coniferous woods, coastal districts. Circumboreal. (Fig. 39.)

Var. *monostachyon* Grev. & Hook. has more incurved leaves about $¾ \times 3¼$ mm. and with more persistent bristles; peduncles 1–5 cm. long, bearing a single spike. Found mostly in the interior but collected near Seward and Juneau in the coast region. This variety is quite distinct so far as Alaska material is concerned, but there are connecting forms found elsewhere.

4. *L. inundatum* L.   Bog Club-moss

Plants small with simple or 1 or 2-forked, short-creeping leafy stems; fertile stems erect, 1–8 cm. tall; leaves of the creeping stems linear-lanceolate and upcurved; leaves of the ascending branches spreading; spike solitary; sporophylls similar to the leaves but with wider ovate base, spreading, usually entire.

Growing in mud, Wrangell—Ore.—Newf.—N. J. Also in Europe and eastern Asia. (Fig. 40.)

5. *L. obscurum* L.   Ground-pine
L. dendroideum Michx.

Main stem creeping underground; aerial branches treelike, 10–35 cm. tall with bushy branches; leaves 8-ranked on the lower branches, 6-ranked on the terminal ones, narrowly lanceolate, spreading but curved upward and usually twisted, acute or mucronate; sporophylls broadly ovate, abruptly acuminate, with scarious, erose margins. Our plant is usually classified as var. *dendroideum* (Michx.) D. C. Eat.

Woods, Alaska distribution scattered, Aleutian Islands, central and southeastern Alaska—Baffinland—Ala.—S. Dak.—Wash. Much used by florists. (Fig. 41.)

6. *L. alpinum* L.   Alpine Club-moss

Main stem creeping on or near the surface of the ground, aerial branches ascending, 2½–11 cm. tall, repeatedly branched, the sterile branches flat with 4-ranked leaves; fertile branches terete with subulate leaves; spikes sessile or nearly so; sporophylls ovate, acute, erose; spores reticulated. Alaska material approaches *L. complanatum* on one hand and *L. sitchense* on the other.

Seward peninsula east and south. Circumboreal. (Fig. 42.)

7. *L. sitchense* Rupr.   Alaska Club-moss

L. sabinaefolium Willd. var. *sitchense* (Rupr.) Fern.

Main stem creeping on or near the surface of the ground, aerial branches several times dichotomous, forming compact tufts 4–8 cm. tall with longer projecting fertile branches; branches terete; leaves of the branchlets 5-ranked, appressed or somewhat spreading, linear, thick, entire, acute; spikes usually sessile, sometimes short-stalked; sporophylls broadly ovate, long-acuminate or subulate, greenish, with scarious, more-or-less erose margins.

Alaska—Lab.—N. Y.—Ore. (Fig. 43.)

8. *L. annotinum* L.   Stiff Club-moss

Main stem creeping on or in moss, up to 4 m. in length; aerial branches 4–35 cm. tall, usually forked 1–4 times; leaves linear-lanceolate, usually serrulate, tipped with a rigid point, spreading or rarely reflexed, upcurved at apex; sporophylls broadly ovate, abruptly acuminate-attenuate. The var. *pungens* (LaPylaie) Desv. is an arctic-alpine form with small, entire, very acute, curved ascending leaves.

Woods, bogs, and alpine meadows, southward from about 68 degrees. Circumboreal. (Fig. 44.)

5. SELAGINELLACEAE (Little Club-moss Family)

Small, leafy, mosslike plants with branching, often prostrate stems and scalelike, 4–6-ranked leaves. Sporangia solitary in the axils of leafy bracts, some containing small, pollen-like spores (microspores), others containing large spores (macrospores) with a roundish base and a triangular-pyramidal apex.

SELAGINELLA Beauv.

Characters of the family. (Diminutive of Selago, ancient name of some Lycopodium).

Bracts thin, spreading, similar to the leaves. .... 1. *S. selaginoides*
Bracts broader, in quadrangular spikes. .... 2. *S. sibirica*

1. *S. selaginoides* (L.) Link. Low Selaginella

Sterile stems prostrate, soft and usually slender, the fertile erect or ascending, 3-8 cm. tall; leaves lanceolate, acute, spreading, sparsely spinulose-ciliate, those of the spike longer, ascending, strongly ciliate; macrospores large, individually visible to the naked eye.

Aleutians, Bering Str., and Wiseman southward and eastward. Circumboreal. (Fig. 45.)

2. *S. sibirica* (Milde) Hieron. Northern Selaginella
S. schmidtii Hieron.

Stems creeping and rooting, forming a dense mat; fertile branches ascending or erect, 1-5 cm. tall; leaves densely imbricated, those of the stem linear-oblong, stiff, about 1½ mm. long with deciduous apical awns about ½ mm. or more long, the margins minutely ciliate; spikes sharply 4-angled, about 1½ mm. thick, the bracts ovate-lanceolate, about 2 mm. long, with short awn.

Dry rocky situations, interior Alaska and Yukon. Also eastern Asia. (Fig. 46.)

6. ISOETACEAE (Quillwort Family)

Small aquatic or marsh plants with short cormlike stem and many crowded subulate or nearly filiform leaves bearing sporangia embedded in their bases. Spores of two kinds, the inner sporangia bearing the microspores, the outer leaves enclosing sporangia with macrospores.

ISOETES L.

The only genus. (Greek, equal at all seasons.)

I. braunii Durieu. Braun's Quillwort
I. braunii Durieu var. *maritima* (Underw.) Pfeiff.
I. echinospora Durieu var. *truncata* Eaton.

Leaves 7-20, erect or spreading, tapering, 2½-5 cm. long; macrospore nearly ½ mm. in diameter, densely covered with broad, often retuse spinules; microspores smooth.

Coastal districts, Aleutians to southeastern Alaska—Greenl.—N. J.—Colo.—Calif. (Fig. 47.)

PHYLUM SPERMATOPHYTA (Seed-bearing Plants)

Plants producing seeds containing young plants in a dormant condition until germination. This seed is the result of the fertilization of the egg-cell of the ovule by a sperm-cell from a pollen-grain. The grains of pollen correspond to the microspore of the heterosporous Pteridophytes while the macrospore is contained within the ovule.

There are probably 150,000 species in existence, and they form the predominant vegetation of the present geological epoch. The diversity and number of species grow progressively greater from the polar regions to the tropics.

Ovules and seed borne on the face of a scale, not enclosed............................
....Class 1. Gymnospermae P.24
Ovules and seed contained in a closed cavity (ovary)...................................
....Class 2. Angiospermae P.41

CLASS 1. GYMNOSPERMAE

Ovules naked, borne on the flat surface of a scale which does not infold to form an ovary, such scale sometimes apparently wanting. Pollen grains dividing at maturity into two or more cells, one of which gives rise to the pollen-tube.

An ancient group which reached its peak in the Triassic geological time; now represented by scarcely 500 species of wide geographic distribution, some of which are of great economic value. Most of the lumber used by mankind is furnished by trees of this group. Only one order (*Coniferales*) of this class is represented in our area. There are two families.

Ovulate flowers without carpellary scales. ....Fam. 1. *Taxaceae* P.24
Ovulate flowers with carpellary scales. ....Fam. 2. *Pinaceae* P.24

1. TAXACEAE (Yew Family)

Evergreen trees or shrubs with linear leaves and dioecious flowers which are axillary and surrounded by bud-scales; the staminate globular, and formed of a few naked stamens; the fertile consisting of an erect ovule developing a fleshy coating.

TAXUS (Tourn.) L.

Branches horizontal or drooping with linear or lanceolate, flat, keeled leaves, revolute on the margins, and persisting 4–5 years. (The classical name.)

T. brevifolia Nutt.   Western Yew

In Alaska reduced to a small tree or shrub not over 10 m. tall. Leaves yellowish-green, 12–16 mm. long, 1–2 mm. wide, acute, 2-ranked by a twist of the flattened and decurrent petiole; fruit red, drupelike.

Extreme southeastern Alaska—Mont.—Calif.

2. PINACEAE (Pine Family)

Resinous trees or shrubs with linear, needle-like or scalelike leaves. Flowers usually monoecious, in scaly aments, the fertile ones becoming cones or berry-like; ovules 2 or more at the base of each fertile scale. All are evergreen except *Larix*. Our species naturally fall into two groups or subfamilies.

1A. Scales of the fertile cones few, opposite (*Cupresseae*).
   1B. Fruit berry-like ....1. *Juniperus*
   2B. Fruit a cone.
      1C. Cone ovoid, its scales oblong......................2. *Thuja*
      2C. Cone globose, its scales peltate. ....3. *Chamaecyparis*

2A. Scales of fertile cones many, alternate (*Abieteae*)
  1B. Leaves in clusters of 2 or more.
    1C. Leaves evergreen. ....4. *Pinus*
    2C. Leaves deciduous ....5. *Larix*
  2B. Leaves solitary.
    1C. Cones erect. ....6. *Abies*
    2C. Cones pendent.
      1D. Leaves flat, blunt. ....7. *Tsuga*
      2D. Leaves quadrangular or thick, acute.8. *Picea*

1. JUNIPERUS (Tourn.) L.

Aromatic trees and shrubs with subulate or scalelike sessile leaves; staminate aments oblong or ovoid, anthers 2–6-celled, each 2-valved; fertile aments of 3–6 fleshy coalescent scales, becoming berry-like, blue fruits, each with 1–6 wingless bony seeds. (The classical name).

Leaves all subulate. ....1. *J. communis*
Leaves mostly scalelike on mature plants......................2. *J. horizontalis*

1. *J. communis* L. var. *montana* Ait.   Low Juniper
  J. sibirica Burgsd.
  J. nana Willd.
  J. communis L. var. *sibirica* Rydb.

A depressed or trailing alpine-arctic shrub forming patches up to 3 m. in diameter. Leaves in whorls of 3, ascending or spreading, 6–10 mm. long, rigid, pungently acute, shining, keeled or strongly convex below, grooved above; staminate aments ovate, 3–6 mm. long; berries globose, 7–9 mm. broad, blue, covered with a white bloom. In some more southern regions this low-growing form intergrades with the upright type.
  Throughout most of Alaska. Circumboreal. (Fig. 48.)

2. *J. horizontalis* Moench.   Creeping Juniper
  J. prostrata Pers.

A prostrate shrub with the stem often rooting. Leaves of young plants subulate, those of the mature stem scalelike, 4-ranked, acute or acuminate; fruits blue, somewhat glaucous, 7–9 mm. in diameter, on short recurved pedicel-like branches.
  Southeastern interior Alaska—Lab.—Newf.—Maine—Iowa—Colo. (Fig. 49.)

2. THUJA L.

Forest trees with frondlike branches, the leaves small and scalelike, appressed, imbricated, opposite, 4-ranked; staminate aments ovate, with 4–6 peltate scales, each bearing 2–4 globose anther-sacs; ovulate aments oblong with 8–12 scales; cones pendulous, their scales thin and flexible. (Ancient name.)

T. plicata D. Don.   Western Red Cedar, Giant Cedar
T. gigantea Nutt.

A large tree with thin, fibrous bark; branchlets bright green and shining above, paler beneath; leaves ovate, short-pointed, about 3 mm. long, obscurely glandular-pitted; cones clustered near the ends of the branches, soon reflexed, 10–14 mm. long; scales leathery. A valuable tree, the wood is soft, brittle, aromatic, light reddish-brown and very durable.

From about 57 degrees N.—Mont.—Calif. (Fig. 50.)

3. CHAMAECYPARIS Spach.

Resembling *Thuja* in general appearance; bark thin, scaly; branchlets 2-ranked in a horizontal plane; leaves scalelike, ovate, acuminate, opposite in pairs; staminate aments oblong; ovulate aments globose, the mature cones with woody, peltate scales, each with a central projection. (Greek, meaning low cypress.)

C. nootkatensis (Lamb.) Spach.   Yellow Cedar, Alaska Cypress

A medium-sized tree with drooping branchlets; leaves hardly glandular, convex or ridged on the back, pointed, appressed except on vigorous shoots; staminate aments about 4 mm. long; cones subglobose, 10–12 mm. broad. The wood is aromatic, sulfur-yellow, fine-grained, and durable.

Along the coast, Prince William Sound to Ore. (Fig. 51.)

4. PINUS (Tourn.) L.

Trees or shrubs with scalelike deciduous primary leaves and needlelike secondary leaves, the secondary leaves borne in clusters of 2–5 terminating short rudimentary branchlets in the axils of the primary leaves and comprising the ordinary foliage which persists for 2–8 years; staminate aments clustered at the base of the seasons growth, forming a distinct zone which remains naked after the aments have fallen; ovulate aments solitary or clustered, borne on twigs of preceding season, composed of many scales and developing into cones the second season, the scales elongating and becoming woody; seeds 2, at the base of the scales, winged above. (The classical Latin name.)

P. contorta Loud.   Lodgepole or Tamarack Pine, Scrub Pine
P. murrayana Balf.

Usually a low scrubby tree or shrub growing in and around muskeags in the coast region. Leaves in 2's, 3–5 cm. long; staminate aments orange-red, about 8 mm. long; cones oblique-ovoid, 3–5 cm. long, usually persisting for several years; wood hard, light reddish-brown, coarse-grained.

Glacier Bay—Calif. The form in Yukon Ter. is the var. *latifolia* Engelm. (Var. *murrayana* Engelm.), an upright-growing, slender tree up to 25 m. tall and distributed in the mountains from the upper Yukon Valley—Colo.—Calif. (Fig. 52.)

5. LARIX (Tourn.) L.

Trees with small, linear, deciduous leaves in fascicles on short, lateral, scaly, budlike branchlets; staminate aments from leafless buds; the ovulate from buds leafy at base, red; cones erect, ovoid, small, with thin scales. (Ancient name.)

L. *laricina* (DuRoi) Koch. American Larch, Tamarack
   L. *americana* Michx.
   L. *alaskensis* Wight.

A small tree, the trunk seldom more than 15 cm. in diameter; leaves 10–20 in a cluster, 15–25 mm. long; cones ovoid, 10–18 mm. long; scales 12–18, suborbicular, thin; wood hard, strong, light brown, resinous, durable.

Wet situations, interior Alaska, Bering Sea—Lab.—Newf.—Mass.—Ill. (Fig. 53.)

6. ABIES (Tourn.) Hill.

Trees with linear, flat, scattered leaves spreading and twisting so as to appear 2-ranked; except on fruiting branches where the leaves are 4-sided and curve upward; staminate aments axillary; ovulate aments lateral; cones erect, subcylindrical or ovoid, their scales deciduous from the persistent axis, thin, incurved at the broad apex. (Ancient name.)

Leaves with stomata on both sides. ....1. *A. lasiocarpa*
Leaves with stomata on lower surface only. ....2. *A. amabilis*

1. *A. lasiocarpa* (Hook.) Nutt. Alpine Fir

A small or medium-sized tree, or at timberline scrubby, with smooth bark except on the oldest and largest trees; branchlets rusty-pubescent; leaves rounded or notched at the apex, grooved on upper side, 20–35 mm. long; scales fan-shaped; wood fine-grained, soft, weak.

Copper river district and southeastern Alaska—Alta.—N. Mex.—Ore. (Fig. 54.)

2. *A. amabilis* (Dougl.) Forbes. Silver Fir, Lovely Fir

A large tree with smooth, gray, white-splotched bark; leaves grooved and green above, whitish beneath, 2–3 cm. long, recurved on the margins, obtuse or notched at apex, erect on the branches by the recurving of those on the lower side; cones oblong, 9–15 cm. long; wood pale brown, hard but weak.

Coast ranges, extreme southeastern Alaska—Ore.

7. TSUGA Carr.

Trees with slender, horizontal, or drooping branches; leaves linear, short-petioled, scattered, appearing 2-ranked by the spreading and twisting of the petioles, jointed to very short sterigmata and falling away on drying; staminate aments axillary, subglobose to ovate; ovulate aments terminal, erect; cones pendulous; scales thin. (Name Japanese.)

Cones small, about 2 cm. long ....1. *T. heterophylla*
Cones larger, about 5 cm. long ....2. *T. mertensiana*

1. *T. heterophylla* (Raf.) Sarg.　　　　　　　　Western Hemlock

A large forest tree up to 60 m. tall and 1½ m. in diameter; branchlets yellowish, pubescent; leaves flat, rounded at the apex, deeply grooved, 8–20 mm. long; staminate aments yellow; ovulate aments purple; cones 16–22 mm. long; scales puberulent; wood pale yellowish-brown, light, hard and strong. Comprises fully 70 per cent of the forest stand in southeastern Alaska.

Kenai Peninsula—Ida.—Ore. (Fig. 55.)

2. *T. mertensiana* (Bong.) Sarg.　　　　　　　　Mountain Hemlock
Hesperopeuce mertensiana (Bong.) Rydb.

A small or medium-sized tree up to 30 m. tall and 9 dm. in diameter, but a mere shrub on muskeags and at timberline; leaves convex or keeled below, grooved above, narrowed toward the base, rounded at the apex, 12–22 mm. long; staminate aments purplish; ovulate aments deep purple; cones sessile, 4–6 cm. long; wood fine-grained, soft, and light.

Cook Inlet—Ida.—Mont.—Calif. (Fig. 56.)

8. PICEA Link.

Forest trees with whorled branches; leaves linear, short, horny-tipped and spreading in all directions, persisting for several seasons, jointed at the base to short persistent sterigmata, falling away in drying; staminate aments axillary; ovulate aments terminal, erect; cones pendulous, their scales numerous, thin, obtuse, persistent. (Name ancient.)

1A. Cones 1½–3 cm. long, persisting for several years 1. *P. mariana*
2A. Cones 4–10 cm. long, falling off at maturity.
　　1B. Leaves quadrangular. ....2. *P. glauca*
　　2B. Leaves rather flat. ....3. *P. sitchensis*

1. *P. mariana* (Mill.) B.S.P.　　　　　　　　　　Black Spruce

A small tree, often scrubby, with pubescent branchlets; leaves stout, generally curved, glaucous, quadrangular, with blunt tip, 6–10 mm. long; cones oval or ovoid; the scales usually with slightly erose margins.

Muskeags and hillsides, southwest Alaska to north of the Arctic Circle—Ungava Bay—Newf.—N. Car.—Wis.—Alta. (Fig. 57.)

2. *P. glauca* (Moench) Voss.　　　　　　　　　　White Spruce
P. canadensis (Mill.) B.S.P.

A medium-sized tree up to 28 m. tall and nearly 1 m. in diameter; branchlets glabrous; leaves rather slender, acute, 12–20 mm. long, bluish-green with more or less bloom; cones cylindric or oblong-cylindric, 3–6 cm. long, the scales thin, entire.

Southwestern Alaska—Noatak River—Ungava Bay—Newf.—Maine —Wis.—Alta. Larger trees of this species furnish most of the lumber sawed in interior Alaska. (Fig. 58.)

3. **P. sitchensis** (Bong.) Carr. Sitka Spruce

Our largest tree, reaching a height of 50 m. and a diameter of 2½ m. or more; branchlets glabrous; leaves acute or acuminate, 15–25 mm. long, keeled on upper surface, rounded or slightly keeled on lower surface, with 2 narrow bands of whitish stomata above and 2 wide bands below; staminate aments dark red; cones cylindrical or narrowly oblong-oval, 5–10 cm. long; scales thin, denticulate above the middle.

Coast region, Kodiak Island and Cook Inlet—Calif. Our most valuable tree. Beside furnishing construction lumber and wood-pulp it is used in airplane construction and for piano sounding boards. (Fig. 59.)

PLATE I

Fig.
1. *Botrychium lunaria*
2. *Botrychium boreale*
3. *Botrychium lanceolatum*
4. *Botrychium silaifolium*
5. *Blechnum spicant*
6. *Struthiopteris filicastrum*
7. *Cryptogramma acrostichoides*
8. *Cryptogramma stelleri*
9. *Adiantum pedatum*
10. *Pteridium aquilinum lanuginosum*
11. *Polypodium vulgare occidentalis*
12. *Woodsia glabella*

PLATE I

PLATE II

FIG. 13. *Woodsia alpina*
14. *Woodsia ilvensis*
15. *Cystopteris fragilis*
16. *Cystopteris montana*
17. *Polystichum lonchites*
18. *Polystichum munitum*
19. *Polystichum braunii*
20. *Dryopteris phegopteris*
21. *Dryopteris linnaeana*
22. *Dryopteris fragrans*
23. *Dryopteris oreopteris*
24. *Dryopteris austriaca*

PLATE II

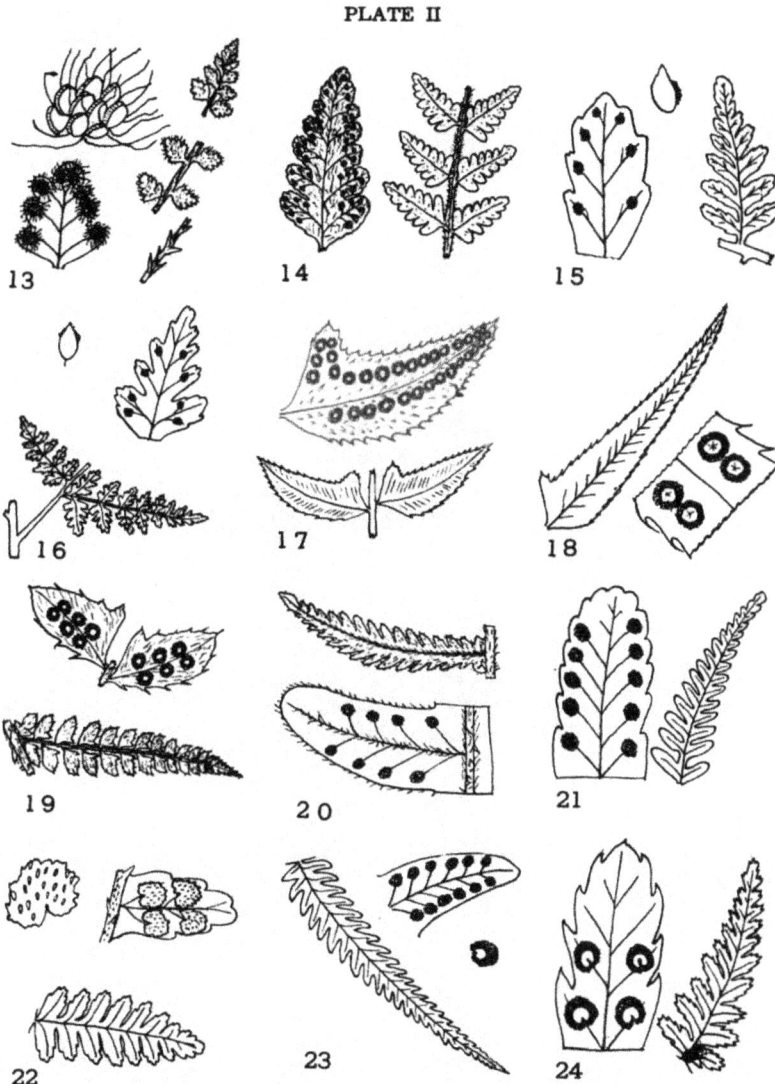

PLATE III

Fig. 25. *Asplenium trichomanes*
26. *Asplenium viride*
27. *Athyrium filix-femina* var. *cyclosorum*
28. *Athyrium alpestre* var. *americanum*
29. *Equisetum arvense*
30. *Equisetum pratense*
31. *Equisetum sylvaticum*
32. *Equisetum palustre*
33. *Equisetum limosum*
34. *Equisetum scirpioides*
35. *Equisetum variegatum*
36. *Equisetum hiemale* var. *californicum*

PLATE III

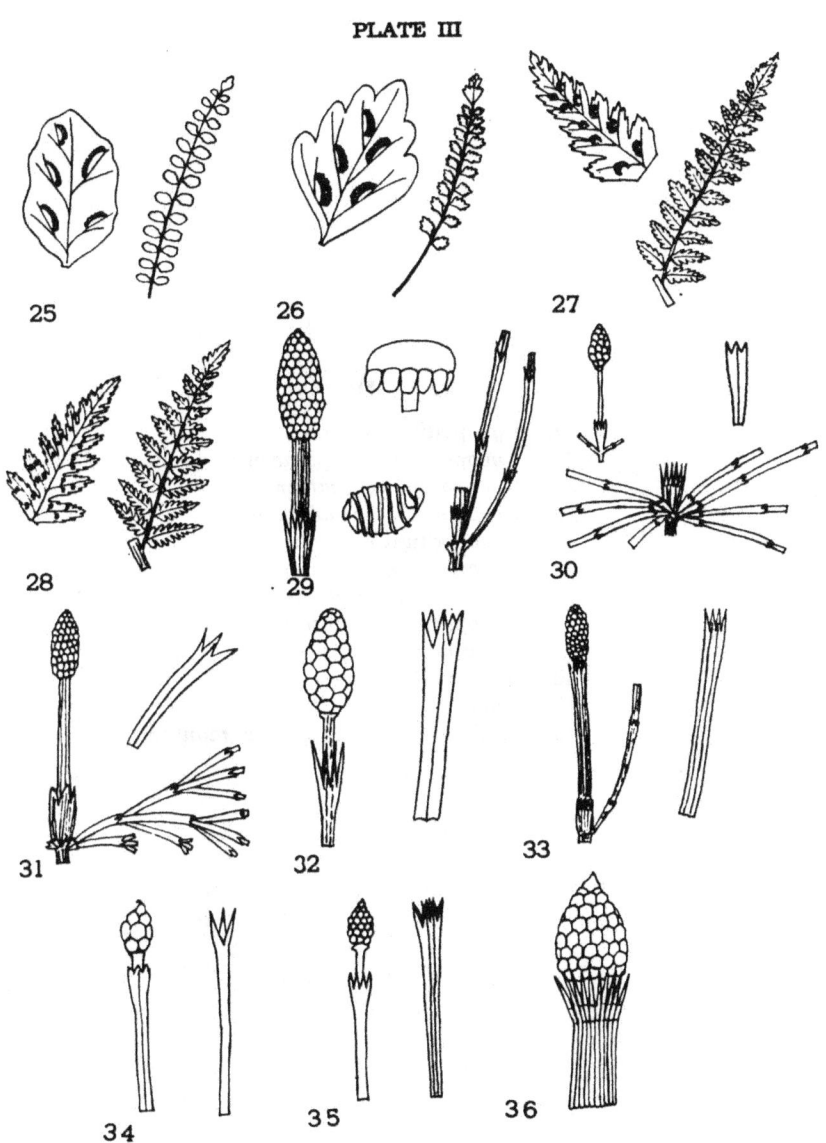

PLATE IV

FIG. 37. *Lycopodium selago*
38. *Lycopodium complanatum*
39. *Lycopodium clavatum*
40. *Lycopodium inundatum*
41. *Lycopodium obscurum*
42. *Lycopodium alpinum*
43. *Lycopodium sitchense*
44. *Lycopodium annotinum*
45. *Selaginella selaginoides*
46. *Selaginella sibirica*
47. *Isoetes braunii*
48. *Juniperus communis* var. *montana*

PLATE IV

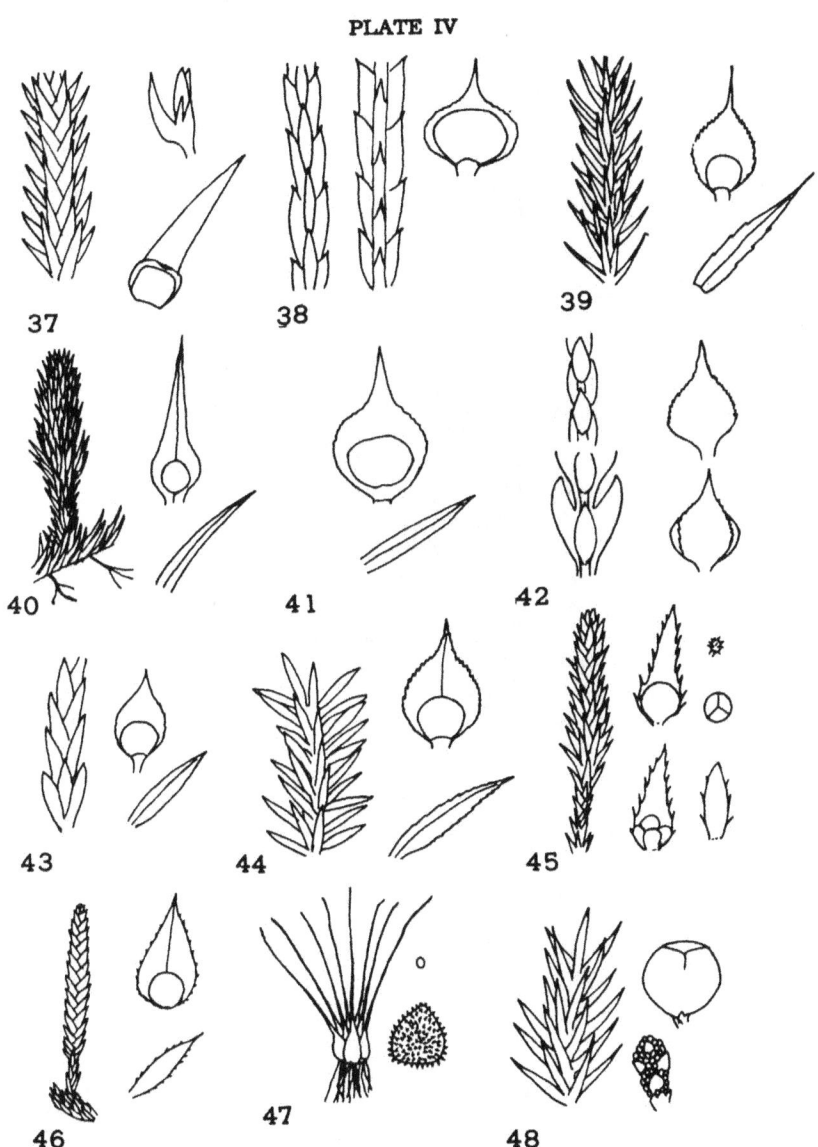

PLATE V

FIG. 49. *Juniperus horizontalis*
50. *Thuja plicata*
51. *Chamaecyparis nootkatensis*
52. *Pinus contorta*
53. *Larix laricina*
54. *Abies lasiocarpa*
55. *Tsuga heterophylla*
56. *Tsuga mertensiana*
57. *Picea mariana*
58. *Picea glauca*
59. *Picea sitchensis*

PLATE V

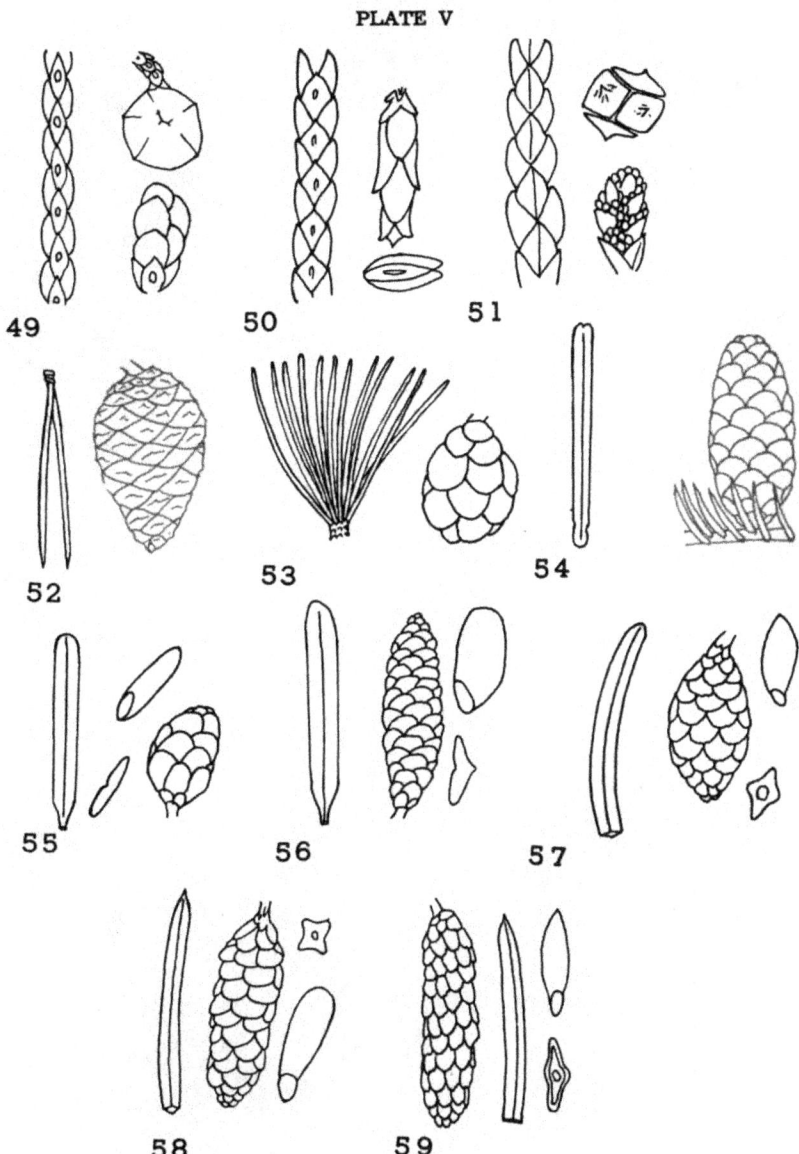

Water arum, *Calla palustris,* page 143.

Class 2. ANGIOSPERMAE

Cotyledons 1, leaves mostly parallel-veined, stems endogenous..................
....Subclass 1. *Monocotyledoneae* P.41
Cotyledons 2, leaves mostly net-veined, stems exogenous............................
....Subclass 2. *Dicotyledoneae* P.181

Subclass 1. MONOCOTYLEDONEAE

1A. Very small, free-floating plants without differentiation into stem and
    leaves ....Family 9. *Lemnaceae* P.143
2A. Plants rooted in the soil.
    1B. Strictly marine plants with ribbon-like leaves.................................
    ....Family 4. *Zosteraceae* P.48
    2B. Not marine but often growing in brackish water.
        1C. Plants submerged, but often with floating leaves.......................
        ....Family 3. *Potamogetonaceae* P.43
        2C. Plants terrestrial or if aquatic, emersed.
            1D. Flowers monoecious, aquatic or marsh plants.
                1E. Flowers in an elongated terminal spike...........................
                ....Family 1. *Typhaceae* P.42
                2E. Flowers in dense spherical heads ....
                ....Family 2. *Sparganiaceae* P.42
                3E. Flowers on a fleshy axis (spadix) and subtended by a
                    large, conspicuous, fleshy bract (spathe).......................
                ....Family 8. *Araceae* P.142
            2D. Flowers perfect.
                1E. Perianth conspicuous.
                    1F. Ovary superior.
                        1G. Styles distinct ....Family 11. *Melanthaceae* P.151
                        2G. Styles united.
                            1H. Herbs with bulbs ....
                            ....Family 12. *Liliaceae* P.153
                            2H. Herbs with rootstocks ....
                            ....Family 13. *Convallariaceae* P.154

2F. Ovary inferior.
    1G. Perianth radial (regular) ....
        ....Family 14. *Iridaceae*   P.157
    2G. Perianth bilateral (irregular) ....
        ....Family 15. *Orchidaceae*   P.158
  2E. Perianth inconspicuous and scaly or absent. (See also *Tofieldia* in *Melanthaceae*.)
    1F. Perianth 6-parted.
      1G. Perianth fleshy, inflorescence a raceme, usually spike-like ....Family 5. *Scheuchzeriaceae* P.48
      2G. Perianth scalelike, inflorescence umbellate or paniculate ....Family 10. *Juncaceae*   P.143
    2F. Perianth absent or represented by 1 or 2 minute scales. Grasslike plants.
      1G. Leaves 2-ranked, fruit a grain..................................
          ....Family 6. *Poaceae*   P.49
      2G. Leaves 3-ranked, fruit an achene ....
          ....Family 7. *Cyperaceae*   P.107

1. TYPHACEAE (Cat-tail Family)

Aquatic or marsh plants; leaves long, linear, flat, striate, sheathing at the base; flowers monoecious, in dense terminal spikes with the staminate part uppermost; perianth of bristles; stamens 2–7; ovary stipitate, 1–2-celled.

TYPHA (Tourn.) L.

The only genus. (Name ancient.)

T. latifolia L.   Common Cat-tail

Stem stout, 1–2.5 m. tall; leaves 6–25 mm. wide; spikes dark brown, the staminate portion lighter than the pistillate, and with bractlets; the pistillate portion without bractlets.

In ponds near Fairbanks. Circumboreal. (Fig. 60.)

2. SPARGANIACEAE (Bur-reed Family)

Marsh or water plants with creeping rootstocks; leaves linear, alternate, clasping at the base; flowers monoecious, in dense, globose heads, the staminate uppermost, the pistillate below, the lower ones peduncled; perianth reduced to a few scales; fruit nutlike.

SPARGANIUM (Tourn.) L.

The only genus (Greek, referring to the ribbon-like leaves).

1A. Peduncles of the upper pistillate heads adnate to the stem (super-axillary).
  1B. Beak very short or lacking...................................1. *S. hyperboreum*
  2B. Beak nearly as long as the achenes.
    1C. Staminate heads remote ....2. *S. simplex*
    2C. Staminate heads approximate ....3. *S. angustifolium*
2A. Pistillate heads all strictly axillary.........................4. *S. minimum*

1. *S. hyperboreum* Laest. Northern Bur-reed

Stem floating and elongated or when growing in mud decumbent and ascending, 1-2 dm. tall; leaves linear, 6-40 cm. long, 1-4 mm. wide, the sheaths somewhat dilated near the base; staminate heads 1 or 2, close to the upper pistillate ones; pistillate heads 2-4, the upper sessile, the lower peduncled, in fruit 8-11 mm. in diameter; achenes ellipsoid.

Generally distributed in our territory except the extreme arctic. Circumboreal. (Fig. 61.)

2. *S. simplex* Huds. Simple-stemmed Bur-reed

Stem rather stout, 4-6 dm. tall, leaves linear, keeled, 4-9 dm. long, 8-15 mm. wide; inflorescence simple; staminate heads 4-8, pistillate heads 2-5, about 15 mm. in diameter at maturity; achenes stipitate, fusiform, 5-6 mm. long, often constricted at the middle; stigma linear.

Extreme southern part of southeastern Alaska. Circumboreal.

3. *S. angustifolium* Michx. Narrow-leaved Bur-reed
S. affine Schniz.
S. multipedunculatum (Morong) Rydb.

Stems floating and elongated or erect and 2-5 dm. tall; leaves 2-6 dm. long, 3-8 mm. wide, dilated and scarious-margined at base, more or less reticulated; staminate heads approximate but distant from the pistillate ones; pistillate heads 2-4, the lower ones peduncled, in fruit 15-20 mm. in diameter; achenes stipitate, fusiform, brown; stigma about 1 mm. long.

Aleutians and Bering Strait—Greenl.—Penn.—Calif. Also N. Europe and Kamtchatka. (Fig. 62.)

4. *S. minimum* (Hartm.) Fr. Small Bur-reed

Stem usually slender and floating, 1-4 dm. long; leaves flat, thin, 2-6 mm. wide, bases of upper ones dilated; pistillate heads 1-3, axillary, the lower sometimes peduncled, less than 1 cm. wide in fruit; achenes ellipsoid to obovoid, sometimes constricted below the middle, short-beaked.

Central Alaska south and east. Circumboreal. (Fig. 63.)

3. POTAMOGETONACEAE (Pondweed Family)

Perennial, mostly fresh-water plants with slender branching stems and floating or submerged leaves or both. Flowers perfect or monoecious, in axillary spikes or clusters; perianth none but flowers sometimes enclosed in hyaline envelopes; stamens 1-4; pistil of 1-4 distinct, 1-celled, 1-ovuled carpels; fruit small druplets.

1A. Flowers perfect.
   1B. Stamens 4, druplets sessile .... 1. *Potamogeton*
   2B. Stamens 2, druplets stipitate .... 2. *Ruppia*
2A. Flowers monoecious, stamen 1 .... 3. *Zannichellia*

1. POTAMOGETON (Tourn.) L.

Leaves alternate or the upper opposite, often of 2 kinds, the submerged thin, pellucid and narrow, the floating broader and coriaceous; stipules present, enclosing the young flower buds; inflorescence spicate, axillary, usually emersed; stamens 4, the connective tissue sometimes becoming perianth-like; carpels 4, distinct; fruit of 4 druplets. (Greek, in allusion to the aquatic habit.)

1A. Leaves of 2 sorts, floating and submerged.
  1B. Submerged leaves without proper blades........1. *P. natans*
  2B. Submerged leaves 2 mm. or more wide.
    1C. Submerged leaves very finely denticulate..
    ....2. *P. gramineus*
    2C. Submerged leaves all entire.
      1D. Submerged leaves ribbon-like ....3. *P. epihydrus*
      2D. Submerged leaves narrowly lanceolate ....
      ....4. *P. alpinus*
2A. Leaves all submerged.
  1B. Stipules free, spike compact.
    1C. Stem flattened, leaves narrow.
      1D. Leaves 9–17-nerved ....5. *P. porsildorum*
      2D. Leaves 1–3-nerved ....6. *P. pusillus*
      3D. Leaves 5-nerved ....7. *P. friesii*
    2C. Stem not conspicuously flattened, leaves broader.
      1D. Leaves with broad blades.
        1E. Leaves half-clasping ....8. *P. praelongus*
        2E. Leaves cordate-clasping ....9. *P. perfoliatus*
      2D. Leaves narrowly lanceolate ....2. *P. gramineus*
  2B. Stipules adnate, spike interrupted.
    1C. Stigmas sessile.
      1D. Leaves filiform ....10. *P. filiformis*
      2D. Leaves narrowly linear ....11. *P. interior*
    2C. Stigmas on a distinct style.
      1D. Leaves blunt ....12. *P. vaginatus*
      2D. Leaves acute ....13. *P. pectinatus*

1. *P. natans* L.   Floating Pondweed

Stems simple or sparingly branched, 6–14 dm. long; floating leaves ovate or elliptical, thick, short-pointed at the apex, rounded or cordate at the base, 4–10 cm. long, 2–5 cm. wide, on rather long petioles; submerged leaves of bladeless petioles and early perishing; stipules 5–10 cm. long, acute, 2-keeled; spike cylindric, dense, 3–5 cm. long; druplets obovoid, 4–5 mm. long; stone 2-grooved on the back.

Coast districts, not common. Circumboreal and nearly cosmopolitan. (Fig. 64.)

2. *P. gramineus* L.   Various-leaved Pondweed
    P. heterophyllus Schreb.

Stems slender, branching, and often very long; floating leaves usually present, oval, pointed at the apex, usually rounded at the base, 1.5–8 cm.

long, 8–28 mm. wide, 9–19-nerved; submerged leaves linear or linear-lanceolate, pellucid, reticulated, 5–15 cm. long, 2–16 mm. wide, 3–9-nerved; peduncles 3–8 cm. long; spikes 2–4 cm. long, many flowered; druplets indistinctly 3-keeled, 2–3 mm. long.

From Bering Str. east and south. Circumboreal. (Fig. 65.)

3. *P. epihydrus* Raf.   Nuttall Pondweed
  P. epihydrus Raf. var. *nuttallii* (Cham. & Schlecht.) Fern.

Stems slender, compressed, 3–18 dm. long; floating leaves opposite, elliptic to obovate, petioled, obtuse, narrowed at the base, 3–8 cm. long, 6–18 mm. wide, many nerved; submerged leaf-blades linear or linear-lanceolate, 2–4 mm. wide, reticulate along the midrib, 5-nerved, the outer nerves nearly marginal; spikes cylindric, many-flowered, 1.5–6 cm. long; druplets round-ovoid, pitted, 3-keeled; style short, apical.

Revillagigedo Island—Labr.—Newf.—Ga.—Colo.—Calif.

4. *P. alpinus* Balbis var. *tenuifolius* (Raf.) Fern.   Northern Pondweed

Plants with a ruddy tinge, simple or branching; floating leaves oblanceolate or spatulate, 5–12 cm. long, sometimes wanting; submerged leaves thin, oblong to linear-lanceolate, 7–30 cm. long; spikes cylindric, 2–4 cm. long; druplets ovoid, lenticular, 2.5–3.5 mm. long, with a sharp middle keel and a short recurved style.

Most of our territory south of the Arctic Circle. Circumboreal. The var. *tenuifolius* is eastern Asiatic and American. (Fig. 66.)

5. *P. porsildorum* Fern.   Porsild Pondweed

Stem simple or branching, 1–6 dm. long; leaves 3.5–9.5 cm. long, 1.5–2 mm. wide, 9–17-nerved, apex rounded, subacute to mucronate; base with 2 prominent glands; stipules subrigid, subpersistent, many-nerved, 1–2.5 cm. long; spikes with 3 or 4 verticils; druplets oblong-ovoid, 3–4 mm. long, 1.5–2 mm. wide, base obliquely truncate.

Known from Buckland River and Takotna to the Mackenzie Delta and James Bay. (Fig. 67.)

6. *P. pusillus* L.   Small Pondweed

Entirely submerged with filiform branched stem; leaves linear, 2–6 cm. long, with strong midrib and usually inconspicuous side veins, 1–1.5 mm. wide, inconspicuous glands at base, the tip often narrowed into a short acumination; peduncles 6–25 mm. long; spikes few-flowered, about 5 mm. long; druplets broadly ovoid, about 2 mm. long, indistinctly 3-keeled. A variable species, of which three varieties of doubtful distinction have been reported from Alaska.

Bering Sea region south and east. Circumboreal. (Fig. 68.)

7. *P. friesii* Rupr.   Fries Pondweed

Stems branching, 4–12 dm. long; leaves linear, 3–6 cm. long, about 2 mm. wide, usually 3-nerved, acute or cuspidate at apex, 2-glandular at base; stipules white, finely nerved, 10–20 mm. long; peduncles often thicker than the stem; mature spikes often somewhat interrupted; druplets with recurved style and usually a shallow pit on sides.

Matanuska. Circumboreal. (Fig. 69.)

8. *P. praelongus* Wulf.   White-stemmed Pondweed

Stems white, flexuous, much-branched, somewhat flattened, up to 25 dm. long; leaves oblong-lanceolate, 5–25 cm. long, 15–30 mm. wide, with 3–5 main nerves; stipules white, scarious, 15–30 mm. long; spikes cylindric, thick, 2–4 cm. long; druplets slightly keeled, 4–5 mm. long.

Atka and Kodiak. Circumboreal.

9. *P. perfoliatus* L.   Clasping-leaved Pondweed
    P. perfoliatus L. var. *gracilis* Fries.
    P. richardsonii (A. Benn.) Rydb.

Stems very leafy; leaves all submerged, thin, lanceolate, with cordate-clasping base, 4–10 cm. long, 8–15 mm. wide; stipules usually conspicuous, often in shreds; peduncles 3–10 cm. long, thickened upward and somewhat spongy; spikes cylindric, 2–3.5 cm. long; druplets obscurely 3-keeled, 3–4 mm. long. This species is represented in central and western Alaska by a near typical form (var. *gracilis* Fries) and in the coastal districts by var. *richardsonii* A. Benn. which has narrower leaves and longer stipules.

Circumboreal. Also northern Africa and southern Australia. (Fig. 70.)

10. *P. filiformis* Pers.   Filiform Pondweed

Stems from running rootstocks, branching, slender above, stouter toward the base; leaves linear-filiform, 5–30 cm. long, less than 1 mm. wide; sheaths 2–3 cm. long; peduncles 4–7 cm. long; spikes interrupted, the verticels 3–20 mm. apart; druplets ovoid, 2–3 mm. long, nearly 2 mm. wide; stigma sessile, forming a broad truncate projection on the druplet.

In most parts of Alaska and Yukon. Circumboreal. (Fig. 71).

11. *P. interior* Rydb.   Interior Pondweed

Stems slender, much-branched; leaves capillary or linear, 3–15 cm. long, about 1 mm. wide, mostly 1-nerved, with acute, pungent apex; adnate portion of stipules 14 mm. or more long, free portion 2–4 mm. long; spikes few-flowered, 15–85 mm. long; druplets obliquely ovoid, 2-grooved on back; stigma subsessile.

Pacific coast districts—Ont.—N. Mex.—Calif.

12. *P. vaginatus* Turcz.   Sheathed Pondweed

Stem compressed, 4–12 dm. long; leaves 0.5–2 mm. wide, up to 3 dm. long; peduncles filiform, 5–10 cm. long; spikes 3–5 cm. long, interrupted; druplets 2.5–3 mm. long, without keel. Related to *P. pectinatus*.

In brackish and salt water, Bering Sea and Pacific coasts. More or less circumboreal.

13. *P. pectinatus* L. Fennel-leaved Pondweed

Stems filiform, much-branched, the branches repeatedly forking; leaves narrowly linear or setaceous, attenuate at the apex, 3–15 cm. long, less than 1 mm. wide; stipular sheaths 1–2 cm. long, adnate one-half their length or more; peduncles filiform, 5–25 cm. long; spikes interrupted with 2–6 verticels; druplets obliquely ovoid, 3–4 mm. long, rounded on the back and with 2 obscure keels.

Matanuska, Cordova, and Circle. Cosmopolitan. (Fig. 72.)

2. RUPPIA L.

Slender, widely-branched water plants with capillary stems and filiform, alternate leaves with membraneous sheaths at the base. Flowers on a capillary spadix-like peduncle which becomes long and coiled in fruit; flowers consisting of 2 sessile anthers and 4 pistils, sessile at first, in fruit long-stipitate; fruit a small obliquely-pointed druplet. (Heinrich Bernhard Rupp was a German botanist.)

Stipular sheaths 15 mm. long..1. *R. spiralis*
Stipular sheaths 20 mm. long ....2. *R. canadensis*

1. *R. spiralis* L. Ditch-grass. Widgeon-grass

Stems much branched, often long; leaves up to 15 cm. long, less than 0.5 mm. wide, and with a sharp tip; stipular sheaths 6–15 mm. long; peduncles in fruit elongating and coiling into a loose spiral; fruit ovoid, about 2 mm. long, obliquely attached.

Salt and brackish water along the coast from St. Paul Island eastward. Cosmopolitan in distribution. (Fig. 73.)

2. *R. canadensis* S. Wats. Western Ditch-grass
   R. lacustris Macoun.

Differs from *R. spiralis* in having stipules 2–4 cm. long, leaves up to 25 cm. long, and generally stouter stems.

Unalaska Island and B. C.—Wash.—Nebr.

3. ZANNICHELLIA L.

Submerged aquatics with capillary, sparsely-branched stems; leaves linear-filiform, 1-nerved; staminate and pistillate flowers in the same axil, the staminate of a single 2-celled anther on a short pedicel-like filament, the pistillate of 2–6 sessile pistils in a cup-shaped involucre; fruit a flattish falcate nutlet with a slender beak, ribbed or toothed on the back. (J. H. Zannichelli was an Italian physician and botanist.)

Z. palustris L. Horned Pondweed

Stems capillary from creeping rhizomes; leaves 2–10 cm. long, 0.5 mm. or less wide, acute at the apex; fruits 2–6 together, 2–4 mm. long, sometimes pedicelled.

In Alaska, known only from the delta of the Buckland River. Cosmopolitan in distribution.

4. ZOSTERACEAE (Eel-grass Family)

Submerged marine plants with creeping rootstocks, flattened branching stems, 2-ranked, ribbon-like leaves, monoecious or dioecious flowers borne on a spadix, enclosed in a spathe, and without perianth but enclosed in a hyaline scale. Staminate flowers consisting of single 1-celled anthers in 2 rows on the spadix which produce filamentous pollen; pistillate flowers of single, 1-celled ovaries composed of two carpels.

Flowers monoecious ....1. *Zostera*
Flowers dioecious ....2. *Phyllospadix*

1. ZOSTERA L.

Marine plants with 2-ranked leaves sheathing at the base, the sheaths with inflexed margins; flowers arranged alternately in two rows on the spadix; pollen threadlike; pistillate flowers fixed on the back near the middle; style elongated; stigma capillary; mature carpels flask-shaped, beaked. (Greek, referring to the ribbon-like leaves.)

Z. marina L.   Eel-grass

Stems branched, arising from a thickish rootstock; leaves ribbonlike, obtuse at apex, 3–15 dm. long, 2–8 mm. wide; spadix 25–60 cm. long, the flowers crowded; at anthesis the anthers escaping and releasing the glutinous, filamentous pollen in the water; fruit strongly 20-ribbed, about 3 mm. long and 1 mm. wide.

Along the coast from Bering Strait south. Circumboreal, but absent from most of the Arctic coasts. (Fig. 74.)

2. PHYLLOSPADIX Hook.

Rootstocks thickened, stems slender, bearing the inflorescence at the summit; leaves linear, sheathing; flowers in spathes, the spadix with a series of short, dilated, foliaceous flaps, which close over the flower; staminate flowers of numerous sessile anthers in 2 rows, producing threadlike pollen; pistillate flowers of single sessile ovaries, tapering into a short style with 2 stigmas; fruit beaked, cordate-sagittate. (Greek, referring to the leaflike appendages of the spathe.)

P. scouleri Hook.   Scouler Surf-grass

Stem winged, 1–4 dm. long; leaves 2–4 mm. wide with 3 primary nerves.

On rocks in the surf, Sitka. Coasts of Japan and B. C.—Calif.

5. SCHEUCHZERIACEAE (Arrow-grass Family)

Marsh herbs with rushlike leaves and small, perfect flowers in spikes or panicles. Perianth 4–6-parted in 2 series; stamens 3–6; anthers 2-celled; carpels 3–6, 1–2-ovuled, more or less united but separating at maturity.

Stems scapose ....1. *Triglochin*
Stems leafy ....2. *Scheuchzeria*

1. TRIGLOCHIN L.

Seaside or marsh herbs with half-round, elongated, linear leaves, sheathing at the base; flowers in long terminal racemes or spikes on a naked scape; stamens 6, the anthers sessile or nearly so; carpels 3–6, 1-celled, 1-ovuled, united at first, at maturity separating from the base upward; stigmas plumose; seed compressed or angular. (Greek, referring to the fruit of the 3-carpelled species.)

Carpels 6, fruit obtuse at base ....1. *T. maritima*
Carpels 3, fruit with subulate base ....2. *T. palustris*

1. *T. maritima* L.   Seaside Arrow-grass

Scape stout, 1–10 dm. tall; leaves 1–6 dm. long, about 3 mm. wide; raceme often 4 dm. long; pedicels 2–4 mm. long, decurrent along the stem, ascending in fruit; fruit 5–6 mm. long, 3–5 mm. wide; carpels triangular, depressed on the back. A form collected near mile 280 on Richardson Highway has fruit only 3–4 mm. long.

Beaches and salt meadows, Kotzebue southward. Occasional in interior from Wiseman south. Circumboreal and to S. America. (Fig. 75.)

2. *T. palustris* L.   Marsh Arrow-grass

Scape slender, 1–5 dm. tall; leaves slender, tapering to a sharp point, 5–30 cm. long; pedicels slender, 2–6 mm. long, erect in fruit; fruit 6–7 mm. long, about 1.5 mm thick, pointed at the lower end.

On very wet soil, Kotzebue and Wiseman south. Circumboreal and in Chile. (Fig. 76.)

2. SCHEUCHZERIA L.

Rushlike, bog plants with creeping rhizomes and erect, leafy stems; leaves elongated, striate, half-round below, flat above, with pore at the apex; flowers small, regular, perfect; perianth 6-parted in 2 series, persistent; stamens 6; anthers linear; ovaries 3, rarely more, separate or connected at the base; stigmas sessile, carpels divergent, 1- or 2-seeded. (Johan Jacob Scheuchzer was a Swiss scientist.)

S. palustris L.   Scheuchzeria

Stems leafy, 1–3 dm. tall; leaves 1–4 dm. long, the upper reduced to bracts; sheaths of the lower leaves up to 1 dm. long; flowers white, segments 1-nerved, 3 mm. long; pedicels 6–20 mm. long, in fruit spreading; follicles 5–9 mm. long.

Extreme southern part of southeastern Alaska. Circumboreal. The American form has longer follicles and styles than the European and has been described as var. *americana* Fern.

6. POACEAE (Grass Family)

Herbs, or in warm climates sometimes woody plants with usually hollow stems (culms) closed at the joints and 2-ranked, parallel-veined leaves, the lower portion forming a sheath enveloping the culm, with an appendage (ligule) at the junction of sheath and blade. Flowers usually perfect, small, with no typical perianth, arranged in spikelets consisting of a shortened axis (rachilla) and 2 to many 2-ranked bracts, the lowest 2 (glumes) empty or rarely obsolete, the succeeding 1 or more (lemmas) bearing a single floret in the axil, and between the floret and the rachilla a second 2-nerved bract (palea); stamens usually 3; pistil of a 1-celled, 1-ovuled ovary with usually 2 styles; fruit a seedlike grain (caryopsis). A large family of cosmopolitan distribution and the most valuable of all plants. Here belong corn, wheat, oats, rye, barley, rice, sugar-cane, bamboo, the sorghums, millet, and most of the hay and forage crops. Indirectly it furnishes most of our meats by furnishing the bulk of the food for all grazing animals.

The family is divided into 2 subfamilies, *Festucoidae* and *Panicoidae*. The latter is not represented in our area. The subfamilies are divided into tribes of which six are represented in our area.

1A. Spikelets distinctly pedicelled, panicles sometimes contracted and spike-like.
  1B. First (lowest) and second florets staminate or neuter......................
     ....1. *Phalarideae*
  2B. Lowest floret perfect, imperfect florets, if any, uppermost.
    1C. Spikelets 1-flowered ....2. *Agrostideae*
    2C. Spikelets 2–many-flowered.
      1D. Lemmas awned on back, glumes longer than the lemmas....
        ....3. *Aveneae*
      2D. Lemmas awnless or with terminal awn, glumes usually shorter than the lemmas ....5. *Festuceae*
2A. Spikelets sessile in spikes.
  1B. Spikes unilateral ....4. *Chlorideae*
  2B. Spikes not unilateral ....6. *Hordeae*

1. Phalarideae

1A. Lower floret staminate; spikelets brown and shining..........................
    .... 1. *Hierochloe*
2A. Lower floret neuter, spikelet green or yellowish.
  1B. Lower florets reduced to small awnless scalelike lemmas; spikelets compressed laterally .... 2. *Phalaris*
  2B. Lower florets consisting of awned, hairy lemmas; spikelets subterete .... 3. *Anthoxanthum*

2. Agrostideae

1A. Lemmas with long terminal awn and closely enveloping the grain......
    .... 4. *Stipa*
2A. Lemmas awnless or short-awned; awn when present dorsal.
  1B. Entire spikelet deciduous at maturity.
    1C. Glumes awnless .... 5. *Alopecurus*
    2C. Glumes awned .... 6. *Polypogon*

2B. Lemmas deciduous above the glumes.
  1C. Glumes awned .... 7. *Phleum*
  2C. Glumes awnless.
    1D. Lemmas 1-nerved .... 8. *Phippsia*
    2D. Lemmas 3-5-nerved.
      1E. Stamen 1; lemma stipitate .... 9. *Cinna*
      2E. Stamens 3, lemmas sessile.
        1F. Lemmas copiously hairy at base....10. *Calamagrostis*
        2F. Lemmas naked or short-hairy at base.
          1G. Glumes longer than the lemmas, spikelets small....
          ....11. *Agrostis*
          2G. Glumes shorter than the lemmas, spikelets large....
          ....12. *Arctagrostis*

3. Aveneae

1A. Spikelets with 1 perfect and 1 staminate floret.
  1B. Lower floret perfect, awnless, upper staminate with hooked awn....
  ....13. *Holcus*
  2B. Lower floret staminate, awned, upper perfect, awnless......................
  ....14. *Arrhenatherum*
2A. Perfect florets 2 or more.
  1B. Lemmas usually awnless.
    1C. Articulation below the glumes, glumes dissimilar........................
    ....15. *Sphenopholis*
    2C. Articulation above the glumes, glumes nearly alike...................
    ....16. *Koeleria*
  2B. Lemmas with twisted awn arising from between 2 terminal teeth
  ....17. *Danthonia*
  3B. Lemmas with dorsal awn.
    1C. Spikelets large, more than 1 cm. long.........18. *Avena*
    2C. Spikelets small, less than 1 cm. long.
      1D. Lemmas keeled, awn arising from above the middle ....
      ....19. *Trisetum*
      2D. Lemmas convex, awn arising from below the middle.............
      ....20. *Deschampsia*

4. Chlorideae

Glumes equal, broad boat-shaped ....21. *Beckmannia*

5. Festuceae

1A. Spikelets nearly sessile in dense one-sided clusters at the ends of the few panicle branches ....22. *Dactylis*
2A. Spikelets not as above.
  1B. Callus barbellate or pilose.
    1C. Panicle erect, the rigid branches often divergent.........................
    ....23. *Dupontia*
    2C. Panicle nodding, the spreading branches capillary.
      1D. Lemmas awned .... 24. *Schizachne*
      2D. Lemmas awnless .... 25. *Colpodium*

2B. Callus naked.
　　1C. Lemmas rounded on back.
　　　　1D. Nerves of the lemmas prominent.
　　　　　　1E. Lemmas long acuminate-pointed................26. *Melica*
　　　　　　2E. Lemmas obtuse ....27. *Glyceria*
　　　　2D. Nerves of lemmas obscure or evident only near the apex.
　　　　　　1E. Lemmas obtuse, awnless.
　　　　　　　　1F. Glumes usually small and shorter than the lemmas....
　　　　　　　　　　..28. *Puccinellia*
　　　　　　　　2F. Glumes usually about as long as the nearest lemma, the lemma usually more or less pubescent...............
　　　　　　　　　　..29. *Poa*
　　　　　　2E. Lemmas acute or obtuse, often awned.
　　　　　　　　1F. Lemmas acute or awned from the apex........................
　　　　　　　　　　..30. *Festuca*
　　　　　　　　2F. Lemmas obtuse, usually awned from below the apex
　　　　　　　　　　..31. *Bromus*
　　2C. Lemmas compressed-keeled.
　　　　1D. Spikelets 1 cm. or more long............................31. *Bromus*
　　　　2D. Spikelets less than 1 cm. long...........................29. *Poa*

6. Hordeae

1A. Spikelets solitary at each node of the rachis.
　　1B. Spikelets placed edgewise to the rachis....................32. *Lolium*
　　2B. Spikelets placed flatwise to the rachis........................33. *Agropyron*
2A. Spikelets 2 or 3 at each node of the rachis.
　　1B. Spikelets 1-flowered ....34. *Hordeum*
　　2B. Spikelets several-flowered ....35. *Elymus*

1. HIEROCHLOE R. Br.

Savastana Schrank.
Torresia Ruiz & Pav.

Perennial, erect, sweet-smelling grasses with small panicles of broad, bronze-colored spikelets; spikelets 3-flowered, the terminal floret perfect, the others staminate; glumes equal, 3-nerved, broad, smooth, acute; staminate lemmas about as long as the glumes, boat-shaped, hispidulous, hairy along the margins; fertile lemma indurate; smooth or nearly so, awnless. (Greek, sacred plus grass.)

1A. Staminate lemmas awned ....1. *H. alpina*
2A. Staminate lemmas awnless.
　　1B. Culm 16 cm. tall or less ....2. *H. pauciflora*
　　2B. Culm 20 cm. long or more ....3. *H. odorata*

1. **H. alpina** (Sw.) Roem. & Schult.　　　　　Alpine Holy-grass
　　Savastana alpina (Sw.) Scribn.

Culms tufted, 1–4 dm. tall, with leafy shoots at the base and short rhizomes; blades 1–2 mm. wide, those of the culm short and wider; panicle

2–4 cm. long; spikelets 5–8 mm. long; glumes glabrous; staminate lemmas ciliate on the margins, the first with a short straight awn, the second with a bent awn, 5–8 mm. long; fertile lemma pubescent near the apex.

Alpine-arctic situations throughout our territory. Circumboreal. (Fig. 77.)

2. H. pauciflora R. Br.   Arctic Holy-grass
Savastana pauciflora (R.Br.) Scribn.

Stems glabrous, erect, simple; basal sheaths overlapping; blades about 1 mm. wide, up to 7 cm. long, involute when dry; stem leaves flat, short and wider, the uppermost almost obsolete; panicle 1–2.5 cm. long, contracted; spikelets few, 3–5 mm. long; glumes smooth and glabrous; stamnate lemmas scabrous, erose-truncate; fertile lemma shorter than the others, obtuse with villous apex.

Arctic regions. Circumpolar. (Fig. 78.)

3. H. odorata (L.) Beauv.   Holy-grass. Sweet-grass
Savastana odorata (L.) Scribn.
Torresia odorata (L.) Hitchc.

Culms 2–6 dm. tall with some leafy shoots and creeping rhizomes; blades 2–6 mm. wide, those of the sterile shoots elongate; panicle 4–12 cm. long, open; spikelets about 5 mm. long; lemmas awnless or nearly so, brown-pubescent.

From about the Arctic Circle south. Circumboreal. (Fig. 79.)

2. PHALARIS L.

Grasses with numerous flat leaves and narrow or spike-like inflorescence; spikelets crowded, laterally compressed, with 1 terminal perfect floret and 2 sterile lemmas below, the rachilla disarticulating above the glumes, the usually inconspicuous sterile lemmas falling close appressed to the fertile floret; glumes equal, boat-shaped, often winged on the keel; fertile lemma coriaceous, enclosing the faintly 2-nerved palea. (Greek, alluding to the shining grain.)

P. arundinacea L.   Reed Canary-grass

Perennial with creeping rhizomes; culms 6–22 dm. tall, glaucous; leaves 6–18 mm. wide, up to 3 dm. long; panicle 7–18 cm. long, the branches spreading during anthesis, later erect; glumes about 5 mm. long, 3-nerved, acute, the keel scabrous; fertile lemma lanceolate, 4 mm. long, with a few appressed hairs; sterile lemmas villous, about 1 mm. long.

Wet places, central Alaska south and east—N. B.—N. Car.—Okla.—Ariz.—Calif. (Fig. 80.)

P. canariensis L., Canary-grass, which furnishes the chief constituent of the bird seed of commerce has been collected a few times, but it is not yet known to be able to maintain itself. It is a glabrous annual with stems branched at the base, 3–9 dm. tall; leaves 4–12 mm. wide; the spikelets broad, imbricate, in a dense, ovoid, headlike panicle; glumes pale with green stripes and a narrowly winged keel. Native of the Mediterranean region.

3. ANTHOXANTHUM L.

Fragrant grasses with spike-like paniculate inflorescence; spikelets with 1 terminal perfect floret, 2 sterile lemmas and unequal glumes; glumes acute or acuminate; sterile lemmas awned from the back, shorter than the glumes but longer than the fertile lemma; fertile lemma awnless; stigmas elongated, plumose. (Greek, referring to the yellowish color of the spikelets of some species.)

A. *odoratum* L.   Sweet Vernal-grass

A tufted perennial 3–6 dm. tall; sheaths shorter than the internodes; leaves 2–5 mm. wide, flat; spike-like panicle greenish yellow-brown, 2–6 cm. long; spikelets 8–10 mm. long; first glume 1-nerved, half as long as the 3-nerved second glume; sterile lemmas subequal, appressed-pilose, the first with a straight awn from near the middle, the second with a long geniculate awn from near the base; fertile lemma smooth and shining, about 2 mm. long.

Established at Unalaska and perhaps elsewhere in Alaska. Native of Eurasia. (Fig 81.)

4. STIPA L.

Tufted perennial grasses; leaves usually convolute; panicles mostly narrow; spikelets 1-flowered, disarticulating above the glumes, the articulation oblique, leaving a bearded, sharp-pointed callus attached to the base of the floret; glumes membranous, acute, acuminate, or aristate; lemma narrow, terete, firm, enclosing the palea and jointed to a usually bent and twisted awn. (Greek, in reference to the feathery awns of the type species.)

Awn 2–2.5 cm. long ....1. *S. columbiana*
Awn 10–15 cm. long ....2. *S. comata*

1. *S. columbiana* Macoun.   Columbia Needle-grass

Stems 3–6 dm. or more tall; sheaths smooth; ligule short; blades 1–3 dm. long, 1–3 mm. wide, mostly involute, those of the stem sometimes flat; panicle 6–20 cm. long, narrow, rather dense, often purplish; glumes about 1 cm. long; lemmas 6–7 mm. long, pubescent; awn twice geniculate, 2–3 cm. long.

Dry plains and meadows, Yukon—Texas—Calif. [Fig. 82 (from a Wyo. specimen).]

2. *S. comata* Trin.   Needle and Thread

Stems 3–6 dm. or more tall; sheaths usually longer than the internodes, smooth or scabrous, the upper one long and inflated, enclosing the base of the panicle; ligule 3–4 mm. long; basal leaves involute-filiform, those of the stem somewhat wider; panicle 1–2 dm. long; glumes 15–20 mm. long with attenuated tips; lemmas 8-12 mm. long, finally brownish with callus about 3 mm. long; awn 10-15 cm. long, very slender, flexuous, indistinctly twice geniculate.

Plains, prairies, and dry hills, Yukon—Ind.—Texas—Calif. [Fig. 83 (from a Wyo. specimen).]

5. ALOPECURUS L.

Ours perennial grasses with flowers in dense spike-like panicles; spikelets 1-flowered, disarticulating below the glumes, compressed; glumes equal, united at the base, ciliate on the keel; lemmas about as long as the glumes, 5-nerved, obtuse, with dorsal awn or point, the margins united at the base; palea none. (Greek, fox plus tail.)

1A. Spikelets 5-6 mm. long ....1. *A. pratensis*
2A. Spikelets 2-4 mm. long.
   1B. Spikelets densely wooly all over.
      1C. Tall-growing, up to 1 m. ....2. *A. glaucus*
      2C. Lower-growing, 15-50 cm. ....3. *A. alpinus*
   2B. Spikelets not densely wooly all over.
      1C. Awn scarcely exceeding the glumes.............4. *A. aequalis*
      2C. Awn exserted 2-3 mm. ....5. *A. geniculatus*

1. *A. pratensis* L.   Meadow Foxtail

Culms erect, 3-10 dm. tall; leaves 2-6 mm. wide; panicle 3-10 cm. long, 7-10 mm. thick; glumes 5-6 mm. long, ciliate on the keel and pubescent on the side nerves; awn exserted, 2-5 mm.

Southeastern Alaska—Labr. and southward. Introduced from Eurasia. (Fig. 84.)

2. *A. glaucus* Less.   Glaucous Foxtail
A. occidentalis Scribn. & Tweedy.

Culms from long-creeping rhizomes, glaucous, up to 1 m. tall; panicle 15-40 mm. long, 8-12 mm. wide. This species has been confused with *A. alpinus* but has longer, more leafy, glaucous culms, longer and more cylindrical panicles, decidedly scabrous leaves.

Bering Sea—Yukon—Alta.—Mont.—Utah—Colo. Asia.

3. *A. alpinus* J. E. Sm.   Mountain Foxtail

Culms from creeping rhizomes, often decumbent at base, 1-6 dm. tall; sheaths glabrous, often inflated, longer than the internodes; blades 3-6 mm. wide; panicle 1-3 cm. long, 7-10 mm. wide; glumes 3-4 mm. long, very wooly; lemmas villous on upper portion, awn attached below the middle, usually exserted. Var. *stejnegeri* (Vasey) Hult. (*A. stejnegeri* Vasey) is characterized by narrow outwardly-curved and very acute glumes which are conspicuously longer than the lemmas. The awn is often longer and the panicle, 12-16 mm. thick. This is the form in the Aleutian Islands. Wet situations, Arctic coast—Central Alaska (Fig. 85.)

4. *A. aequalis* Sobol. Short-awned Foxtail

Culms erect or spreading, 12–60 cm. long; leaves 1–4 mm. wide; panicle cylindric, 2–7 cm. long, 3–5 mm. wide; spikelets about 2 mm. long; awn scarcely exserted or exserted up to 1 mm.; anthers about 1 mm. long.

In water or wet soil, Aleutians, Bering Sea, and central Alaska east and south. Circumboreal. (Fig. 86.)

5. *A. geniculatus* L. Marsh Foxtail

Similar to *A. aequalis*. Culms often rooting at the nodes; spikelets about 2.5 mm. long, the tip purple; anthers about 1.5 mm. long; awn of lemma about twice as long as the spikelet, exserted and giving the panicle a bristly appearance.

Introduced into southeastern Alaska. Native of Eurasia. (Fig. 87.)

6. POLYPOGON Desf.

Mostly decumbent annual grasses with spike-like panicles; spikelets 1-flowered, disarticulating below the glumes, leaving a short-pointed callus attached; glumes equal, awned from the tip; lemma shorter than the glumes, hyaline, usually with a short, straight awn. (Greek, many plus beard, referring to the bristly inflorescence.)

P. monspeliensis (L.) Desf. Rabbit-foot Grass. Annual Beard-grass

Culms erect from a usually decumbent base, 15–50 cm. tall, sometimes depauperate or taller; leaves flat, 4–12 cm. long, 2–6 mm. wide; inflorescence dense, 2–15 cm. long, 1–2 cm. thick; glumes hispidulous, about 2 mm. long; awn 6–8 mm. long; lemma scarcely 1 mm long, its awn 0.5–1.5 mm. long.

Sparingly introduced as a weed. Native of Europe. (Fig. 88.)

7. PHLEUM L.

Perennial grasses with flat leaves; inflorescence a dense, cylindric, spike-like panicle; spikelets 1-flowered, compressed, disarticulating above the glumes; glumes equal, persistent, keeled, abruptly mucronate or awned; lemma shorter, hyaline, truncate, denticulate; palea narrow, nearly as long as the lemma. (Greek, a kind of reed.)

Panicle 1.5–3 times as long as broad ....1. *P. alpinum*
Panicle several times as long as broad...2. *P. pratense*

1. *P. alpinum* L. Mountain Timothy

Culms 15–50 cm. tall from a decumbent base; leaf-blades 2–10 cm. long, 3–8 mm. wide; inflorescence 1–5 cm. long, 7–12 mm. thick; glumes 3–4 mm. long exclusive of the awns which are about 2 mm. long.

Aleutian and Pribilof Islands and Pacific Coast regions of Alaska. Circumboreal and in South America. Our form averages taller and has shorter awns and more inflated upper sheaths than the European plant and has been described as var. *americanum* Fournier. (Fig. 89.)

2. **P. pratense** L. Common Timothy

Culms erect, 5–10 dm. tall from a more or less swollen base; leaf-blades 5–25 cm. long, 4–8 mm. wide; inflorescence cylindric, 3–10 cm. long, 5–8 mm. thick; glumes 3–4 mm. long, scabrous, ciliate on the keel; awns about 1 mm. long.

A native of Eurasia and widely introduced in America, being cultivated for hay and pasture. Spontaneous in all Pacific Coast sections of Alaska and to some extent in the interior. (Fig. 90.)

8. PHIPPSIA R. Br.

Low annual tufted grass; leaves flat; panicle narrow; spikelets 1-flowered, disarticulating above the glumes; glumes unequal, minute, the first sometimes wanting; lemma thin, slightly keeled, 3-nerved, abruptly acute; palea somewhat shorter than the lemma. (John Constantine Phipps, 1744–1792, was an Arctic navigator.)

P. algida (Soland.) R. Br. Phippsia

Glabrous; culms 3–15 cm. long; leaves soft, narrow, with boat-shaped tips; panicles 5–30 mm. long; spikelets 1–1.5 mm. long; grain oblong, enclosed in the lemma and palea.

Arctic Coast to St. Paul Island. Circumpolar. (Fig. 91.)

9. CINNA L.

Tall perennials with broad flat leaves and numerous spikelets in large, often nodding panicles; spikelets 1-flowered, the rachilla forming a stipe below the floret and produced beyond the palea as a minute bristle; glumes nearly equal, keeled, acute; lemma similar, nearly as long, 3-nerved, short-awned or awn-pointed; palea shorter, 1-nerved. (Greek name for some grass.)

C. latifolia (Trev.) Griseb. Slender Reed-grass

Culms 5–15 dm. tall; leaf-blades 5–15 mm. wide; panicle 15–30 cm. long, the branches in verticils, capillary, flexuous, often drooping; spikelets about 4 mm. long; glumes hispidulous; lemma nearly as long as the glumes, hispidulous toward the apex; awn short.

Alaska Range and Pacific Coast districts of Alaska—Labr.—Newf.—N. Car.—Ill.—N. Mex.—Calif. (Fig. 92.)

10. CALAMAGROSTIS Adans.

Erect perennial grasses with panicled inflorescence; spikelets 1-flowered, the rachilla prolonged beyond the palea as a short, usually hairy bristle; glumes nearly equal, persistent, acute or acuminate; the lemma shorter with a basal ring of long hairs and a dorsal awn. Some of the species of this genus are very variable. This variability has resulted in much confusion regarding species and varieties. Some hybridization may have taken place. (Greek, signifying reed-grass.)

1A. Awn comparatively long, bent or geniculate, exserted.
  1B. Low-grown, less than 3 dm. tall.
    1C. Awn fixed at or below the middle of the lemma..........................
    ....1. C. deschampsioides
    2C. Awn fixed near the apex of the lemma....2. C. holmii
  1B. Taller, stout grasses, 3-15 dm. tall.
    1C. 4-8 dm. tall, awn long-exserted ....3. C. purpurescens
    2C. 6-15 dm. tall, awn but slightly exserted..4. C. nutkaensis
2A. Awn not exserted, callus hairs as long or nearly as long as the lemma.
  1B. Panicle open, spreading......................................5. C. canadensis
  2B. Panicle narrow or condensed.
    1C. Awn fixed at or above the middle of the lemma..........................
    ....6. C. neglecta
    2C. Awn fixed below the middle of the lemma.
      1D. Leaves and panicle stiff ....7. C. inexpansa
      2D. Leaves and panicle soft ....8. C. lapponica

1. *C. deschampsioides* Trin.

Stems slender, erect, or decumbent at base, sheaths glabrous; leaves narrow, glabrous, mostly clustered at base, more or less involute, at least when dry; panicle erect, ovate, shining, 2-5 cm. long; spikelets 4-5 mm. long; glumes nearly equal, the lower 1-nerved, the upper 3-nerved, acute; lemma nearly equaling the glumes, bidentate, 5-nerved, awned at or below the middle; palea about equaling the lemma, bidentate; awn slightly longer than the lemma.

Western coast of Alaska, Eurasia, west coast of Hudson Bay. (Fig. 93.)

2. *C. holmii* Lange.   Holm Reed-grass

Relatively low-grown; upper ligules rudimentary; panicle open, small, rather dense; resembles *C. deschampsioides*, but the callus hairs are relatively shorter and the short, bent, or twisted awn is fixed near the apex of the lemma.

St. Paul Island, northeastern Asia, Nova Zembla, Arctic Siberia.

3. *C. purpurescens* R. Br.   Purple Reed-grass
   C. yukonensis Nash.
   Deyeuxia purpurescens (R. Br.) Schult.

Stems tufted, mostly 4-6 dm. tall; sheaths scabrous; leaves 2-4 mm. wide, flat or involute, scabrous; panicle dense, usually pinkish or purplish, somewhat spike-like, 2-12 cm. long; glumes 5-8 mm. long; lemma as long as the glumes, 4-toothed at apex, awned from near the base; hairs of the callus about one-third as long as the lemma.

Ssp. *arctica* (Vasey) Hult. [var. *arctica* (Vasey) Kearney] (*C. vaseyi* Beal) is a form growing up to 1 m. tall with hyaline abruptly pointed glumes. It occurs from the Alaska Peninsula to Japan.

Alaska—Victoria Land—Baffin Land—Greenl.—Que.—S. Dak.—Colo.—Calif.—N. Asia. (Fig. 94.)

4. **C. nutkaensis** (Presl) Steud.   Pacific Reed-grass
    C. *aleutica* Trin.

Culms stout, 7–15 dm. tall; leaf-blades elongate, flat, becoming involute, gradually narrowed into a long point; panicle usually purplish, narrow, 12–30 cm. long, the branches stiffly ascending; glumes 5–7 mm. long, acuminate; lemma nearly as long as the glumes, indistinctly nerved; awn from below the middle, slightly geniculate, scarcely as long as the lemma; callus hairs and rachilla scarcely half as long as the lemma.

Along the coast, Aleutian Islands to Calif. (Fig. 95.)

5. **C. canadensis** (Michx.) Beauv.   Bluejoint

Culms tufted, 6–15 dm. tall, with numerous creeping rhizomes; leaf-blades elongate, flat, 2–8 mm. wide, scabrous; panicle 8–25 cm. long, open, usually purplish, the branches spreading; glumes 3–4 mm. long, acute, more or less scabrous; lemma nearly as long as the glumes, thin in texture, the awn delicate and extending to or slightly beyond its tip; callus hairs abundant, fully as long as the lemma.

The ssp. *langsdorfii* (Link) Hult. [var. *scabra* (Presl) Hitchc.] [C. *langsdorfii* (Link) Trin.] is the form most abundant in the coast districts. It has spikelets 4.5–6 mm. long, firm glumes which are hispid-ciliate on the keel, and the culms may attain a length of 23 dm.

Alaska—Greenl.—N. Car.—Kans.—Ariz.—Calif. (Fig. 96.)

6. **C. neglecta** (Ehrh.) Gaertn.   Narrow Reed-grass

Resembling C. *inexpansa* but averages smaller. Culms 3–8 dm. tall; leaves smooth or nearly so, narrow, often filiform; panicle 5–10 cm. long; spikelets 3–4 mm. long; glumes often nearly smooth except on the keel; callus hairs long but distinctly shorter than the lemma.

Var. *borealis* (Least.) Kearney differs from the type in its low growth, broad and short, flat leaves, and dark, purplish very acute glumes.

Bering Sea—Greenl.—Maine—Wis.—Utah—Ore. The variety is circumpolar.

7. **C. inexpansa** A. Gray.   Northern Reed-grass
    C. *hyperborea* Lange.

Culms often scabrous below the panicle, 6–12 dm. tall, rhizomes present; leaf-blades loosely involute, scabrous, 2–4 mm. wide; panicle narrow, dense, 5–20 cm. long, the branches erect and spikelet-bearing from the base; glumes 3–4 mm. long, scabrous, abruptly acuminate; lemma about as long as the glumes, scabrous, green on the back with purplish tip; awn short, attached about the middle of the lemma or below; some of the rachilla hairs about reaching to the tip of the lemma.

Meadows and marshes, central Alaska—Greenl.—Maine—Mo.—N. Mex.—Calif. (Fig. 97.)

8. **C. lapponica** (Wahl.) Hartm.   Lapland Reed-grass
    C. *alaskana* Kearney

Culms 4–9 dm. tall; leaves and panicle soft; panicle narrow, not very compact, often nodding; glumes purplish on back, brownish at the tip, more or less evenly covered with scattered scabrous pubescence and a few comparatively large spinelike hairs on the middle of the keel; awn fixed near the base of the lemma, reaching about to the top of the glumes, longest callus-hairs as long as the lemma.

Bering Sea—Yukon—Mack.—Eurasia.

11. AGROSTIS L.

Delicate to moderately tall tufted grasses with paniculate inflorescence; spikelets small, numerous, 1-flowered; glumes obtuse, usually shorter and thinner than the glumes, awned or awnless; often hairy on the callus; palea small or obsolete. (Greek, referring to the field habitat of many of the species.)

1A. Palea evident, 2-nerved, at least half as long as the lemma.
  1B. Rachilla prolonged beyond the palea (Podagrostis).
    1C. Spikelets 3 mm. long .... 1. *A. aequivalvis*
    2C. Spikelets 2 mm. long .... 2. *A. thurberiana*
  2B. Rachilla not prolonged.
    1C. Branches of panicle naked at the base ...... 3. *A. tenuis*
    2C. Branches of panicle or some of them floriferous from base.
      1D. Panicles contracted, the branches appressed........................
        .... 4. *A. palustris*
      2D. Panicles open, the branches ascending.
        1E. Stems decumbent at base, rhizome wanting......................
          .... 5. *A. stolonifera*
        2E. Stems erect, rhizome present............ 6. *A. alba*
2A. Palea obsolete or minute.
  1B. Panicle narrow, contracted, branches spikelet-bearing from the base .... 7. *A. exerata*
  2B. Panicle open.
    1C. Panicle very diffuse .... 8. *A. scabra*
    2C. Panicle open but not diffuse.
      1D. Lemma awnless .... 9. *A. idahoensis*
      2D. Lemma awned.
        1E. Awn straight ....10. *A. alaskana*
        2E. Awn geniculate ....11. *A. borealis*

1. *A. aequivalvis* Trin.   Northern Bent-grass
Podagrostis aequivalvis (Trin.) Scribn. & Merr.

Culms tufted, ascending from a spreading base, 2–6 dm. tall; leaves flat, 1–3 mm. wide; panicle usually purplish, 5–15 cm. long, the branches slender; spikelets 3 mm. or more long; glumes about equal, minutely scabrous beneath the tip of the keel; lemma about as long as the glumes, the palea nearly as long as the lemma; prolongation of the rachilla one-fifth to one-half as long as the lemma.

Western Aleutians south to Calif.

2. **A. thurberiana** Hitchc.   Thurber redtop
Podagrostis thurberiana (Hitchc.) Hult.

Culms in small tufts, erect, 2-4 dm. tall; leaves about 2 mm. wide; panicles rather narrow, lax, 5-9 cm. long; spikelets about 2 mm. long, lemmas a little shorter than the glumes; palea about two-thirds as long as the lemma; prolongation of the rachilla short, hairy.

Aleutians—Central Alaska—Mont.—Colo.—Calif. (Fig. 98.)

3. **A. tenuis** Sibth.   Colonial Bent. Rhode Island Bent

Culms slender, erect, 2-5 dm. tall; stolons short; leaves 1-3 mm. wide; panicle 3-10 cm. long, open, delicate; glumes 2 mm. long or less, sometimes slightly scabrous on the keel near the apex.

Introduced into southeastern Alaska. Native of Europe.

4. **A. palustris** Huds.   Dense-flowered Bent
A. maritima Lam.

Culms tufted, erect, usually with a decumbent base, 2-4 dm. tall; leaves erect, rough on both surfaces, 4-8 cm. long, 1-3 mm. wide; panicle dense and compact, 3-10 cm. long; spikelets crowded, acute at both ends, lanceolate when closed, about 2 mm. long, on hispidulous pedicels thickened at apex; glumes acute, hispidulous on the upper part of the keel; lemma hyaline, palea one-half to two-thirds as long as the lemma.

Along the coast, Hyder—Calif. Also east coast and in Europe. (Fig. 99.)

5. **A. stolonifera** L.

Culms ascending, 2-7 dm. tall, with decumbent base and with stolons; leaves flat, 2-4 mm. wide, light or glaucous-green and scabrous; panicle 5-15 cm. long, pale or purple, somewhat open, some of the branches spikelet-bearing to the base; glumes acute, glabrous except the scabrous keel, 2-2.5 mm. long; lemma shorter than the glume, usually awnless, palea more than half as long as the lemma.

Central Alaska—N. Jer.—Ore. Also northern Europe. This form may be native. (Fig. 100.)

6. **A. alba** L.   Redtop

Culms 3-15 dm. tall, erect or decumbent at base, with strong creeping rhizomes; leaves flat, 3-10 mm. wide, panicle reddish, 5-30 cm. long, lower branches verticillate, spreading; glumes acute, 2.5-3 mm. long; lemmas rarely awned.

Often used in lawn and pasture mixtures. Introduced. Native of Europe.

7. **A. exarata** Trin.   Spike Redtop

Culms usually tufted, 2-12 dm. tall; ligule prominent, leaves flat, 1-8 mm. wide; panicle narrow, from somewhat open to spike-like, 10-25 cm. long; glumes acuminate or awn-pointed, 2.5–4 mm. long, scabrous, especially on the keel; lemma about 2 mm. long, often bearing an awn; palea about 0.5 mm. long.

Moist soil, common and variable, Aleutians—Man.—Nebr.—N. Mex.—Mex. and eastern Asia. (Fig. 101.)

Var. *purpurescens* Hult. 3–4 dm. tall, inflorescence dark purple. May be a hybrid with *A. alaskana*.

8. *A. scabra* Willd.   Ticklegrass
    A. hiemalis Auct.

Culms tufted, 2–8 dm. long; leaves mostly basal, the blades usually narrow or even setaceous; panicle large and diffuse, 2–5 dm. long; spikelets crowded at the ends of the branchlets; glumes acute or acuminate, usually purplish; lemma shorter than the glumes, awnless or rarely awned; palea none. At maturity the panicle branches spread widely, and the whole panicle breaks away and rolls before the wind.

Var. *geminata* (Trin.) Hult. (*A. geminata* Trin.) Caespitose perennial; culms 1–3 dm. long; glumes about 3 mm. long; lemma usually awned. Alaska—Calif.

Var. *aristata* Hult. Spikelets about 2.4 mm. long; awns geniculate, exserted; panicle rather contracted. Central and southern Alaska.

Alaska—Newf.—Fla.—Mex.—Calif. (Fig. 102.)

9. *A. idahoensis* Nash.   Idaho Bent-grass

Culms slender, tufted, 1–3 dm. tall; leaves mostly basal, narrow; panicle loosely spreading, 5–10 cm. long, the flexuous branches capillary and minutely scabrous; spikelets 1.5–2 mm. long; lemma awnless; palea minute.

Fairbanks—Wash.—Mont.—N. Mex.—Calif.

10. *A. alaskana* Hult.   Alaska Redtop
    A. melaleuca Am. auct.

Culms tufted, 1–5 dm. tall; panicles 5–20 cm. long, open, the branches in whorls; glumes dark purple, 2.5–3 mm. long, acute, smooth except the scabrous keel; lemmas white, about 2 mm. long, with a short, straight, variable awn; palea none. The var. *breviflora* Hult. has shorter and broader lemmas.

Coastal districts, Aleutians to southeastern Alaska. (Fig. 103.)

11. *A. borealis* Hartm.   Red Bent-grass

Culms tufted, 1–4 dm. tall; leaves mostly basal, 5–10 cm. long, 1–3 mm. wide; panicle 4–12 cm. long, the lower branches whorled and usually spreading; glumes 2.5–3 mm. long, acute, the lower usually slightly longer and more acute than the upper; lemma slightly shorter than the glumes, awned, the awn exserted; palea obsolete or nearly so.

Aleutians and Bering Sea Coast eastward. Circumboreal. (Fig. 104.)

12. ARCTAGROSTIS Griseb.

Perennial grasses; leaves flat; panicle contracted; spikelets 1-flowered; glumes unequal, membranous, acute; lemma longer than the glumes, obtuse; palea obtuse, 2-nerved; the lemma and palea strongly hispidulous. (Latin, Arctic and Agrostis.)

Spikelets 3–4.6 mm. long ....1. *A. latifolia*
Spikelets 2–3 mm. long ....2. *A. poaeoides*

1. *A. latifolia* (R.Br.) Griseb.   Arctagrostis

Culms 3–12 dm. tall, ligule prominent; blades 4–30 cm. long, 4–14 mm. wide; panicle rather narrow, somewhat open, 7–28 cm. long; glumes unequal, smooth, acute; lemma densely hispidulous, appearing acute in side view; palea similar to the lemma. A variable species, the typical form, seldom exceeding 5 dm. in height with purple spikelets 4 mm. or more long, is common in the Bering Sea and Arctic regions.

Var. *arundinacea* (Trin.) Griseb. [*A. arundinacea* (Trin.) Beal] is usually more than 5 dm. tall; spikelets usually less than 4 mm. long; usually purplish. This is the form common in central and southern Alaska. (Fig. 105.)

Var. *angustifolia* (Nash) Hult. (*A. angustifolia* Nash) has long, narrow, very flexible panicles with short branches bearing greenish spikelets and long, lax, bluish-green leaves. Dr. Hulten believes this form to be a hybrid between *A. latifolia* var. *arundinacea* and *A. poaeoides*.

This species is circumboreal.

2. *A. poaeoides* Nash.

Somewhat tufted and with running, branched rootstocks; culms 6–9 dm. tall, erect; sheaths striate, shorter than the internodes; ligule prominent; blades rough on both surfaces, 5–8 mm. wide; panicle about 15 cm. long, the main axis smooth, the pedicels rough; spikelets numerous, 2–3 mm. long; first glume 1-nerved, two-thirds as long as the 3-nerved second glume; lemma short, broad; the glumes, lemma, and palea all thin and translucent.

Central Alaska—Sask.—Man.

13. HOLCUS L.

Perennial grasses; leaf-blades flat; panicles contracted; spikelets 2-flowered, the pedicel disarticulating below the glumes, the rachilla curved and somewhat elongated below the first floret, not prolonged beyond the second floret; glumes about equal, longer than the lemmas; upper lemma bearing a short awn. (Old Latin name for a grass.)

H. lanatus L.   Velvet Grass

Plant grayish velvety-pubescent; culms erect, 3–7 dm. tall; leaves 4–8 mm. wide; panicles 6–15 cm. long, pale, purple-tinged; spikelets about 4 mm. long; glumes villous, hirsute on the nerves, the second broader than the first, 3-nerved; lemmas smooth and shining; awn hooklike.

A European grass sometimes cultivated and naturalized in southeastern Alaska. (Fig. 106.)

14. ARRHENATHERUM Beauv.

Tall perennial grasses; lower floret staminate, the upper perfect, the rachilla disarticulating above the glumes and produced beyond the upper floret; glumes rather broad and papery, the first 1-nerved, the second longer and 3-nerved, about the full length of the spikelet; lemmas 5-nerved, hairy on the callus, the lower bearing a short, minute, straight awn from near the tip; panicle rather dense. (Greek, for masculine plus awn, referring to the awned staminate floret.)

A. elatius (L.) Mert. & Koch. Tall Oat-grass

Culms erect, smooth, 10–15 dm. tall; leaves flat, scabrous on both sides, 3–8 mm. wide; panicle shining, narrow, 1–3 dm. long, the short branches verticillate; spikelets 7–8 mm. long; lemmas scabrous.

Introduced in southeastern Alaska. Native of Europe. (Fig. 107.)

15. SPHENOPHOLIS Scribn.

Slender perennials; leaves flat; panicles narrow, shining; spikelets 2- or 3-flowered, the pedicel disarticulating below the glumes; the rachilla produced beyond the upper floret; glumes unlike, the first narrow, acute, 1-nerved, the second broadly ovate, 3–5-nerved, somewhat coriaceous, the margin scarious; lemmas scarcely nerved, awnless; palea hyaline. exposed. (Greek, wedge plus scale.)

S. intermedia (Rydb.) Rydb. Slender Wedgegrass

Culms tufted, erect, 3–10 dm. tall; panicle erect and spike-like, somewhat interrupted and lobed; 5–20 cm. long; spikelets 2.5–3.5 mm. long.

Occurs at Manly Hot Springs. B. C.—Newf.—Fla.—Ariz. (Fig. 108.)

16. KOELERIA Pers.

Tufted grasses; leaf-blades narrow; panicle spike-like; spikelets 2–4-flowered, the rachilla disarticulating above the glumes and between the florets, prolonged beyond the perfect floret and often bearing a rudimentary floret at the tip; first glume narrow, 1-nerved, somewhat shorter than the second, which is wider and 3–5-nerved; lemmas somewhat scarious and shining, a little longer than the glumes, acute or short-awned. (George Ludwig Koeler was a German botanist.)

Culms and sheaths glabrous ....1. *K. yukonensis*
Culms and sheaths pubescent ....2. *K. cairnesiana*

1. *K. yukonensis* Hult. Yukon Koeler-grass

Densely caespitose; culms 20–25 cm. tall, glabrous; basal leaves filiform, 0.2–0.3 mm. wide, glabrous, acute, ashy-green; cauline leaves wider, involute; spikelets about 7 mm. long, 3–5-flowered, long-pedicelled; glumes of about equal length; lemmas 5–7 mm. long, 5-nerved.

Known only from the upper Yukon district.

2. **K. cairnesiana** Hult.   Cairnes Koeler-grass

Culms 25–30 cm. tall, upper part pilose, slender; basal leaves filiform, about 0.5 mm. wide, glabrous; cauline leaves about 1 mm. wide, ciliate; panicles about 25 mm. long, dense, violet-tinged; spikelets 2-flowered, short-stipitate; glumes wide-lanceolate, scarious; lemmas about 3.5 mm. long, sparingly long-pilose, hyaline-margined.

Known only from the upper Yukon region.

17. DANTHONIA DC.

Tufted perennials; panicles spike-like; spikelets several-flowered; rachilla pubescent, extending beyond the florets; glumes about equal, broad, papery, acute, usually exceeding the florets; lemmas rounded on the back, the apex bifid, the lobes acute, often extending into slender awns, with a stout, flat, twisted, geniculate awn arising between them. (Etienne Danthione was a French botanist.)

Glumes pilose on the back .... 1. *D. spicata*
Glumes glabrous on the back .... 2. *D. intermedia*

1. *D. spicata* (L.) Beauv.   Poverty Oat-grass

Culms 2–5 dm. tall, slender; sheaths pubescent at the mouth; blades filiform to 2 mm. wide; panicle 2–5 cm. long, often 1-sided; spikelets 3–10; glumes 10–12 mm. long, acute; lemmas 4–5 mm. long, sparsely villous except the 2-toothed summit, the teeth acuminate or subsetaceous, 1.5–2.5 mm. long, flat.

Southeastern Alaska—Newf.—Fla.—N. Mexico—Ore. (Fig. 109.)

2. *D. intermedia* Vasey   Timber Oat-grass

Culms 1–5 dm. tall; blades subinvolute, or those of the stem flat; panicle 2–5 cm. long, few-flowered, purplish; branches appressed; glumes 12–16 mm. long, appressed-pilose along the margin below and on the callus, the teeth acuminate, aristate-tipped; awn 7–8 mm. long.

Extreme northwestern B. C.—Labr.—W. Newf.—Mich.—N. Mex.—Calif. Also Kamtchatka.

18. AVENA (Tourn.) L.

Annual or perennial grasses; spikelets usually large, in open or contracted panicles, 2–several-flowered, the rachilla bearded, disarticulating above the glumes and between the florets; glumes about equal, longer than the first floret and usually exceeding the upper one; lemmas indurate, except near the tip, 5–9-nerved, bidentate at tip, bearing a dorsal, bent, and twisted awn which is much reduced in the cultivated oat. (Old Latin name for the oat.)

A. fatua L.   Wild Oat

Annual; culm 3–12 dm. tall; leaves numerous, the blades flat, 1–3 dm. long, 5–15 mm. wide; panicle loose and open, 1–3 dm. long, the rachilla and lower part of the lemmas with brownish hair; lemmas 12–20 mm. long; awn 15–40 mm. long.

Introduced in southeastern Alaska. (Fig. 110.)

A. sativa L., the cultivated oat, has usually smooth lemmas, 2-flowered spikelets, and the awn straight or wanting. It is occasionally found along roadsides.

19. TRISETUM Pers.

Tufted perennial grasses; leaf-blades flat; inflorescence a spike-like, contracted, or open panicle; spikelets usually 2–3, rarely 4- or 5-flowered, the rachilla extending beyond the florets; glumes unequal, acute, entire at the apex, awnless, persistent; lemmas 2-toothed at the apex, the teeth acuminate and often bristle- or awn-pointed; awn often twisted and inserted below the apex on the back of the lemma. (Latin, referring to the awn and the two sharp teeth of the lemma.)

1A. Glumes nearly equal .... 1. *T. spicatum*
2A. Second glume much longer than the first.
   1B. Panicle lax or drooping .... 2. *T. cernuum*
   2B. Panicle dense .... 3. *T. sibiricum*

1. *T. spicatum* (L.) Richt.   Downy Oat-grass
   T. subspicatum Beauv.
   T. alaskanum Nash.

Culms densely tufted, 15–50 cm. tall; sheaths and usually the blades puberulent; panicle dense and spike-like, often interrupted at the base, pale or more often purplish, 5–15 cm. long; spikelets 2–3-flowered, 5–6 mm. long; glumes somewhat unequal, acute or acuminate; lemmas 4–5 mm. long, the teeth setaceous; awn 5–6 mm. long, bent and twisted.

An arctic-alpine plant of cosmopolitan distribution and exceedingly variable. (Fig. 111.)

2. *T. cernuum* Trin.   Nodding Trisetum

Culms 6–12 dm. tall; leaves thin, flat, 1–2 dm. long, 4–8 mm. wide; panicle open, lax, drooping, 1–3 dm. long, the branches verticillate, flexuous, spikelet-bearing toward the ends; spikelets 6–12 mm. long, 2–3-flowered; lemmas 5–6 mm. long, the teeth setaceous, the awn twice as long as the lemmas.

Woods, southeastern Alaska—Ida.—Calif. (Fig. 112.)

3. *T. sibiricum* Rupr.   Siberian Trisetum

Culms erect, simple, smooth, 3–6 dm. tall; sheaths glabrous; leaves sometimes sparingly hairy; panicle contracted, 5–10 cm. long, brownish, shining; spikelets 5–10 mm. long, mostly 3-flowered; second glume much longer than the first; lemmas brown, about 5 mm. long; awn twisted and curved, 4–7 mm. long.

An Asiatic species. The plant here described and figured is the arctic form which is found on the tundra in the northern part of the Bering Sea region. (Fig. 113.)

20. DESCHAMPSIA Beauv.

Mostly perennial grasses; panicles contracted or open; spikelets shining, pale or purplish, 2-flowered, the rachilla articulated above the glumes and prolonged beyond the florets; glumes nearly equal, persistent, keeled, acute; lemmas thin, almost hyaline, 2-4-toothed at the apex, bearded at the base, bearing a slender awn at or below the middle. (J. C. A. Loiseleur-Deslongchamps was a French physician and botanist.)

1A. Glumes extending beyond the lemmas.
   1B. Annual ....1. *D. danthonioides*
   2B. Perennial.
      1C. Panicle very narrow ....2. *D. elongata*
      2C. Panicle spreading ....3. *D. atropurpurea*
2A. Upper lemma extending to or beyond the glumes.
   1B. Awn exserted, geniculate, twisted ....4. *D. flexuosa*
   2B. Awn included or slightly exserted, nearly straight.
      1C. Panicle open.
         1D. Spikelets 6-7 mm. long ....5. *D. beringensis*
         2D. Spikelets 3-5 mm. long ....6. *D. caespitosa*
      2C. Panicle narrow ....7. *D. holciformis*

1. *D. danthonioides* (Trin.) Munro.   Annual Hair-grass
   D. calycina Presl.

Stems slender, erect, 15-60 cm. tall; leaves few, short, narrow; panicle open, 7-20 cm. long, the capillary branches usually in twos, ascending, naked below; glumes 4-8 mm. long, 3-nerved, smooth except on the keel; lemmas smooth, shining, 2-3 mm. long, those of the base floret and the rachilla pilose; awns geniculate, 4-6 mm. long.

Probably introduced; Alaska—Mont.—Lower Calif. Also in Chile.

2. *D. elongata* (Hook.) Munro.   Slender Hair-grass

Culms densely tufted, erect, slender, 3-12 dm. tall; leaves 1-1.5 mm. wide, those of the basal tuft filiform; panicle very narrow, 1-3 dm. long, the capillary branches appressed; glumes 3-nerved, 4-6 mm. long, equaling or exceeding the florets; lemmas 2-3 mm. long, pilose at base, finely toothed at apex; rachilla pilose; awn inserted near the base of the lemma, 4-6 mm. long.

Alaska—Wyo.—Calif.—Mexico. Also in Chile. (Fig. 114.)

3. *D. atropurpurea* (Wahl.) Scheele.   Mountain Hair-grass
   Vahlodea atropurpurea (Wahl.) Fr.

Culms loosely tufted, purplish at base, 3-8 dm. tall; leaves flat, 2-6 mm. wide; panicle loose, open, 5-10 cm. long; spikelets mostly purplish, broad; glumes about 5 mm. long, broad, exceeding the florets; lemmas about 2.5 mm. long; awn attached at about the middle, bent. Our plant differs from the eastern American and European form in having wider leaves, upper part of inflorescence hairy, and shorter callus hairs which are one-third to one-half as long as the lemma. It was described as var. *paramushirensis* Kudo (ssp. *paramushirensis* Hult.) (var. *latifolia* Scribn.).

Var. *patentissima* has large panicle branches in verticels of 4–6, the lower up to 15 cm. long and bearing spikelets only on the distal fourth of their length.

The Pacific form occurs in the coastal districts of eastern Asia around the Bering Sea and south to Calif. (Fig. 115.)

4. *D. flexuosa* (L.) Trin.   Wavy Hair-grass

Stems densely tufted, erect, slender, 3–8 dm. tall; leaves numerous, mostly in a basal tuft, the sheaths scabrous, the blades involute, slender or setaceous, flexuous; panicle open, nodding, 5–12 cm. long, spikelets 4–5 mm. long, purplish or bronze; glumes 1-nerved, acute; lemmas scabrous, the callus hairs about 1 mm. long, awn attached near the base; awn geniculate, twisted, 5–7 mm. long.

Attu and Sitka—Greenl.—N. Car.—Okla. Also Eurasia, S. Am., E. Africa.

5. *D. beringensis* Hult.   Bering Hair-grass
D. bottnica Am. auct. not Trin.

Culms tufted, 3–12 dm. tall; leaves 1.5–4 mm. wide; panicle 7–20 cm. long, open, nodding, the branches scabrous; spikelets often 3-flowered, 6–7 mm. long; glumes about reaching to the top of the second floret, narrow, acute; lemmas 4–5 mm. long with long hairs at the base; palea scabrous. Hybridizes with *D. caespitosa* giving rise to intermediate forms.

Eastern Asia—Aleutian and Pribilof Islands—Ore. (Fig. 116.)

6. *D. caespitosa* (L.) Beauv.   Tufted Hair-grass
D. alaskana L. & M.

Culms in dense tufts, erect, 2–12 dm. tall; panicle 1–3 dm. long, loose, open, nodding; spikelets 3–5 mm. long, pale or purple-tinged, the rachilla joint half the length of the flower floret; glumes acute, 3–4.5 mm. long, lemmas 2.5–3.5 mm. long; awn about as long as, or slightly exceeding the lemma. A very variable species of circumboreal distribution. (Fig. 117.)

Hulten recognizes two forms as sufficiently distinct from the type to be entitled to recognition. Both are usually less than 3 dm. tall. Var. *glauca* (Hartm.) Sam. (*D. curtifolia* Scribn.) has filiform basal leaves which are bluish-green, the florets are very small, and the awn is fixed close to the base of the lemma.

Ssp. *orientalis* Hult. Basal leaves broad, flat, usually yellowish-green. This is the form found along the Arctic and Bering Sea Coasts.

7. *D. holciformis* Presl.   California Hair-grass

Culms caespitose, 5–12 dm. tall, relatively robust; leaves mostly basal, tightly folded or involute, firm, the lower ones long; panicle 10–25 cm. long, rather dense, purplish to brownish; spikelets 6–8 mm. long; glumes and lemmas scaberulous, the glumes about equaling or a little shorter than the spikelet; lemmas awned from below the middle; awns erect, exceeding the spikelet.

Along the coast, Ketchikan—Calif. (Fig. 118.)

21. BECKMANNIA Host.

Rather tall, more or less erect annuals; leaves flat; inflorescence consisting of numerous short, appressed or ascending spikes in a narrow somewhat interrupted panicle; spikelets 1–2-flowered, on one side of a slender rachis, falling off entire; glumes equal, saccate, 3-nerved; lemmas narrow, 5-nerved, about as long as the glumes. (Johann Beckmann, 1739-1811, taught natural history in St. Petersburg, now Leningrad.)

B. *syzigachne* (Steud.) Fern.   Slough-grass
B. *erucaeformis* Am. Auct.

Culms 3–10 dm. long; leaf-blades 4–8 mm. wide; panicle 10–25 cm. long; spikes crowded, 1–2 cm. long; spikelets 1-flowered, 3 mm. long; glumes transversely wrinkled, and with a deep keel; lemma with an acuminate apex protruding above the glumes.

Growing in mud or water, Alaska—Man.—Ill.—N. Mex.—Calif. Also in N. Y., Ohio, Asia. (Fig. 119.)

22. DACTYLIS L.

Tall perennial grass; panicle contracted, the spikelets crowded at the ends of the branches in unilateral clusters; spikelets 3–5 flowered; glumes unequal, acute, hispid–ciliate on the keel; lemmas compressed-keeled, mucronate, 5-nerved, ciliate on the keel. (Greek, finger, referring to the stiff branches of the panicle.)

D. *glomerata* L.   Orchard Grass

Culms 6–12 dm. tall, arising from large, dense tussocks; leaves flat, elongate, 2–8 mm. wide; panicles 5–20 cm. long, the branches spreading in anthesis, appressed at maturity, the rachis hispid; lemmas about 6 mm. long, mucronate or short-awned.

Introduced, native of Europe. (Fig. 120.)

23. DUPONTIA R. Br.

An arctic perennial grass; leaf-blades flat; panicle narrow; spikelets 2–4-flowered; glumes extending beyond the lemmas, membranous; lemmas membranous, entire, with a tuft of hair at the base. (J. D. Dupont was a French botanist.)

D. *fischeri* R. Br.   Dupontia

Culms smooth, erect, simple, 12–50 cm. tall; sheaths overlapping; blades 3–15 cm. long, 2–4 mm. wide; panicle usually contracted, 4–12 cm. long; spikelets mostly 2-flowered, 6–8 mm. long; glumes thin, generally acute, the first 1-nerved and usually shorter than the second which is usually 3-nerved; lemmas 4–6 mm. long, 1-nerved or obscurely 3-nerved. The typical form with blunt-tipped glumes and subsericeous lemmas is a low-growing plant of the Arctic Coast. Ssp. *psilosantha* (Rupr.) Hult. (*D. psilosantha* Rupr.) is a taller growing form with acute glumes and glabrous lemmas. Intermediate forms occur where both types are found.

Arctic and Bering Sea coasts. (Fig. 121.)

24. SCHIZACHNE Hack.

Rather tall perennials with simple culms and open, rather few-flowered panicles; spikelets several-flowered, disarticulating above the glumes and between the florets; glumes unequal, 3-nerved and 5-nerved; lemmas lanceolate, strongly 7-nerved, long-pilose on the callus, awned from just below the teeth of the strongly bifid apex; palea with softly pubescent submarginal keels, the hairs longer toward the summit. (Greek, schizein, to split, plus achne, chaff, referring to the teeth of the lemma.)

S. purpurescens (Torr.) Swallen.   False Melic
Avena striata Michx.

Culms erect from a loosely tufted, decumbent base, 5–10 cm. tall; blades flat, narrowed at the base, 1–5 mm. wide; panicles about 1 dm. long, the branches more or less drooping, bearing 1 or 2 spikelets; spikelets 20–25 mm. long; glumes purplish; lemmas about 1 cm. long, the awn as long or longer than the lemma.

Woods, southern Alaska—Newf.—Penn.—S. Dak.—Mont.—N. Mex.—B. C.—Siberia—Japan. (Fig. 122.)

25. COLPODIUM Trin.

Annual or perennial grasses; leaf-blades flat or almost setaceous; panicles diffuse, pyramidal, the branches capillary; spikelets 1–6 flowered, often colored, the rachilla disarticulating above the glumes and between the florets; glumes membranous or hyaline, 1–3-nerved or nerveless, obtuse or rather acute, unequal; lemmas with texture of the glumes, broad, obtuse, more or less 5-nerved, the lateral ones short or almost obsolete; palea almost as long as the lemma, hyaline.

Leaves 5–8 mm. broad ....1. *C. fulvum*
Leaves 1 mm. or less broad ....2. *C. wrightii*

1. *C. fulvum* (Trin.) Griseb.
   Arctophila fulva (Trin.) Rupr.

A stout perennial, 2–9 dm. tall, rarely taller; culms and leaves smooth; blades flat, pungent-pointed or sometimes obtuse, 5–25 cm. long, 5–8 mm. wide; panicle open, ovoid, 8–15 cm. long, the branches drooping and bearing spikelets on the outer half; spikelets pedicellate, ovate or oblong, 5–6 mm. long, 4–6-flowered; first glume 1-nerved, second glume 3-nerved and about as long as the lemma; lemmas 3–5-nerved, 3–4 mm. long. The Arctic form of this species has the branches of the panicle ascending and the spikelets are smaller, often only 1 or 2-flowered. This is the var. *effusum* (Lange) Polunin. There are intermediate forms.

Shallow water or mud, throughout most of our territory. Circumpolar. (Fig. 123.)

2. *C. wrightii* Scribn. & Merr.

Poa wrightii (Scribn. & Merr.) Hitchc.

Densely caespitose perennial, glabrous, 3–5 dm. tall; basal leaves rather short, linear, involute, about 5 cm. long, 1 mm. or less wide, those of the culm shorter; panicles open, purplish, 4–9 cm. long, the branches glabrous, the lower ones usually in pairs, spreading or ascending; spikelets 3- or 4-flowered, 6–8 mm. long; glumes unequal, the first 1.5–2.5 mm. long, the second 2.5–3.5 mm. long, obtuse, 3-nerved; lemmas lanceolate, rather obtuse, 4.5–5 mm. long, quite prominently 5-nerved, appressed silky-pubescent on the back toward the base, glabrous above.

Seward Peninsula and eastern Asia.

26. MELICA L.

Moderately tall perennial grasses with the base of the culm often swollen into a corm; spikelets 2–several-flowered, the rachilla disarticulating above the glumes and between the fertile florets, prolonged beyond the perfect florets and bearing at the apex 2 or 3 smaller empty lemmas each enclosing the one above; glumes somewhat unequal, thin, papery, scarious-margined, 3–5-nerved, sometimes nearly as long as the lower lemma; lemmas convex, membranous or rather firm, scarious-margined, usually awnless. (Italian for a sorghum, from the Greek, mel, honey.)

M. subulata (Griseb.) Scribn. Alaska Onion-grass

Bromus subulatus Griseb.

Culms 6–12 dm. tall, mostly bulbous at the base; leaves thin, usually 2–5 mm. wide; panicle usually narrow, the branches appressed or sometimes spreading; spikelets narrow, 15–20 mm. long, loosely several-flowered; glumes narrow, obscurely nerved; lemmas prominently 7-nerved, narrowed to an acuminate point, awnless, the nerves more or less pilose-ciliate.

Apparently rare, Unalaska—Wash.—Mont.—Wyo.—Calif. [Fig. 124. (From a Wash. specimen).]

27. GLYCERIA R. Br.

Mostly perennial aquatic or marsh grasses with flat leaves and paniculate inflorescence; spikelets few–many-flowered, subterete or slightly compressed, the rachilla disarticulating above the glumes and between the florets; glumes unequal, short, usually scarious; lemmas broad, convex on the back, firm, scarious at the apex, 5–9-nerved; palea 2-keeled. (Greek, sweet, the seed of the type species being sweet.)

1A. Spikelets linear, more than 1 cm. long.

    1B. Lemmas glabrous between the nerves.................1. *G. borealis*

    2B. Lemmas scaberulous between the nerves ....2. *G. leptostachya*

2A. Spikelets less than 1 cm. long.
   1B. Lemmas appearing to be 5-nerved ....3. *G. pauciflora*
   2B. Lemmas plainly 7-nerved.
      1C. Lemmas less than 2 mm. long ....4. *G. striata*
      2C. Lemmas more than 2 mm. long.
         1D. Culms usually more than 1 m. tall ....5. *G. grandis*
         2D. Culms less than 1 m. tall ....6. *G. pulchella*

1. *G. borealis* (Nash) Batch.   Northern Manna-grass
   Panicularia borealis Nash.

Culms glabrous, 6–15 dm. tall; sheaths smooth or slightly scabrous, keeled; blades flat, or usually folded, 1–2 dm. long, 2–4 mm. or more wide; panicle narrow, 2–4 dm. long, the branches and slender pedicels appressed; spikelets narrow, 10–15 mm. long, 6–12-flowered; glumes about 1.5 and 3 mm. long; lemmas 3–4 mm. long, 7-nerved, smooth except on the scabrous nerves.

In shallow water, central Alaska—Newf.—Conn.—Iowa—N. Mex.—Calif. (Fig. 125.)

2. *G. leptostachya* Buckl.   Davy Manna-grass

Culms 1–2 m. tall, rather stout or succulent; leaves flat, scaberulous on the upper surface, 4–10 mm. wide; panicle 2–6 dm. long, narrow with ascending branches; spikelets 1–2 cm. long, 8–14-flowered, often purplish; lemmas firm, broadly rounded toward the apex, about 3 mm. long, scaberulous both on the nerves and in between.

Wrangell—central Calif. (Fig. 126.)

3. *G. pauciflora* Presl.   Weak Manna-grass
   Panicularia pauciflora (Presl.) Kuntze.

Culms 5–12 dm. long, from a decumbent, rooting base; leaves thin, flat, scabrous, mostly 10–15 cm. long, 5–15 mm. wide; panicle 10–25 cm. long, the branches usually more or less flexuous, the spikelets crowded on the upper half; spikelets 4–6 mm. long, 4–8-flowered; glumes short, broad; lemmas scabrous, about 2 mm. long, rounded and somewhat erose at the summit, prominently 5-nerved, the 2 marginal nerves short and inconspicuous.

Central Pacific Coast region of Alaska—S. Dak.—Colo.—N. Mex.—Calif. (Fig. 127.)

4. *G. striata* (Lam.) Hitchc. ssp. *stricta* (Scribn.) Hult.
                                                    Fowl Manna-grass
   Panicularia nervata (Willd.) Kuntze var. *stricta* Scribn.

Culms 3–5 dm. tall, erect; blades 5–15 cm. long, 2–4 mm. wide; panicle about 1 dm. long; spikelets about 3 mm. long, 4–6-flowered; glumes about 0.5 and 1 mm. long; lemmas 1.5–2 mm. long, prominently 7-nerved, usually purplish, scarious tip inconspicuous; palea about as long as the lemma.

Central Alaska—Labr.—Newf.—N. Hamp.—Iowa—N. Mex.—Ariz.—

northern Calif. The type form is found in eastern America and extends to northern Florida.

5. **G. grandis** S. Wats. American Manna-grass
 G. maxima (Hartm.) Holmb. ssp. *grandis* (S. Wats) Hult.
 Panicularia americana (Torr.) MacM.

Culm stout, glabrous, 1–2 m. tall; blades flat, 15–35 cm. long, 6–15 mm. wide; panicle 2–4 dm. long, very compound and spreading; spikelets 5–8 mm. long, 4–7-flowered; glumes scarious; lemmas purplish, about 2.5 mm. long.

Central Alaska—Pr. Edward Isl.—Tenn.—N. Mex.—Nev.—eastern Ore. (Fig. 128.)

6. **G. pulchella** (Nash) K. Schum.

Culms 4–6 dm. tall, stout, smooth; leaves crowded, blades 15–30 cm. long, 2.5–5 mm. wide, long-acuminate; panicle open, 15–30 cm. long, naked toward the base; spikelets 5–6 mm. long, 4–6-flowered; glumes brownish or purplish, scarious-margined, obtuse, much shorter than the lemmas; lemmas usually purplish, with broad hyaline margins above, strongly but minutely hispidulous, prominently 7-nerved, about 3 mm. long.

Central Alaska—Mack.—Alta.—B.C.

28. PUCCINELLIA Parl.

Spikelets several-flowered, usually terete or only slightly flattened, the rachilla disarticulating above the glumes and between the florets; glumes unequal; lemmas rounded on the back, usually 5-nerved, scarious and often erose at the tip; palea nearly equaling the lemma. Our species are tufted perennials with narrow or open panicles. Closely related to Poa and Glyceria, the species often being listed under one or the other of those genera. The treatment of the genus here followed is that of Mr. Jason R. Swallen in a paper recently published in the Journal of the Washington Academy of Sciences. (Puccinelli was an Italian botanist.)

1A. Anthers 1.8–2 mm. long .... 1. *P. phryganodes*
2A. Anthers less than 1.5 mm. long.
   1B. Panicle branches distinctly scabrous.
      1C. Anthers 0.3–0.5 mm. long .... 2. *P. hauptiana*
      2C. Anthers 0.7–1.5 mm. long.
         1D. Lemmas 3–4 mm. long, anthers 1.3–1.5 mm. long.................
         .... 3. *P. grandis*
         2D. Lemmas 2–3 mm. long, anthers less than 1 mm. long...........
         .... 4. *P. borealis*
   2B. Panicle branches glabrous or only very sparsely scabrous.
      1C. Lemmas 3.5–4 mm. long, anthers 1.2–1.5 mm. long.
         1D. Panicle branches ascending, elongate.... 5. *P. glabra*
         2D. Panicle branches stiffly spreading or reflexed.
            1E. Spikelets 2- or 3-flowered .... 6. *P. triflora*
            2E. Spikelets 4–8-flowered .... 7. *P. andersoni*

2C. Lemmas about 3 mm. long or less.
   1D. Anthers 0.3-0.6 mm. long.
      1E. Lemmas about 3 mm. long, culms up to 30 cm. tall............
            .... 8. *P. alaskana* .. ..
      2E. Lemmas 2-2.5 mm. long, culms less than 2 dm. tall............
            .... 9. *P. paupercula*
   2D. Anthers mostly 0.8-1 mm. long.
      1E. Palea longer than the lemma ....13. *P. kamtschatica*
      2E. Palea as long as or slightly shorter than the lemma.
         1F. Panicle branches slender, usually closely appressed....
              ....12. *P. nutkaensis*
         2F. Panicle branches stout, stiffly spreading or reflexed, naked in the lower half.
            1G. Culms stout, erect from a decumbent base............
                ....10. *P. hulteni*
            2G. Culms relatively slender, erect or ascending, densely tufted ....11. *P. pumila*

1. *P. phryganodes* (Trin.) Scribn. & Merr.   Creeping Alkali-grass

Culms 5-15 cm. tall and in addition creeping decumbent culms resembling stolons, up to 4 dm. long; leaves short, rarely more than 4 cm. long and 1 mm. wide; panicle 1-4 cm. long, barely exserted, with comparatively few spikelets; spikelets 5-8 mm. long, 3-6-flowered; second glume nearly as long as the lemmas; lemmas 3-4 mm. long. Easily distinguished from the other species by the stolon-like culms and the long anthers. Fruits more freely in the northern parts of its range.

Saline or brackish flats along the coast. circumpolar. (Fig. 129.)

2. *P. hauptiana* Krecz.   Haupt Alkali-grass

Densely tufted; culms 2-6 dm. tall; leaf-blades 1.5-2 mm. wide, loosely involute; panicle 4-16 cm. long, the branches ascending, spreading or reflexed; spikelets 3-5 mm. long,. 3-6-flowered; first glume 1 mm. long; second glume 1.5 mm. long, mostly obtuse, laciniate; lemmas 2 mm. long or less, obtuse, laciniate.

Siberia—Alaska—Alta. (Fig. 130.)

3. *P. grandis* Swallen.   Large Alkali-grass

Culms 4-9 dm. tall, erect or geniculate at the lower nodes; leaf-blades 0.5-3 mm. wide, flat or involute on drying; panicles 1-2 dm. long, the branches at first appressed but later spreading; spikelets appressed, tinged with purple, 8-15 mm. long, 5-12-flowered; first glume 2-3 mm. long, subacute; second glume more than 3 mm. long, obtuse; lemmas 3-4 mm. long, sparsely pilose at the base.

Salt marshes, southwestern Alaska—Calif. (Fig. 131.)

4. *P. borealis* Swallen.   Northern Alkali-grass

Culms 3-7 dm. tall, erect from a decumbent base; leaf-blades 1-2 mm. wide, flat, glabrous below, scabrous above; panicles 1-2 dm. long, the slender branches ascending to reflexed, in rather distant fasicles of 2–5; spikelets 4–6 mm. long, 4–6-flowered, usually purplish; first glume more than 1 mm. long; second glume less than 2 mm. long; lemmas 2–3 mm. long, minutely erose-ciliolate; anthers 0.6–0.7 mm. long.

Along coast and rivers, Alaska and Yukon. (Fig. 132.)

5. *P. glabra* Swallen. Smooth Alkali-grass

Culms 25–40 cm. tall, erect or decumbent at the base; leaf-blades 1.5–3 mm. wide, flat or involute toward the tip, glabrous; panicles mostly 1–2 dm. long, the glabrous branches 4–10 cm. long, naked at the base; spikelets appressed, pale or purplish, 8–10 mm. long, 5–7-flowered; first glume 2–3 mm. long; second glume 3–4 mm. long, obtuse, minutely ciliolate; lemmas 3.5–4 mm. long, rather thin and shining, the nerves obscure.

Tidal flats, Alaska Peninsula—Kenai Peninsula—Kodiak Isl.

6. *P. triflora* Swallen. Three-flowered Alkali-grass

Culms densely tufted, erect, 45–60 cm. tall; leaf-blades 1–1.5 mm. wide, soft, glabrous, flat or becoming involute; panicles 15–20 cm. long, the glabrous branches in rather distant fasicles of 2–5, naked at the base; spikelets appressed, deeply tinged with purple, 5–7 mm. long, 2 or 3-flowered; first glume 1.5–3 mm. long; second glume 2.5–4 mm. long; lemmas 3.5–4 mm. long, the nerves evident.

Shores of Cook Inlet.

7. *P. andersoni* Swallen. Anderson Alkali-grass

Culms densely tufted, 15–40 cm. tall, erect from a decumbent base, shorter culms arising from the sides; leaf-blades 1–3 mm. wide, flat and soft; panicle 4–10 cm. long, the branches ascending or spreading, slightly scabrous, bearing 1–3 or up to 5 appressed spikelets, spikelets up to 1 cm. long, 4–8-flowered; first glume about 2 mm. long, acute; second glume 2.5–3 mm. long, acute; lemmas 3–3.5 mm. long, acute, erose, sparsely pilose at the base and lower part of the prominent nerves.

Known only from Point Lay on the Arctic coast. (Fig. 133.)

8. *P. alaskana* Scribn. & Merr. Alaska Alkali-grass

Culms 6–30 cm. tall; leaf-blades 1–2 mm. wide, shorter than, to longer than the culms, usually soft and flat; panicle contracted. 2–9 cm. long; spikelets 4–6 mm. long, 3–5-flowered; first glume 1–1.5 mm. long, acute; second glume 2–2.5 mm. long, obtuse; lemmas 2.5–3 mm. long, rather prominently 5-nerved, appressed ciliolate on the nerves below.

Along the coast, Aleutian Islands and Bering Sea region. (Fig. 134.)

9. *P. paupercula* (Holm) Fern. & Weath. Arctic Alkali-grass

Culms 5–20 cm. tall, longer than the leaves; leaf-blades 0.5–1 mm. wide, rather stiff and often curved; panicles 1–7 cm. long, few-flowered; spikelets 4–8 mm. long, 3–5-flowered; first glume 1–1.5 mm. long; second glume 1.5–2 mm. long; lemmas 2–2.5 mm. long, elliptic to ovate, erose.

Aleutians and eastern Asia through arctic America. (Fig. 135.)

10. *P. hulteni* Swallen. Hulten Alkali-grass

Culms 3–4 dm. tall; leaf-blades 0.5–2.5 mm. wide, mostly involute; panicle 6–14 cm. long, the branches ascending, spreading, or reflexed; spikelets 5–6 mm. long, 3–5-flowered; first glume 1.5–2 mm. long, acute, slightly keeled; second glume 2–2.5 mm. long, slightly keeled and with strong lateral nerves; lemmas 2.5–2.8 mm. long, minutely toothed.

Beaches, Kodiak Island—southeastern Alaska. (Fig. 136.)

11. *P. pumila* (Vasey) Hitchc. Small Alkali-grass

Culms erect or somewhat decumbent at the base, 1–3 dm. tall; leaf-blades flat, scaberulous, 1–2.5 mm. wide; panicles 2.5–15 cm. long, the short branches usually stiffly ascending to reflexed, bearing only one to a few spikelets; spikelets 5–7 mm. long, 3–6-flowered; first glume 1.5–2.5 mm. long; second glume 2.5–3 mm. long; lemmas about 3 mm. long, rather abruptly narrowed toward the apex, the nerves conspicuous, sparsely pubescent on the callus.

Along the coast, Kodiak Isl.—Vancouver Isl. (Fig. 137.)

12. *P. nutkaensis* (Presl) Fern. & Weath. Pacific Alkali-grass

Culms 2–6 dm. tall, usually erect; leaf-blades 1–2 mm. wide, flat or loosely involute; panicle narrow, 5–20 cm. long, the few slender branches appressed; spikelets 6–9 mm. long, 4–9-flowered; glumes about 1.5 mm. and 2.5 mm. long; lemmas about 3 mm. long, narrowed to an obtuse apex which is erose or minutely fimbriate.

Along the coast, Aleutian Islands—Calif. (Fig. 138.)

13. *P. kamtschatica* Holmb. var. *sublaevis* Holmb.

Culms erect or somewhat decumbent at the base, 12–25 cm. tall; leaf-blades rather soft, smooth, flat or involute on drying, 2 mm. or less wide; panicle 4–10 cm. long, the branches narrowly ascending or later spreading, spikelet-bearing on the upper half; spikelets 3–4 mm. long, 3- or 4-flowered; first glume about half as long as the lemmas; second glume much broader and obtuse; lemmas about 2 mm. long, obtuse, glabrous.

Cold wet soil, Kamtchatka—southeastern Alaska.

29. POA L.

Grasses with contracted or open paniculate inflorescence and narrow, flat, folded, or involute leaves with a boat-shaped tip. Spikelets 2–6-flowered, flat, the rachilla disarticulating above the glumes and between the florets, the uppermost floret rudimentary; glumes acute, keeled, the first usually 1-nerved, the second 3-nerved; lemmas somewhat keeled, awnless, membranous, often scarious at the tip, 5-nerved, the nerves sometimes pubescent. (Greek for a grass.)

1A. Low grasses, usually 3 dm. tall or less.
  1B. Plants annual .... 1. *P. annua*
  2B. Plants perennial.
    1C. Spikelets 2–4-flowered.

1D. Stolons present.
    1E. Lower branches of panicle usually in twos............................
    ....12. *P. arctica*
    2E. Lower branches of panicle in twos–fours...........................
    ....13. *P. irrigata*
2D. Plants tufted, not stoloniferous.
    1E. Coma at base of lemma copious........15. *P. leptocoma*
    2E. Coma at base of lemma lacking.
        1F. Panicle branches slender, capillary ....
        ....19. *P. brachyanthera*
        2F. Panicle branches erect or appressed.
            1G. Spikelets 5–6 mm. long ....24. *P. glauca*
            2G. Spikelets 4–5 mm. long ....25. *P. rupicola*
2C. Spikelets 3–5–9-flowered.
    1D. Plants with stolons.
        1E. Plants dioecious .... 3. *P. confinis*
        2E. Plants with perfect flowers.
            1F. Culm flat with sharp edges .... 2. *P. compressa*
            2F. Culm terete or slightly flattened.
                1G. Panicle nodding, lemmas 6 mm. long....................
                .... 6. *P. turneri*
                2G. Panicle erect, lemmas shorter.............................
                ....23. *P. komarovii*
    2D. Plants tufted, not stoloniferous.
        1E. Spikelets but little compressed, much longer than wide
        ....27. *P. sandbergii*
        2E. Spikelets decidedly compressed.
            1F. Leaves about 1 mm. wide..........20. *P. abbreviata*
            2F. Leaves 2–5 mm. wide.
                1G. Panicle compact ....21. *P. alpina*
                2G. Panicle open ....18. *P. merrilliana*
2A. Medium and tall perennial grasses more than 3 dm. high.
    1B. Plants with stolons.
        1C. Spikelets 2–4-flowered.
            1D. Glumes 6–8 mm. long, nearly as long as the spikelet..............
            .... 8. *P. macrocalyx*
            2D. Glumes decidedly shorter than the spikelet ....
            .... 4. *P. laxiflora*
        2C. Spikelets 3–5–9-flowered.
            1D. Culm strongly flattened, 2-edged .... 2. *P. compressa*
            2D. Culm terete or slightly flattened.
                1E. Lemmas 5–7 mm. long.
                    1F. Leaves 4–8 mm. wide .... 7. *P. eminens*
                    2F. Leaves narrower.
                        1G. Lemmas lanate-pubescent on lower part.
                            1H. Panicle nodding, spikelets green ....
                          .... 6. *P. turneri*

　　　　　　2H. Panicle usually erect, spikelets violet or grayish .... 5. *P. lanata*
　　2E. Lemmas shorter.
　　　　1F. Lemmas lanate-pubescent on lower part ....
　　　　　　.. 9. *P. norbergii*
　　　　2F. Lemmas smooth or scabrate, not lanate between the nerves.
　　　　　　1G. Panicle branches 2–4 (mostly 4) in lowest whorl
　　　　　　..11. *P. eyerdamii*
　　　　　　2G. Panicle branches 3–5 (mostly 5) in lowest whorl
　　　　　　...10. *P. pratensis*
2B. Plants tufted, without stolons.
　　1C. Spikelets 2–4-flowered.
　　　　1D. Lemmas about 4 mm. long, spikelets more than 5 mm. long.
　　　　　　1E. Spikelets much flattened ....15. *P. leptocoma*
　　　　　　2E. Spikelets but little flattened ....26. *P. hispidula*
　　　　2D. Lemmas about 3 mm. long or less, spikelets less than 5 mm. long.
　　　　　　1E. Panicle erect, narrow ....24. *P. glauca*
　　　　　　2E. Panicle open.
　　　　　　　　1F. Lemmas glabrous or the keel slightly pubescent........
　　　　　　　　..14. *P. trivialis*
　　　　　　　　2F. Lemmas pubescent on the keel and marginal nerves.
　　　　　　　　　　1G. Glumes about as long as the first lemma..................
　　　　　　　　　　...16. *P. nemoralis*
　　　　　　　　　　2G. Glumes shorter than the first lemma......................
　　　　　　　　　　..17. *P. palustris*
　　2C. Spikelets 3–7-flowered.
　　　　1D. Spikelets 8–10 mm. long, 4–7-flowered............................
　　　　　　..29. *P. ampla*
　　　　2D. Spikelets 6–8 mm. long, 3–5-flowered.
　　　　　　1E. Lemmas about 4 mm. long.............28. *P. canbyi*
　　　　　　2E. Lemmas about 5 mm. long...............22. *P. stenantha*

1. *P. annua* L.　　　　　　　　　　　　　　　　Annual Bluegrass

Tufted, often decumbent, rooting at the nodes and forming mats; culms 5–25 cm. tall; leaf-blades soft, flat, lax, 1–3 mm. wide; panicle open, 3–7 cm. long; spikelets crowded, 3–6-flowered, about 4 mm. long; lemmas 2.5–3 mm. long, not webbed at base, 5-nerved, the nerves more or less pubescent on the lower half.

A weed, probably introduced from Europe but widespread and common in Alaska. (Fig. 139.)

2. *P. compressa* L.　　　　　　　　　　　　　　Canada Bluegrass

Culms 15–60 cm. tall, pale bluish-green, solitary or a few together, decumbent at the base, strongly flattened, with long creeping rhizomes; leaves rather short, 1–4 mm. wide; panicle narrow, 3–7 cm. long, the short branches usually in pairs; spikelets 4–7 mm. long, 3–9-flowered;

glumes 2–3 mm. long; lemmas firm, 2–3 mm. long, the keel and marginal nerves sparingly pubescent; web at base scant or wanting.

Roadsides, central and southeastern Alaska. Introduced. Native of Eurasia. (Fig. 140.)

3. *P. confinis* Vasey   Dune Bluegrass

Plants dioecious, the two kinds similar; culms often geniculate at the base, usually less than 15 cm. tall, sometimes much taller; sheaths and involute blades smooth, firm, narrow; panicle narrow, contracted, 1–3 cm. long; spikelets 4–5 mm. long, 3–4-flowered; glumes unequal; lemmas 3 mm. long, scaberulous, sparsely webbed at base, the nerves faint.

Sand dunes and sandy meadows near the coast, B. C.—Calif. Reported as growing in southeastern Alaska.

4. *P. laxiflora* Buckl.   Loose-flowered Bluegrass
    P. leptocoma elatior Scribn. & Merr.

Culms scabrous, 10–15 dm. tall; sheaths slightly scabrous, leaves lax, 2–4 mm. wide; panicle loose, open, 10–15 cm. long, nodding or drooping; lower branches in whorls of 3 or 4; spikelets 5–6 mm. long, 3–4-flowered; lemmas about 4 mm. long, webbed at base, sparsely pubescent on lower part of nerves.

Southeastern Alaska—western Ore.

5. *P. lanata* Scribn. & Merr.   Lanate Bluegrass

Culms erect, 25–40 cm. tall, from creeping rootstocks; leaves glaucous, scabrous, rather rigid, 2–4 mm. wide, acute and hooded at the apex; panicle open, 5–12 cm. long, the branches usually in twos; spikelets ovate, acute, purplish or brownish, 8–10 mm. long, 3–6-flowered; glumes acute, 3-nerved, scabrous on the keel; lemmas 6–7 mm. long, with broad hyaline margins, 5-nerved, obtuse, densely webby on the lower half, strigose above. Viviparous forms are frequent.

Aleutian Islands—Lake Athabasca—B. C. (Fig. 141.)

6. *P. turneri* Scribn.   Turner Bluegrass

Culms leafy, 2–4 dm. tall, from creeping rhizomes; leaves 3–5 mm. wide, 4–8 cm. long; panicles 6–9 cm. long, plumose and nodding; spikelets 7–10 mm. long, usually 3-flowered; glumes long and narrow; lemmas acute, about 6 mm. long, copiously pubescent on the keel and lower part of lateral nerves, less so between the nerves; coma at base copious. A beautiful grass.

Kenai Peninsula and Aleutian Islands. (Fig. 142.)

7. *P. eminens* Presl.   Large-flowered Spear-grass
    P. glumaris Trin.
    P. trinii Scribn. & Merr.

Culms glaucous, 4–10 dm. tall from creeping rootstocks; leaves thick, 4–8 mm. wide; panicle dense, 1–2 dm. long, contracted; spikelets 10–14 mm. long, 3–6-flowered; glumes up to 1 cm. long; lemmas 5–6 mm. long, laciniate at the hyaline tip, pubescent at the base and lower part of the midrib and nerves.

All of our beaches except the high Arctic—Vancouver Island. Also Labr.—Que. (Fig. 143.)

8. *P. macrocalyx* Tr. & Mey.   Large-glumed Bluegrass

Culms smooth, up to 8 dm. tall; leaves flat, 2–5 mm. wide; panicle open, 1–2 dm. long, the lower branches in whorls of 3–5, spikelet-bearing near the ends; spikelets 6–9 mm. long, 2–4-flowered; glumes 3-nerved, narrow and long-acuminate, 6–8 mm. long, reaching to the apex of the second lemma; lemmas 5–6 mm. long, webbed at base, densely white-hairy on keel and marginal nerves.

Prince William Sound—Aleutian Islands—eastern Asia. (Fig. 144.)

9. *P. norbergii* Hult.   Norberg Bluegrass

Stoloniferous; culms 5–7 dm. tall, glabrous; culm leaves 3 or 4, 6–8 cm. long, 3.5–4 mm. wide, the upper surface minutely scaberulous, lower surface and margins scabrous; glumes 1-nerved, glabrous, glaucous, with wide hyaline margins, the apex tinged violet, the lower lanceolate, the upper ovate to acute lanceolate-ovate; lemmas minutely and densely scaberulous, sparsely long-pilose below, the keel and lateral nerves pilose for two-thirds their length.

Known only from Hoonah.

10. *P. pratensis* L.   Kentucky Bluegrass

Culms erect, 3–10 dm. tall, from creeping rhizomes; leaves flat or folded, 2–4 mm. wide, the basal often elongated; panicle open, the branches in fascicles of 3–5, ascending or spreading, naked below; spikelets 3–5-flowered, 3–6 mm. (mostly 4–5 mm.) long; lemmas about 3 mm. long, copiously webbed at base, silky pubescent on the keel and marginal nerves, the intermediate nerves prominent but glabrous. A very variable species and giving rise to many forms, some of which may be hybrids. The cultivated form has probably been introduced but the var. *alpigena* E. Fries [*P. alpigena* (E. Fr.) Hartm.] is native and in the far north has given rise to viviparous forms. On the average it is not so tall as the typical form, the culms arise singly and do not form mats, the leaves are narrower and the spikelets purplish. The var. *angustifolia* (L.) Kunth has been collected at Seward. It has basal shoots with long, narrow, involute leaves and was probably introduced.

The entire species is circumboreal. (Fig. 145.)

11. *P. eyerdamii* Hult.   Eyerdam Bluegrass

Rhizomes long-creeping; culms 5–7 dm. tall, slender, glabrous; leaves about 2 mm. wide, the margins and under surface smooth, the upper surface minutely scaberulous; panicles 10–15 cm. long; glumes 1–2-nerved, glabrous with hyaline margins and scaberulous keel; lemmas minutely scaberulous, webbed at base, the lower two-thirds of the keel and one-third of the lateral nerves white-pilose, glabrous between; intermediate nerves indistinct, anthers 1.4–1.9 mm. long.

Kodiak Island and Prince William Sound region.

12. P. *arctica* R. Br.   Arctic Bluegrass
P. *cenisia* All.

Culms loosely tufted, erect from a decumbent base, 1–3 dm. tall; leaves mostly basal, flat or folded, 1–4 mm. wide, a single culm-leaf at about the middle of the culm; panicle open, 5–10 cm. long, the lower branches usually 2, spreading or even reflexed; spikelets 2–4-flowered, 5–8 mm. long; lemmas densely villous on keel and marginal nerves, pubescent on lower part between the nerves; webbing at base very variable. A form with short runners and with long cobweb-like coma at the base of the lemmas is the ssp. *williamsii* (Nash) Hult. (P. *williamsii* Nash). Another form with long slender culm and involute leaves has been described as ssp. *longiculmis* Hult. There are also viviparous forms.

Circumpolar. (Fig. 146.)

13. P. *irrigata* Lindm.

Stoloniferous; culms 12–40 cm. tall, solitary or a few at the ends of the stolons, glaucesent; leaves clustered on innovations at the base of the culm, less than 1 dm. long, culm-leaf short; panicle small, rather lax; spikelets 3.5–6 mm. long, 2–3-flowered, the short peduncles scabrous; glumes subequal, usually acuminate; lemmas with cobwebby base, the keel and lateral nerves pubescent; the keel of the glumes scaberulous and incurved toward the apex.

Central and southeastern coast of Alaska, probably introduced. Described from Sweden. Range not definitely known. (Fig. 147.)

14. P. *trivialis* L.   Rough Bluegrass

Culms erect from a decumbent base, 3–10 dm. tall, scabrous, at least toward the summit; leaves 2–4 mm. wide; panicle 6–15 cm. long, open, the branches spreading or ascending; spikelets 2 or sometimes 3-flowered, about 3 mm. long, lemmas about 2.5 mm. long, glabrous except the slightly pubescent keel and the prominent coma at the base, nerves prominent.

Aleutian Islands—southeastern Alaska—Newf.—Va.—S. Dak.— northern Calif. Introduced from Europe. (Fig. 148.)

15. P. *leptocoma* Trin.   Bog Bluegrass
P. *paucispicula* Scribn. & Merr.

Culms solitary or a few together, smooth, rather lax, 2–6 dm. tall; leaves flat, flaccid, 2–4 mm. wide; panicle lax, 5–10 cm. long; spikelets narrow, 5–6 mm. long, 2–4-flowered; glumes narrow, acuminate; lemmas 3.5–4.5 mm. long, narrowly lanceolate, acuminate. Var. *scabrinervis* Hult. has the keel and nerves of the lemmas scaberulous and the tuft at the base lacking.

Boggy places, southeastern Alaska—Utah—northern Mex.—Calif. (Fig. 149.)

16. *P. nemoralis* L. Wood Bluegrass

Culms tufted, glabrous, 3-7 dm. tall; leaves rather lax, 1-2 mm. wide; panicle 4-12 cm. long, the branches spreading; spikelets 3-5 mm. long, 2-5-flowered; glumes narrow, sharply acuminate; lemmas 2-3 mm. long, faintly 5-nerved, sparsely webbed at base, silky-pubescent on keel and marginal nerves below.

Aleutian Islands—southeastern Alaska. Circumboreal. May have been introduced in our territory. (Fig. 150.)

17. *P. palustris* L.
P. triflora Gilib.
P. crocata Michx.

Culms loosely tufted, glabrous, with decumbent, flattened, purplish base, 3-15 dm. tall; leaves 1-4 mm. wide; panicle open, nodding, yellowish-green or purplish, 1-3 dm. long; spikelets 3-5 mm. long, 2-4-flowered; glumes acute; lemmas 2.5-3 mm. long, usually bronzed at the tip, webbed at base, villous on the keel and marginal nerves, intermediate nerves faint.

Wet or moist soil, Aleutian Islands and central Alaska south and east. Circumboreal. (Fig. 151.)

18. *P. merrilliana* Htichc. Merrill Bluegrass
P. glacialis Scribn. & Merr. not Stapf.

Densely caespitose, glabrous, 2-3 dm. tall; leaves rather broad, thin, flat, glabrous, ascending, pale green, 4-6 cm. long, 3-4 mm. wide; panicles 3-9 cm. long, the branches flexuous with 2 or 3 spikelets near the ends; spikelets about 7 mm. long, 5-flowered, broadly lanceolate; glumes unequal, acute, the first 3 mm. long, the second 1 mm. longer, 3-nerved; lemmas acute, 5-nerved, 4-5 mm. long, with very few hairs on keel and marginal nerves, not webbed at base or only slightly so.

Southeastern Alaska. (Fig. 152.)

19. *P. brachyanthera* Hult.

Caespitose, about 1 dm. tall; culm leaves 2 or 3, 1-3 cm. long, 1-1.5 mm. wide; panicle branches in twos, glabrous, bearing spikelets at the ends; spikelets 2-5-flowered; glumes about 2.5 mm. long, glabrous, wide-lanceolate, acute; lemmas with lateral nerves and keel short white-ciliate; palea as long as the lemma; anthers about 0.5 mm. long.

Aleutian Islands to Copper River.

20. *P. abbreviata* R. Br. Low Spear-grass

Culms from close tufts, 15 cm. tall or less, erect, smooth; leaves crowded at the base, about 1 mm. wide; panicle contracted, 15-25 mm. long, branches short and erect; spikelets about 5 mm. long, 3-5-flowered; glumes acute, smooth; lemmas about 3 mm. long, obtuse, strongly pubescent all over.

Occurs on the Arctic Archipelago and has been reported from Alaska.

21. *P. alpina* L.   Alpine Bluegrass

Culms erect from a rather thick vertical crown, 1–3 dm. tall; leaves mostly basal, short, 2–5 mm. wide; panicle rather compact, 2–7 cm. long; spikelets broad, purplish, 5–6 mm. long, 3–5-flowered; glumes broad, acute, scabrous on the keel; lemmas 3–4 mm. long, villous on the keel and lateral nerves.

Alpine-arctic, Bering Strait east and south. Circumboreal. (Fig. 153.)

22. *P. stenantha* Trin.   Narrow-flowered Bluegrass
P. acutiglumis Scribn.

Culms tufted, 3–7 dm. tall; ligule prominent, as much as 5 mm. long; leaves flat or slightly involute, mostly basal; panicle lax, 5–13 cm. long, the branches in twos or threes; spikelets 3–5-flowered, 6–8 mm. long; glumes 3-nerved; lemmas about 5 mm. long, copiously pubescent on lower part of keel and marginal nerves, sparsely pubescent or glabrous between, intermediate nerves faint. Often the spikelets produce growing plants thus forming the var. *vivipara* Trin.

Aleutian Islands—Mont.—Colo.—Ore. (Fig. 154.)

23. *P. komarovii* Roshew.   Komarov Bluegrass

Subterranean stolons curved; culms not over 3 dm. tall, the base surrounded by a cylinder of hyaline sheaths; leaves relatively broad; panicle erect or nearly so, short-pyramidal, green or the scales brown-tipped; lemmas webbed at base and often with straight hairs between the veins below. Has the appearance of *P. alpina*. Produces viviparous forms.

Eastern Asia—Aleutian Islands—Arctic coast—southern Alaska.

24. *P. glauca* Vahl.   Glaucous Spear-grass

Culms tufted, erect, rigid, 15–50 cm. tall; uppermost leaf usually below the middle of the culm; leaves usually short, 1–2 mm. wide; panicle 3–8 cm. long, the branches erect or ascending; spikelets 2–4-flowered; 5–6 mm. long; glumes 3-nerved, glabrous, rough on the upper part of the keel; lemmas 3–4 mm. long, strongly pubescent on lower part of keel and marginal nerves, slightly pubescent on the base of the faint intermediate nerves, not webbed at base.

A common grass in most parts of our territory. Circumboreal. (Fig. 155.)

25. *P. rupicola* Nash.   Timberline Bluegrass

Culms tufted, erect, 10–25 cm. tall; leaf-blades erect or ascending, involute, 1–5 cm. long, 0.5–1.5 mm. wide; panicle 2–5 cm. long, purplish, the short branches ascending or appressed; spikelets 4–5 mm. long, 2–4-flowered; glumes 3-nerved, 2.5–3 mm. long; lemmas 3 mm. or more long, villous on the lower part of keel and marginal nerves, sometimes a few hairs on the internerves, no coma at the base.

Buffalo range of central Alaska, Mont.—N. Mex.—Calif. (Fig. 156.)

26. *P. hispidula* Vasey. Hispid Bluegrass

Tufted, culms 3-7 dm. tall; leaves 2-3 mm. wide; panicle somewhat contracted, 3-15 cm. long; spikelets about 3-flowered, 6 mm. long; glumes prominently nerved, lanceolate; lemmas narrow, acute, about 4 mm. long, the keel and marginal nerves with white lanate hairs, intermediate nerves sometimes slightly lanate, the intervening space scabrous with fine short hairs. The var. *aleutica* Hult. is a dwarf form with narrow leaves and small spikelets found in exposed spaces in the Aleutian Islands. This species also produces viviparous forms. It is probably responsible for the reports of *P. gracillima* Vasey from Alaska.

Bering Island and the Aleutians—southeastern Alaska. (Fig. 157.)

27. *P. sandbergii* Vasey. Sandberg Bluegrass

Culms up to 3 dm. or more tall form a dense tuft of short basal foliage; leaves soft, flat, folded or involute, about 2 mm. wide; panicle narrow, 2-10 cm. long, the branches short, erect or ascending; spikelets 5-7 mm. long; glumes about 4 mm. long; lemmas pubescent on the lower half, especially on the keel and near the margin.

As *P. secunda* this species has been reported as found in Yukon Territory—Nebr.—N. Mex.—Calif.

28. *P. canbyi* (Scribn.) Piper. Canby Bluegrass

Culms tufted, erect, smooth, 5-12 dm. tall; leaves scabrous above, leaves 1-2 mm. wide; panicle narrow, often compact, 10-15 cm. long; spikelets 6-8 mm. long, 3-5-flowered; glumes unequal, acute; lemmas about 4 mm. long, more or less crisp-pubescent on lower part of back.

Sandy or dry ground, Yukon—Que.—Isle Royal—Minn.—Colo.—Ariz.—eastern Ore.

29. *P. ampla* Merr. Big Bluegrass

Tufted; culms 8-12 dm. tall; leaves 1-3 mm. wide; panicle narrow, 10-15 cm. long, usually dense; spikelets 4-7-flowered, 8-10 mm. long; lemmas rounded on the back, smooth or minutely scaberulous, 4-5 mm. long.

Skagway and Yukon—Mont.—N. Mex.—Calif.

30. FESTUCA L.

Mostly tufted perennial grasses with paniculate inflorescence and 2-several-flowered spikelets with the rachilla disarticulating above the glumes and between the florets, the uppermost floret reduced; glumes narrow, acute, unequal; lemmas rounded on the back, more or less awned or sometimes awnless; palea scarcely shorter than the lemma. (An old Latin name for a weedy grass.)

1A. Basal leaves 3-10 mm. wide, flat, lax.
   1B. Lemmas awnless or very short-awned ....1. *F. elatior*
   2B. Lemmas with awns 5-20 mm. long ....2. *F. subulata*

2A. Basal leaves narrow, folded, or involute.
  1B. Low-growing; less than 3 dm. tall ....3. *F. brachyphylla*
  2B. Culms 3–5 dm. tall ....4. *F rubra*
  3B. Culms 5–10 dm. tall ....5. *F. altaica*

1. *F. elatior* L.   Meadow Fescue

Culms smooth, 5–12 dm. tall; leaves flat, 3–8 mm. wide, somewhat scabrous above; panicle mostly erect, 1–2 dm. long, contracted after flowering, branches spikelet-bearing nearly to the base; spikelets usually 6–8-flowered, 8–15 mm. long; glumes about 3 mm. and 4 mm. long; lemmas coriaceous, 5–7 mm. long, the apex hyaline, rarely short-awned. The var. *arundinacea* (Schreb.) Wimm. (*F. arundinacea* Schreb.) is a tall-growing form with usually 4–5-flowered spikelets.

Introduced at several places in Alaska. Native of Eurasia. (Fig. 158.)

2. *F. subulata* Trin.   Bearded Fescue

Culms 4–12 dm. tall; leaves flat, thin, green above, scabrous on both sides, 1–3 dm. long, 3–10 mm. wide; panicle loose, open, drooping, 15–40 cm. long, the branches in twos or threes, at length spreading or reflexed; spikelets loosely 3–5-flowered, 7–12 mm. long; lemmas somewhat keeled, scabrous toward the apex, 5–7 mm. long, attenuate into a scabrous awn 5–20 mm. long.

Woods, southeastern Alaska—Wyo.—Utah—northern Calif. (Fig. 159.)

3. *F. brachyphylla* Schult.   Alpine Fescue

Densely tufted; culms 10–25 cm. tall; leaves narrow, involute, 2–7 cm. long; panicle narrow and spike-like, 2–7 cm. long; spikelets 2–6-flowered, often purplish; lemmas 3–4 mm. long; awn scabrous, 2–3 mm. long. The type form has the leaves glabrous or nearly so. Ssp. *saximontana* (Rydb.) Hult. is a somewhat taller form with scabrous leaves, occurring in the interior.

Alpine-arctic situations throughout our territory. Circumboreal. (Fig. 160.)

4. *F. rubra* L.   Red Fescue

Culms more or less tufted, erect from a decumbent base, 3–10 dm. tall; leaves soft, smooth, usually involute, 7–15 cm. long; panicle 4–20 cm. long, usually contracted and narrow; spikelets 4–6-flowered, often purplish; lemmas 5–7 mm. long, smooth to scabrous or villous; awn 1–4 mm. long. A very variable group from which species, subspecies, and varieties have been described. Geographical races can be distinguished, but some of the characters such as the amount and nature of the pubescence of the lemma, do not follow the variation of other characters. The following names have been used for species, subspecies, and varieties, and the plants reported under these names in the genus *Festuca* refer to this species: *aucta, arenaria, barbata, glabrata, kitaibeliana, lanuginosa, megastachya, mutica, richardsonii, subvillosa*.

A common and widely distributed species in our territory. Circumboreal. (Fig. 161.)

5. *F. altaica* Trin. Rough Fescue

Plants forming dense tussocks; culms erect, smooth, 3-10 dm. tall; leaves narrow, involute, 15-30 cm. long; panicle loose and open, 1-2 dm. long; spikelets 10-15 mm. long, 3-5-flowered, usually suffused with purple; glumes unequal, nearly smooth; lemmas ovate, attenuate, finely and densely scaberulous, 7-11 mm. long, usually with a short awn.

Nearly throughout our territory and in eastern Asia. (Fig. 162.)

F. duriuscula L. [*F. ovina* L. var. *duriuscula* (L.) Koch], Hard Fescue, a native of Europe naturalized in America, has been found in Alaska. It resembles *F. brachyphylla* but is taller and has wider and firmer leaves.

F. megalura Nutt., the Western Six-weeks Fescue, a native of B. C.—Baja, Calif., has been found introduced in our area. It grows 2-6 dm. tall; has narrow panicles 7-20 cm. long, with appressed branches; spikelets 4 or 5-flowered; lemmas linear-lanceolate, scabrous on the back toward the apex, ciliate on the upper half and with awns 8-18 mm. long.

31. BROMUS L.

Spikelets several to many-flowered, the rachilla disarticulating above the glumes and between the florets; glumes unequal, acute, the first 1-3-nerved, the second 3-5-nerved; lemmas convex or keeled on the back, 5-9-nerved, 2-toothed at the apex, sometimes awnless but usually awned from between the teeth; palea shorter than the lemma; sheaths closed; leaf-blades flat; inflorescence a panicle of large spikelets. All species described here have been collected in Alaska, but it is not known if *B. brizaeformis*, *B. marginatus*, *B. racemosus*, and *B. secalinus* are able to maintain themselves. (Greek, an ancient name of the oat.)

1A. Lemmas compressed-keeled.
  1B. Spikelets glabrous or slightly pilose.
    1C. Panicle branches elongate, drooping............1. *B. sitchensis*
    2C. Panicle branches shorter, erect....................2. *B. aleutensis*
  2B. Spikelets densely pilose ....3. *B. marginatus*
2A. Lemmas rounded on the back, not compressed-keeled.
  1B. Perennials.
    1C. Creeping rhizomes present.
      1D. Lemmas glabrous ....4. *B. inermis*
      2D. Lemmas pubescent, at least near the margin ....
      .... 5. *B. pumpellianus*
    2C. Creeping rhizomes wanting.
      1D. Lemmas glabrous ....6. *B. ciliatus*
      2D. Lemmas pubescent ....7. *B. pacificus*
  2B. Annuals.
    1C. Lemmas rounded above, teeth short.

    1D. Panicle contracted, rather dense.
       1E. Lemmas glabrous .... 8. *B. racemosus*
       2E. Lemmas pubescent .... 9. *B. mollis*
    2D. Panicle open, the branches spreading.
       1E. Awn short or wanting .... 10. *B. brizaeformis*
       2E. Awn well developed.
          1F. Sheaths glabrous .... 11. *B. secalinus*
          2F. Sheaths pubescent .... 12. *B. commutatus*
  2C. Lemmas narrow, the teeth long .... 13. *B. tectorum*

1. *B. sitchensis* Trin.   Alaska Brome Grass

Perennial; culms smooth, 10–18 dm. tall, sheaths smooth; leaves smooth beneath, sparsely pilose above, 6–12 mm. wide; panicle large, lax, drooping, 25–35 cm. long; spikelets 2–3.5 cm. long, 4–12-flowered; lemmas scabrous, often hairy toward the base, about 12 mm. long; awn 5–10 mm. long.

Near the coast, southeastern Alaska—Wash. (Fig. 163.)

2. *B. aleutensis* Trin.   Aleutian Brome Grass
B. sitchensis Trin. var. *aleutensis* (Trin.) Hult.

Culms 5–10 dm. tall; leaves 5–10 mm. wide; panicle erect with stiffly ascending branches; spikelets 3–7-flowered; lemmas often 15 mm. long; and awn nearly 1 cm. long.

Near the coast, Aleutian Islands to Wash. (Fig. 164.)

3. *B. marginatus* Nees.   Large Mountain Brome Grass

Short-lived perennial; culms 6–12 dm. long, sheaths pilose; panicle erect, rather narrow, 1–2 dm. long; spikelets 25–35 mm. long, 7- or 8-flowered; glumes scabrous or scabrous-pubescent; lemmas coarsely pubescent, ovate-lanceolate, 11–14 mm. long; awn 4–7 mm. long.

Introduced, native of B. C.—S. Dak.—N. Mex.—Calif. (Fig. 165.)

4. *B. inermis* Leyss.   Smooth Brome

Culms erect, 6–12 dm. tall, from creeping rhizomes; leaves smooth, 5–10 mm. wide; panicle 1–2 dm. long, erect with whorled branches; spikelets 20-25 mm. long; first glume 4–5 mm. long, second glume 6–8 mm. long; lemmas 9–12 mm. long, obtuse, glabrous or scabrous, emarginate, usually awnless, occasionally with an awn 1 or 2 mm. long.

A cultivated grass from Europe that has become established in some places in our territory. (Fig. 166.)

\5. *B. pumpellianus* Scribn.   Arctic Brome-grass

Culms 5–12 dm. tall, with creeping rhizomes; leaves 1–2 dm. long, 5-10 mm. wide, smooth beneath, scabrous or pubescent above; panicle 1–2 dm. long, rather narrow, with short, erect, or ascending branches; spikelets 2–3 cm. long, 7–11-flowered; lemmas 10–12 mm. long, 5–7-nerved, pubescent along the margin and across the back at the base; awn 2–3 mm. long. Var. *arcticus* (Shear) Porsild (*B. arcticus* Shear). Panicle purplish, the branches spreading at flowering time, erect or ascending at maturity; spikelets 2–4.5 cm. long, 6–14-flowered; glumes and lemmas coarsely pubescent, the lemmas 5-nerved, 12–14 mm. long. Var. *villosissimus* Hult. Glumes and lemmas densely villous-gray; leaves and sheaths often also villous.

Northern Bering Sea region—Ida.—Black Hills—Colo. (Fig. 167.)

6. B. *ciliatus* L. Fringed Brome
    B. *richardsonii* Link

Culms moderately robust, 7–12 dm. tall; sheaths often more or less pubescent; leaves up to 1 cm. wide; panicle 15–25 cm. long, open, the branches drooping; spikelets 2–3 cm. long, 5–10-flowered; lemmas nearly glabrous on the back, pubescent along the lower half to three-quarters of the margins, about 12 or 13 mm. long; awn 3–5 mm. long.

Central Alaska—Newf.—N. Jer.—Tenn.—Texas—Calif.—northern Asia. (Fig. 168.)

7. B. *pacificus* Shear. Pacific Brome-grass

Culms stout, erect, 10–15 dm. tall, pubescent at the nodes; sheaths more or less retrosely pilose; leaves sparsely pillose above, 8–14 mm. wide; panicle very open, 10–25 cm. long, the branches slender, drooping; spikelets 20–25 mm. long, pubescent; lemmas 11–12 mm. long; awn 4–6 mm. long.

Along the coast, southeastern Alaska—Ore. (Fig. 169.)

8. B. *racemosus* L. Smooth-flowered Soft Cheat

Resembling B. *mollis* but the panicle usually more open and lemmas glabrous or scabrous.

An European annual species adventitive in central Alaska and Yukon.

9. B. *mollis* L. Soft Chess

Softly pubescent throughout; culms erect, 2–8 dm. tall; panicle contracted, the branches erect or ascending, 5–10 cm. long; glumes broad; lemmas broad with hyaline margins and tip, obtuse, 7-nerved, bidentate, 7–9 mm. long; awn 4–8 mm. long. This species has been reported as B. *hordaceus* L.

An introduced annual weed, native of Europe. (Fig. 170.)

10. B. *brizaeformis* Fisch. & Mey. Rattlesnake Grass

Culms 3–6 dm. tall; sheaths and blades pilose-pubescent; panicle 5–15 cm. long, lax, secund, nodding; spikelets 15–25 mm. long, about 1 cm. wide, flat; lemmas about 1 cm. long, very broad, inflated, smooth, with broad scarious margins, awnless or nearly so.

An European species that has been collected at Seward and Nome.

11. B. *secalinus* L. Chess or Cheat

Culms erect, 3–6 dm. tall; sheaths smooth, panicle nodding, 7–12 cm. long, the lower branches 3–5, unequal, drooping; spikelets 1–2 cm. long,

6-8 mm. wide; lemmas 6-8 mm. long, the margin strongly involute at maturity; awns usually 3-5 mm. long.

Often a weed in fields of grain. Native of Europe.

12. B. *commutatus* Schrad. Hairy Chess

Resembles B. *secalinus*, but the sheaths are pilose with short retrose hairs; lemmas at maturity are less plump and the awn straight and usually longer.

An introduced weedy grass that is native of Europe.

13. B. *tectorum* L. Downy Chess

Culms 3-6 dm. tall, glabrous; sheaths and blades more or less pubescent; panicle 6-15 cm. long, open, the branches slender and drooping, somewhat one-sided; spikelets nodding, 12-20 mm. long exclusive of awns; glumes villous; lemmas villous or pilose, 10-12 mm. long, the teeth 2-3 mm. long; awn straight, 12-18 mm. long.

An introduced weed that is becoming a pest in some localities. Native of Europe. (Fig. 171.)

Secale cereale L., the cultivated rye, has infrequently been found along roadsides and in old fields where it sometimes persists for several years. It is doubtful if it can maintain itself indefinitely.

Triticum aestivum L., the common wheat, like rye, is sometimes found along roadsides. It cannot be considered as really established and therefore a part of our flora.

32. LOLIUM L.

Spikelets several-flowered, solitary, sessile, placed edgewise to the rachis, one edge fitting into the alternate concavities; first glume wanting (except in terminal spikelet), the second outward, 3-5-nerved, equaling or exceeding the second floret; lemmas rounded on back, 5-7-nerved. (Lolium, an old Italian name for darnel.)

Glume shorter than the spikelet.
    Lemmas awned ....1. L. *multiflorum*
    Lemmas awnless or nearly so ....2. L. *perenne*
Glume as long as or longer than the spikelet ....3. L. *tremulatum*

1. L. *multiflorum* Lam. Italian Ryegrass

A short-lived perennial with culms 4-10 dm. tall; spikes 8-30 cm. long, flat; spikelets 10-20-flowered, up to 25 mm. long; lemmas 7-8 mm. long, at least part of them awned.

Introduced. Native of Europe. (Fig. 172.)

2. L. *perenne* L. English Ryegrass

Culms 3-6 dm. tall; spikelets 6-10-flowered; lemmas 5-7 mm. long, awnless. Often used in lawn grass mixtures.

Introduced. Native of Europe.

3. **L. temulentum** L. Darnel

Annual, culms 6–9 dm. tall; leaves 3–6 mm. wide; spike 15–20 cm. long; glume about 25 mm. long, as long as or longer than the 5–7-flowered spikelet.

Has been collected near Dawson and at St. Michael, but it is doubtful if it has become established. Native of Europe.

33. AGROPYRON Gaertn.

Perennial grasses; leaves flat or involute; inflorescence a terminal spike; spikelets several-flowered, usually solitary, compressed, placed flatwise at each joint of the rachis, the rachilla disarticulating above the glumes and between the florets; glumes firm; lemmas convex on the back, rigid, 5–7-nerved, acute or awned at the apex; palea nearly as long as the lemma.

A difficult genus with many variable forms probably due to hybridization. (Greek, wild and wheat, referring to their growth in wheat fields.)

```
1A. Plants with creeping rhizomes
    1B. Glumes rigid, tapering to a short awn............1. A. smithii
    2B. Glumes not rigid, acute or abruptly short-awned.
        1C. Lemmas glabrous ............................2. A. repens
        2C. Lemmas pubescent ..........................3. A. yukonense
2A. Plants tufted, without creeping rhizomes.
    1B. Lemmas awnless or awn-tipped only.
        1C. Nodes of culm finely appressed-pilose ........4. A. alaskanum
        2C. Nodes of culm glabrous.
            1D. Lemmas pubescent.
                1E. Spikelets very narrow ......................5. A. sericeum
                2E. Spikelets comparatively broad ..........6. A. latiglume
            2D. Lemmas glabrous ............................7. A. trachycaulum
    2B. Lemmas awned.
        1C. Awns straight ................................8. A. subsecundum
        2C. Awns divergent ..............................9. A. spicatum
```

1. **A. smithii** Rydb. Western Wheat-grass

Culms usually glaucous, 3–6 dm. or more tall; leaves firm, stiff, scabrous; striate, 2–4 mm. wide, sharp-pointed, becoming involute on drying; spikes 6–15 cm. long, the rachis scabrous on the angles; spikelets rarely 2 at a node, 5–10-flowered, 1–2 cm. long; glumes rigid, tapering into a short awn, faintly nerved, 7–12 mm. long; lemmas about 1 cm. long, acuminate, mucronate or short-awned.

Central Alaska, probably introduced from the western states. (Fig. 173.)

2. **A. repens** (L.) Beauv. Quackgrass

Culms 5–15 dm. tall, from long-jointed running rootstocks; leaves flat, smooth beneath, rough above, mostly 5–10 mm. wide; spike 5–15 cm. long, the rachis scabrous on the margins; spikelets 3–7-flowered, 10–15 mm. long; glumes strongly 3–7-nerved, acute or awn-pointed; lemmas about 8 mm. long, acute or awned, the awn when present may approach that of the lemma in length.

Central Alaska south and east. A native of Eurasia and extensively introduced in North America. (Fig. 174.)

3. *A. yukonense* Scribn. & Merr.   Yukon Wheat-grass

Culms glabrous, 4–8 dm. tall, from creeping rhizomes; leaves 2–6 mm. wide, flat or involute; spikes 5–12 cm. long; spikelets closely imbricate, 4–8-flowered, 10–15 mm. long; glumes acute, 3-nerved, pilose; lemmas villous, about 8 mm. long, acute or short-awned.

Upper and central Yukon River valley. (Fig. 175.)

4. *A. alaskanum* Scribn. & Merr.   Alaska Wheat-grass

Culms glabrous, erect, 4–9 dm. tall, the nodes pubescent; leaves flat or involute, 4–7 mm. wide, scabrous on both surfaces; spike 6–10 cm. long, rather slender; spikelets sometimes 2 at a node, 15–20 mm. long, 4–6-flowered, exceeding the scabrous internodes of the rachis; glumes 6–8 mm. long, oblanceolate; lemmas 8–10 mm. long exclusive of the short awn, lanceolate, hispid with short stiff hairs along the sides but the pubescence variable. The arctic variety *arcticum* Hult. has the glumes and lemmas hispid to pilose.

Matanuska—Arctic coast—upper Yukon district. (Fig. 176.)

5. *A. sericeum* Hitchc.

Culms tufted, 4–12 dm. tall; leaves flat, rather long, mostly 4–9 mm. wide; spikes 6–25 cm. long; spikelets remote to loosely imbricated, usually narrow, 3–7-flowered, 15–22 mm. long; glumes mostly 3-nerved, acute or short-awned, glabrous or slightly pubescent, 6–15 mm. long; lemmas short-villous, acuminate or short-awned, 8–12 mm. long; rachilla pubescent.

Nome—Matanuska—upper Yukon valley. (Fig. 177.)

6. *A. latiglume* (Scribn. & Sm.) Rydb.
A. violaceum (Hornem.) Lange var. *latiglume* Scribn. & Sm.

Culms loosely tufted, curved or geniculate below, 2–7 dm. tall; leaves flat, rather short, 3–5 mm. wide; spike 3–7 cm. long; spikelets imbricate, 10–18 mm. long, 3–5-flowered; glumes rather broad, flat or rounded, 9–12 mm. long; lemmas pubescent, acute or short-awned, 7–11 mm. long.

Most of our territory; circumpolar. (Fig. 178.)

7. *A. trachycaulum* (Link) Hitchc.   Slender Wheat-grass
    A. angustiglume Nevski.
    A. pauciflorum (Schwein.) Hitchc.
    A. tenerum Vasey

Culms tufted, 6–12 dm. tall; leaves mostly 2–4 mm. wide; spikes slender, 5–20 cm. long; spikelets from rather remote to closely imbricate,

12–15 mm. long, 2–5-flowered; glumes firm, acute to awn-pointed, 10–12 mm. long; lemmas 8–10 mm. long, acute or short-awned. Very variable and our most common species.

Above the Arctic Circle in Alaska—Labr.—W. Va.—Mo.—N. Mex.—Calif. (Fig. 179.)

8. *A. subsecundum* (Link) Hitchc.   Bearded Wheat-grass
    A. caninum Am. auct. not (L.) Beauv.
    A. richardsonii Schrad.

Culms tufted, erect, 3–10 dm. tall; leaves scabrous, flat, 3–8 mm. wide; spike erect or slightly nodding, 6–15 cm. long, rather dense; spikelets 12–15 mm. long, 3–5-flowered; glumes unequal, acuminate or awn-pointed, 4–7-nerved; lemmas scabrous, 8–12 mm. long, with a nearly straight awn 1–2 cm. long.

Central Alaska—Newf.—Md.—Nebr.—N. Mex.—Calif. (Fig. 180.)

9. *A. spicatum* (Pursh.) Scribn. & Sm.   Bluebunch Wheat-grass

Culms tufted, rigid, 6–10 dm. tall; leaves 1–4 mm. wide; spike slender, 8–15 cm. long; spikelets distant, often shorter than the internodes of the rachis, 3–8-flowered; glumes acute but not awned; lemmas 8–10 mm. long, terminating in a bent awn 10–25 mm. long, rachilla scabrous.

Central Alaska—Mich.—N. Mex.—Calif. (Fig. 181.)

34. HORDEUM L.

Leaf-blades flat; inflorescence a terminal spike; spikelets 1-flowered, usually in threes at each joint of the rachis, the middle spikelet sessile and perfect, the lateral usually pedicelled and imperfect; glumes narrow, often subulate and awned, rigid, standing in front of the spikelet; lemmas lanceolate, rounded on the back, tapering to a usually long awn. (Latin name for barley.)

Awn less than 12 mm. long .... 1. *H. brachyantherum*
Awn 15–35 mm. long .... 2. *H. caespitosum*
Awn 4–7 cm. long .... 3. *H. jubatum*

1. *H. brachyantherum* Nevski.   Meadow barley
    H. boreale Scribn. & Sm. not Gavdoger.
    H. nodosum Auct. in part.

Tufted perennial; culms 5–10 dm. tall; leaves 4–8 mm. wide, scabrous; spikes slender, 2–8 cm. long; glumes all setaceous, 8–15 mm. long; lemma of central spikelets 7–8 mm. long; awn exceeding the glumes; lemmas of lateral spikelets considerably reduced.

Coastal regions of Alaska and northeastern Asia. Also Labr. and Newf. (Fig. 182.)

2. *H. caespitosum* Scribn.   Bobtail barley

Culms 3–10 dm. tall; glumes and awns 15–35 mm. long. In nearly all characters this form is intermediate between H. brachyantherum and H. jubatum and since it is found only in the coastal sections where both these species occur, it is probably a hybrid between them. (Fig. 183.)

3. H. jubatum L. Squirrel-tail barley

Tufted perennial; leaves 2–4 mm. wide, scabrous; spikes nodding, 5–10 cm. long; glumes awnlike, 4-7 cm. long; lemma of central spikelet 6–8 mm. long with awn as long as the glumes; lemmas of lateral spikelets reduced almost to a short awn.

Open ground, all of Alaska except the Arctic—Labr.—Newf.—Md.—Mo.—Mex. Also Asia. (Fig. 184.)

Hordeum vulgare L., the annual, cultivated barley, like other grains, is sometimes found adventitive along roadsides.

35. ELYMUS L.

Perennials with spicate inflorescence; spikelets 2–6-flowered, usually 2 but sometimes 1 or 3 at each node of the rachis, the rachilla disarticulating above the glumes and between the florets; glumes equal, forming an apparent involucre to the cluster, rigid, narrow to subulate; lemmas oblong to lanceolate, rounded on the back, 5-nerved, usually awned; palea a little shorter than the lemma, 2-keeled. (Greek, an ancient name for a kind of barley.)

1A. Culms from creeping rhizomes.
  1B. Spikelets 10–15 mm. long ....1. *E. innovatus*
  2B. Spikelets 12–25 mm. long.
    1C. Glumes nearly as long as the spikelet ....2. *E. mollis*
    2C. Glumes much shorter than the spikelet ....3. *E. aleuticus*
2A. Culms tufted, no creeping rhizomes.
  1B. Lemmas awnless or nearly so ....4. *E. virescens*
  2B. Lemmas awned.
    1C. Awns curved, divergent ....5. *E. canadensis*
    2C. Awns straight.
      1D. Rachis tardily disjointing ....6. *E. macounii*
      2D. Rachis continuous.
        1E. Lemmas glabrous or scabrous ....7. *E. glaucus*
        2E. Lemmas sparsely long-hirsute on the edge ....
        ....8. *E. hirsutus*

1. *E. innovatus* Beal. Downy Rye-grass

Culms erect from horizontal rhizomes, 4–9 dm. tall; leaves rather rigid, flat or involute, 2–8 mm. wide; spike rather dense, 4–10 cm. long; spikelets 10–15 mm. long, 3–6-flowered; the narrow glumes and the lemmas densely purplish or grayish villous, the lemmas 8–10 mm. long with awns 1–4 mm. long.

Grassy flats, central Alaska—mouth of the Mackenzie River—S. Dak. —Wyo.—B. C. (Fig. 185.)

2. **E. mollis** Trin. Beach Ryegrass
   E. arenarius L. ssp. mollis (Trin.) Hult.
   E. villosissmus Scribn.

   Culms stout, glaucous, erect, from a widely creeping rhizome, 6–20 dm. tall; sheaths smooth, leaves smooth or scabrous above, 7–12 mm. wide; often involute on drying; spikes erect, dense, thick, soft, 7–25 cm. long; glumes scabrous or pubescent, 12–24 mm. long, acuminate, nearly as long as the spikelet; lemmas scabrous to villous-pubescent, acuminate or mucronate. Attu baskets are made of the fibers from the leaf of this species. In the far north it becomes much dwarfed.

   Sandy beaches, Alaska—Calif.; with closely related species, circumboreal. (Fig. 186.)

3. **E. aleuticus** Hult. Aleutian Rye-grass

   Culms 6–7 dm. tall, arising from elongated creeping rhizomes; ligule 0.7 mm. long, ciliate, spikes 10–15 cm. long; spikelets 3–5-flowered, about 25 mm. long; glumes lanceolate, 3–5-nerved, about 1 cm. long, scarious wing-margined, acute, sparsely pilose; lemmas 15–20 mm. long, acute, 5-nerved, short-pilose; awn 2–4 mm. long.

   Known only from Atka Island.

4. **E. virescens** Piper Pacific Rye-grass
   E. howellii Scribn. & Merr.

   Somewhat tufted; culms 3–12 dm. tall; sheaths smooth; leaves flat, 5–15 mm. wide, minutely scabrous; spike 6–16 cm. long; spikelets few-flowered; glumes strongly nerved, pointed or awn-tipped; lemmas 10–12 mm. long, scabrous toward the sharp-pointed or short-awned apex.

   Woods, southeastern Alaska—Calif. (Fig. 187.)

5. **E. canadensis** L. Canada Rye-grass

   Culms erect, tufted, smooth, 7–15 dm. tall; leaves 4–20 mm. wide; scabrous; spike 1–3 dm. long, nodding; spikelets 3–5-flowered; glumes narrow, scabrous; lemmas 8–14 mm. long, strongly nerved above, scabrous-hirsute; awns divergently curved, 2–3 cm. long.

   Along the Alaska Railroad—Que.—N. Car.—Texas—Ariz.—Calif. Probably introduced in our area. (Fig. 188.)

6. **E. macounii** Vasey. Macoun Rye-grass

   Culms densely tufted, erect, slender, 5–10 dm. tall; sheaths smooth; leaves usually scabrous on both sides, 2–5 mm. wide; spike slender, erect or somewhat nodding, 4–12 cm. long; spikelets imbricate, appressed, 1–3-flowered, about 1 cm. long; glumes narrow, scabrous, awned; lemmas 8–10 mm. long, scabrous and somewhat hirsute toward the apex; awns 1–2 cm. long.

   Central Alaska—Alta.—Minn.—Nebr.—N. Mex.—Calif. (Fig. 189.)

7. **E. glaucus** Buckley. Western Rye-grass

   Culms tufted, often bent at the base, 6–15 dm. tall; leaves more or less scabrous on both sides, 6–15 mm. wide; spike erect or somewhat nodding, 5–20 cm. long; glumes lanceolate, 8–15 mm. long, with prominent scabrous nerves; lemmas 7–10 mm. long, scabrous toward the apex and with awns 8–20 mm. long.

Southeastern Alaska—Ont.—Mich.—Mo.—N. Mex.—Calif. (Fig. 190.)

8. *E. hirsutus* Presl. Northern Rye-grass
   E. borealis Scribn.

Culms rather weak, 5–14 dm. tall; leaves lax, 4–13 mm. wide, somewhat scabrous beneath, sparsely pilose above; spike loosely flowered, nodding, 10–18 cm. long; spikelets about 15 mm. long; glumes strongly nerved, awned; lemmas long-hirsute on the margins toward the summit, sometimes coarsely pubescent on the back; awn up to 2 cm. long.

Coastal sections. Alaska—Ore. (Fig. 191.)

Pleuropogon sabinii R. Br. is a small grass found in the arctic regions of Canada and Eurasia and may be expected in the most northerly parts of Alaska and the Yukon. It is 15 cm. or less tall; leaves 1–5 cm. long or when growing in water longer; spikelets 2–6, about 1 cm. long on spreading or reflexed pedicels, 5–8-flowered; glumes small, unequal, scarious at the somewhat lacerate tip; lemmas 4–5 mm. long, 7-nerved, the midvein sometimes excurrent as a sharp point; keels of the palea winged on lower half, bearing awnlike appendages near the middle.

PLATE VI

60. *Typha latifolia* L. Inflorescence and flowers.
61. *Sparganium hyperboreum* Laest. Inflorescence, achenes, and perianth scales.
62. *Sparganium angustifolium* Michx. Achenes and scales.
63. *Sparganium minimum* (Hartm.) Fr. Inflorescence, achene, and scale.
64. *Potamogeton natans* L. Leaves.
65. *Potamogeton gramineus* L. Leaves, fruit, and nutlet.
66. *Potamogeton alpinus* Balbis. Leaves and nutlet.
67. *Potamogeton porsildorum* Fern. Leaf, section of leaf, and nutlet.
68. *Potamogeton pusillus* L. Leaf, fruit, and nutlet.
69. *Potamogeton friesii* Rupr. Leaf, tip of leaf, and nutlet.
70. *Potamogeton perfoliatus* L. Leaf and nutlet.
71. *Potamogeton filiformis* Pers. Leaf, fruit, and nutlet.
72. *Potamogeton pectinatus* L. Leaf, fruit, and nutlet.
73. *Ruppia spiralis* L. Leaves, fruit, and nutlet.
74. *Zostera marina* L. Inflorescence, tip of leaf, and nutlet.
75. *Triglochin maritima* L. Leaf, flower, and fruit.
76. *Triglochin palustris* L. Flowers and fruits.
77. *Hierochloe alpina* (Sw.) Roem. & Schult. Spikelet and lemmas.
78. *Hierochloe pauciflora* R. Br. Spikelet.
79. *Hierochloe odorata* (L.) Beauv. Spikelet.

FLORA OF ALASKA

PLATE VI

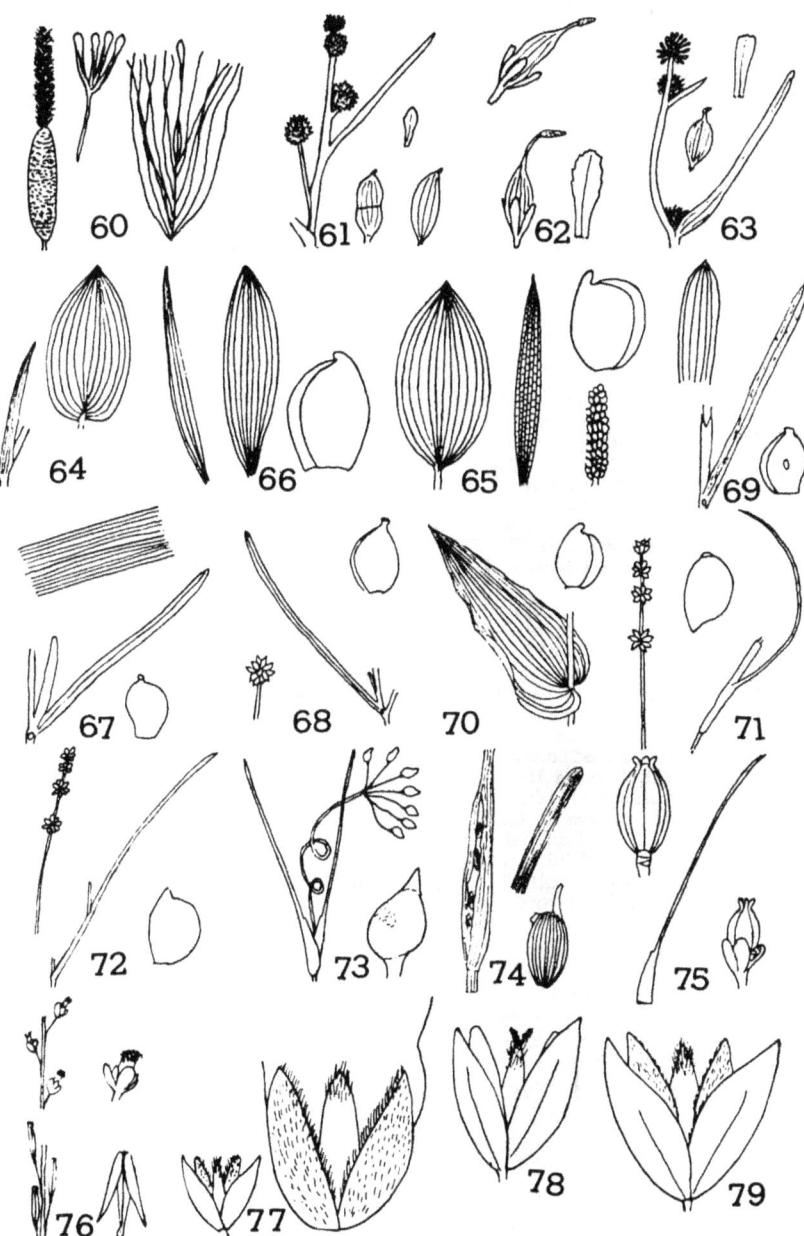

PLATE VII

80. *Phalaris arundinacea* L. Spikelet and lemma.
81. *Anthoxanthum odoratum* L. Glumes, sterile lemmas, and fertile lemmas.
82. *Stipa columbiana* Macoun. Lemma.
83. *Stipa comata* Trin. Lemma.
84. *Alopecurus pratensis* L. Spikelet.
85. *Alopecurus alpinus* J. E. Sm. Spikelet and lemma.
86. *Alopecurus aequalis* Sobol. Spike, spikelet, and lemma.
87. *Alopecurus geniculatus* L. Spikelet.
88. *Polypogon monspeliensis* (L.) Desf. Spikelet and lemma.
89. *Phleum alpinum* L. Spikelet and floret.
90. *Phleum pratense* L. Spikelet and lemma.
91. *Phippsia algida* (Soland.) R. Br. Glumes and fruiting floret.
92. *Cinna latifolia* (Trev.) Griseb. Spikelet.
93. *Calamagrostis deschampsioides* Trin. Spikelet and floret.
94. *Calamagrostis purpurescens* R. Br. Spikelet and lemma.
95. *Calamagrostis nutkaensis* (Presl) Steud. Spikelet and lemma.
96. *Calamagrostis canadensis* (Michx.) Beauv. Glumes and lemma.
97. *Calamagrostis inexpansa* A. Gray. Glumes and lemma.
98. *Agrostis thurberiana* Hitchc. Glumes and floret.
99. *Agrostis palustris* Huds. Glumes and floret.
100. *Agrostis stolonifera* L. Glumes and floret.
101. *Agrostis exarata* Trin. Glumes and lemma.
102. *Agrostis scabra* Willd. Spikelet and lemma.
103. *Agrostis alaskana* Hult. Glumes and lemma.
104. *Agrostis borealis* Hartm. Glumes and lemma.

FLORA OF ALASKA

PLATE VII

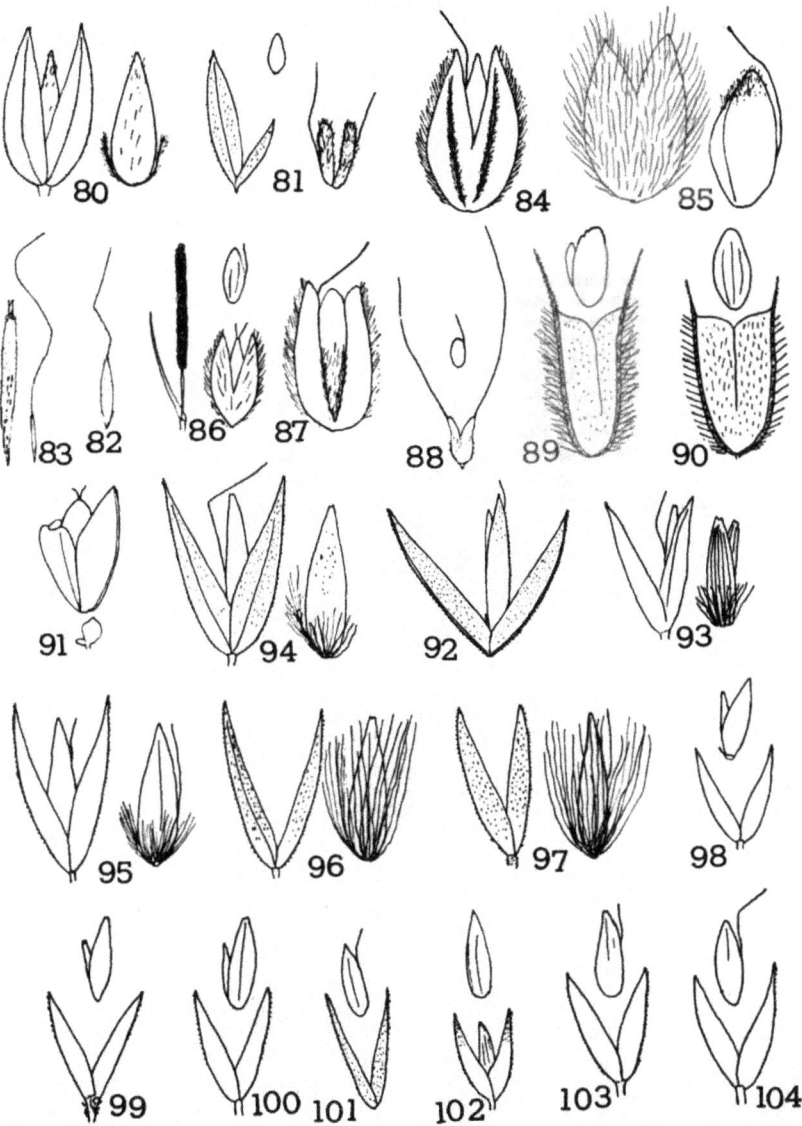

PLATE VIII

105. *Arctagrostis latifolia* (R. Br.) Griseb. Spikelet.
106. *Holcus lanatus* L. Glumes and florets.
107. *Arrhenatherum elatius* (L.) Mert. & Koch. Spikelet.
108. *Sphenopholis intermedia* (Rydb.) Rydb. Spikelet.
109. *Danthonia spicata* (L.) Beauv. Spikelet and lemma.
110. *Avena fatua* L. Spikelet.
111. *Trisetum spicatum* (L.) Richt. Spikelet and lemma.
112. *Trisetum cernuum* Trin. Spikelet.
113. *Trisetum sibiricum* Rupr. Spikelet.
114. *Deschampsia elongata* (Hook.) Munro. Spikelet and lemma.
115. *Deschampsia atropurpurea* (Wahl.) Scheele. Spikelet and lemma.
116. *Deschampsia beringensis* Hult. Spikelet and lemma.
117. *Deschampsia caespitosa* (L.) Beauv. Spikelet and lemma.
118. *Deschampsia holciformis* Presl. Spikelet.
119. *Beckmannia syzigachne* (Steud.) Fern. Spikelet and floret.
120. *Dactylis glomerata* L. Spikelet.
121. *Dupontia fischeri* R. Br. Spikelet and floret.
122. *Schizachne purpurescens* (Torr.) Swallen. Spikelet and lemma.
123. *Colpodium fulvum* (Trin.) Griseb. Spikelet.
124. *Melica subulata* (Griseb.) Scribn. Spikelet and lemma.
125. *Glyceria borealis* (Nash) Batch. Spikelet and lemma.
126. *Glyceria leptostachya* Buckl. Spikelet and lemma.
127. *Glyceria pauciflora* Presl. Spikelet.
128. *Glyceria grandis* S. Wats. Spikelet.
129. *Puccinellia phryganodes* (Trin.) Scribn. & Merr. Spikelet.
130. *Puccinellia hauptiana* Krecz. Spikelet.
131. *Puccinellia grandis* Swallen. Spikelet.
132. *Puccinellia borealis* Swallen. Spikelet.
133. *Puccinellia andersoni* Swallen. Spikelet.
134. *Puccinellia alaskana* Scribn. & Merr. Spikelet.
135. *Puccinellia paupercula* (Holm) Fern. & Weath. Spikelet.
136. *Puccinellia hulteni* Swallen. Spikelet.
137. *Puccinellia pumila* (Vasey) Hitchc. Spikelet.

PLATE VIII

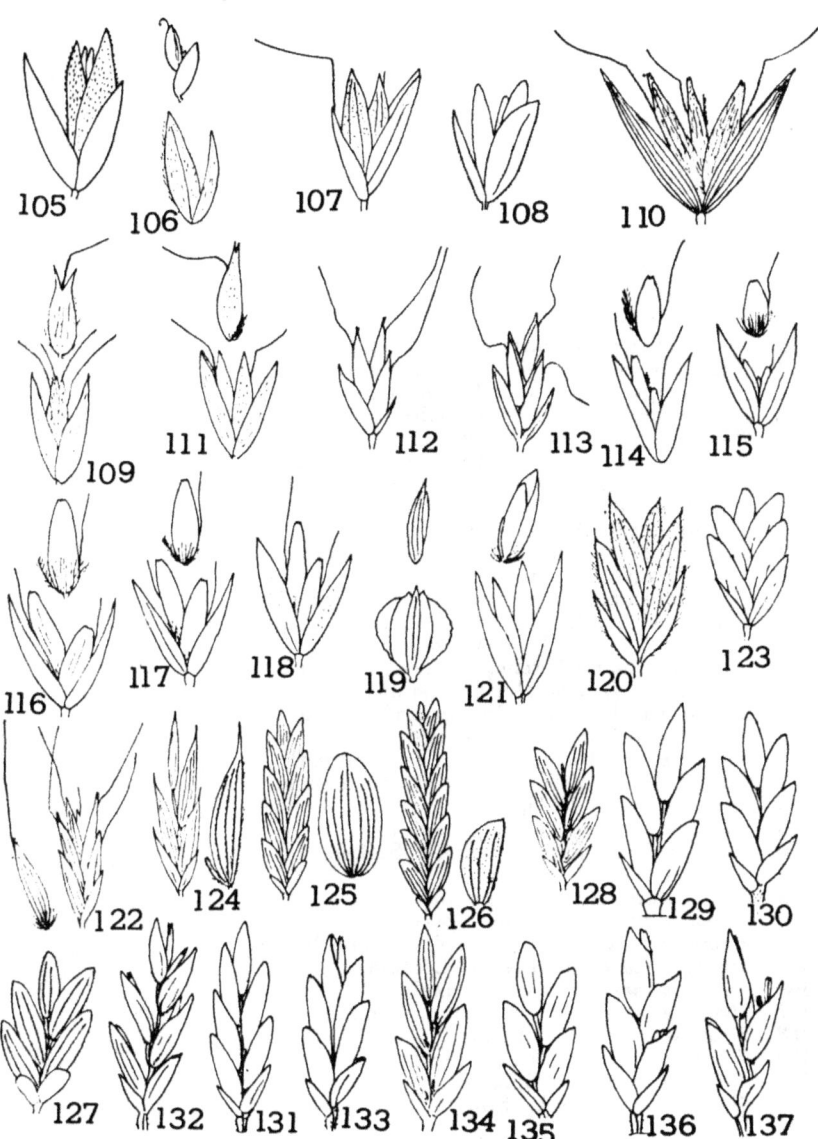

PLATE IX

138. *Puccinellia nutkaensis* (Presl) Fern. & Weath. Glumes and florets.
139. *Poa annua* L. Spikelet and lemma.
140. *Poa compressa* L. Spikelet and lemma.
141. *Poa lanata* Scribn. & Merr. Glumes and lemma.
142. *Poa turneri* Scribn. Floret.
143. *Poa eminens* Presl. Spikelet and lemma.
144. *Poa macrocalyx* Tr. & Mey. Spikelet and lemma.
145. *Poa pratensis* L. Glumes and a lemma.
146. *Poa arctica* R. Br. Lemma.
147. *Poa irrigata* Lindm. Glumes and lemmas.
148. *Poa trivialis* L. Glumes and lemma.
149. *Poa leptocoma* Trin. Glumes and lemma.
150. *Poa nemoralis* L. Glumes and lemma.
151. *Poa palustris* L. Glumes and lemma.
152. *Poa merrilliana* Hitchc. Spikelet and lemma.
153. *Poa alpina* L. Spikelet and lemma.
154. *Poa stenantha* Trin. Glumes and lemma.
155. *Poa glauca* Vahl. Spikelet and lemma.
156. *Poa rupicola* Nash. Glumes and lemma.
157. *Poa hispidula* Vasey. Spikelet and lemma.
158. *Festuca elatior* L. Spikelet.
159. *Festuca subulata* Trin. Spikelet
160. *Festuca brachyphylla* Schult. Spikelet.
161. *Festuca rubra* L. Spikelet.
162. *Festuca altaica* Trin. Spikelet.
163. *Bromus sitchensis* Trin. Spikelet.
164. *Bromus aleutensis* Trin. Spikelet and lemma.
165. *Bromus marginatus* Nees. Lemma.
166. *Bromus inermis* Leyss. Spikelet.
167. *Bromus pumpellianus* Scribn. Lemma.
168. *Bromus ciliatus* L. Lemma.
169. *Bromus pacificus* Shear. Floret.
170. *Bromus mollis* L. Spikelet.
171. *Bromus tectorum* L. Spikelet.

FLORA OF ALASKA

PLATE IX

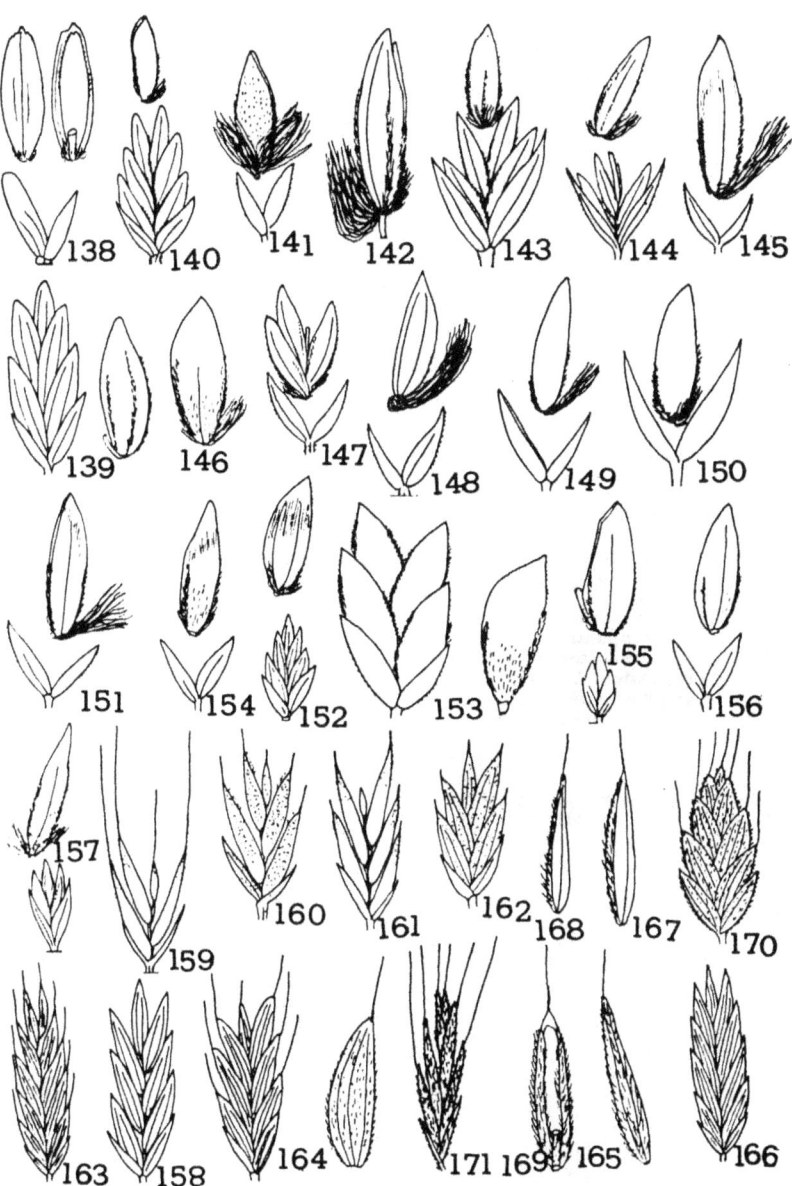

PLATE X

172. *Lolium multiflorum* Lam. Spikelet.
173. *Agropyron smithii* Rydb. Spikelet.
174. *Agropyron repens* (L.) Beauv. Spikelet, floret, and awned lemma.
175. *Agropyron yukonense* Scribn. & Merr. Spikelet.
176. *Agropyron alaskanum* Scribn. & Merr. (a) Spikelet. (b) Var. *arcticum* Hult.
177. *Agropyron sericeum* Hitchc. Spikelet and floret.
178. *Agropyron latiglume* (Scribn. & Sm.) Rydb. Spikelet.
179. *Agropyron trachycaulum* (Link) Hitchc. Spikelet and floret.
180. *Agropyron subsecundum* (Link) Hitchc. Spikelet.
181. *Agropyron spicatum* (Pursh) Scribn. & Sm. Spikelet.
182. *Hordeum brachyantherum* Nevski. Node of 3 spikelets.
183. *Hordeum caespitosum* Scribn. Node of 3 spikelets.
184. *Hordeum jubatum* L. Node of spikelets.
185. *Elymus innovatus* Beal. Lemma.
186. *Elymus mollis* Trin. Spikelet.
187. *Elymus virescens* Piper. Spikelet.
188. *Elymus canadensis* L. Spikelet and lemma.
189. *Elymus macounii* Vasey. Spikelet.
190. *Elymus glaucus* Buckl. Spikelet.
191. *Elymus hirsutus* Presl. Spikelet.

PLATE X

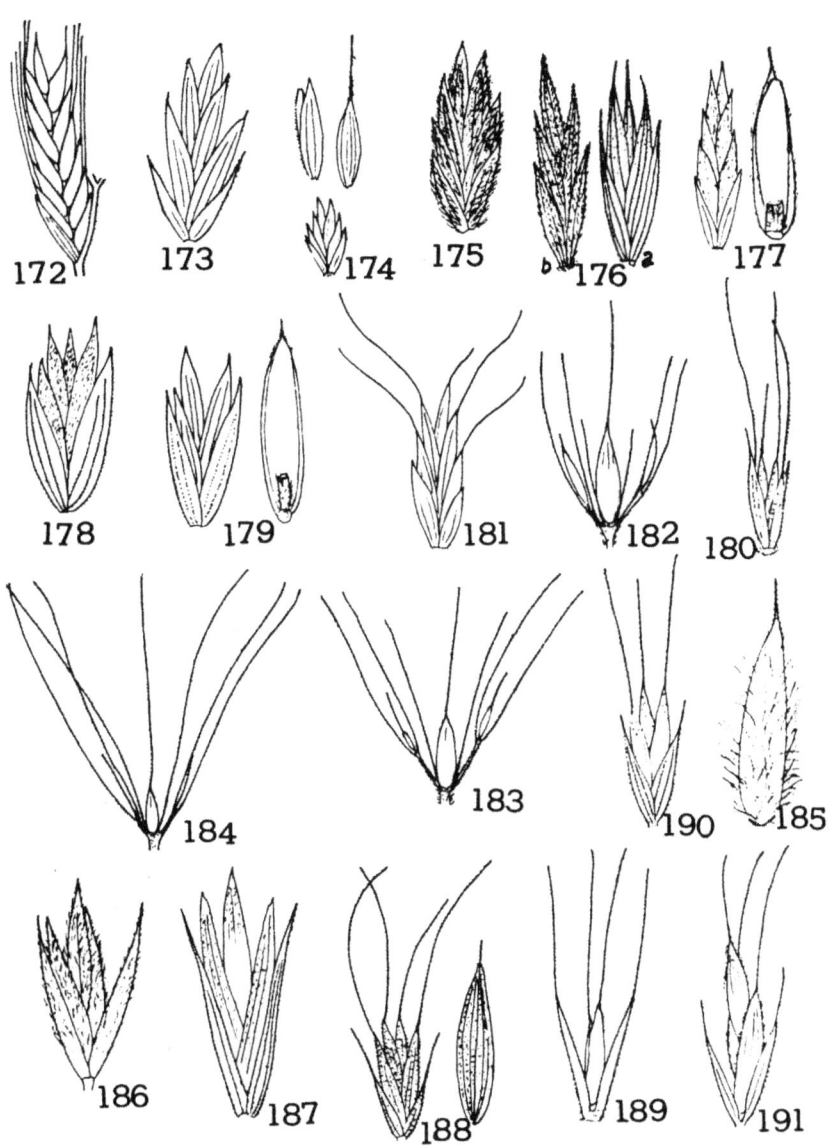

7. CYPERACEAE (Sedge Family)

Grass-like or rush-like plants with usually solid stems and three-ranked leaves with closed sheaths and narrow blades. Flowers in spikes or spikelets, in the axil of two-ranked or spirally arranged scales; perianth composed of bristles, a sac-like organ (perigynium), or wanting; stamens and styles two or three; anthers two-celled, basifixed; ovary one-celled, one-ovuled; fruit an achene.

1A. Fertile flowers all perfect, sometimes staminate flowers present.
   1B. Base of style persistent as a tubercle.
      1C. Basal empty scales several..................................1. *Rynchospora*
      2C. Basal empty scales not more than two or three..2. *Eleocharis*
   2B. Base of style not persistent.
      1C. Bristles six to many, much elongated.................3. *Eriophorum*
      2C. Bristles few, short ....4. *Scirpus*
2A. All flowers unisexual.
   1B. Pistillate flower partly enwrapped in the scale....5. *Kobresia*
   2B. Pistillate flowers enclosed in a sac ....6. *Carex*

1. RYNCHOSPORA Vahl.

Leafy perennials with erect stems; leaves narrow, flat or involute; spikelets clustered, ovoid, oblong, or fusiform; scales imbricate, thin, one-nerved, usually mucronate by the excurrent midvein; perianth of one to twenty-four, mostly six, barbed or scabrous bristles; achenes lenticular, capped by the persistent base of the style or the whole style. (Greek, referring to the beak-like tubercle.)

R. alba (L.) Vahl.   White Beaked-rush

Stem slender, glabrous, 10–25 cm. tall; leaves bristle-like, 1 mm. or less wide; spikelets several or numerous, 4–6 mm. long, in one to four dense heads; scales light-colored, acute; bristles nine to fifteen, downwardly barbed, about as long as the achene and the tubercle.

Southeastern Alaska; has an interrupted circumboreal distribution. (Fig. 192.)

2. ELEOCHARIS R. Br.

Rush-like tufted plants growing in water or wet places; leaves reduced to mere sheaths or rarely the lower blade-bearing; inflorescence an erect, terminal spikelet; perianth of one to twelve, usually retrosely barbed, bristles; base of styles persistent, forming a tubercle at the summit of the achene. (Greek, referring to the growth of most species in marshy ground.)

1A. Culms low, 3-15 cm. long.
   1B. Achene reticulate ....1. *E. acicularis*
   2B. Achene rough ....2. *E. nitida*
2A. Culms taller, 12-150 cm. long.
   1B. Achenes finely papillose ....3. *E. kamtschatica*
   2B. Achenes smooth.
      1C. Tubercle more than one-half as wide as the achene...................
      ....4. *E. uniglumis*
      2C. Tubercle less than one-half as wide as the achene.
         1D. Tubercle conical-triangular ....5. *E. palustris*
         2D. Tubercle cap-like ....6. *E. mamillata*

1. *E. acicularis* (L.) R. & S.   Needle Spike-rush
    Scirpus acicularis L.

Stems filiform, grooved, obscurely four-angled, 3-10 cm. tall; spikelet 3-6 mm. long; three- to ten-flowered; scales oblong, pale green, usually with a brown band on each side of the midvein; bristles three or four, short and fugacious; achenes pale, obscurely three-angled, with intermediate ribs; tubercle conic, about one-fourth as long as the achene.

From Seward Peninsula south and east. Circumboreal. (Fig. 193.)

2. *E. nitida* Fern.   Slender Spike-rush
    E. tenuis (Willd.) Schult.
    Scirpus nitidus (Fern.) Hult.

Perennial by slender rootstocks; culms slender, tufted, four-angled, striate, 2-8 cm. tall; tips of upper sheaths whitish; spikelets 2.5-4.5 mm. long, 1.5-2.5 mm. wide; scales ovate or ovate-oblong, the tips obtuse, purplish-brown with greenish midrib and narrow, scarious margins; bristles two to four, shorter than the achene, fugacious, or wanting; achene trigonous, very minutely roughened, 1 mm. or less long, tubercle conic, short, acute.

Western Pacific district and Ottawa Valley—Newf.—N. S.—N. Hamp.

3. *E. kamtschatica* (C. A. Mey.) Kom.   Kamchatka Spike-rush
    Scirpus kamtschaticus C. A. Mey.

Stems erect, 1-4 dm. tall; spikelet ovoid, 6-15 mm. long; scales ovate, purplish-brown with reddish midvein; bristles about as long as the achene and tubercle; achene greenish-yellow, about 1.5 mm. long, finely papillose; tubercle nearly as large as the achene, cap-like.

Eastern Asia and the Bering Sea and Aleutian regions—Southeastern Alaska. (Fig. 194.)

4. **E. uniglumis** (Link) Schult. One-bracted Spike-rush
   Scirpus uniglumis Link.

   Stoloniferous and loosely caespitose; culms slender, 5–70 cm. tall, reddish at the base; spikelet 5–15 mm. long, 2–6 mm. thick, five- to thirty-flowered; basal scale roundish, completely clasping the base of the spikelet; fertile scales castaneous or purplish, firm, lustrous, 3–5 mm. long; achenes obovoid, yellowish or darker, tubercle conic-ovoid, one-half to two-thirds as wide as the achene.

   Collected at Circle and at Hyder, a variable, circumboreal species. (Fig. 195.)

5. **E. palustris** (L.) R. & S. Creeping Spike-rush
   Scirpus palustris L.

   Stems erect, striate, 3–15 dm. tall; spikelet ovoid-cylindric, 8–20 mm. long, many-flowered; scales brown with scarious margins; bristles usually four, longer than the achene and tubercle; achene lenticular, smooth, yellowish; tubercle conic-triangular, flattened, 0.25–0.5 as long as the achene.

   Central Alaska south and east; circumboreal. (Fig. 196.)

6. **E. mamillata** Lindb. f. Pale Spike-rush
   Scirpus mamillatus Lindb. f.

   Resembling *E. palustris;* culms 2–12 dm. tall, pale; spikelet subcylindric to lanceolate, 1–3 cm. long, 2–5 mm. thick, many-flowered, acute, scales narrowly ovate, obtuse to subacute, appressed, 2–4 mm. long; achenes yellow or pale brown; tubercle yellow, small, cap-like.

   Pacific Coast districts; circumboreal.

3. ERIOPHORUM L.

Bog plants with erect stems and linear leaves or the upper one or two reduced to bladeless sheaths; spikes terminal, solitary and capitate, or several in an involucrate umbel; scales spirally imbricated; flowers perfect; perianth of soft capillary bristles which are much exerted at maturity; achenes three-angled, oblong, ellipsoid, or obovoid. (Greek, wool-bearing.)

1A. Spike solitary.
  1B. Bristles six .... 1. *E. alpinum*
  2B. Bristles numerous.
    1C. Plants stoloniferous.
      1D. Anthers 0.5–1 mm. long, bristles white ....
      .... 2. *E. scheuchzeri*
      2D. Anthers longer.
        1E. Middle scales blunt with broad hyaline margins ....
        .... 3. *E. chamissonis*
        2E. Middle scales acute with narrow hyaline margins ....
        .... 4. *E. medium*

  2C. Plants densely tufted, no stolons.
    3D. Scales gray, translucent .... 5. *E. vaginatum*
    4D. Scales grayish or greenish-black, not translucent.
      1E. Plant 6–20 cm. tall .... 6. *E. callitrix*
      2E. Plants 3–7 dm. tall .... 7. *E. brachyantherum*
 2A. Spikes more than one.
  1B. Leaf-blades triangular-channeled throughout ....
  .... 8. *E. gracile*
  2B. Leaf-blades flat below the middle.
    1C. Midrib of the scales prominent to the very tip ....
    .... 9. *E. viridi-carinatum*
    2C. Midrib of scales not prominent at the tip ....
    .... 10. *E. angustifolium*

1. *E. alpinum* L.   Alpine Cotton-grass

  Stems scattered or somewhat tufted, triangular, 10–25 cm. tall; leaves subulate, 6–20 mm. long, borne near the base, lower sheaths often bladeless; involucral bract blunt-subulate, shorter than the spike; spike small, erect; glumes yellowish-brown with slender midvein; bristles six, white, flat, crisped, 10–20 mm. long; achene obovate, apiculate.
  Cook Inlet—central Alaska—Hudson Bay—Conn.—Mich.—B. C. (Fig. 197.)

2. *E. scheuchzeri* Hoppe.   White Cotton-grass

  Stems slender, 2–6 dm. tall; sheaths all blade-bearing except the uppermost one; blades filiform, channeled, shorter than the culm; spike erect, globose at maturity; bristles numerous, white, or in drying often yellowish, 15–30 mm. long; achenes narrowly oblong, acute with a subulate beak, scarcely 2 mm. long.
  Throughout our area; circumboreal. (Fig. 198.)

3. *E. chamissonis* C. A. Mey.   Russet Cotton-grass
  E. russeolum Fries.

  Culms triangular, 3–7 dm. tall; upper sheaths somewhat inflated. This species closely resembles *E. scheuchzeri* but is of taller growth, the scales are broader with wide, hyaline margins, the achene is broader and narrowed at the base, the numerous bristles 2–4 cm. long and usually of a russet-brown color, although a pale form occurs. This pale form which is usually nearly white is the only one found in the Bering Sea region and on the Arctic Coast. It has been described as var. *albidum* Fern. (var. *leucothrix* (Blomg.) Hult.
  Throughout our area; circumboreal. (Fig. 199.)

4. *E. medium* Anders.

  This name is applied to plants forming a connecting link between *E. chamissonis* and *E. scheuchzeri*. It is probably a hybrid of these two species and occurs where both parent species are found, but according to Hultén it does not occur in regions where *E. scheuchzeri* alone is found. In

E. scheuchzeri the anthers are 1 mm. or less in length, in *E. chamissonis* they are 2-3 mm. long, and in *E. medium* they are 1-2 mm. long. The bristles are tinged with brown.

5. *E. vaginatum* L.   Niggerheads, Sheathed Cotton-grass

Densely tufted, forming "niggerheads"; culms stiff, obtusely triangular, 2-5 dm. tall; leaves filiform, triangular, channeled; upper sheaths inflated; spike oblong, 1-3 cm. long; glumes ovate-lanceolate, acuminate, thin, mostly hyaline; anthers 2-3 mm. long; bristles white or slightly dingy, 10-16 mm. long; achene narrowly ovoid, scarcely apiculate. Ssp. *spissum* (Fern.) Hult. (*E. spissum* Fern.) spikes ovoid to subglobose, the rachis 6-10 mm. long compared to 9-15 in the type form, anthers 1-2 mm. long, achenes broadly ovoid. This is the form found in the Bering Sea and Arctic regions.

The typical form from central Alaska eastward; circumboreal. (Fig. 200.)

6. *E. callitrix* Cham.   Arctic Cotton-grass

Culms low, 6-20 cm. tall; usually only one sheath which is close to the base and often bears a short blade; leaves rigid, spreading, the blades forming an angle with the sheath; scales nearly uniform in color; bristles pure white.

Northeastern Asia and Bering Sea islands—Baffin Island—E. Greenl.— N. Newf.

7. *E. brachyantherum* Trautv.   Close-sheathed Cotton-grass
E. opacum Am. Auct.

Culms 3-7 dm. tall, from dense tussocks; basal leaves elongate, continuous with the sheath; uppermost sheath scarcely inflated; scales dark, ovate-lanceolate or the inner linear-lanceolate, acuminate; bristles white or slightly tinged brown, 1-2 cm. long; achenes obovate-oblong, smooth, conspicuously apiculate.

Throughout most of our territory; circumboreal. (Fig. 201.)

8. *E. gracile* Koch.   Slender Cotton-grass

Culms slender, smooth, terete, 3-6 dm. tall; sheaths all blade-bearing, the blades narrowly linear, not over 2 mm. wide; spikes two to six, some of them on slender, drooping, pubescent peduncles; scales ovate with prominent midribs; bristles white, 15-25 mm. long; achenes linear-oblong, about 2.5 mm. long.

Central Alaska east and south; circumboreal. (Fig. 202.)

9. *E. viridi-carinatum* (Engelm.) Fern.   Thin-leaved Cotton-grass

Similar in appearance to *E. angustifolium;* leaves thin, flat, black at the base; spikes usually numerous, up to thirty; peduncles finely hairy, elongated or short; scales ovate-lanceolate, the midvein extending to the tip and sometimes excurrent; achene oblong-ovoid; bristles white or slightly yellowish.

Sphagnum bogs, Cook Inlet region and B. C.—Hudson Bay—Newf.—N. Y.—Ohio—Wyo.

10. *E. angustifolium* Roth.   Tall Cotton-grass

Culms smooth, obtusely triangular above 3–7 dm. tall; leaf-blades more than 3 mm. wide; bracts two to four, often blackish at the base; spikes two to twelve, in a terminal umbel; peduncles smooth; scales ovate-lanceolate, purple-green or brown; bristles white or tawny, up to 3 cm. long; achenes nearly black, sharp-pointed, about 2.5 mm. long. Forms of this species have been reported as *E. polystachyon* L.

Common throughout our territory; circumboreal. (Fig. 203.)

4. SCIRPUS L.

Ours all perennials of swamps or wet places; leaves grass-like or in some species reduced to sheaths; spikelets solitary, clustered, or umbellate, the inflorescence usually subtended by one or more leafy bracts, often appearing lateral; scales arranged spirally, the lower often empty; flowers perfect; perianth of one to six usually barbed or pubescent bristles; styles and stamens two or three. (Latin name for the bulrush.)

1A. Spikelet small, solitary, terminal.
   1B. None of the sheaths leaf-bearing ....1. *S. pauciflorus*
   2B. One or more of the sheaths leaf-bearing...............2. *S. caespitosus*
2A. Spikelets normally more than one.
   1B. Spikelets few, appearing lateral ....3. *S. americanus*
   2B. Spikelets several.
      1C. Spikelets spicate ....4. *S. rufus*
      2C. Spikelets umbellate ....5. *S. pacificus*
   3B. Spikelets numerous.
      1C. Sheaths bladeless, culms terete ....6. *S. validus*
      2C. Plant leafy, culms triangular ....7. *S. microcarpus*

1. *S. paucifloris* Light.   Few-flowered Club-rush

Similar in appearance to the common *C. caespitosus* but less densely tufted; culms three-angled; upper sheath truncate, without trace of a leaf; no involucral bract; bristles two to six, hispid.

Known from Manly Hot Springs and B. C.—Que.—N. Y.—Calif.

2. *S. caespitosus* L. var. *callosus* Bigel.   Tufted Club-rush

S. caespitosus L. ssp. *austriacus* (Pella) Achers. & Graebn.

Culms slender, densely tufted, 1–3 dm. tall; basal sheaths numerous, the upper one bearing a short blade; spikelet 4–5 mm. long, glumes yellowish-brown; bristles six, smooth, longer than the acute achene.

Central Alaska southward; circumboreal. (Fig. 204.)

3. *S. americanus* Pers.   Three-square

Culms sharply triangular, erect, 3–12 dm. tall; leaves 1–3, linear, keeled, shorter than the culm; spikelets one to seven, oblong-ovoid, acute, 8–15 mm. long, appearing as if lateral; bract 2–10 cm. long; glumes broadly ovate, brown, often emarginate or two-cleft; awned; bristles two to six, barbed.

Circle Hot Springs and B. C.—Newf.—Bermuda—S. Am.—Calif.—Europe. (Fig. 205.)

4. *S. pacificus* Britt.　　　　　　　　　　　　　　　　Pacific Bulrush

Culms leafy, stout, sharply three-angled with flat sides, 5–8 dm. tall; leaves 1 cm. or less wide, the longer often as long as the culm; bracts two to five, some of them longer than the inflorescence; spikelets ovoid, 1–2 cm. long, usually densely clustered; scales brown-tipped with a recurved awn; bristles shorter than the achene; achene light brown, about 2.5 mm. wide, nearly 4 mm. long.

Saline marshes along the coast, Anchorage—s. Calif. (Fig. 206.)

5. *S. rufus* (Huds.) Schrad.　　　　　　　　　　　Red Club-Rush

Culms in small clusters from slender rootstocks, erect, 8–30 cm. tall; leaves narrow, channeled, up to 15 cm. long, the lower reduced; spikelets reddish-brown, few-flowered, ovoid-oblong, 5–7 mm. long, in a terminal two-ranked spike 1–2 cm. long; bract 5–25 mm. long; scales lanceolate, acute, one-nerved; bristles one to six, shorter than the achene, deciduous.

Matanuska—N. W. Terr.—Newf.—N. S.—James Bay. Also N. Europe. (Fig. 207.)

6. *S. validus* Vahl.　　　　　　　　　　　　　　　　Great Bulrush
S. lacustris Am. Auct.

Culms stout, terete, smooth, spongy, 1–3 m. tall, 1–2 cm. thick, sheathed below; spikelets 5–12 mm. long, numerous in a compound cluster; scales ovate to suborbicular, reddish-brown, with strong midrib; achenes gray, plano-convex, about 1.5 mm. by 2 mm., bristles four to six, downwardly barbed.

Cook Inlet region—Newf.—West Indies—Calif. (Fig. 208.)

7. *S. microcarpus* Presl.　　　　　　　　　　　Small-fruited Bulrush

Culms stout and leafy, 6–15 dm. tall; leaves 7–18 mm. wide, up to 1 m. long, rough-margined; spikelets very numerous in a very compound inflorescence, ovoid-oblong, acute, 3–4 mm. long; scales greenish; bristles four, barbed, longer than the smooth whitish achene.

Western Pacific Coast of Alaska—Newf.—Conn.—Calif. (Fig. 209.)

5. KOBRESIA Willd.

Slender arctic and mountain sedges; culms erect, leafy below; spikelets very small, one- or two-flowered, in our species arranged in spikes; stamens three; perianth bristles and perigynium wanting; ovary oblong, narrowed into the style; stigmas two or three, linear; achenes obtusely three-angled, sessile. (von Kobres was a naturalist of Augsberg, Germany.)

Spike one ....1. *K. myosuroides*
Spikes more than one ....2. *K. simpliciuscula*

1. **K. myosuroides** (Vill.) Fiori & Paol. Bellard Kobresia
   K. bellardii (All.) Degland.

   Culms tufted, very slender, 1–4 dm. tall, longer than the leaves; leaves near the base, 2–20 cm. long, 0.25–0.5 mm. wide, acicular; spike bractless, 1–3 cm. long, 2–4 mm. in diameter, the terminal spikelet staminate, the lateral ones with one staminate and one pistillate flower; scales 2–3 mm. long; achenes about 2.5 mm. long, 1 mm. wide.

   Arctic and alpine; circumpolar. (Fig. 210.)

2. **K. simpliciuscula** (Wahl.) Mack.

   Culms and leaves similar to *K. myosuroides;* spikes three to ten, 3–8 mm. long, 1.5–2.5 mm. wide, in a head 10–35 mm. long, which sometimes appears spike-like; terminal spikelets staminate, the lateral androgynous or pistillate and one-flowered; achenes about 3 mm. long, 0.5 mm. wide.

   Bering Sea and Alaska Range regions; circumpolar.

6. CAREX L.

Perennial grass-like sedges with mostly triangular stems (culms) and three-ranked leaves, the upper (bracts) subtending the spikes or wanting; plants monoecious or sometimes dioecious; spikes one-many, either wholly staminate, wholly pistillate, or producing both staminate and pistillate flowers in different ends of the same spike; flowers solitary in the axils of scales; perianth none; staminate flowers of three (rarely two) stamens with filiform filaments; pistillate flowers of a single pistil with a style and two or three stigmas, forming an achene enclosed in a sac (perigynium) through the orifice of which the stigmas protrude; achenes triangular, lenticular, or plano-convex and enclosed in the perigynium or rarely rupturing it. (Greek, to cut, on account of the sharp leaves.)

A vast genus, well represented in our area. The division of genus here adopted is that of Kukenthal which is much easier to use though more artificial than that adopted for the American species by Mackenzie. The illustrations are of glume, perigynium and achene. They are not drawn to any particular scale but the parts illustrated are in proportion for that species.

1A. Spike single, terminal ....*Primocarex*
2A. Spikes two or more.
  1B. Spikes sessile, bisexual ....*Vignea*
  2B. Spikes peduncled, usually unisexual, sometimes bisexual.
    1C. Stigmas two .... *Eucarices distigmaticae*
    2C. Stigmas three ....*Eucarices tristigmaticae*

Primocarex

1A. Pistillate scales persistent.
  1B. Stigmas two.
    1C. Spike androgynous.
      1D. Perygynia with rounded base .... 3. *C. capitata*

2D. Perigynia tapering to a stipulate base.
   1E. Beak scabrous, leaves filiform .... 1. *C. nardina*
   2E. Beak smooth, leaves flat .... 2. *C. jacobi-peteri*
2C. Spike unisexual .... 4. *C. gynocrates*
2B. Stigmas three.
   1C. Perigynia lanceolate with long beak.
      1D. Leaves flat, rhizomes long .... 6. *C. anthoxanthea*
      2D. Leaves canaliculate, rhizomes short..... 7. *C. circinata*
   2C. Perigynia with short beak or beakless.
      1D. Spike unisexual .... 5. *C. scirpoidea*
      2D. Spike androgynous.
         1E. Perigynia beakless, flat .... 8. *C. leptalea*
         2E. Perigynia with short beaks, trigonous.
            1F. Perigynia obovate.
               1G. Leaves filiform, plant tufted 9. *C. filifolia*
               2G. Leaves keeled or flat, plant
                   with creeping rhizomes ....10. *C. rupestris*
            2F. Perigynia ovate ....11. *C. obtusata*
2A. Pistillate scales early deciduous.
   1B. Spike densely flowered, only lower perigynia reflexed.
      1C. Stigmas two ....12. *C. pyrenaica*
      2C. Stigmas three ....13. *C. nigricans*
   2B. Spike few-flowered, perigynia all reflexed in age.
      1C. Perigynia 6–7 mm. long ....14. *C. pauciflora*
      2C. Perigynia 4–5 mm. long ....15. *C. microglochin*

Vignea

1A. Spikes androgynous.
   1B. Stigmas two.
      1C. Rhizome long, creeping.
         1D. Perigynia not wing-margined.
            1E. Spikes densely aggregated, perigynia inflated ....16. *C. maritima*
            2E. Spikes distinct, perigynia not inflated.
               1F. Rootstock slender, leaves 1–1.5 mm. wide ....17. *C. stenophylla*
               2F. Rootstock stout, leaves 1.5–3 mm. wide ....19. *C. praegracilis*
         2D. Perigynia wing-margined ....18. *C. chordorrhiza*
      2C. Rhizome short, plants tufted.
         1D. Leaves 4–8 mm. wide ....20. *C. stipata*
         2D. Leaves 1–2.5 mm. wide ....21. *C. diandra*
   2B. Stigmas three ....22. *C. macrocephala*
2A. Spikes gynaecandrous.
   1B. Margins of perigynia winged.
      1C. Spikes aggregated into a dense head.
         1D. Bracts leaf-like, exceeding the head....23. *C. athrostachya*

2D. Bracts shorter than the head.
   1E. Perigynia very conspicuous ....24. *C. macloviana*
   2E. Perigynia not conspicuous ....26. *C. phaeocephala*
2C. Spikes not aggregated into a head.
   1D. Perigynia lanceolate ....27. *C. crawfordii*
   2D. Perigynia ovate.
      1E. Beak of perigynia flattened and serrulate to the tip ....28. *C. aenea*
      2E. Beak of perigynia terete, not serrulate toward the tip ....25. *C. praticola*
2B. Margins of the perigynia not winged.
   1C. Perigynia white-puncticulate, beak short.
      1D. Plants tufted, lacking stolons.
         1E. Spikes two to four, congested.
            1F. Leaves 2 mm. wide, flat ....29. *C. lachenalii*
            2F. Leaves narrower.
               1G. Culms scabrous, scales yellowish-brown.
                   1H. Perigynia distinctly nerved ....31. *C. neurochleana*
                   2H. Perigynia almost nerveless ....30. *C. heleonastes*
               2G. Culms glabrous or nearly so, scales darker.
                   1H. Perigynia many-nerved, 1.5 mm. wide ....32. *C. pribylovensis*
                   2H. Perigynia few-nerved, narrow ....33. *C. glareosa*
         2E. Spikes four to eight, the lower ones distant.
            1F. Beaks of the perigynia scabrous on the margins.
               1G. Beak and part of perigynia. with a distinct hyaline suture38. *C. brunnescens*
               2G. Beak and perigynia without such suture.
                   1H. Perigynia 1.5–1.8 mm. long ....37. *C. bonanzensis*
                   2H. Perigynia 2–3 mm. long..35. *C. canescens*
            2F. Beaks of perigynia smooth.
               1G. Perigynia about 3 mm. long..34. *C. mackenziei*
               2G. Perigynia much smaller ......36. *C. lapponica*
   2D. Plants loosely tufted, stolons present.
      1E. Spikes androgynous ....39. *C. disperma*
      2E. Spikes gynaecandrous.
         1F. Spikes aggregated at top of culm ....40. *C. tenuiflora*
         2F. Spikes at some distance from each other ....41. *C. loliacea*
2C. Perigynia not white-puncticulate, beaks long.
   1D. Perigynia broadest near base.
      1E. Perigynia 2.5–4 mm. long ....42. *C. stellulata*

    2E. Perigynia 4–4.5 mm. long ....43. *C. phyllomanica*
  2D. Perigynia tapering toward base ....44. *C. laeviculmis*

Eucarices distigmaticae

1A. Beaks with truncate mouths ....98. *C. physocarpa*
2A. Perigynia short-beaked or beakless.
  1B. Lowest bract long-sheathing.
    1C. Sheath 2–4 mm. long with black auricles..45. *C. bicolor*
    2C. Sheaths longer, without black auricles.
      1D. Perigynia white-papillose, dry ....47. *C. garberi*
      2D. Perigynia not or only slightly papillose, fleshy ....46. *C. aurea*
  2B. Lowest bract sheathless or nearly so.
    1C. Lowest bract shorter than the inflorescence.
      1D. Aphyllopodic, runners present ....48. *C. bigelowii*
      2D. Phyllopodic, runners absent ....49. *C. lugens*
    2C. Lowest bract as long as the inflorescence or longer.
      1D. Spikes ovate, congested at the top of the culm ....65. *C. enanderi*
      2D. Spikes cylindrical or prolonged, the upper ones staminate.
        1E. Culms with lower leaves blade-bearing (phyllopodic).
          1F. Perigynia nerved, ovate.
            1G. Scales acute, spikes slender..50. *C. kelloggii*
            2G. Scales blunt, spikes thicker..51. *C. hindsii*
          2F. Perigynia rounded.
            1G. Scales strongly nerved ....52. *C. kokrinensis*
            2G. Scales not strongly nerved.53. *C. aquatilis*
        2E. Culms with lower leaves not blade-bearing (aphyllopodic).
          1F. Normally high-growing plants.
            1G. Pistillate spikes usually erect, long and narrow ....54. *C. sitchensis*
            2G. Pistillate spikes drooping, rather short and thick ....57. *C. lyngbyei*
          2F. Comparatively low-growing, 3 dm. tall or less.
            1G. Low-growing, spikes few-flowered ....55. *C. subspathacea*
            2G. Medium-low, spikes many-flowered ....56. *C. ramenskii*

Eucarices tristigmaticae

1A. Beaks with truncate mouths.
  1B. Bracts sheathless or nearly so.
    1C. Lower bract foliaceous.
      1D. Terminal spike staminate.
        1E. Spikes more or less approximate....60. *C. stylosa*

2E. Spikes distant.
  1F. Scales long-aristate ....68. *C. macrochaeta*
  2F. Scales short-aristate ....84. *C. magellanica*
  3F. Scales blunt or acute.
    1G. Perigynia ciliate on the margins.
      1H. Pistillate scales cuspidate ....69. *C. karaginensis*
      2H. Pistillate scales merely acute ....88. *C. atrofusca*
    2G. Perigynia smooth on the margins.
      1H. Culms aphyllopodic.
        1I. Scales with midveins reaching the apex and sometimes excurrent..71. *C. spectabilis*
        2I. Scales with midveins obsolete toward the apex ....70. *C. montanensis*
      2H. Culms phyllopodic.
        1I. Spikes linear, long and narrow ....72. *C. nesophila*
        2I. Spikes oblong, thick and short ....73. *C. podocarpa*
2D. Terminal spike gynaecandrous.
  1E. Pistillate scales awned or cuspidate.
    1F. Spikes sessile ....59. *C. buxbaumii*
    2F. Spikes distinctly peduncled........61. *C. gmelini*
  2E. Pistillate scales not awned or cuspidate.
    1F. Perigynia 5 mm. long, spikes six to ten ....67. *C. mertensii*
    2F. Perigynia shorter, spikes three to six.
      1G. Perigynia nerved, sparsely spinulose on margins ....65. *C. enanderi*
      2G. Perigynia not spinulose on margins.
        1H. Rootstocks long, leaves smooth ....62. *C. leiophylla*
        1I. Scales purplish-black with conspicuous hyaline margins.
          1J. Culms slender, perigynia subinflated 58. *C. norvegica*
          2J. Culms stiff, perigynia flat ....63. *C. albo-nigra*
        2I. Scales lacking distinct hyaline margins.
          1J. Culms scabrous, spikes linear ....66. *C. atratiformis*
          2J. Culms glabrous, spikes oblong--ovoid ....64. *C. atrata*
2C. Lower bract scale-like.
  1D. Perigynia glabrous ....77. *C. supina*

2D. Perigynia pubescent.
   1E. Lower pistillate spikes on elongated subradical peduncles.
      1F. Loosely caespitose, rootstocks thin .... 74. *C. deflexa*
      2F. Densely caespitose, rootstocks stout .... 75. *C. rossii*
   2E. Subradical pistillate spikes absent.. 76. *C. peckii*
2B. Bracts with distinct sheaths.
   1C. Perigynia pubescent .... 78. *C. concinna*
   2C. Perigynia glabrous ....
      1D. Leaves 0.2–1 mm. wide, canaliculate or involute.
         1E. Lowest bract bladeless .... 80. *C. eburnea*
         2E. Lowest bract with a setaceous blade .... 79. *C. glacialis*
      2D. Leaves broader, flat (or canaliculate).
         1E. Pistillate spikes more or less densely flowered, drooping.
            1F. Sheath of lowest bract long, tubiform .... 85. *C. laxa*
            2F. Sheath of lowest bract short, spathiform.
               1G. Lowest bract leaf-like.
               2G. Lowest bract subulate.
                  1H. Pistillate scales obtuse.
                     1I. Pistillate spikes two to ten-flowered .... 81. *C. rariflora*
                     2I. Pistillate spikes ten to twenty-five flowered.. 82. *C. pluriflora*
                  2H. Pistillate scales cuspidate or mucronate .... 83. *C. limosa*
         2E. Pistillate spikes loosely flowered, erect.
            1F. Perigynia nearly beakless .... 86. *C. livida*
            2F. Perigynia long-beaked .... 87. *C. vaginata*
2A. Beak with bidentate mouth, the teeth sometimes rather indistinct.
   1B. Leaves not septate-nodulose.
      1C. Perigynia flat, ciliate-serrulate on the margins.
         1D. Perigynia rounded at the base, about as long as the scales.
            1E. Pistillate scales cuspidate .... 69. *C. karaginensis*
            2E. Pistillate scales merely acute .... 88. *C. atrofusca*
         2D. Perigynia tapering at the base, longer than the scales .... 89. *C. misandra*
      2C. Perigynia trigonous, not serrulate or ciliate on the margins.
         1D. Spikes on capillary peduncles, drooping.
            1E. Terminal spike gynaecandrous ...... 91. *C. krausei*
            2E. Terminal spike staminate.
               1F. Leaves setiform, involute .... 92. *C. williamsii*
               2F. Leaves flat .... 90. *C. capillaris*
         2D. Spikes short on short peduncles, erect.

1E. Beak of perigynia as long as the
       body, curved ....93. *C. flava*
   2E. Beak of perigynia short, erect ....94. *C.* ~~oederi~~ *viridula*
 2B. Leaves septate-nodulose.
   1C. Teeth of beak subulate, 1 mm. long............100. *C. atherodes*
   2C. Teeth of beak shorter.
     1D. Pistillate spikes 1–2 cm. long.
       1E. Leaves flat, channeled toward the
           base ....99. *C. membranacea*
       2E. Leaves involute ....96. *C. rotundata*
     2D. Pistillate spikes 5–7 cm. long.
       1E. Perigynia horizontal ....97. *C. rhyncophysa*
       2E. Perigynia ascending ....95. *C. rostrata*

1. *C. nardina* Fr.   Hepburn Sedge
   C. hepburnii Boott.
   Densely caespitose; culms 2–15 cm. tall, slender, wiry, not exceeding the leaves; leaves setaceous, stiff, erect or recurving, about 0.25 mm. wide; spike 5–15 mm. long, bractless; scales reddish-brown with straw-colored center; perigynia five to fifteen, 3.5–4.5 mm. long, biconvex or plano-convex, light-colored with some brown at the apex, sharp-edged, serrulate above; achenes lenticular or triangular, brown, apiculate; stigmas two or three.
   Central Alaska—Alta.—Colo.—Wash. (Fig. 211.)

2. *C. jacobi-peteri* Hult.   Anderson Sedge
   Plants caespitose; culms 3–10 cm. tall, usually curved; leaves longer than the culm, flat, 1–1.5 mm. wide, usually curved; spikelet 4–11 mm. long, without bracts; scales acute or acuminate, brownish with greenish midrib, as long or nearly as long as the perigynia; perigynia decidedly stipitate, about 2.5 mm. long, brownish at tip; achenes lenticular, about 1.5 mm. long.
   Known only from Tin City. (Fig. 212.)

3. *C. capitata* L.   Capitate Sedge
   Loosely caespitose; rootstocks slender, ascending obliquely, culms 10–35 cm. tall, erect; leaf-blades 0.5 mm. or less wide, filiform, involute; spike 4–10 mm. long, bractless; scales brown with hyaline apex and margins, the staminate narrower, more acute and lighter colored; perigynia six to twenty-five, 2–3 mm. long, plano-convex, sharp-edged, broad-margined; achenes yellowish-brown, lenticular.
   Bering Strait region through central Alaska. Distribution circumpolar and in S. Am. (Fig. 213.)

4. *C. gynocrates* Wormskj.   Northern Bog Sedge
   Stoloniferous, stolons long, 1 mm. thick; culms 4–30 cm. long, stiff, obtusely triangular; leaves 0.5 mm. wide, involute, stiff; spike staminate, pistillate or androgynous, 5–15 mm. long, brownish with hyaline margins;

perigynia four to ten, 3-3.5 mm. long, ascending, spreading, or reflexed, often curved toward the tip, yellowish or dark, finely nerved dorsally, serrulate above, hyaline at the mouth; achenes lenticular, 1.5 mm. long, yellowish-brown, shining.

Throughout most of our area—Greenl.—N. Y.—Colo.—B. C.—Siberia. (Fig. 214.)

5. *C. scirpoidea* Michx.   Northern Single-spike Sedge
C. stenochleana (Holm.) Mack.

Rootstocks creeping, dark reddish-purple; culms 1-5 dm. tall, stiff, roughened above; leaves 1-3 mm. wide, flat or canaliculate; spike dioecious, 1-3 cm. long, 3-7 mm. thick, often with a leaf-like bract 3-50 mm. long 5-50 mm. below the spike; pistillate scales brownish or blackish, with hyaline margins and lighter center, often more or less hairy on the back and with ciliate margins; perigynia compressed-triangular, dark-colored, short white-pubescent; achenes 1.5-2 mm. long, yellowish-brown.

A variable species of circumpolar distribution found throughout our area. (Fig. 215.)

6. *C. anthoxanthea* Presl.

Rootstocks rather long, scaly; culms 5-35 cm. tall, roughened above; leaves 1.5-2.5 mm. wide, erect or recurved; spike usually pistillate, sometimes androgynous or staminate, bractless; lower scales cuspidate or awned, the upper obtuse, chestnut-brown with one- to three-nerved lighter or greenish center; perigynia four to fourteen, about 4 mm. long, 1.5 mm. wide, compressed triangular, yellowish-green, many-nerved, achenes about 1.5 x 1 mm., triangular.

Grassy banks, Aleutian and Pribylof Islands.—B. C. (Fig. 216.)

7. *C. circinata* C. A. Mey.   Coiled Sedge

Densely caespitose; culms 5-20 cm. long, erect or more often recurved; leaves about 0.5 mm. wide, involute-filiform, curved, stiff; spike androgynous, 15-30 mm. long, bractless; lowest scale cuspidate, the upper obtuse, reddish-brown with hyaline apex and margins; perigynia 4.5-6 mm. long, narrow, erect-ascending, obscurely compressed-triangular, straw-colored with some reddish-brown below the hyaline-tipped beak, serrulate; achene 2.5-3 mm. long, obtusely triangular.

Near the coast, Aleutian Islands—Wash. (Fig. 217.)

8. *C. leptalea* Wahl.   Bristle-stalked Sedge

Caespitose; culms filiform, 1-5 dm. tall; leaves very narrow; spike androgynous, 4-15 mm. long, 2-3 mm. thick, bractless; staminate flowers few-many, their scales connate below; perigynia one to ten, 2.5-5 mm. long, thick, yellowish or light green, striate; lowest scale cuspidate, the upper usually obtuse; achenes 1.5-2 mm. long, triangular with concave sides below, yellowish or brownish, shining.

Bogs, central Alaska—Labr.—Fla.—Texas—Colo. (Fig. 218.)

9. *C. filifolia* Nutt. Thread-leaved Sedge

Densely caespitose, culms slender, stiff, 8–30 cm. tall; leaves acicular, involute, stiff, 3–20 cm. long, 0.25–0.5 mm. wide; spike 1–3 cm. long, bractless, the upper half staminate; scales usually obtuse, light reddish-brown with broad hyaline margins; perigynia five to fifteen, 3–35 mm. long, obtusely triangular, dull whitish or straw-colored, darker above, obscurely 2-ribbed; beak truncate, hyaline; achenes 2.25–3 mm. long, triangular.

Yukon—Man.—N. Mex.—eastern Ore.

10. *C. rupestris* All. Rock Sedge

Loosely caespitose and stoloniferous; culms 4–15 cm. tall, wiry; leaves 1–3 mm. wide, spreading or recurving, canaliculate, stiff; spike 1–2 cm. long, bractless; scales thin, chestnut-brown with hyaline margins and lighter center; perigynia three to eight, 3–4 mm. long, triangular, greenish straw-colored tinged brownish, shining, two-keeled; achenes 2.25 mm. long, triangular, dark chestnut-brown, short-apiculate.

Alpine-arctic; circumpolar. Rare in our area.

11. *C. obtusata* Lilj.

Rootstocks long-creeping, slender, purplish-black; culms 6–20 cm. tall, scattered, or two or three together; leaves channeled, 1–1.5 mm. wide; spike 5–12 mm. long, bractless; scales acuminate or cuspidate, thin, light brownish with hyaline margins and lighter midvein; perigynia one to six, 3–3.5 mm. long, dark chestnut or blackish brown, shining; beak obliquely cut, bidentulate, hyaline-tipped; achene 1.75 mm. long, triangular and with prominent ridges, light yellowish-brown.

Central Alaska—Man.—S. Dak.—N. Mex.—B. C.—Eurasia. (Fig. 219.)

12. *C. pyrenaica* Wahl. Pyrenean Sedge
C. pyrenaica Wahl ssp. *micropoda* (C. A. Mey) Hult.
C. micropoda C. A. Mey.

Densely caespitose; culms slender, 3–25 cm. tall; leaves 0.25–1.5 mm. wide, channeled; spike androgynous, 5–20 mm. long, 3–5 mm. thick, bractless; scales blackish-chestnut to straw-color, with hyaline margins; staminate flowers inconspicuous; perigynia ten to many, brownish above, lighter at the base, shining; achenes 1.25–1.5 mm. long, light brown.

Aleutian and Pribylof Islands eastward; circumboreal. (Fig. 220.)

13. *C. nigricans* C. A. Mey. Blackish Sedge

Loosely caespitose; culms 5–30 cm. tall, striate, rather stiff; leaves 1.5–2 mm. wide; spike androgynous, 8–15 mm. long, bractless; staminate scales persistent, reddish-brown, becoming straw-colored; pistillate scales deciduous, dark brown; staminate flowers conspicuous; perigynia several to fifty, 3.5–4 mm. long, exceeding the scales, jointed to the rachis, deflexed at maturity; compressed-triangular, yellowish to brownish, the orifice hyaline; achenes 1.5–2 mm. long, triangular, yellowish-brown.

Aleutian and Commander Islands—Colo.—Calif. (Fig. 221.)

14. *C. pauciflora* Lightf.   Few-flowered Sedge

Rootstocks long, slender; culms 5–60 cm. tall, stiff; leaves 0.75–1.5 mm. wide, involute or channeled; spike androgynous, bractless; scales acutish, light-colored, the pistillate early deciduous; perigynia one to six, 6–7 mm. long, soon reflexed, light green, soon becoming straw-color or brownish, finely striate; achene about 2 mm. long, triangular with concave sides near the base, usually convex above; stigmas three, short.

Muskegs, Pacific Coast of Alaska; circumboreal. (Fig. 222.)

15. *C. microglochin* Wahl.   False Uncinia

Rootstocks long, slender; culms 5–25 cm. tall, stiff; leaves about 0.5 mm. wide, involute, light green with blunt tip; spike androgynous, 7–14 mm. long, bractless; scales light chestnut brown, sometimes with lighter margins and center; perigynia three to twelve, 4–6 mm. long, about 1 mm. wide, bright brownish-green or straw-color, finally reflexed, the orifice oblique; achenes about 2.5 mm. long, triangular, yellowish-brown.

Bering Sea through central Alaska; circumpolar. (Fig. 223.)

16. *C. maritima* Gunner.   Curved Sedge
C. incurva Lightf.

Rootstocks long, forking; culms solitary or a few together, 2–16 cm. long, stiff, usually more or less curved; leaves 2–10 cm. long, 1–2 mm. wide, involute above, erect or recurved-spreading; spikes four to twelve, in a dense head 6–12 mm. long, bractless; staminate flowers inconspicuous; perigynia 3.25–4 mm. long, longer than the scales, plano-convex, slightly inflated, shaded light yellowish-brown to brownish-black, sharp-edged, sparingly serrulate on and near the beak; achenes 1.5 mm. long, lenticular, brownish.

Near the coast and in tundra, arctic—southeastern Alaska; circumpolar. (Fig. 224.)

17. *C. stenophylla* Wahl ssp.*eleocharis* (Bailey) Hult.
                                               Involute-leaved Sedge
C. eleocharis Bailey.

Rootstocks long, slender, culms one or a few together, 3–20 cm. tall, slender, stiff; leaves 1–1.5 mm. wide, involute above; spikes few, aggregated into a head 7–15 mm. long; bracts ovate, cuspidate; scales slightly exceeding the perigynia, brownish with wide hyaline margins; perigynia one to eight to a spike, 2.5–3 mm. long, slightly elevated and serrulate near and along the beak; achenes lenticular, about 1.75 x 1.5 mm.

Yukon—Slave Lake—Man.—Iowa—N. Mex.—E. Ore. The full species is circumboreal.

18. *C. chordorrhiza* Ehrh.   Creeping Sedge

Old culms prostrate, producing fertile culms 1–3 dm. tall terminally and from upper nodes, sterile culms from the lower nodes; leaves about 1 mm. wide, slightly scabrous, canaliculate; acute or acuminate, light brown with hyaline margins and lighter center; perigynia 2.5–3.5 mm.

long, concealed by the scales, thick plano-convex, brownish, shining, strongly nerved, sharp-edged; achenes lenticular, 1.75–2 mm. x 1.25 mm., thick, brownish, punctate.

Collected on Buckland River; circumboreal. (Fig. 225.)

19. *C. praegracilis* W. Boott. Clustered Field-sedge

Rootstocks long, stout, black; culms 20–75 cm. tall, roughened above; leaves 1.5–3 mm. wide, flat or channeled; spikes five to fifteen, 4–8 mm. x 4–6 mm. in a head 1–5 cm. long; bracts none or one or two; scales acute or cuspidate, nearly concealing the perigynia, dull brownish with hyaline margins; perigynia plano-convex, smooth, dull blackish with age, 3–4 mm. long, 1.5 mm. wide, nerved dorsally, the margins sharp; beak about 1 mm. long, serrulate, obliquely cut and hyaline at orifice; achenes 1.25 mm. long, lenticular.

Yukon—Sask.—Man.—Kans.—Mex.—L. Calif.

20. *C. stipata* Muhl. Awl-fruited Sedge

Caespitose; culms 3–12 dm. tall, sharply triangular, erect, weak, flattened in drying; leaves 4–10 mm. wide, flat, flaccid, serrulate on the margins near the apex; spikes many, yellowish-brown, in a compound head 3–10 cm. long, 10–25 mm. thick; lowest bract setiform, up to 5 cm. long, or lacking; scales acuminate or cuspidate, brownish or hyaline; staminate flowers inconspicuous; perigynia 4–5 mm. long, plano-convex, thick, yellowish, strongly nerved, sharp-edged; beak 2–2.5 mm. long, serrulate, tipped reddish-brown; achenes 1.5–2 x 1.25–1.75 mm., plump.

Eastern Asia—coast of Alaska—Newf.—N. C.—Calif. (Fig. 226.)

21. *C. diandra* Schk. Lesser Panicled Sedge

Caespitose; culms 3–7 dm. tall, stiff, roughened on the edges; leaves 1–2.5 mm. wide; spikes many, in a brownish head 2–5 cm. long; bracts short, subulate, often absent; scales acute or cuspidate, brownish with hyaline margins and lighter midrib; staminate flowers inconspicuous; perigynia 2–2.75 mm. long, strongly biconvex, brown, shining, few-nerved dorsally, sharp-edged and serrulate above; achenes lenticular.

Wet meadows, central Alaska east and south. Circumboreal and in New Zealand. (Fig. 227.)

22. *C. macrocephala* Willd. ssp. *anthericoides* (Presl.) Hult.

Large-headed Sedge

Perpendicular rootstocks from long, deep, horizontal ones; culms 15–35 cm. tall, stiff, stout; leaves 4–8 mm. wide, the margins minutely serrulate; heads 4–6 x 2.5–5 cm., composed of numerous scarcely distinguishable spikes about 1.5 cm. long; bracts variable, sometimes highly developed; scales acuminate to awned, striate, brownish with green center and hyaline margins; perigynia 10–15 x 4–6 mm., thick, shining, strongly nerved, the margins winged and serrulate; beak 4–7 mm. long, bidentate; achenes 4 x 2.5 mm., triangular, constricted in the middle.

Along the coast, Alaska—Calif. Main species in eastern Asia.

23. **C. athrostachya** Olney. Slender-beaked Sedge

Caespitose with short rootstocks; culms 5–60 cm. tall; leaves 1.5–3 mm. wide; spikes 4–20, ovoid, 4–10 mm. long, in a head 1–3 cm. long; lower bracts elongated and exceeding the head, dilated and hyaline-margined at the base; scales acute to cuspidate, brownish with hyaline margins and green center; staminate flowers inconspicuous; perigynia ascending, ovate-lanceolate, thin, 3–4 mm. long, wing-margined, serrulate above; achenes lenticular; stigmas two.

Southeastern Alaska—Sask.—Colo.—Calif. (Fig. 228.)

24. **C. macloviana** d'Urv. ssp. **pachystachya** (Cham.) Hult.
Thick-headed Sedge
C. pachystachya Cham.

Densely caespitose, culms 3–10 dm. tall, striate; leaves 2–4 mm. wide, flat; spikes four to twelve, 5–8 x 4–6 mm., in a dense head 10–25 mm. long; bracts scale-like or the lower awned; scales acute, brown or blackish, often with lighter midrib; staminate flowers inconspicuous; perigynia six to twenty to a spike, 4.5–6.5 mm. long, appressed, plano-convex, wing-margined, serrulate, light-colored; beak brownish, bidentulate; achenes 1.5–2 mm. long, lenticular, yellowish-brown; stigmas two.

Central Alaska—Greenl.—Que.—Colo.—Calif. (Fig. 229.)

25. **C. praticola** Rydb. Meadow Sedge

Caespitose; culms 2–7 dm. tall; leaves 1–3.5 mm. wide, flat, light green; spikes two to seven, 6–16 x 4–6 mm., in a flexuous head 15–50 mm. long; bracts scale-like, the lowest often cuspidate; scales acutish,, tinged reddish-brown with silvery-hyaline margins; staminate flowers inconspicuous; perigynia six to twenty to a spike, 4.5–6.5 mm. long, appressed, plano-convex, wing-margined, serrulate, light-colored; beak brownish, bidentulate; achenes 1.5–2 mm. long, lenticular, yellowish-brown; stigmas two.

Central Alaska—Greenl.—Que.—Colo.—Calif. (Fig. 230.)

26. **C. phaeocephala** Piper. Mountain Hare Sedge

Caespitose with densely matted rootstocks; culms 1–3 dm. tall, stiff; leaves 1.5–2 mm. wide, canaliculate or involute; spikes two to five, occasionally up to seven, 6–12 x 5–8 mm., in a head 12–25 mm. long; lowest bract sometimes developed; scales acute, covering the perigynia, dark with hyaline margins and lighter midvein; staminate flowers conspicuous; perigynia 4–6 mm. long, oblong-ovate, plano-convex, brownish, strongly veined dorsally, wing-margined, minutely serrulate; beak 1 mm. long; achenes 1.5 x 1 mm., lenticular, brownish.

Reported from Glacier Bay—B. C.—Alta.—Colo.—Calif.

27. **C. crawfordii** Fern. Crawford Sedge

Densely caespitose; culms 1–6 dm. tall, stiff; leaves 1–3 mm. wide; spikes three to twelve, densely-flowered, in a head 12–25 mm. long; lower bracts setaceous; scales acute or acuminate, light brown with greenish center; staminate flowers inconspicuous; perigynia numerous, about 4 mm. long, thin, distended over the achene, brownish, winged, serrulate above; beak bidentate, reddish-brown at the tip; achenes about 1 mm. long, lenticular with prominent beak.

Central Alaska—Newf.—N. Jer.—Mich.—Wash. (Fig. 231.)

28. *C. aenea* Fern.   Fernald Hay-Sedge

Caespitose; culms 3–12 dm. tall, nodding; leaves 2–4 mm. wide, flat, weak; spikes four to ten, 7–25 x 5–7 mm., in a flexuous, moniliform or loose head 35–70 mm. long; lower bracts cuspidate, the upper scale-like; scales acute or acuminate, dull or yellowish brown with hyaline margins and three-ribbed green center; perigynia 4–5 x 2 mm., nearly concealed by the scales, concavo-convex, dull green or brownish, nerved dorsally, delicately serrulate above; achenes 2 x 1.5 mm., dull yellowish-brown.

Circle Hot Springs—Labr.—Newf.—N. Y.—S. Dak.—B. C. (Fig. 232.)

29. *C. lachenalii* Schk.   Arctic Hare's-foot Sedge
C. bipartata All.
C. lagopina Wahl.

Loosely caespitose, rootstocks short, brownish; culms 5–30 cm. tall, slender, erect or curving; leaves 1–3 mm. wide; spikes two to five, dark brown, 5–10 mm. long in a head 1–2 cm. long; bracts scale-like; scales obtuse, keeled, chestnut-brown with hyaline margins and yellowish-brown center; perigynia 2–3.5 mm. long, appressed-ascending, plano-convex, several-nerved, sharp-edged; achenes about 1.5 mm. long.

Arctic-alpine situations; circumpolar. (Fig. 233.)

30. *C. heleonastes* Ehrh.   Hudson Bay Sedge

Loosely caespitose with long slender rootstocks; culms slender, stiff, 15–35 cm. tall; leaves 1–2 mm. wide, flat or involute; spikes two to four, 4–7 x 4–6 mm. in a head 8–18 mm. long; bract scale-like, sometimes cuspidate; scales thin, keeled, reddish-brown with narrow hyaline margins and lighter center; perigynia five to ten to a spike, 2.5–3 mm. long, 1.25 mm. wide, plano-convex, thick, blunt-edged, faintly nerved; beak 0.5 mm. long, cleft dorsally; achenes lenticular, 1.5 x 1 mm.

Known from Kusilof; circumboreal but local.

31. *C. neurochlaena* Holm.   Northern Clustered Sedge

Caespitose in small clumps; rootstocks very slender; culms scabrous, slender, weak, often curved, 15–25 cm. long; leaves canaliculate, 0.75–1.5 mm. wide; spikes two to four, the terminal one gynaecandrous, 7–12 mm. long, the lower usually pistillate and shorter; scales distinctly hyaline-margined; perigynia distinctly nerved.

Yukon and N. W. Territories.

32. *C. pribylovensis* Macoun.   Pribylof Sedge

Loosely caespitose; culms 2–4 dm. tall, stiff or slightly flexuous; leaves 1–2.5 mm. wide, flat, thickish; spikes three to four, the terminal gynaecandrous, 7-12 mm. long, the lateral usually pistillate and shorter, in a head 12-20 mm. long; scales keeled, deep brown with wide hyaline margins and straw-colored center; perigynia ten to thirty to a spike, 2.5-3 x 1.5 mm., light yellowish-green; achenes lenticular, about 2 mm. long.

Aleutian Islands and islands in Bering Sea. (Fig. 234.)

33. *C. glareosa* Wahl.   Weak Clustered Sedge

Loosely caespitose; rootstocks long, slender; culms 10-25 cm. tall, smooth, brownish; leaves 0.5-1.5 mm. wide, canaliculate; spikes two or three, the terminal gynaecandrous, 7-12 mm. long, 2 mm. wide, the lateral pistillate and shorter, in a head 12-19 mm. long; bracts usually scale-like; scales thin, keeled, brownish with hyaline margins; perigynia narrow, about 3.5 mm. long, plano-convex, brownish above, lighter below; achenes lenticular, nearly filling the perigynia.

Coastal regions; circumpolar. (Fig. 235.)

34. *C. mackenziei* Kretch.   Norway Sedge
    C. norvegica Willd. not Retz.

Rootstocks long, slender; culms 10-45 cm. tall, smooth; leaves 1-3 mm. wide, flat, thin, soft, yellowish-green; spikes three to six, the terminal gynaecandrous, 1-2 cm. long, the lateral gynaecandrous or pistillate, 5-15 mm. long, in a head 15-55 mm. long; bracts scale-like, the lowest often setaceous-pointed; scales light reddish-brown with hyaline margins and lighter center; perigynia five to twenty to a spike 2.5-3.3 mm. long, plano-convex, thick, glaucous-green, white-puncticulate, striate; achenes about 2 mm. long, lenticular, filling the perigynia.

Along the coast; circumboreal. (Fig. 236.)

35. *C. canescens* L.   Silvery Sedge

Caespitose, the rootstocks short; culms 2-8 dm. tall, erect; leaves flat, 2-4 mm. wide, shorter than the culm; spikes four to eight, 3-12 mm. long, in a cluster 2-15 cm. long; bracts scale-like, the lowest often prolonged into a bristle; scales hyaline with greenish center and somewhat brown-tinged when mature; perigynia 1.8-2.8 mm. long, plano-convex, gray-green or yellowish-brown, white-puncticulate, rough or minutely serrulate near the apex, brownish-tinged at mouth; achenes lenticular, 1.5 mm. long, yellowish-brown.

Common in swamps and bogs; circumboreal. (Fig. 237.)

36. *C. lapponica* O. F. Lang.   Lapland Sedge

Resembles *C. canescens* but is less distinctly tufted, the culms and leaves are more slender, the spikes are smaller and fewer-flowered, and the perigynia are smooth and not serrulate on the margins.

Bering Sea and central Yukon regions; circumpolar and more arctic in distribution than *C. canescens*.

37. *C. bonanzensis* Britt.   Yukon Sedge

Caespitose; rootstocks long, slender; culms 25-45 cm. tall, stiff, with concave sides; leaves 2-3 mm. wide, flat; spikes about seven, 5-14 x 4 mm., the lower distant; lowest bract 15–30 mm. long, the upper scale-like; scales thin, keeled; staminate flowers conspicuous in terminal spike; perigynia small, about 1.5 x 1 mm., exceeding the scales, dark straw-colored, white-puncticulate, strongly nerved dorsally, sharp-edged; achenes about 1.25 mm. long.

Yukon and Siberia.

38. *C. brunnescens* (Pers.) Poir.   Brownish Sedge

Caespitose, rootstocks short; culms 7–70 cm. tall, slender, lax; leaves long, 1–2.5 mm. wide, roughened toward the apex; spikes five to ten, mostly gynaecandrous, scattered, the lateral 3–7 mm. long, the terminal up to 13 mm. long; lowest bract prolonged, the upper scale-like; scales white-hyaline with greenish center and usually more or less tinged with brown; perigynia 2–2.5 mm. long, appressed-ascending, plano-convex, greenish or brownish, puncticulate, nerved, finely serrulate above; beak bidentate; achenes lenticular.

Pacific coastal and Alaskan Range districts; circumboreal. (Fig. 238.)

39. *C. disperma* Dewey.   Soft-leaved Sedge

Loosely caespitose; culms 1–6 dm. tall, slender, weak; leaves 0.75–2 mm. wide, soft, thin; spikes two to four, in a cluster 15–25 mm. long; bracts setaceous, less than 1 cm. long; scales acuminate or mucronate, white-hyaline with greenish midrib; staminate flowers one or two, inconspicuous, perigynia one to six to a spike, 2.25–3 mm. long, biconvex, light or yellowish-green, often darker with age, finely nerved; achenes about 1.75 mm. long, lenticular, brownish-yellow, shining, filling the perigynia.

Yukon valley and Pacific coastal districts; circumboreal. (Fig. 239.)

40. *C. tenuiflora* Wahl.   Sparse-flowered Sedge

Loosely caespitose; culms 15–60 cm. tall, slender; leaves 0.5–2 mm. wide, soft, flat or canaliculate; spikes two to four, 4–9 x 3–6 mm., whitish, in a head 6–12 mm. long; bracts scale-like or the lower prolonged; scales obtuse with a three-nerved center; staminate flowers inconspicuous; perigynia three to fifteen in a spike, 3–3.5 mm. long, concealed by the scales, greenish-white, obscurely nerved; achenes about 2 x 1.5 mm., lenticular, light brown.

Sphagnum bogs, not common; circumboreal. (Fig. 240.)

41. *C. loliacea* L.

Loosely caespitose with long, slender stolons; culms 15–40 cm. long, slender, weak; leaves 0.5–2 mm. wide, flat, soft; spikes two to five, in a cluster 1–3 cm. long, lowest bract up to 8 mm. long, the upper scale-like; scales keeled, thin, hyaline; staminate flowers inconspicuous; perigynia three to eight to a spike, 2.5–3 mm. long, thick, plano-convex, light green, white-puncticulate, finely ribbed, beakless; achenes lenticular, 1.75 mm. long.

Central Alaska—Alta.—B. C.—also Eurasia. (Fig. 241.)

42. *C. stellulata* Good.   Little Prickly Sedge

Caespitose; culms wiry, 15–35 cm. tall; leaves 1–2 mm. wide, flat or canaliculate; spikes two to four, the terminal gynaecandrous, the lateral usually pistillate, in a head 1–3 cm. long; lower bract often cuspidate, the upper scale-like; scales light brown with wide hyaline margins and green midrib; perigynia three to ten to a spike, 2.5–3.25 mm. long, exceeding the scales, spreading, thick, nerved dorsally, sharp-edged, serrulate toward and on the bidentate beak; achene about 1.5 mm. long, lenticular, yellowish-brown. Reports of *C. echinata* Murr., *C. leersii* Willd. and *C. muricata* L. from Alaska all refer to this species.

Aleutian Islands; probably more or less circumboreal. (Fig. 242.)

43. *C. phyllomanica* W. Boott.   Coastal Stellate Sedge

Caespitose from slender creeping rootstocks; culms 2–6 dm. tall; leaves 1.75–2.75 mm. wide, flat or canaliculate; spikes three to four, burlike, in a head 15–25 mm. long, the terminal gynaecandrous, the lateral often pistillate; lowest bract setaceous, the upper scale-like; scales obtuse, light brown with hyaline margins and green center; perigynia eight to fifteen to a spike, 3.75–4.5 mm. long, plano-convex, thick, light-colored, striate, tapering into a serrulate beak; achenes about 2 mm. long, yellowish.

Southern Alaska—Calif. (Fig. 243.)

44. *C. laeviculmis* Meinsh.   Smooth-stemmed Sedge

Caespitose with short, slender rootstocks; culms slender, 3–7 dm. tall; leaves 1–2 mm. wide, flat, weak; spikes three to eight, the terminal gynaecandrous, the lateral pistillate, the upper approximate, the lower distant, 3–10 mm. long; lowest bract 15–50 mm. long, the upper reduced; scales ovate, hyaline with conspicuous green midrib; perigynia three to ten to a spike, 2.5–4 mm. long, light or yellowish-green, sharp-edged, serrulate above, tawny-tipped; achenes 1.25–1.75 mm. long, lenticular, brownish.

Seward Peninsula—Mont.—Calif. (Fig. 244.)

45. *C. bicolor* All.   Two-color Sedge

Loosely caespitose and stoloniferous; culms 5–20 cm. tall, roughened above; leaves 3–6 cm. long, 1–2.5 mm. wide; spikes two to five, the terminal gynaecandrous, the lateral pistillate, 5–10 mm. long; lower bract short-sheathing, leaf-like; the upper scale-like; scales obtuse or mucronate, dark with yellowish-green center; perigynia 2–2.5 long, appressed-ascending, glabrous, glandular-roughened, ribbed; achenes lenticular, yellowish-brown, puncticulate.

Southern half of our area; circumboreal. (Fig. 245.)

46. *C. aurea* Nutt.   Gold-fruited Sedge

Loosely caespitose and stoloniferous; culms 5–55 cm. tall; leaves 2–4 mm. wide; terminal spike staminate, 3–10 mm. long, occasionally with a few perigynia; lateral spikes three to five, pistillate, 4–20 mm. long, the lowest on nearly basal peduncles 3–8 cm. long; bracts leaf-like, sheathing; scales light reddish-brown with hyaline margins and a wide light center; perigynia 2–3 mm. long, flattened-oval, translucent, fleshy, puncticulate, coarsely ribbed; achenes lenticular, brownish, minutely puncticulate.

Central Alaska—Newf.—Pa.—Nebr.—N. Mex.—Calif. (Fig. 246.)

47. *C. garberi* Fern. ssp. *bifaria* Fern.   Garber Sedge
C. hassei Am. auct.

Loosely caespitose and stoloniferous; culms 5–70 cm. tall; leaves 2–4 mm. wide, flat above, channeled below; terminal spike gynaecandrous or staminate, 6–20 mm. long; lateral spikes three to five, pistillate, 7–25 mm. long, the lower on long, rough peduncles; lower bracts short-sheathing; scales brown with hyaline margins and prominent light center; perigynia 2.5–3 mm. long, flattened-suborbicular, whitish, minutely granular; achenes lenticular, 1.5 mm. long, brown, puncticulate.

Southern half of our area—Alta.—B. C. also near mouth of St. Lawrence R. (Fig. 247.)

48. *C. bigelowii* Torr.   Bigelow Sedge
C. concolor R. Br.

Stoloniferous, the stolons horizontal or ascending; culms 1–4 dm. tall, rather stout and stiff; leaves 2–8 mm. wide, flat; terminal spike staminate, 5–25 mm. long; lateral spikes one to six, pistillate or the upper androgynous, 5–30 mm. long; lowest bract leaf-like, black-auricled, the upper reduced and scale-like; scales brownish-black with narrow hyaline margins and lighter midrib; perigynia 2.5–3.5 mm. long, biconvex, light green, usually purplish-black blotched above, two-ribbed, short-beaked, the orifice entire; achenes lenticular, 1.5–2 mm. long.

Most of our area; circumboreal. (Fig. 248.)

49. *C. lugens* Holm.

Densely caespitose; culms 2–5 dm. tall; leaves 1–2.5 mm. wide, channeled and with revolute margins; terminal spike staminate, 10–25 mm. long; lateral spikes two to three, pistillate or occasionally one of them androgynous, 8–25 mm. long; lowest bract 5–30 mm. long, black-auricled, the upper reduced to auricles; scales blackish with lighter margins and midrib; perigynia 1.5–2.5 mm. long, appressed plano-convex, straw-color below, dark above, beak short, purple-black; achenes lenticular, dark, filling the perigynia.

Bering Sea region—Mack. (Fig. 249.)

50. *C. kelloggii* W. Bott.   Kellogg Sedge

Caespitose with very short ascending stolons; culms slender, 2–7 dm. tall, leaves 1.5–3 mm. wide; terminal spike staminate, 1–4 cm. long; lateral spikes three to five, pistillate, 15–35 mm. long, lowest bract leaf-like, the upper reduced; scales dark with hyaline margins and light center; perigynia numerous, appressed-ascending, 1.5–3 mm. long, flattened biconvex, light green, granular, the beak usually black-tipped; achenes 1 mm. long, lenticular, blackish.

Pacific Coast regions—Alta.—Colo.—Calif. (Fig. 250.)

51. *C. hindsii* C. B. Clarke.   Hinds Sedge

Caespitose with short or long branching rootstocks; culms 1–5 dm. tall; leaves 1.5–3 mm. wide; terminal spike staminate, 15–35 mm. long;

lateral spikes pistillate, 1–4 cm. long; lowest bract leaf-like, the upper reduced; scales purplish-black with lighter center; perigynia numerous, 2–3.5 mm. long, flattened biconvex, yellowish-green, ribbed, two-edged, papillate; beak usually black-tipped; achenes lenticular, brownish-black, 1.5 mm. long, granular.

Near the coast, Aleutian Islands—Calif. (Fig. 251.)

52. *C. kokrinensis* Porsild.   Kokrines Mountain Sedge

Loosely caespitose; culms 25–35 cm. tall, erect, exceeding the leaves, somewhat flattened; leaves about 2 mm. wide, flat; spikes cylindrical, 1–2 cm. long, erect, usually four, the terminal gynaecandrous, the lateral pistillate but generally with a few staminate flowers at the apex, the upper three closely aggregated; uppermost bract equaling, the lower exceeding the inflorescence; scales black with conspicuous greenish midvein reaching to the apex; perigynia flattened on one side, nerveless, pale grayish-green, smooth; beak very short. May be a hybrid.

Kokrines Mountains.

53. *C. aquatilis* Wahl.   Water Sedge

Rootstocks sending out long horizontal scaly stolons; culms caespitose, 3–7 dm. tall, slender, sharply triangular, reddened at the base; leaves 2–5 mm. wide; staminate spikes one or two, slender, 1–5 cm. long; pistillate or androgynous spikes two to four, sessile or short-peduncled, 1–4 cm. long; bracts leaf-like, the lower exceeding the culm; scales obtuse to acuminate, blackish or reddish-brown, one-nerved with light center; perigynia about 2.5 mm. long and half as wide, biconvex, puncticulate, glandular-dotted, two-ribbed; achenes lenticular, stigmas two.

A circumboreal species found in most of our area. (Fig. 252.)

54. *C. sitchensis* Prescott.   Sitka Sedge

Caespitose; rootstocks short, creeping; culms 25–125 cm. tall, reddish-brown at base; leaves 3–9 mm. wide, flat with revolute margins or channeled toward the base; terminal one to four spikes staminate, 2–8 cm. long; lower three to five spikes pistillate or androgynous, 2–9 cm. long, erect or the lowest drooping on slender peduncles; bracts leaf-like; scales usually acute, longer than the perigynia; perigynia fifty to one hundred fifty to a spike, 2.5–3.5 x 1.25–2 mm., plano-convex, sharp-edged, achenes 1.5–2 mm. x 1 mm., brownish, loosely enveloped.

Along streams and lakes, southwestern Alaska—Calif. (Fig. 253.)

55. *C. subspathacea* Wormskj.   Hoppner Sedge

Culms 3–20 cm. tall, stiff, smooth, usually curved, arising from elongated, horizontal rootstocks; leaves 1–2.5 mm. wide, flat, but involute toward the apex; lower bracts foliaceous, rather spathe-like; terminal spike staminate, 5–15 mm. long; lateral spikes pistillate, 5–15 mm. long; scales dark with hyaline margins and prominent light center; perigynia five to fifteen to a spike, about 3 mm. long, white puncticulate.

An arctic, circumpolar species of coastal marshes. (Fig. 254.)

56. *C. ramenskii* Komarov. Ramenski Sedge

Culms 1-3 dm. tall, stiff, from horizontal, creeping rootstocks; leaves 1.5-4 mm. wide, firm, flat or revolute toward the tip; upper one or two spikes staminate, 8-20 mm. long; lower two or three spikes pistillate or the upper of these androgynous, 1-3 cm. long; lower bracts leaf-like, scales dark, ovate, one-nerved, blunt; perigynia 2.5-3 mm. long, short-beaked or beakless, achenes about 2 mm. long, brownish. Var. *caudata* Hult. has the pistillate scales with awns 1-3 mm. long.

Coastal regions, Kenai Peninsula—Arctic and northeastern Asia. (Fig. 255.)

57. *C. lyngbyei* Hornem. ssp. *cryptocarpa* (C. A. Mey.) Hult.

Lyngbye Sedge

C. cryptocarpa C. A. Mey.

Long stoloniferous; culms 2-10 dm. tall, purple-red or brownish at base; leaves 2-12 mm. wide, flat with revolute margins; upper one to three spikes staminate, lower two to four spikes pistillate or androgynous, 15-80 mm. long, many-flowered, pendulous on slender peduncles; lower bracts leaf-like, often exceeding the inflorescence, the upper reduced; scales lanceolate, acuminate, exceeding the perigynia, brownish or blackish with light center; perigynia 2.5-3.5 mm. long, biconvex, glaucous-green or brownish, puncticulate; achenes about 2.5 mm. long, lenticular.

Common in brackish soil along the coast except the high arctic. The species is circumboreal. (Fig. 256.)

58. *C. norvegica* Retz. ssp. *inferalpina* (Wahl.) Hult.

C. angarae Steud.

Caespitose; culms rather slender, 2-6 dm. tall; leaves flat, with roughened and often revolute margins, 2-4 mm. wide; spikes two to four, the terminal gynaecandrous, the lateral pistillate, 4-8 mm. long, 3-5 mm. thick; lower bract often leaf-like; scales ovate, 1.5-2.5 mm. long, dark with rather narrow hyaline margins; perigynia obtusely triangular, yellowish-green, 2-3 mm. long; achenes triangular, about 1.75 mm. long.

In all our area except the arctic. The species is circumboreal. (Fig. 257.)

59. *C. buxbaumii* Wahl. Buxbaum Sedge

Loosely caespitose with long, slender, horizontal stolons; culms 25-100 cm. tall; leaves 1.5-4 mm. wide, flat and keeled, with revolute margins and channeled toward the base; spikes two to five, the terminal gynaecandrous, 1-4 cm. long, the lateral pistillate, 5-20 mm. long; bracts dark-auricled, the upper reduced; scales acuminate or aristate, dark with light center; perigynia 2.5-4 mm. long, triangular-biconvex, glaucous-green, shaded brownish, papillose; achenes 1.75 x 1.5 mm., brownish, triangular with rounded angles.

Southern half of Alaska; circumboreal. (Fig. 258.)

60. *C. stylosa* C. A. Mey. Variegated Sedge

Caespitose; culms 15-50 cm. tall, slender; leaves 1.5-3 mm. wide, flat with revolute margins or channeled toward the base; terminal spike staminate or with a few perigynia, 1–2 cm. long; lateral spikes two or three, pistillate, 7–18 mm. long; scales obtuse to acute, very dark, with hyaline margins and lighter midrib; perigynia 2.5–3.5 mm. long, yellowish-brown, tinged purplish-black, papillose above; achenes 1.5 x 1.25 mm., brownish, triangular.

Bering Strait—Greenl.—Newf.—Wash. Also eastern Asia. (Fig. 259.)

61. *C. gmelini* Hook. & Arn.   Gmelin. Sedge

Caespitose; rootstocks short, stout; culms 1–6 dm. tall, purplish-red at base; leaves 1.5–4 mm. wide, flat with revolute margins or channeled toward the base; spikes three to six, the terminal gynaecandrous or staminate, the lateral pistillate, 1–3 cm. long; lowest bract leaf-like, the upper reduced; scales dark with hyaline margins, light center and spiny-cuspidate tip; perigynia 4–5 mm. long, yellowish-brown, purple-tipped; achenes 1.75–2 mm. long.

Seashores, Norton Sound—B. C. and the Asiatic coast. (Fig. 260.)

62. *C. leiophylla* Mack.   Carcross Sedge

Loosely caespitose; rootstocks long, slender; culms 25–35 cm. tall, nodding; leaves 2–3.5 mm. wide, flat or channeled and with revolute margins; spikes four or five, the terminal gynaecandrous, the lateral pistillate, in a dense head 25 mm. x 12–16 mm.; scales acute, purplish-brown, with slender midvein and hyaline margins at apex; perigynia ten to twenty to a spike, 4 x 2 mm., inflated, straw-colored blotched purple; achenes 2.25 x 1.5 mm., triangular.

Known only from Carcross, Yukon.

63. *C. albo-nigra* Mack.   Black and White-scaled Sedge

Caespitose; culms 1–3 dm. tall, stiff; leaves 2.5–5 mm. wide, flat with revolute margins; spikes usually three, the terminal gynaecandrous, 10–12 mm. long, the lateral shorter and pistillate; lowest bract brownish-tinged and subsheathing at the base; scales purplish-black with white hyaline margins and apex; perigynia 3–3.5 x 2 mm., flattened, purplish-black, granular, the beak bidentate; achenes 1.25 x 0.75 mm., triangular with concave sides, light yellowish-brown.

Central Alaska and Wash.—Alta.—Colo.—Ariz.—Calif.

64. *C. atrata* L.   Black-scaled Sedge

Caespitose; culms 15–50 cm. tall, usually nodding above; leaves 2–8 mm. wide; spikes three to seven, the terminal gynaecandrous, the lateral pistillate, 1–2 cm. long, the lower nodding on slender peduncles; lowest bract leaf-like, the upper reduced; scales obtuse to acute, mostly brownish-black with lighter midrib and often lighter margins and tip; perigynia appressed, flattened but distended by the achene, papillose, brown-spotted, the beak dark, emarginate; achenes about 2 mm. long, triangular, yellowish-brown. Ssp. *atrosquama* (Mack.) Hult. is the more common form. It has shorter scales and the beak of the perigynia is shorter and broader than the type.

A circumboreal arctic-alpine species found in central and southeastern Alaska. (Fig. 261.)

65. *C. enanderi* Hult. Enander Sedge

Loosely caespitose with long rhizomes; culms 10–25 cm. tall, stiff; leaves 1.5–2 mm. wide; spikes three to five, oblong, densely flowered, the lower long-peduncled; terminal spike gynaecandrous, the lateral pistillate; pistillate scales dark purplish without hyaline margins or tip but with conspicuous green midrib; perigynia densely puncticulate, distinctly nerved, sparsely ciliate-serrulate on the margins, almost beakless, stipitate; stigmas two or sometimes three.

Known from Skagway and Akutan.

66. *C. atratiformis* Britt. Black Sedge

Loosely caespitose; culms 2–10 dm. tall, roughened above; leaves 2.5–5 mm. wide, flat with revolute margins; spikes three to six, the terminal gynaecandrous, the lateral pistillate with occasionally a few staminate flowers at the base, 10–25 x 4–6 mm., the lower nodding on slender peduncles; scales acute to cuspidate, dark reddish-brown to black with hyaline margins; perigynia ten to thirty to a spike, 2.5–3 x 1.5–1.75 mm., flattened, purplish-brown or straw-colored below, puncticulate, the beak bidentate; achenes 1.5–1.75 x 0.75 mm., silvery-black, shining.

Yukon—Labr.—Newf.—Maine—Mich.—Alta.

67. *C. mertensii* Prescott. Mertens Sedge

Caespitose, with short stolons; culms 3–12 dm. tall, sharply triangular, rough; leaves 4–8 mm. wide, flat with revolute margins; spikes five to ten, 1–4 cm. long, the uppermost strongly staminate at the base, the lateral sparingly so, drooping on slender peduncles; the lower two or three bracts leaf-like; scales mostly acute, shorter than the perigynia, dark with light center; perigynia numerous, flat, thin, distended over the achene, light brownish, often dark-spotted near the apex, 4.5–6 mm. long; achenes triangular, silvery-brown, about 2 mm. long. Our most beautiful species of Carex.

Central Alaska—Mont.—Calif. Common in the coast regions. (Fig. 262.)

68. *C. macrochaeta* C. A. Mey. Alaska Long-awned Sedge

Loosely caespitose with densely matted rootstocks; culms 2–6 dm. tall, purplish-red at base; leaves 2–4 mm. wide, flat with revolute margins; terminal spike staminate, 15–25 mm. long; lateral spikes two to four, pistillate, 1–3 cm. long, erect or drooping on slender peduncles; lower bract leaf-like, the upper reduced; scales dark with light whitish midrib excurrent as a serrulate awn 2–12 mm. long; perigynia 4.5–6 mm. long, smooth, obscurely nerved, straw-color or blotched or brownish, the beak dark; achenes 2–2.5 mm. long, triangular, brownish.

Near the coast, Aleutian and Pribylof Islands—Calif.—eastern Asia. One of our commonest species. (Fig. 263.)

69. *C. karaginensis* Meinsh. Karaginsk Island Sedge

Caespitose and short-stoloniferous; culms 15–80 cm. tall, slender; leaves 3–5 mm. wide; terminal spike staminate, 15–20 mm. long; lateral spikes pistillate; lower bracts foliaceous, the upper scale-like; scales oblong-lanceolate, obtuse or short-cuspidate, dark purplish-black or brownish; perigynia compressed-triangular, broad, the margins ciliate-serrulate, the beak emarginate or shallowly bidentate.

St. Matthew Island—northeastern Asia.

70. *C. montanensis* Bailey. Montana Sedge

Loosely caespitose; rootstocks long, slender; culms 1-5 dm. tall, stiff, somewhat nodding above; leaves 2-4 mm. wide, flat, firm; terminal spike staminate, 7-25 mm. long, sometimes with a smaller one at the base; lateral spikes two to four, pistillate or sometimes androgynous, 1-2 cm. x 4-6 mm., drooping or erect; bracts black-auricled; scales acute to obtuse, nearly black with inconspicuous or obsolete, lighter midrib; perigynia about 4 x 2 mm., glandular, straw-color with darker shadings; achenes 1.5 x 0.75 mm., brownish, triangular with concave sides, long-stipitate.

Eastern Asia across Alaska—Mack.—Mont.—Ida. (Fig. 264.)

71. *C. spectabilis* Dewey. Showy Sedge

Rootstocks stout, short-branching, matted; culms 25-90 cm. tall, slender; leaves 2-5 mm. wide, flat with revolute margins; terminal and occasionally the second spike staminate, 8-20 mm. long; pistillate spikes one to four, 1-4 cm. long; scales cuspidate, blackish with thick, lighter or whitish midrib, the pistillate with hyaline margins; perigynia 4-5 mm. long, flattened, light green, blotched, glandular-roughened; achenes 2.5 x 1 mm., triangular, light brown.

Asia—Bering Sea and Pacific regions of Alaska—Alta.—Mont.—Calif. (Fig. 265.)

72. *C. nesophila* Holm. Bering Sea Sedge

Stoloniferous, the stolons ascending; culms 1-4 dm. long, stiff; leaves 2.5-6 mm. wide, flat, with revolute margins, stiff; terminal spike staminate, 1-2 cm. long; lateral spikes two to five; pistillate 3-35 mm. long, erect on stiff peduncles; scales obtuse or acute to cuspidate, purplish-black with whitish midvein and occasionally hyaline margins or tip; perigynia 3-4.5 mm. long, flattened-triangular, light-colored with darker shadings, the faces three-nerved; achene nearly 2 mm. long, triangular, yellowish-brown.

Coast regions of Bering Sea—Kenai Peninsula. (Fig. 266.)

73. *C. podocarpa* R. Br. Short-stalk Sedge

Caespitose; culms 15-60 cm. tall; leaves 2.5-6 mm. wide, flat with revolute margins, the lower ones much reduced, bright green; terminal one or two spikes staminate, 1-3 cm. long; pistillate spikes two to six, 7-25 mm. long, 4-6 mm. thick; bracts dark-auricled, the upper reduced; scales usually acute, purplish-black with light midrib, some with hyaline margins; perigynia 2-4 mm. long, flattened, light-green, purplish-blotched, two-ribbed, the beak bidentulate; achenes 1.75 mm. long, triangular, light brown.

Central Alaska—Mack.—Mont.—Wyo.—Ore. (Fig. 267.)

74. *C. deflexa* Hornem. Northern Sedge

Caespitose; rootstocks branching, slender; culms 2-24 cm. tall, very slender; leaves short, 1-2 mm. wide, flat above, channeled toward the base, thin; terminal spike staminate, 2-5 x 0.5-1 mm.; lateral spikes two to four, pistillate, 2-6 x 3 mm., the lowest nearly basal on capillary peduncles; scales acute to cuspidate, brown with hyaline margins and lighter or green center; perigynia two to eight to a spike, about 2.5 x 1 mm., green, short-pubescent, ciliate-serrulate on the bidentate beak; achenes 1.5 mm. long, triangular.

Head of Yukon R.—L. Athabaska—Greenl.—Newf.—N. Y.—Mich.—Man. (Fig. 268.)

75. *C. rossii* Boott. Ross Sedge

Caespitose; culms 5-30 cm. tall, roughened above; leaves 1-2.5 mm. wide, channeled above, thin, firm; terminal spike staminate, 3-15 mm. long; lateral spikes two or three in the inflorescence near the top, usually with one or two from near the base on long slender peduncles, 3-5 mm. long; scales obtuse to cuspidate or awned; perigynia three to fifteen to a spike, 3-4.5 x 1 mm., pale green, short-pubescent, stipitate, with rather long ciliate-serrulate, bidentate beak; achenes triangular with concave sides.

Eastern Alaska and Yukon—L. Athabaska—Mich.—S. Dak.—Colo.—Calif. (Fig. 269) (from Colo.).

76. *C. peckii* Howe. Peck Sedge

Caespitose and stoloniferous; culms 15-65 cm. tall; leaves 1-1.5 mm. wide, flat; terminal spike staminate, 1-13 mm. long, inconspicuous; lateral spikes pistillate, 4-8 x 4 mm. in an inflorescence 8-20 mm. long; scales obtuse to mucronate, reddish-brown with hyaline margins, lighter center and green roughish midvein; perigynia three to twelve to a spike, about 3.5 x 1 mm., grayish- or yellowish-green, hirsute-pubescent, two-ridged, stipitate; beak obliquely cut, bidentulate; achenes about 2 x 1 mm., triangular with convex sides, yellowish-brown.

Yukon—Que.—N. B.—N. Jer.—Wis.—B. C.

77. *C. supina* Willd. ssp. *spaniocarpa* (Steud.) Hult. Weak Arctic Sedge
C. spaniocarpa Steud.

Caespitose and stoloniferous; culms 5-30 cm. tall; leaves 1-1.5 mm. wide, channeled, stiff, roughened, especially toward the attenuate apex; terminal spike staminate, 6-25 mm. long; lateral spikes one to three pistillate, 4-12 x 4 mm.; bracts scale-like; scales reddish-brown with wide hyaline margins and lighter center; perigynia four to fifteen to a spike or the one immediately below the terminal spike reduced to one to four, 2.5-3.5 mm. long, hard, brownish, shining; beak with hyaline orifice, achenes yellowish-brown, 2 x 1.5 mm.

Central Alaska—Baffin Land—Greenl.—Minn. (Fig. 270.)

78. *C. concinna* R. Br. Low Northern Sedge

Caespitose; rootstocks slender, often long; culms 5-20 cm. long, slender, erect or incurved; leaves 2-2.5 mm. wide; terminal spike staminate, 2–3 mm. long, very narrow; lateral spikes two or three, 4–8 mm. long, all crowded at the end of the culm; bracts reduced to sheaths; scales obtuse, reddish-brown, the pistillate with hyaline margins and lighter midrib, ciliate and hairy; perigynia about 3 mm long, obtusely triangular, light-colored, two-ribbed, hairy, the beak dark-colored; achenes triangular.

Dry soil, central Alaska—Newf.—Que.—Mich.—Colo.—B. C. (Fig. 271.)

79. *C. glacialis* Mack.   Glacier Sedge

Very densely caespitose; culms 3–15 cm. long, wiry, stiff; leaves 2–4 cm. 1–1.5 mm., flat at base, recurved, triangular and channeled above, stiff; terminal spike staminate, 2–6 x less than 1 mm.; lateral spikes one to three, pistillate, 2–5 mm. long, the entire inflorescence 7–20 mm. long; lowest bract loose, short-tubular, often prolonged into a cusp not over 15 mm. long; scales dark with hyaline margins; perigynia one to five to a spike, about 2.5 mm. long, yellowish-green below, dark above, the beak hyaline-tipped; achenes about 1.75 mm. long.

Alpine-arctic situations, Nome eastward; circumpolar. (Fig. 272) (from Newf.).

80. *C. eburnea* Boott.   Bristle-leaved Sedge

Caespitose; rootstocks long, slender; culms 10–35 cm. tall; leaves 0.5 mm. wide; setaceous, involute, firm, often recurved-spreading; terminal spike staminate, 4–8 mm. long; lateral spikes two to four, pistillate, 2–6 x 2 mm., on setaceous peduncles 10–25 mm. long; bracts tubular, truncate; scales whitish with green midrib, often tinged yellowish-brown; perigynia two to six to a spike, 2 x 1 mm., light green or brownish, shining puncticulate, finely nerved, the beak short and hyaline at the orifice; achenes about 2 x 0.75 mm., brown, granular, jointed with the bulbous-thickened base of the style.

Chitina R.—Great Bear L.—Newf.—Va.—Tenn.—Mo.—B. C.

81. *C. rariflora* (Wahl.) J. E. Smith.   Loose-flowered Alpine Sedge

Loosely stoloniferous; culms 10–35 cm. tall; leaves 1.5–2.5 mm. wide, the lower very short; terminal spike staminate, 6–15 mm. long, narrow; lateral spikes one or two, 6–15 x 3.5–5 mm.; bracts colored at the base, the lowest usually short-sheathing; scales brownish to blackish, the pistillate darker than the staminate; perigynia two to twelve to a spike, 3–4 mm. long, glaucous-green, two-edged, dark around the orifice; achenes about 2 mm. long, blackish, triangular.

Arctic and central Alaska east; circumpolar. (Fig. 273.)

82. *C. pluriflora* Hult.
C. stygia Auct.

Rhizomes long-stoloniferous, dark or purplish-black; culms 2–5 dm. tall; leaves about equaling the culm, 2–4 mm. wide, flat, roughened toward the attenuate apex; terminal spike staminate, 1–2 cm. long; lateral spikes one or two, about 13 x 6 mm., on long capillary peduncles; staminate scales reddish-brown with hyaline margins; pistillate scales blackish, acute to abruptly cuspidate; perigynia ten to twenty to a spike, 4–4.5 mm. long, glaucous-green or whitish, later turning brown, papillate, strongly nerved. beakless.

Near the coast, Bering Sea regions—Wash. (Fig. 274.)

83. C. limosa L.   Shore Sedge

Long-stoloniferous; culms 15–50 cm. tall, rather slender; terminal spike staminate, 1–3 cm. x 2.5 mm.; lateral spikes pistillate or occasionally androgynous, 10–25 x 5–8 mm., drooping on slender peduncles; lowest bract up to 6 cm. long with dark auricles, the upper reduced; scales acute to cuspidate, brownish; perigynia 2.5–4 mm. long, compressed-triangular, glaucous-green, papillate, prominently nerved; achenes about 2.25 mm. long, brown, triangular.

Bering Sea across central Alaska; circumboreal. (Fig. 275.)

84. C. magellanica Lam.   Bog Sedge
    C. paupercula Michx.

Loosely caespitose, with long or short rootstocks; culms 1–4 dm. tall; leaves 2–4 mm. wide, flat with revolute margins; terminal spike staminate or occasionally gynaecandrous, 7–15 mm. long; lateral spikes one to four, usually all pistillate but sometimes gynaecandrous, 4–20 mm. long, drooping on slender peduncles; lowest bract leaf-like, slightly sheathing at the base; scales usually acuminate or cuspidate, brownish, with or without greenish center and tip; perigynia 2.5–3 mm. long, compressed triangular, pale or glaucous-green, papillate, prominently nerved; achene about 2 mm. long.

Seward Peninsula east and south; circumboreal and in S. Am. and the Falkland Islands. (Fig. 276.)

85. C. laxa, Wahl.   Weak Sedge

Stoloniferous; culms 1–4 dm. tall; slender, weak; leaves 1–2.5 mm. wide, terminal spike staminate, often with a few perigynia, 1–2 cm. long; pistillate spikes one to three, 1–2 cm. long, drooping on capillary peduncles; lowest bract leaf-like with long sheath, scales obtuse, brownish, with rather wide hyaline margins and lighter center; perigynia much as in C. limosa.

An old world species reported in America only from Mile 172–174 along the Richardson Highway.

86. C. livida (Wahl.) Willd.   Livid Sedge

Rootstocks long, slender; culms 1–5 dm. tall; leaves 3 mm. or less wide, glaucous-green, involute, thickened, stiff; terminal spike staminate or with a few perigynia, 7–30 mm. long; pistillate spikes one to three, 10–20 x 5 mm., the lower sometimes subradical; bracts leaf-like, the lower sheathing; scales purplish with hyaline margins and greenish center, the pistillate wider than the staminate; perigynia 2.25–4.5 mm. long, obscurely triangular, glaucous-green, puncticulate, two-keeled, beakless; achenes triangular, about 2.5 mm. long, brownish-black.

Southwestern Alaska—Labr.—Newf.—N. Jer.—Ida.—Calif. (Fig. 277.)

87. *C. vaginata* Tausch. Sheathed Sedge
C. saltuensis Bailey.

Producing long, horizontal, yellowish-brown stolons; culms 16–80 cm. tall; leaves 1.5–5 mm. wide, flat or channeled toward the base; terminal spike staminate, 1–2 cm. long; lateral spikes two or three, pistillate, 8–20 x 3–5 mm,. the lower ones on long peduncles; bracts with sheaths up to 3 cm. long; scales purplish-brown with hyaline margins and three-nerved light center; perigynia three to twenty to a spike, about 4 mm. long, longer than the scales, yellowish-green or brown, puncticulate, the beak tinged with purple; achenes 2.5–3 mm. long, triangular with concave sides, yellowish.

Alaska Range northward; circumpolar. (Fig. 278.)

88. *C. atrofusca* Schk. Dark-brown Sedge

Caespitose and stoloniferous; culms 1–3 dm. tall, obtusely triangular, usually nodding; leaves clustered at the base, usually less than 1 dm. long but sometimes longer, 2–4 mm wide; terminal spike gynaecandrous, the lateral spikes two to three, 8–18 mm. long, drooping on slender peduncles; lowest bract long-sheathing; scales black with somewhat lighter margins and midribs; perigynia 4–5 mm. long, dark with lighter base and hyaline at the tip of the beak, granular, two-ribbed, slightly serrulate above; achenes conspicuously stipitate and apiculate.

Bering Sea eastward; circumpolar. (Fig. 279.)

89. *C. misandra* R. Br. Short-leaved Sedge

Caespitose; culms slender, 1–3 dm. tall; leaves 1.5–3 mm. wide, canaliculate below, thickish, stiff, long-attenuate; terminal spike gynaecandrous, drooping; lateral spikes two or three, pistillate, 7–20 x 4–6 mm.; lowest bract long-sheathing, the sheath tight, tinged purplish, the blade short; perigynia 4–6 mm. long, 1 mm. wide, flattened-triangular, dark above, light-colored below, ciliate-serrulate on the margin above, two-edged, the beak with a hyaline tip; achenes about 2 x 0.75 mm., brownish.

Arctic, Bering Sea and Pacific coastal districts. Distribution interrupted circumpolar. (Fig. 280.)

90. *C. capillaris* L. Hair-like Sedge

Caespitose; culms 1–6 dm. long, very slender, erect, spreading or decumbent; leaves 0.75–2.5 mm. wide; terminal spike staminate, 4–8 mm. long; lateral spikes two or three, pistillate, 5–15 mm. long, on slender drooping peduncles; bracts sheathing, tubular; scales ovate, hyaline-margined, shorter than the perigynia; perigynia 2.5–3 mm. long, obtusely triangular, somewhat inflated, greenish-brown, two-ribbed, the beak hyaline-tipped; achenes 1.5 mm. long, triangular.

Throughout most of Alaska—Greenl.—Maine—N. Mex.—Nev.—B. C. (Fig. 281.)

91. C. krausei Boeck. Krause Sedge
C. capillaris var. nana (Cham.) Kuk.

Caespitose; culms 3–30 cm. tall; leaves nearly as long as the culms, sometimes longer; terminal spike gynaecandrous. Resembles C. capillaris but is of much lower growth, the terminal spike is gynaecandrous, the perigynia have shorter beaks which are finely spinulose on the margins.

Central and southeastern Alaska—Yukon. Range probably more extensive.

92. C. williamsii Britt. Williams Sedge

Caespitose; culms 3–30 cm. tall, slender; leaves 2–8 cm. long, 0.25–0.75 mm. wide, canaliculate, minutely serrulate toward the base; terminal spike staminate 2–6 x 0.5–1 mm.; lateral spikes three to five, pistillate, 4–10 x 2.5 mm.; lowest bract tubular-sheathing; scales obtuse or the staminate mucronate; perigynia three to nine to a spike, 2.5–3.5 mm. long, narrow, greenish, puncticulate, the beak with hyaline orifice; achenes 1.5 x 0.5 mm.

Bering Sea region—Labr.

93. C. flava L. Yellow Sedge

Caespitose; culms 1–8 dm. tall, stiff; leaves 3–5 mm. wide, flat or canaliculate at the base, thickish; terminal spike staminate or with a few perigynia, 5–20 mm. long; lateral spikes two to five, pistillate or sometimes androgynous, 7–18 x 10 mm.; bracts leaf-like, the lowest with sheath 2–20 mm. long; scales reddish-brown with hyaline margins and lighter center; perigynia 4.5–6 x 1.25–2 mm., spreading or deflexed, yellowish-green or yellow, puncticulate, ribbed; beak 2–3 mm. long, serrulate, the teeth of the bidentate beak tinged with red; achenes 1.5 x 1 mm.

Yakutat Bay—Hudson Bay—Labr. .

94. C. viridula Michx. Green Sedge
C. oederi Retz. var. viridula (Michx.) Kuk.

Caespitose; culms 6–30 cm. tall, stiff; leaves 1–3 mm. wide, canaliculate, thickish; terminal spike staminate, 3–15 mm. long; lateral spikes two to six, pistillate, 5–10 x 4–7 mm.; bracts leaf-like, the lowest with sheath 4–18 mm. long; scales shaded brownish; perigynia fifteen to thirty to a spike, 2–3 mm. long, spreading, yellowish-green, puncticulate, ribbed, the beak scarcely half as long as the body; achenes 1.25 x 1 mm., blackish.

Southeastern Alaska—Sask.—Utah—N. Mex.—Calif. Also in eastern America and eastern Asia. (Fig. 282.)

95. C. rostrata Stokes. Beaked Sedge

Caespitose and long-stoloniferous; culms 3–12 dm. tall, tinged red at the base; leaves 2–12 mm. wide, flat above and with revolute margins, septate-nodulose; upper two to five spikes staminate, 1–7 cm. long, narrow.

lower two to five spikes pistillate or some of them androgynous, 1–8 cm. x 4–8 mm., bracts leaf-like; scales brownish with hyaline margins and lighter center; perigynia 4–8 mm. long, inflated, yellowish-green to brown, puncticulate, strongly nerved, the beak bidentate; achenes 2 mm. long, yellowish-brown.

All of our area except the high arctic; circumboreal. (Fig. 283.)

96. *C. rotundata* Wahl.   Round-fruited Sedge
     C. melozitensis Porsild.

Loosely caespitose and with long stolons; culms 15–45 cm. tall; leaves 1–3 mm. wide, involute; terminal spike staminate, 10–25 mm. long, often with one or two smaller ones at the base; lateral spikes one or two, pistillate, 8–25 x 6–9 mm.; lowest bract leaf-like, 1–3 cm. long; staminate scales brown, the pistillate purplish-black, both with hyaline apex and lighter midrib; perigynia 3–3.5 mm. long, inflated, straw-colored, tinged brownish, puncticulate; achenes 2.25 mm. long, grayish-brown.

Bering Sea and Alaska Range north and east; circumpolar. (Fig. 284.)

97. *C. rhyncophysa* C. A. Mey.
     C. laevirostris (Blytt) Fr.

Caespitose and with long stolons; culms 4–10 dm. tall, stout; leaves 6–15 mm. wide, flat above, channeled below, firm, strongly septate-nodulose; upper two to four spikes staminate, 2–6 cm. long, lower two to five spikes pistillate, 25–75 x 9–12 mm.; lower bracts leaf-like, sheathless or nearly so; staminate scales obtusish, pistillate scales acute or acuminate, both tinged reddish-brown with hyaline margins and lighter center; perigynia 4.5–7 mm. long, inflated, greenish straw-color, coarsely nerved, long-beaked; achenes 2 x 1.5 mm., triangular with concave sides below.

Matanuska—Yukon—Mack.—Eurasia. (Fig. 285.)

98. *C. physocarpa* Presl.

Rootstocks long, slender; culms 2–8 dm. tall; leaves 1.5–5 mm. wide; terminal spike staminate, 2–4 cm. long, often with one or two shorter ones at the base; lateral spikes one to three, pistillate, 15–35 x 6–12 mm., spreading or drooping on slender peduncles; lowest bract leaf-like; scales purplish-black with lighter midrib and prominently hyaline tips; perigynia 4–5 mm. long, dull grayish-yellow, usually dark tinged above; achenes 1.5–2 mm. long, lenticular, yellow, continuous with the bent style. This species forms hybrids with *C. rostrata* (*C. utriculata* Cov. & Wight).

Common in our area except the Arctic and extends eastward to Mack. and south to Colo. & Utah. Also in eastern Asia. (Fig. 286.)

99. *C. membranacea* Hook.   Fragile Sedge
     C. membranopacta Bailey.

Caespitose and stoloniferous, culms 15–50 cm. tall, stiff; leaves 3–5 mm. wide, flat, often with revolute margins; terminal spike staminate, 10–25 mm. long, sometimes with one or two smaller ones at the base; pistillate spikes one to three, sessile to short-peduncled, 12–30 mm. long, 7–9 mm.

thick; bracts leaf-like, the upper reduced; scales dark, often blackish, usually with hyaline margins and lighter center; perigynia 3.5–4.5 mm. long, inflated, fragile, dark-tinged above, the beak bidentate; achenes about 1.5 mm. long.

Throughout most of our area—Ellsmereland—Greenl.—Ungava. Also in Asia. (Fig. 287.)

100. *C. atherodes* Spreng.   Awned Sedge

Loosely caespitose and stoloniferous; culms 5–15 dm. tall; leaves 3–12 mm. wide, flat, thin, septate-nodulose, sparsely hairy beneath toward the base, the sheaths soft-hairy; upper two to six spikes staminate, 4–10 cm. long; lower two to four spikes pistillate or androgynous, 5–12 cm. x 8–15 mm.; bracts leaf-like, the lowest sheathing; scales rough-aristate, ciliate; perigynia 7–12 x 2 mm., inflated, yellowish-green or brownish, strongly ribbed, the beak bidentate with long, spreading teeth; achenes 2.5 x 1.25 mm., triangular, yellowish-brown.

Yukon—Mack.—St. Lawrence R.—N. Jer.—Mo.—Utah.—Ore.

In addition to the species here described several other species of *Carex* have been reported from our area but their occurrence needs confirmation.

8. ARACEAE (Arum Family)

Perennials with basal, reticulate-veined, petioled leaves and the flowers in a dense, fleshy spike borne on a spadix subtended or enclosed by a large foliaceous or colored bract (spathe). Perianth of scale-like parts or none; flowers in ours perfect; stamens four to ten; fruit a berry or utricle.

Flowers with a perianth, spathe yellow..................................1. *Lysichitum*
Flowers without a perianth, spathe white ....2. *Calla*

1. LYSICHITUM Schott.

Swamp plants with short, thick rootstocks, large fleshy roots and large, fleshy leaves, with the petioles sheathing at the base; spadix at first enveloped by the spathe, later exerted; perianth of four segments; stamens four; filaments flat; anthers two-celled; ovary conical, two-celled, two-ovuled; stigma depressed; fruit fleshy. Also spelled Lysichiton. (Greek, base and tunic, referring to the spathe.)

L. americanum Hult. & St. J.   Skunk Cabbage

Large tropical-looking swamp plants with elliptic or oblong-lanceolate leaves, a large blade having actually measured more than 13 dm. long and 7 dm. wide; spathe yellow, oblong-lanceolate, acute, the open part 1–3 dm. long; spadix 7–12 cm. long in flower, elongating in fruit, on a stout, fleshy peduncle.

Pacific coastal district of Alaska—Mont.—Calif. (Fig. 288.)

2. CALLA L.

A bog or aquatic herb; rootstocks creeping, acrid; leaves broadly ovate-cordate; spathe white, ovate-lanceolate, acuminate with a long point;

spadix cylindric, densely covered with flowers; stamens about six; anthers with two divaricate sacs; ovary ovoid, one-celled; ovules six to nine; berries depressed-obconic. (Ancient name.)

C. palustris L. Water Arum

Petioles 1–3 dm. long; leaf-blades 6–12 cm. long, 4–9 cm. wide, cuspidate at the apex, cordate at the base; scapes as long as the petioles; spathe 4–7 cm. long, the mucronation often 6–8 mm. long; fruiting spadix 2–3 cm. long and half as wide.

Shallow water, southwestern and central Alaska eastward; circumboreal. (Fig. 289.)

9. LEMNACEAE L. (Duckweed Family)

Minute stemless and leafless perennial plants with thallus-like body, floating on fresh water; roots one or more from the lower surface; inflorescence of one or more monoecious flwers borne on the edge of the upper surface, staminate flower of a single stamen with two to four pollen-sacs; pistillate flower of a single flask-like pistil; fruit a one to six seeded utricle. The simplest and smallest of flowering plants, propagating mostly by budding.

LEMNA L.

Fronds disk-like or oval; rootlet solitary, without fibrovascular tissues; anthers dehiscent transversely. Not often found in flower. (Greek, in allusion to the swamp habitat.)

Fronds long-tailed, mostly submerged .... 1. *L. trisulca*
Fronds short-stalked or sessile, floating .... 2. *L. minor*

1. *L. trisulca* L. Ivy-leaved Duckweed

Fronds usually submerged with several generations attached to each other, oblong, or oblong-lanceolate, 6–10 mm. long, 2–3 mm. wide, obscurely three-nerved and denticulate at the apex, often without rootlets.

Cook Inlet district, circumboreal. (Fig. 290.)

2. *L. minor* L. Lesser Duckweed

Fronds 2–4 mm. wide, rounded to obovate-oblong, symmetrical, green or rarely reddish or purplish-tinged, obscurely three-nerved, and often a row of papillae on the midrib; fruit symmetrical, subturbinate; seed deeply and unevenly 12–15-ribbed. (Fig. 291.)

10. JUNCACEAE (Rush Family)

Annual or mostly perennial grass-like herbs; flowers perfect; regular inconspicuous, sepals and petals each three, similar and scale-like; stamens six or three, rarely four or five; anthers introrse; pistil of three united carpels; ovary one or three-celled; stigmas three, filiform; capsule loculicidal; seed three to many, usually reticulated or ribbed and often tailed.

Leaf-sheaths open, capsule many-seeded .... 1. *Juncus*
Leaf-sheaths closed, seeds three .... 2. *Luzula*

1. JUNCUS (Tourn.) L.

Mostly glabrous perennial swamp plants, with or without leaves, leafsheaths with free margins, the blades terete, gladiate, grass-like or channeled; flowers subtended by a bract, sometimes also by two bractlets; capsules many-seeded. Our illustrations show the mature capsule surrounded by the perianth, seed, and the bractlets when present. (Latin, *jungo*, to bind, in allusion to the use of these plants for withes.)

1A. Inflorescence appearing lateral due to the prolongation of the lower bract.
  1B. Bract about as long as the stem .... 1. *J. filiformis*
  2B. Bract shorter than the stem.
    1C. Involucral bract short, 0.5–3 cm. long .... 2. *J. drummondii*
    2C. Involucral bract longer.
      1D. Stems tufted .... 3. *J. effusus*
      2D. Stems in rows from horizontal rhizomes.
        1E. Involucral bracts 2–5 cm. long .... 4. *J. arcticus*
        2E. Involucral bracts longer.
          1F. Anthers about 2 mm. long .... 6. *J. ater*
          2F. Anthers shorter .... 5. *J. balticus*
2A. Inflorescence appearing terminal.
  1B. Leaves gladiate .... 7. *J. ensifolius*
  2B. Leaves terete, convolute, or channeled.
    1C. Flowers borne separately.
      1D. Low annual .... 8. *J. bufonius*
      2D. Tall perennial .... 9. *J. macer*
    2C. Flowers in one or more compact heads.
      1D. Heads normally one.
        1E. Heads many-flowered .... 13. *J. mertensianus*
        2E. Heads 1–5 flowered.
          1F. Stem with leaf in its middle part ....10. *J. stygius*
          2F. Stem with only basal leaves.
            1G. Involucral bract overtopping the head ....11. *J. biglumis*
            2G. Involucral bract shorter ....12. *J. triglumis*
      2D. Heads normally more than one.
        1E. Leaves septate, terete.
          1F. Heads globose ....16. *J. nodosus*
          2F. Heads conical with erect or ascending flowers.
            1G. Perianth 3.5–5 mm. long ....14. *J. oreganus*
            2G. Perianth 2–3 mm. long ....15. *J. alpinus*
        2E. Leaves not septate.
          1F. Capsule truncate or depressed at the apex ....17. *J. falcatus*
          2F. Capsule rounded or acute at the apex.
            1G. Capsule acutish, pale, about double the length of the perianth .... 19. *J. leucochlamys*

2G. Capsule obtusish, dark, only
slightly longer than the
perianth ....18. *J. castaneus*

1. *J. filiformis* L. Thread Rush

Stems tufted from creeping rootstocks, 1–6 dm. tall, about 1 mm. thick; basal leaf-blades filiform rudiments; lower leaf of inflorescence erect and often longer than the stem; inflorescence several-flowered, spreading; perianth about 3 mm. long, the segments lanceolate, acute; stamens six, about half as long as the perianth; capsule green, barely pointed, as long as or shorter than the perianth; seeds obliquely oblong, pointed at one or both ends.

St. George Isl.—southeastern and central Alaska; circumboreal. (Fig. 292.)

2. *J. drummondii* E. Mey. Drummond Rush

Stems densely tufted, 1–4 dm. tall, from matted rootstocks; leaf-sheaths bladeless or with mere rudiments; inflorescence one to five, but mostly three-flowered; lower bract 10–25 mm. long; flowers with a pair of brown bractlets at the base; perianth about 6 mm. long, the segments lanceolate, acute or acuminate with broad brown margins; stamens six, about half the length of the perianth; seeds caudate at both ends.

Aleutians—Alta.—N. Mex.—Calif. (Fig. 293.)

3. *J. effusus* L. Bog Rush

Stems tufted, 5–14 dm. tall; basal leaf-blades reduced to short filiform rudiments; inflorescence many-flowered, usually dense and congested, 2–5 cm. long; lowest bract of the inflorescence 5–25 cm. long, perianth 2–3 mm. long, the segments lanceolate, acuminate; stamens three; capsule obovoid, about as long as the perianth; seed with short points, two or three times as long as broad.

Along the coast, southeastern Alaska—Calif. (Fig. 294.)

4. *J. arcticus* Willd. Arctic Rush

Stems arising at close intervals from creeping rootstocks and somewhat caespitose, 12–35 cm. tall; sheaths leafless or with a small mucronation; inflorescence three to ten flowered; lowest bract 4–20 cm. long; perianth 4–5 mm. long, the outer segments somewhat longer than the inner and darker brown, all with light center; stamens shorter than the perianth, the anther about the same length as the filament; capsule dark and shining, about the same length as the perianth; seed somewhat irregular in shape, about 0.5 mm. long. Most of our forms belong to the ssp. *alaskanus* Hult. with more lax inflorescence and more acute inner segments of the perianth.

Arctic and Bering Sea coasts eastward; circumpolar. (Fig. 295.)

5. *J. balticus* Willd. Baltic Rush

Stems arising from creeping rootstocks, 2–8 dm. tall; sheaths leafless or with a slender mucronation; main floral bract 6–20 cm. long; perianth 4–5 mm. long, the segments lanceolate, the outer more acute than the inner, purplish-brown; anther as long as or longer than the filament; capsule brown and shining, narrowly ovoid, conspicuously mucronate, fully as long as the perianth; seed striate. The typical form occurs in southeastern Alaska but is not so common as the ssp. *sitchensis* (Engelm.) Hult., with anthers about as long as the filaments and the inner perianth segments more narrowly scarious-margined, which occurs over most of our area south of the Arctic Circle.

Whole species circumboreal; the ssp. in eastern Asia and Alaska. (Fig. 296.)

6. *J. ater* Rydb.   Mountain Rush

Stems slender, about 2 mm. thick; flowers 5–25; anthers about 2 mm. long on very short filaments. Resembles *J. balticus* and its variety but the stem is more slender, the inflorescence is more lax, the lower floral bract averages shorter, the inner perianth segments are more acute with narrower hyaline margins and the anthers are longer.

Central Alaska—Mont.—Nebr.—N. Mex.—Calif.

7. *J. ensifolius* Wiks.   Dagger-leaved Rush

Stems flattened, leafy, erect, 2–6 dm. tall, arising from creeping rootstocks; leaves flattened laterally and equitant, 7–25 cm. long, 3–6 mm. wide; heads one to seven, very dense; perianth dark brown, the segments lanceolate, acuminate, about 3 mm. long; stamens three, two-thirds as long as the perianth; capsule dark brown, obtuse below the mucronation, slightly exceeding the perianth.

Eastern Asia—Aleutian Islands—Alta.—Utah—Calif. (Fig. 297.)

8. *J. bufonius* L.   Toad Rush

Low, profusely branched annual 5–25 cm. tall; leafy below, the leaves narrow and involute, the lower up to 5 cm. long, the upper short; flowers greenish, inserted singly on the branches and in the axils of the leaves; perianth 4–6 mm. long, the inner segments shorter and less attenuate than the outer; capsule shorter than the perianth; seeds broadly oblong but variable.

Nearly cosmopolitan but may be introduced in our area. (Fig. 298.)

9. *J. macer* S. F. Gray.   Slender Rush

Stems wiry, tufted, 2–6 dm. tall; leaves from one-half to nearly as long as the stem, flat; bracts two or three, leaf-like, at least one of them longer than the open inflorescence; flowers greenish, aggregated at the top of, or scattered along the branches of the panicle; perianth segments lanceolate with scarious margins, acute, 3–4.5 mm. long; capsule broadly ovoid, obscurely triangular.

Introduced in southeastern Alaska. Has been confused with *J. tenuis* Willd. (Fig. 299.)

10. *J. stygius* L. ssp. *americanus* (Buch.) Hult.   Moor Rush

Stems 7–15 cm. tall, erect, one- to three-leaved below; leaves erect or ascending, the sheaths nerved and auriculate; heads usually one, sometimes more, one to four flowered; lowest bract usually exceeding the flowers; perianth 3-5 mm. long, the parts about equal; capsule 6-8 mm. long, spindle-shaped, pale brown, acute, mucronate, few-seeded; seeds with a thick coat forming thick tails, the total length 2.5-3 mm.

Central Alaska—Labr.—Newf.—Gt. Lakes region. The typical form in Europe and western Asia. (Fig. 300.)

11. *J. biglumis* L. Two-flowered Rush

Stems one to few, from branched rootstocks, erect, 25-200 mm. tall; leaves one to five, all basal; longest bract of the inflorescence foliose, 5-25 mm. long; flowers one to four, usually two, perianth 3-4 mm. long, dark brown, the segments about equal, obtuse; stamens equaling the perianth; capsule exceeding the perianth, retuse at apex with three-keeled shoulders; seed 1 mm. long or more, brown with broad white tails.

Western and central Alaska—Ellsmereland—Greenl.—Hudson Bay—Labr.—arctic Eurasia. (Fig. 301.)

12. *J. triglumis* L. Three-flowered Rush
J. albescens (Lge.) Fern.

Stems tufted, 5-15 cm. tall; leaves one to five, all basal with clasping, auriculate sheaths and narrow terete blades 1-7 cm. long; inflorescence a cluster of one to five, usually three, flowers; the lowest two or three bracts divergent, usually brown and membranous, perianth about 4 mm. long; stamens nearly as long as the perianth; capsule about equaling the perianth, mucronate; seed less than 2 mm. long including the tails.

Bering Strait east and south; circumboreal. (Fig. 302.)

13. *J. mertensianus* Bong. Mertens Rush

Stems tufted, erect, 8-30 cm. tall, 1-1.5 mm. thick; leaves two or three on the stems, occasionally overtopping the stem; inflorescence a dense head 8-15 mm. broad, capitate, the lower bract usually longer than the head; perianth nearly black, about 4 mm. long, stamens nearly equaling the perianth, the anthers much shorter than the filaments; capsule scarcely as long as the perianth.

Alpine meadows, Japan, Aleutian Islands and central Alaska—Colo.—Calif. (Fig. 303.)

14. *J. oreganus* S. Wats. Oregon Rush

Stems tufted from slender, matted rootstocks, 1-2 dm. tall; stem leaves two to four, the sheaths with conspicuous hyaline margins and auricles; heads usually two to four, small, few- several-flowered; perianth segments nearly equal, narrowly lanceolate, acute, brown; stamens half as long as the perianth, the filaments longer than the anthers; capsule often twice as long as the perianth, acute mucronate, dark brown; seeds ribbed and cross-lined.

Bogs near the coast, Kodiak Island—Ore.

15. *J. alpinus* Vill. Richardson Rush
J. richardsonianus Schult.

Stems erect, 15-50 cm. tall from creeping rootstocks; one or two-leaved, leaf-blades terete or slightly compressed, septate; flowers in panicled heads of two to twelve; perianth 2-2.5 mm. long, the inner segments shorter than the outer; capsule ovoid-oblong, longer than the perianth, straw-colored or light brown; seed apiculate, acute or acuminate at the base. Except in the Pacific coast and the Aleutian Islands districts our forms belong to the ssp. *nodulosus* (Wahl.) Lindm. (*J. nodulosis* Wahl.) which has some of the flowers peduncled in the loose heads, whereas in the type all flowers are sessile.

Central Alaska southward; circumboreal. (Fig. 304.)

16. *J. nodosus* L. Knotted Rush

Stems 15–60 cm. tall, arising from thickenings of slender rootstocks; leaves erect, conspicuously septate-nodulose; inflorescence of one to thirty heads 7–12 mm. in diameter, six to twenty flowered; perianth 3–4 mm. long, the inner segments longer than the outer; capsule acutely triangular, long-pointed, usually longer than the perianth; seed acute below, apiculate above, reticulate.

Manly Hot Springs and southeastern Alaska—Newf.—Va.—N. Mex.—Nev. (Fig. 305.)

17. *J. falcatus* E. Mey. ssp. *sitchensis* (Buch.) Hult. Sickleleaved Rush
J. falcatus var. *sitchensis* Buch.

Stems 8–30 cm. tall from slender creeping rootstocks; basal leaves grass-like, from two-thirds as long to longer than the stems, 1.5–3 mm. wide; stem leaves one or two; heads one to six, the lowest bract foliaceous; perianth about 4 mm. long, minutely roughened, the segments about equal, the outer minutely mucronate, brown or with green midrib; capsule slightly retuse, as long as or longer than the perianth; seeds reticulate, with a light-colored coat.

Eastern Asia—Aleutian Islands—southeastern Alaska. The type form from B. C.—Calif., Japan and Tasmania. (Fig. 306.)

18. *J. castaneus* J. E. Smith. Chestnut Rush

Stoloniferous; stems erect, 1–5 dm. tall, more or less leafy; leaves 4–12 cm. long, tapering from an involute tubular base to a slender channeled apex; heads one to four, three- to twelve-flowered; perianth brown, about 5 mm. long, the segments lanceolate, acute; capsule dark brown, paler toward the base, longer than the perianth; seeds with long, light brown tails.

Found in most of our area; circumboreal. (Fig. 307.)

19. *J. leucochlamys* Zing. & Kretch.

Resembles *J. castaneus* but is taller, has from two to twelve heads, the capsules are longer and more acute, being about twice as long as the perianth; the perianth and capsule are much lighter in color, being a pale brown.

Eastern Asia, known in America only from Matanuska.

2. LUZULA DC.

Glabrous or sparingly pubescent plants with leaf-bearing stems, the leaves grass-like with closed sheaths; inflorescence umbel-like, capitate,

or spike-like; flowers always subtended by usually dentate or lacerate bractlets; stamens six; ovary one-celled, developing into a three-seeded capsule. (Latin, *lux*, light, suggested by leaves of a species shining with dew. *Juncoides* (Dill.) Adans.)
1A. Inflorescence umbel-like ....1. L. rufescens
2A. Inflorescence an open panicle with usually solitary flowers.
   1B. Bractlets lacerate and abundantly ciliate.......... 2. L. wahlenbergii
   2B. Bractlets less lacerate and glabrous or with a
      single cilium ....3. L. parviflora
3A. Inflorescence of one to several spike-like or head-like glomerules of flowers.
   1B. Flowers in a dense, drooping, spike-like panicle..4. L. spicata
   2B. Flowers in glomerules or sometimes in an erect spike in number six.
      1C. Leaves involute or channeled, with purplish bases.
         1D. Inflorescence widely branched with
            curved branches ....5. L. arcuata
         2D. Inflorescence spike-like or sparsely
            branched with straight branches ....6. L. hyperborea
      2C. Leaves flat with brownish bases.
         1D. Bract shorter than the inflorescence.........7. L. nivalis
         2D. Bract longer than the inflorescence.......8. L. multiflora

1. *L. rufescens* Fisch.   Hairy Wood-Rush

Stoloniferous and somewhat caespitose; stems slender, 1–3 dm. tall; leaves mostly basal, 1–3 mm. wide, slightly webbed when young; inflorescence umbellate, some of the pedicels often reflexed, rarely bearing more than one flower; perianth 2–3 mm. long; capsule acute, longer than the perianth; seed with a conspicuous irregular caruncle. Reports of *Luzula* (or *Juncoides*) *carolinae, japonica, pilosa* or *saltuensis* from Alaska all refer to this species.

Eastern Asia extending to east central Alaska. (Fig. 308.)

2. *L. wahlenbergii* Rupr.   Wahlenberg Wood-Rush

Stems caespitose, erect, 10–35 cm. tall; leaves mostly basal, usually not more than 3 mm. wide; inflorescence diffuse, the capillary branches curved; flowers solitary but often two to four approximate; bractlets lacerate or ciliate; perianth 2.25–3 mm. long, the segments acute, brown with hyaline tips; capsule about equaling the perianth, ovoid; seed about 1.5 mm. long, brown, attached to the placentas by white fibers. Along the southern Alaska coast this species approaches *L. parviflora.*

Seward Peninsula east and south; circumboreal. (Fig. 309.)

3. *L. parviflora* (Ehrh.) Desv.   Small-flowered Wood-Rush
   J. parviflorum (Ehrh.) Cov.

Stems erect, terete, 3–6 dm. tall; leaves 5–15 cm. long, 4–12 mm. wide, glabrous except for a few hairs at the mouth of the sheath; inflorescence a nodding, compound panicle, the flowers usually borne singly on slender pedicels; perianth 2-2.5 mm. long, the segments lanceolate, acute, green or brown; capsule ovoid, dark, slightly exceeding the perianth; seeds ellipsoid, brown, with cottony fibers at lower ends. The ssp. *divaricata* (S. Wats.) Hult., has larger, more spreading panicle with lighter colored flowers. This and the preceding species sometimes form hybrids.

Kotzebue Sound—Baffinland south; circumpolar. (Fig. 310.)

4. *L. spicata* (L.) DC.   Spiked Wood-Rush
J. spicatum (L.) Kuntze.

Stems tufted, erect, 1-3 dm. tall; leaves 4-12 cm. long, 1-3 mm. wide, with sharp apex and sparingly webby; inflorescence spike-like, usually nodding; bractlets ovate-lanceolate, acuminate, more or less webby, often equaling the perianth; perianth dark brown with lighter margins; capsule acute, shorter than the perianth; seeds brown with light base, about 1 mm. long.

Central Alaska southward; circumpolar. (Fig. 311.)

5. *L. arcuata* Wahl.   Alpine Wood-Rush
J. arcuatum (Wahl.) Kuntze.

Caespitose and short-stoloniferous; stems 8-20 cm. tall, slender; leaves 1-3 mm. wide, usually some of them curved, canaliculate, the apex subulate; inflorescence paniculate, the branches curved and spreading; perianth brown, 2-2.5 mm. long, the parts about equal; capsule slightly shorter than the perianth; apiculate; seed attached with a tuft of white fibers.

Seward Peninsula—Wash., and in Eurasia. (Fig. 312.)

6. *L. hyperborea* R. Br.   Northern Wood-Rush
L. confusa Lindeb.
J. hyperborium (R. Br.) Sheldon.

Stems tufted, erect, 1-3 dm. tall; leaves erect, 1-3 mm. wide, sparingly ciliate at the mouth of the sheath, sharp-pointed; inflorescence of a single head or two- or three-branched, the branches erect or curved; lowest bract short-foliose, the upper bracts and the bractlets fimbriate; perianth 2-2.5 mm. long, the parts brown, paler above; capsule somewhat shorter than the perianth, ovoid.

Arctic and central Alaska; circumpolar. (Fig. 313.)

7. *L. nivalis* (Laest.) Beurl.   Snow Wood-Rush
J. arcticum Am. auct.

Stems tufted, 5-10 cm. tall, erect; leaves 2-4 mm. wide, usually less than 8 cm. long; inflorescence an ovate, spike-like cluster about 1 cm. long; perianth 2 mm. long or less. Known from Little Diomede Island and southeastern Alaska. The common form is the var. *latifolia* Kjellm. Stems 8-25 cm. tall; leaves 3-5 mm. wide; inflorescence with one to three slender, erect or curved branches; bractlets with hyaline, more or less laciniate margins, one-half to two-twirds as long as the perianth; perianth 2.5-3 mm. long, its parts equal; capsule as long as the perianth or shorter; seed brown with dark tip and light base.

A circumpolar species. (Fig. 314.)

8. L. *multiflora* (Retz.) Lej. Many-flowered Wood-Rush
   L. *campestris* Am. auct.
   J. *campestris* Am. auct.

Stems tufted, 1-5 dm. tall; leaves ciliate on the margins, webbed at the mouth of the sheath; inflorescence of four to ten, eight to sixteen-flowered heads, the branches erect or ascending, sometimes congested; lowest bract foliose, longer than the inflorescence, bractlets often entirely hyaline; perianth 2.5-3.5 mm. long, brown with hyaline margins; capsule somewhat shorter than the perianth; seed with white caruncle at the base. Var. *frigida* (Buch.) G. Sam. Stems 10-15 cm. tall; leaves 2-4 mm. wide, inflorescence dense. Var. *kobayasii* (Satake) G. Sam. Leaves up to 1 cm. wide. Ssp. *comosa* (E. Mey.) Hult. High growing; bract large; bractlets strongly ciliate; perianth yellowish.

A variable, circumboreal species. (Fig. 315.)

11. MELANTHACEAE (Bunch-flower Family)

Leafy-stemmed or scapose perennials with elongated or bulb-like rootstocks; leaves alternate or basal; flowers perfect, dioecious, or polygamous in racemes or panicles; sepals and petals distinct or nearly so; stamens six, often partly adnate to the base of the sepals and petals; anthers versatile; pistil composed of three united carpels; ovary three-celled; styles three; fruit a septicidal capsule. By many authors this group is considered to be only a tribe or subfamily of *Liliaceae*.

1A. Plants with bulbs ....3. *Zygadenus*
2A. Plants with rootstocks.
   1B. Leaves narrow ....1. *Tofieldia*
   2B. Leaves broad ....2. *Veratrum*

1. TOFIELDIA Huds.

Rootstocks short with numerous fibrous roots; leaves two-ranked, linear or equitant; flowers small, in a terminal spike-like raceme, subtended by a small involucre of three, more or less, united bractlets below the perianth; sepals and petals nearly equal, persistent, glandless; capsule septicidal to the base; seeds numerous. (Tofield was a botanist of Yorkshire, England.)

1A. Stems viscid-pubescent above ....1. *T. occidentalis*
2A. Stems glabrous, scapiform.
   1B. Flowers greenish ....2. *T. pusilla*
   2B. Flowers purplish ....3. *T. coccinea*

1. *T. occidentalis* S. Wats. Western Tofieldia
   T. intermedia Rydb.

Stems 15-20 cm. tall, viscid above with black, stalked glands; leaves 5-25 cm. long, 2-6 mm. wide; racemes 15-50 mm. long, rather dense, the pedicels usually three together; flowers greenish-yellow, the sepals ovate, 4-5 mm. long, the petals narrower and slightly longer; capsule ovoid, 5-7 mm. long. Most of the Alaskan material belongs to the form described as

T. intermedia but the characters on which that form is based are not constant.

Southern Alaska—Sask.—Calif. (Fig. 316.)

2. *T. pusilla* (Michx.) Pers.   Scotch Asphodel, False Asphodel
T. palustris Am. auct.

Stems tufted, 6–18 cm. tall; leaves 2–10 cm. long, gladiate; raceme 6–25 mm. long, usually rather dense; flowers yellowish-green on short pedicels; sepals and petals obovate, much shorter than the oblong-globose, beaked capsule.

A circumpolar species occurring in most of our area. (Fig. 317.)

3. *T. coccinea* Richards.   Northern Asphodel
T. nutans Willd.

Stems tufted, 4–10 cm. tall; basal leaves gladiate, 2–6 cm. long, 2–4 mm. wide; raceme short and dense; flowers short-pedicelled or nearly sessile, tinged with purple; petals and sepals obovate, nearly as long as the depressed globose, minutely beaked, dark purple capsule.

A circumpolar species occurring almost throughout our area. (Fig. 318.)

2. VERATRUM (Tourn.) L.

Tall, stout, poisonous perennials; rootstocks stout; leaves broad, strongly veined and plaited; flowers polygamous, in large panicles; sepals and petals each three, nearly equal; stamens six; anthers cordate; fruit a many-seeded, slightly inferior capsule; seeds flat, winged. (Ancient name of the Hellebore.)

Flowers yellowish-green .... 1. *V. eschscholtzii*
Flowers whitish inside .... 2. *V. album*

1. *V. eschscholtzii* A. Gray.   American White Hellebore
V. viride Ait. in part.

Stems 10–25 dm. tall; leaves broadly round-oval to ovate-lanceolate, narrower toward the inflorescence, glabrous above, pubescent, often densely so, below, up to 3 dm. long and 2 dm. wide, the base sheathing the stem; inflorescence a large panicle with drooping branches; sepals and petals oblanceolate, greenish, 8–10 mm. long, about twice as long as the stamens; capsule 10–12 mm. long. May be only a variety of *V. viride*.

Bering Sea—Mont.—Calif. (Fig. 319.)

2. *V. album* L. ssp. *oxysepalum* (Turcz.) Hult. European White Hellebore

Resembling the preceding species but not so tall and with an upright, spike-like inflorescence, the branches short and ascending; flowers whitish, on very short pedicels; lower surface of leaves glabrous or pubescent on the veins only. An old world species found on Attu Island and on Seward Peninsula extending toward central Alaska.

3. ZYGADENUS Michx.

Glabrous or obscurely scabrous perennials; bulbs membranous-coated; leaves linear, mainly basal; flowers in terminal racemes, perfect or polygamous; petals and sepals withering-persistent, sometimes adnate to the base of the ovary, bearing one or two glands just above the narrowed base; anthers cordate or peltate, one-celled; capsule three-lobed, three-celled; seed angled. (Greek, yoke and gland, referring to the pair of glands in some species.)

Z. *elegans* Pursh. Glaucus Zygadenus
Z. *chloranthus* Richards
Anticlea elegans (Pursh.) Rydb.

Stems glabrous, 3–6 dm. tall; basal leaves 1–3 dm. long, 5–15 mm. wide, slightly keeled, the few stem leaves shorter; bracts lanceolate, rather large, often purplish; flowers greenish-white; sepals and petals obovate or oval, obtuse, nearly 1 cm. long; gland obcordate; capsule ovoid, about 15 mm. long.

Most of Alaska—Gt. Bear Lake—Sask.—Minn.—Wash. (Fig. 320.)

12. LILIACEAE (Lily Family)

Caulescent or scapose perennials, our species from bulbs; leaves various; flowers solitary or clustered; sepals and petals each three, similar and petaloid, distinct or partly united; stamens six; anthers two-celled; pistil of three united carpels; ovary superior, three-celled; styles united; fruit a loculicidal capsule.

Flowers umbelloid ....1. *Allium*
Flowers usually solitary, white ....2. *Lloydia*
Flowers usually several, dark ....3. *Fritillaria*

1. ALLIUM (Tourn.) L.

Scapose, bulbous plants with characteristic odor; leaves fleshy, usually narrowly linear but sometimes flat and broad, mostly basal; stems simple, erect; flowers in a terminal umbel, subtended by three membranous bracts; petals and sepals free or partly united at the base, one-nerved; stamens adnate to the base of the sepals and petals; styles filiform, usually deciduous; seeds black, one or two in each cell of the capsule. (Latin name of garlic.)

Leaves narrowly linear ....1. *A. sibiricum*
Leaves broad ....2. *A. victoralis*

1. *A. sibiricum* L. Wild Chives

Bulbs small, narrowly ovoid, clustered, membranous-coated; leaves very narrow, 10–35 cm. long; scapes 3–6 dm. tall, bearing a capitate umbel of rose-purple flowers; sepals and petals about 1 cm. long with dark midrib; capsule obtusely three-lobed, about half as long as the perianth. This species is closely related to the garden chives (*A. schoenoprasum* L.) and is sometimes rated as only a variety of it.

Widely distributed in Alaska and Yukon—Newf.—N. Y.—Wyo.—Ore.—Sib. (Fig. 321.)

2. **A. victoralis** L. ssp. **platyphyllum** Hult.

A vigorous form growing up to 75 cm. tall; leaves one or two, the blade thin and flat, elliptical or ovate, up to more than 2 dm. long and nearly 1 dm. wide, the sheath enclosing the base of the scape; flowers white; bulbs with reticulate cover.

An Asiatic form found on Attu Island. (Fig. 322.)

2. LLOYDIA Salisb.

Dwarf herb with tunicated bulb arising from a creeping rootstock; leaves very narrow and grass-like; flowers in our form usually reduced to one; sepals and petals nearly alike; stamens basi-fixed, dehiscent by marginal slits; ovary three-celled. (George Lloyd was an English naturalist.)

L. serotina (L.) Wats.   Alp Lily

Bulbs small, covered with a grayish fibrous coat; stem slender, erect, 5–15 cm. tall; basal leaves 8–15 cm. long, about 1 mm. wide; stem leaves few, 1–4 cm. long, wider; flowers creamy white, about 1 cm. long, purple-veined and tinged with rose on the back; capsule ovoid, many-seeded, about 8 mm. long.

A circumpolar species widely distributed in Alaska. (Fig. 323.)

3. FRITILLARIA (Tourn.) L.

Simple leafy-stemmed plants from bulbs with thick scales; flowers campanulate, large, nodding; sepals and petals nearly equal, deciduous, each with a nectiferous pit at the base; anther attached at the base; style slender, three-cleft; seeds numerous, flat, winged. (Latin, chess-board, from the checkered marking of the perianth of some of the species.)

F. camtchatcensis (L.) Ker.   Indian Rice. Black Lily

Bulb of several larger scales subtended by numerous rice-like bulblets; stems stout, simple, 3–6 dm. tall; leaves mostly in two or three whorls with a few scattered ones near the top, lanceolate, blunt, 5–9 cm. long; flowers one to six, dark-wine color, often almost black, tinged greenish-yellow on the outside, 18–30 mm. long; pod obtusely angled, 2–3 cm. long.

Along the coast, eastern Asia—Aleutian and Pribylof Islands—western Ore. (Fig. 324.)

13. CONVALLARIACEAE (Lily-of-the-valley Family)

Perennials with simple or branched rootstocks; leaves alternate or basal; flowers perfect, in axillary or terminal racemes or panicles or sometimes solitary; sepals and petals two, or more commonly three each, distinct or partly united; stamens four or six; gynoecium superior, of two or three united carpels; ovary two or three celled, styles united; fruit a berry. This group is often included in *Liliaceae*.

1A. Leaves all basal ....1. *Clintonia*
2A. Stem more or less leafy.
   1B. Perianth four-parted ....2. *Maianthemum*

2B. Perianth six-parted.
   1C. Flowers in a terminal raceme or panicle ....3. *Smilacina*
   2C. Flowers axillary.
      1D. Perianth rotate ....4. *Kruhsea*
      2D. Perianth campanulate ....5. *Streptopus*

1. CLINTONIA Raf.

Perennials with creeping rootstocks; leaves basal, broad, many-nerved; flowers borne on scapes; sepals and petals similar and petaloid; stamens six; anthers versatile, ovary two- or three-celled; style slender; berry ovoid or nearly globose. (Named for DeWitt Clinton, governor of New York.)

C. uniflora (Schultz) Kunth.   Blue-bead

Leaves two to four, oblanceolate, more or less villous beneath, 1–2 dm. long, 3–6 cm. wide, acute at both ends; scape shorter than the leaves; flower white, campanulate, the sepals and petals about 2 cm. long; berry about 1 cm. long, five- to 10-seeded. The scape is occasionally two-flowered.

Woods, southeastern Alaska—Mont.—Calif. (Fig. 325.)

2. MAIANTHEMUM Weber.

Low perennials with slender rootstocks; leaves usually two or three, broad, many-nerved; flowers white, small, in a terminal raceme; sepals and petals each two, distinct and spreading; stamens four; anthers versatile; stigmas two; ovary two-celled; fruit a globose berry with one or two seeds. (Greek, May and flower, referring to the season of flowering.) *Unifolium* Haller.

M. dilitatum (Wood) Nels. & Macb.   Deerberry
   M. bifolium DC. var. *kamtschaticum* (Gmel.) Jeps.
   U. dilitatum (Wood) Howell.
   U. eschscholtzianum (Anders. & Bess.) Wight.

Stems 15–40 cm. tall, glabrous; leaves broadly cordate to sagittate, acuminate, 5–15 cm. long, 3–10 cm. wide, or those of the sterile stems up to 15 cm. wide; racemes many-flowered, the pedicels 2–4 mm. long, spreading, often fascicled; sepals and petals 2–3 mm. long, becoming reflexed, style stout; berry spotted, becoming red on drying.

Woods, eastern Asia—Aleutian Islands—southeastern Alaska—Ida.—Calif. (Fig. 326.)

3. SMILACINA Desf.

Perennials with slender, creeping rootstocks; stems leafy; flowers in terminal racemes or panicles; sepals and petals white or greenish-white, distinct or nearly so; stamens six; anthers introrse; ovary three-celled; style short, stigma three-lobed; berry globose. (Name a diminutive of Smilax.) *Vagnera* Adans.

Inflorescence a panicle ....1. *S. racemosa*
Inflorescence a raceme ....2. *S. stellata*

1. *S. racemosa* (L.) Desf.   Wild or False Spikenard
V. racemosa (L.) Morong.

Rootstocks fleshy; stem somewhat angled, 3–10 dm. tall; leaves oblong-lanceolate, sessile or short-petioled, 7–20 cm. long, pubescent below with short, stiff hairs, the margins minutely ciliate; panicle densely many-flowered, 4–10 cm. long; sepals and petals oblong, about 2 mm. long; berries speckled with purple, 4–6 mm. in diameter.

Hyder—N. S.—Geo.—Ariz.—B. C. (Fig. 327.)

2. *S. stellata* (L.) Desf.   Star-flowered Solomon's Seal
V. stellata (L.) Morong.

Stem more or less flexuous above, 3–5 dm. long; leaves sessile, pubescent beneath, lanceolate, many-nerved, 5–15 cm. long, 2–4 cm. wide; racemes 3–7 cm. long, several-flowered; sepals and petals about 6 mm. long; berries 7–10 mm. in diameter.

Cook Inlet district—Mont.—Utah—Calif. (Fig. 328.)

4. KRUHSEA Regel.

A low, glabrous perennial with slender rootstocks; flowers solitary, extra-axillary, the perianth rotate, deeply wine-colored with greenish reflexed tips; stamens six, the filaments very short; ovary three-celled, becoming a bright red, globose berry. (Named for Dr. Kruhse of Siberia.)

K. streptopoides (Ledeb.) Kearney.   Kruhsea
Streptopus streptopoides (Ledeb.) Frye & Rigg.

Stem simple, 3–15 cm. tall; leaves four to eight, sessile, ovate-lanceolate, acute, 25–50 mm. long; flowers one to five, on recurved pedicels scarcely 1 cm. long; sepals and petals 2–2.5 mm. long.

Woods, eastern Asia, central and southeastern Alaska—Ida.—Wash. (Fig. 329.)

5. STREPTOPUS Michx.

Leafy perennials; leaves thin, many-nerved, sessile or clasping; flowers usually solitary, extra-axillary; peduncles slender, twisted or bent above the middle; sepals and petals nearly alike, recurved, deciduous, petals keeled; stamens six; anthers sagittate, extrorse; ovary three-celled, stigma three-lobed; berry globose or oval, many-seeded. (Greek, twisted-stalk, referring to the peduncles.)

Leaves clasping, flowers greenish ....1. *S. amplexifolius*
Leaves sessile, flowers pinkish ....2. *S. roseus*

1. *S. amplexifolius* (L.) DC.   Cucumber-root. Clasping Twisted-stalk

Rootstock short, stout, horizontal, with thick fibrous roots; stems usually branched, 3–10 dm. tall; leaves ovate-lanceolate, acuminate, strongly clasping, glaucous beneath, 5–12 cm. long; flowers campanulate, sepals and petals 8–12 mm. long, attenuate; berry ellipsoid, 10–15 mm. long, yellowish-white to light red.

Most of our area south of the Arctic Circle; circumboreal. (Fig. 330.)

2. *S. roseus* Michx. ssp. *curvipes* (Vail) Hult.

Simple-stemmed Twisted-Stalk

Rootstocks slender with fibrous roots; stem usually simple, 1-4 dm. tall; leaves sessile, oblong-lanceolate, acuminate, 4-10 cm. long, the margins more or less short-ciliate; sepals and petals pinkish, 5-7 mm. long, glandular-pubescent on the inner surface; peduncles glandular, not geniculate; fruit red, globose, 7-9 mm. in diameter.

Woods, southeastern Alaska and northwestern B. C.—Wash. Other forms in eastern U. S. and Canada. (Fig. 331.)

14. IRIDACEAE (Iris Family)

Perennials arising from bulbs or commonly from rootstocks; leaves equitant, often grass-like; flowers bracted, perfect, regular or nearly so, sepals and petals three each, often dissimilar but both colored; stamens three, opposite the sepals; gynoecium of three-united carpels with three-celled, inferior ovary and three distinct styles; fruit a loculicidally dehiscent capsule.

Flowers more than 5 cm. wide ....1. *Iris*
Flowers less than 25 mm. wide ....2. *Sisyrinchium*

1. IRIS (Tourn.) L.

Perennials with sword-shaped or linear leaves; flowers large and showy in few-flowered terminal clusters; perianth in ours blue, the sepals spreading or recurved, the petals smaller and erect or ascending, the tube prolonged beyond the ovary; the three styles are petal-like and arch over the stamens; capsule elongated, three or six angled with numerous seeds. (Greek, rainbow, referring to the colored flowers.)

I. setosa Pall.   Wild Iris, Flag
I. arctica Eastw.

Stems densely tufted, 35-70 cm. tall; basal leaves linear-lanceolate, 2-5 dm. long, 5-12 mm. wide; stem leaves two or three, the stem usually with one branch; perianth blue varying to lavender and purple, the sepals 5-6 cm. long, copiously veined, petals cuspidate; style branches large, crested; capsule oblong, 3-4 cm. long. Var. *platyrincha* Hult. has the petals broad, flat, dilated below and constricted in the middle. Ssp. *interior* (E. Anderson) Hult. is the form found in the interior. It differs from the coast form in having narrower and less arched leaves, and in more scarious and somewhat violet-colored bracts shorter than the peduncles.

Siberia—Alaska—Labr.—Newf.—Maine. (Fig. 332.)

2. SISYRINCHIUM L.

Grass-like, tufted perennials, usually with winged stems; flowers usually blue, borne in a few-flowered terminal umbel subtended by a pair of erect green bracts; perianth-tube short or none; sepals and petals similar, spreading, aristulate; capsule globose or ovoid, three-valved, dehiscent. (Name used by Theophrastus for a plant allied to Iris.)

S. littorale Greene.   Blue-eyed Grass

Stems 2–4 dm. tall, prominently winged; leaves 8–25 cm. long, less than 5 mm. wide; umbels three to six flowered, sepals and petals about 11 mm. long; capsules subglobose, 6–8 mm. long, thick-walled.

Wet soil. Pacific Coast Districts of Alaska and B. C. (Fig. 333.)

15. ORCHIDACEAE (Orchid Family)

Perennials with corms, bulbs, or tuberous roots; leaves simple, entire, sheathing, often reduced to scales; flowers irregular, perfect, bracted, solitary or in racemes or spike; sepals three, similar or nearly so, the lower two sometimes united; petals three, the lateral ones alike, the third (lip) differing, usually very markedly so, often spurred, usually inferior by the twisting of the ovary; functional stamens two in Cypripedium, one in the other genera, adnate to the pistil and forming a column, two-celled and containing two or more waxy or powdery pollinia which are usually stalked and attached at the base to a viscid gland; style often terminating in a beak (rostellum) at the base of the anther or between its sacs; ovary inferior, usually long and twisted, three-angled, one-celled with three parietal placentae and numerous ovules; seeds vary numerous and minute. A very large and interesting family of plants. Many tropical species are epiphytes, attached to the limbs of trees but not parasitic, deriving nutriment from the air and moisture alone, being assisted in this by the symbiotic fungi living in the roots, corms or bulbs. The flowers are so constructed that they are dependent on insects for pollination, the head of the insect when reaching for the nectar comes in contact with the bases of the pollinia which adhere to its head, and when visiting the next flower the pollinia adhere to the sticky stigma and the process is repeated.

1A. Fertile stamens two, lip saccate .... 1. *Cypripedium*
2A. Fertile stamen one.
   1B. Pollinia caudate at the base, attached to a viscid gland.
      1C. Gland enclosed in a pouch-like fold, lip three-lobed .... 2. *Orchis*
      2C. Gland surrounded by a thin membrane, lip toothed at the apex .... 3. *Coeloglossum*
      3C. Gland naked, lip entire.
         1D. Sepals one-nerved, plants with corms ...... 7. *Piperia*
         2D. Sepals three to five nerved, plants with rootstocks.
            1E. Stem leafy .... 6. *Limnorchis*
            2E. Stems scapiform, leaves one or two, basal.
                1F. Leaves two, ovary straight .... 4. *Lysias*
                2F. Leaf one, ovary curved .... 5. *Lysiella*
   2B. Pollinia not caudate at the base.
      1C. Pollinia granulose or powdery.
         1D. Leaves basal, usually white-reticulate ...... 10. *Peramium*
         2D. Leaves green, borne on stem.
            1E. Leaves alternate, spike twisted .... 8. *Spiranthes*
            2E. Leaves two, opposite .... 9. *Listera*

  2C. Pollinia smooth or waxy.
    1D. Plants with coralloid roots ....13. *Corallorrhiza*
    2D. Plants with corms.
      1E. Lip flat, flowers in a raceme ....11. *Malaxis*
      2E. Lip saccate, flowers usually one ....12. *Calypso*

1. CYPRIPEDIUM L.

  Glandular-pubescent herbs with coarse, fibrous roots; leaves large and broad, somewhat plaited, many-nerved; flowers in ours solitary, showy; sepals spreading or two of them united; lip a large inflated sac; column incurved, concealed by a recurving petaloid sterile stamen; capsule ribbed. (Greek, Venus and shoe.)
1A. Leaves two, lip pink ....1. *C. guttatum*
2A. Leaves several.
  1B. Flowers large, sepals longer than lip ....2. *C. montanum*
  2B. Flowers small, sepals shorter than lip ....3. *C. passerinum*

1. *C. guttatum* Sw.   Pink Ladies' Slipper

  Rootstock horizontal; the two-leaved stems 12–24 cm. tall; leaves lanceolate, 6–11 cm. long; lip 18–25 mm. long, pink spotted purplish and veined, longer than the sepals; capsule glandular-pubescent, strongly ribbed, reflexed.
  Aleutian Islands—Yukon Valley—Great Bear Lake—Great Slave Lake and Eurasia. (Fig. 334.)

2. *C. montanum* Dougl.   Mountain Ladies' Slipper

  Stems 3–7 dm. tall; leaves ovate to broadly lanceolate, 8–16 cm. long, 4–8 cm. wide, abruptly acuminate; sepals lanceolate, 4–6 cm. long, the petals similar but narrower; lip oblong, 2–3 cm. long, white veined with purple; capsule erect or nearly so, 2–3 cm. long.
  Glacier Bay, the Stickine River, and the Lewes River in Yukon—B. C.—Sask—Wyo.—Calif.

3. *C. passerinum* Richards.   Northern Ladies' Slipper

  Stems 1–3 dm. tall, often retrosely villous; leaves oval or lanceolate, 6–15 cm. long; lip about 15 mm. long, whitish with purplish spots inside; sepals 10–15 mm. long, the lower slightly two-cleft; capsule upright.
  Woods, Seward Peninsula—Yukon valley—Alta.—Ont. (Fig. 335.)

2. ORCHIS (Tourn.) L.

  Sepals distinct, spreading; petals narrower than the sepals; lip spurred; anther of two pollen masses, slightly diverging, prolonged into a slender stalk attached to a small gland which is enclosed in a pouch. (Ancient name.)
Stem leafy ....1. *O. aristata*
Stem scapose with one basal leaf ....2. *O. rotundifolia*

1. *O. aristata* Fisch.   Rose-purple Orchis

  Stems 12–40 cm. tall, from thick, fusiform, forked tubers; leaves obovate to lanceolate, up to 12 cm. long and 25 mm. wide; lower bracts leaf-like; flowers several to many, rose-purple, in a spike-like raceme 4–10 cm. long; lip very broad, the middle lobe acute; sepals and petals acuminate to aristate; spur large.

Eastern Asia—Aleutian Islands—Nunivak Island—Pr. William Sound. (Fig. 336.)

2. O. rotundifolia Pursh.   Round-leaved Orchis

Stem slender, 8–16 cm. tall, arising from a rootstock with fibrous roots; leaf solitary, near the base, oval to orbicular, 2–6 cm. long; spike two to six flowered; flowers 12–15 mm. long, subtended by bracts, sepals elliptic, pink, 6–7 mm. long, petals narrower; lip white, purple-spotted, the large middle lobe notched at the apex; spur slender, curved, shorter than the lip.

Central Alaska—Greenl.—Maine—N. Y.—Minn.—Alta.—B. C. (Fig. 337.)

3. COELOGLOSSUM Hartm.

Tubers two to three cleft; leaves alternate; flowers in terminal spikes; sepals distinct, converging, forming a hood; lateral petals narrow, erect; lip obtuse, two or three toothed at the apex, prolonged below into a sac-like spur. Included by some authors in Habenaria. (Latin, Heaven-tongue.)

C. viride (L.) Hartm.   Long-bracted Orchid
   H. viridis var. interjecta Fern.

Leaves ovate, obovate or lanceolate; bracts linear-lanceolate, the lower ones usually about twice as long as the flowers; sepals ovate-lanceolate, petals narrow; lip 5–8 mm. long, oblong or somewhat cuneate. The typical form growing only 6–15 cm. tall with two or three stem-leaves is found on Seward Peninsula and Central Alaska. Much more frequent is the ssp. bracteatum (Muhl.) Hult. (C. bracteatum (Muhl.) Parl. (H. bracteata (Muhl.) R. Br.) found in the Pacific Coast districts. It grows 1–4 dm. tall and has three to five stem leaves.

The species is circumboreal. (Fig. 338.)

4. LYSIAS Salisb.

Plants with fleshy rootstocks or tubers; leaves two, near the base, broad; flowers in a terminal spike, greenish or white; sepals distinct, large, spreading, the upper one broadly cordate, the lateral ones obliquely ovate, lateral petals small and narrow; lip entire, narrow, prolonged at the base into a slender spur. Included in Habenaria by some authors, in Platanthera by others. (Lysias was an Attic orator.)

L. orbiculata (Pursh.) Rydb.   Large Round-leaved Orchid
   H. orbiculata (Pursh.) Torr.
   P. orbiculata (Pursh.) Lindl.

Scapes 3–4 dm. tall with one or two lanceolate bracts above the basal leaves; leaves 7-12 cm. long, 4-8 cm. wide; sepals 7-8 mm. long; lip linear, 8-10 mm. long; spur 15-20 mm. long; anther-sacs large and prominent.

Rare, southeastern Alaska—Newf.—S. Car.—Ill.—Wash. (Fig. 339.)

5. LYSIELLA Rydb.

Small plants with scapiform stems; leaf solitary, basal; flowers greenish-yellow; upper sepal round-ovate, erect, surrounding the column, lateral sepals reflexed-spreading; lip entire, linear-lanceolate, deflexed; spur slightly curved, shorter than the ovary; capsule ovoid. (Diminutive of *Lysias*.) Often included in Habenaria or Platanthera.

L. obtusata (Pursh.) Rydb.   Small Northern Bog Orchid
  H. obtusata (Pursh.) Rich.
  P. obtusata (Pursh.) Rich.

Stem slender, glabrous, 8-25 cm. tall; leaf 5-10 cm. long; flowers four to twelve, about 1 cm. long; spur slender, about as long as the lip and nearly as long as the ovary.

Widely distributed in Alaska—Labr.—Newf.—N. Y.—Colo.—B. C. (Fig. 340.)

6. LIMNORCHIS Rydb.

Leafy-stemmed plants with fusiform, root-like tubers; flowers small, greenish or white, borne in a terminal spike; upper sepal ovate to suborbicular, erect, three to seven nerved; lateral sepals linear to ovate-lanceolate, usually three nerved; lateral petals erect, usually lanceolate and three nerved; lip entire, reflexed, from linear to rhombic lanceolate, or orbicular in one species; column short and thick; anther-sacs parallel. This group is included in Habenaria by some authors and in Platanthera by others. Some forms are hard to place and there has been much confusion regarding the species. It seems evident that several of the species hybridize readily resulting in much natural variation. No two writers seem to agree on the limits of the species and the same name has been used in literature for different forms. (Greek, marsh and orchid.)

1A. Floral parts about 1 mm. long .... 1. *L. chorisiana*
2A. Floral parts much longer.
  1B. Spur at least twice as long as the lip .... 2. *L. behringiana*
  2B. Spur one-half to one and one-half as long as the lip.
    1C. Lip linear.
      1D. Spike rather dense .... 3. *L. convallariaefolia*
      2D. Spike lax .... 4. *L. stricta*
    2C. Lip distinctly broader at the base.
      1D. Lip obtusely triangular .... 5. *L. hyperborea*
      2D. Lip linear with dilated base .... 6. *L. dilitata*

1. *L. chorisiana* (Cham.) new comb.   Choriso Bog-orchid
  H. chorisiana (Cham.)
  P. chorisiana (Cham.) Rchb.

Stems 10-15 cm. tall; tuber elongated-fusiform, leaves two at or near the base, 25–40 mm. long, 8–20 mm. wide, with usually a bract-like leaf on the stem above; bracts lanceolate, longer than the flowers; flowers very small, the parts scarcely more than 1 mm. long; capsule about 5 mm. long.

Not common, eastern Asia, Aleutian Islands, and on Douglas Island. (Fig. 341.)

2. *L. behringiana* Rydb.   Bering Bog-orchid
   P. behringiana (Rydb.) Tatew. & Kobay.

Stems 10–15 cm. tall; tubers elongate fusiform; main leaf about 5 cm. long, 15–20 mm. wide, and usually two lanceolate smaller ones; spike 3–4 cm. long; bracts linear-lanceolate, the lowest about twice as long as the flowers; flowers purplish; lip about 5 mm. long; spur fully 1 cm. long. Hultén considers this as only a dwarf form of *Platanthera tipuloides* (L.f.) Lindl. of Asia.

In America known only from Attu Island.

3. *L. convallariaefolia* (Fisch.) Rydb.

Stems 2–6 dm. tall; tuber fusiform, moderately elongated; leaves 4–6 cm. long, 10–22 mm. wide, the lower obtuse, the upper acute; spikes 5–12 cm. long; flowers greenish or sometimes whitish; lip linear; spur about equaling the lip, linear or clavate. This species seems to hybridize with *L. dilitata* and *L. hyperborea*. Var. *dilitatoides* is probably such a hybrid. It is a rather robust plant with whitish flowers and the lip dilated at the base.

Eastern Asia—Aleutian Islands—Cook Inlet. (Fig. 342.)

4. *L. stricta* (Lindl.) Rydb.   Slender Bog-orchid
   H. saccata Greene.
   P. stricta Lindl.

Stems 2–10 dm. tall; lower leaves lanceolate, obtuse, upper leaves smaller, acute; spikes 1–3 dm. long, lax; bracts linear-lanceolate, the lower much longer than the flowers; flowers greenish, upper sepal erect, ovate, 4–5 mm. long; spur saccate, shorter than the lip.

Common in the Pacific coastal districts of Alaska—Alta.—Colo.—Calif. (Fig. 343.)

5. *L. hyperborea* (L.) Rydb.   Northern Bog-orchid
   H. hyperborea (L.) R. Br.
   P. hyperborea (L.) Lindl.

Stem 15–50 cm. tall; lower leaves oblanceolate, obtuse, upper leaves lanceolate and acute; spike rather dense, 4–10 cm. long; flowers light green, upper sepal ovate, 3–4 mm. long; lip triangular-ovate, obtuse, 3–5 mm. long; spur clavate, curved, about equaling the lip.

Central Alaska—Labr.—Iceland—Pa.—Colo.—S. C. (Fig. 344.)

6. *L. dilitata* (Pursh.) Rydb.   White Bog-orchid
   H. dilitata (Pursh.) Hook.
   P. dilitata (Pursh.) Lindl.

Stems 2–8 dm. tall; tubers elongated-fusiform; lower leaves lanceolate, often obtuse, the upper narrower and acute, 8–20 mm. wide; flowers white; lip linear with distinctly dilated base, the margins papillose; spur linear, about the same length as the lip. The above description applies to the more typical forms. Var. *angustifolia* Hook. (*L. leptoceratatis* Rydb.) Leaves not over 8 mm. wide; tuber slender and elongated; inflorescence rather lax; spur one to two times as long as the lip. Var. *chlorantha* Hult. Base of lip dilated, the margins sparsely papillose; flowers greenish; leaves wide. Var. *leucostachys* (Lindl.) Ames. More robust; inflorescence dense; flowers white, very fragrant; tuber thicker and less elongated; spur fully one-half longer than the lip. This is the form known as Wild Hyacinth.

Pacific Coast districts of Alaska; across the continent in some of its forms. (Fig. 345.)

7. PIPERIA Rydb.

Stems arising from spherical or broadly ellipsoid tubers; leaves few, near the base, usually withering at or shortly after anthesis; flowers small, spicate; upper sepal erect, the lateral spreading; lateral petals free, oblique; lip linear-lanceolate to ovate, concave, united with the base of the lower sepals; anther-sacs parallel. Often included in Habenaria or Platanthera. (Charles V. Piper was a botanist of Washington State and the U. S. Dept. of Agriculture.)

P. unalaschensis (Spreng.) Rydb.   Alaska Piperia
   H. unalaschensis (Spreng.) Wats.
   Platanthera unalaschensis (Spreng.) F. Kurtz.

Stem slender, 3–5 dm. tall; basal leaves oblanceolate, the largest 10–15 cm. long, stem leaves bract-like; spike lax, 1–3 dm. long; flowers greenish, 8–14 mm. long; sepals and petals 2–4 mm. long, upper sepal ovate; lip oblong, obtuse; spur narrow, rather longer than the lip.

Unalaska, southeastern Alaska—Que.—Colo.—Calif. (Fig. 346.)

8. SPIRANTHES L. C. Richards

Herbs with tuberous-thickened or fleshy-fibrous roots; leaves alternate or mostly basal; flowers in twisted spikes, white or cream-colored, small, spurless, sepals and petals in ours more or less united or connivant into a hood; lip concave, small, dilated at the reflexed apex; column oblique, arched; pollinia two; stigma with a beak. (Name from the spiral arrangement of the flowers.) (*Ibidium* Salisb.) (*Gyrosrachys* Pers.)

S. romanzoffiana Cham. & Schleicht.   Hooded Ladies' Tresses
   I. romanzoffianum (C. & S.) House.

Stems 7–30 cm. tall; lower leaves linear to lanceolate, 6–15 cm. long; spikes dense, 4–10 cm. long, the flowers in three spiral rows; bracts often longer than the flowers; lip oblong, broad at the base, contracted above the dilated, cusped apex.

Wet soil, Unalaska—Labr.—Newf.—N. Y.—Colo.—Calif., and in Ireland. (Fig. 347.)

9. LISTERA R. Br.

Slender woodland plants; rootstocks with fleshy-fibrous roots; leaves two, opposite, near the middle of the stem; flowers small, greenish or purplish, spurless, in terminal racemes; sepals and lateral petals similar, spreading or reflexed; lip longer than the sepals; pollinia two, united to a minute gland; capsule ovoid or obovoid. (Martin Lister was an English naturalist.) (*Ophrys* (Tourn.) L.)

1A. Lip narrow, deeply cleft ....4. *L. cordata*
2A. Lip broad, slightly cleft or notched at apex.
   1B. Lip with auricles ....1. *L. borealis*
   2B. Lip without auricles but with small teeth at base.
      1C. Ovary glandular ....2. *L. convallariodes*
      2C. Ovary glabrous ....3. *L. caurina*

1. *L. borealis* Morong.   Northern Twayblade
   O. borealis (Morong) Rydb.

Stems 6–15 cm. tall; leaves 10–35 mm. long, rather firm, elliptic ovate, obtuse, borne above the middle of the stem; flowers two to six, 10–12 mm. long; lip 7–8 mm. long, oblong-cuneate, the lobes at the apex obtuse and without mucro; column 3–4 mm. long.

Central Alaska—Mack.—Colo. (Fig. 348.)

2. *L. convallarioides* (Sw.) Torr.   Broad-leaved Twayblade
   O. convallarioides (Sw.) Rydb.

Stems 10–25 cm. tall, glandular-pubescent above the leaves; leaves broadly oval or suborbicular, obtuse or very short-cuspidate, 3–6 cm. long; flowers greenish-yellow on short, slender, bracted pedicels; sepals linear-lanceolate; lip broadly cuneate, 7–10 mm. long with two obtuse lobes at the apex and a mucro between; ovary glandular and pubescent.

Woods, Aleutian Islands and B. C.—Newf.—Mass.—N. Mex. (Fig. 349.)

3. *L. caurina* Piper   Western Twayblade
   O. caurina (Piper) Rydb.

Stems 1–3 dm. tall; leaves short-elliptic to ovate, 3–7 cm. long; lip 4–7 mm. long, cuneate, retuse with a blunt mucro in the sinus; ovary glabrous. Resembles *L. convallarioides* in appearance but is a more slender plant, has narrower leaves, longer pedicels and smaller flowers. The pedicels are two to four times as long as the bracts. This is the common species in southeastern Alaska.

Alaska—Mont.—Ore. (Fig. 350.)

4. *L. cordata* (L.) R. Br.   Heart-leaved Twayblade
   O. cordata L.

Stem slender and delicate, 1–2 dm. tall, glabrous except just above the leaves; leaves cordate-reniform, mucronate, 15–35 mm. long and about as wide; racemes four to twenty flowered, bracts minute, pedicels about 2 mm. long; flowers greenish or purplish, sepals and petals about 2 mm.

long; lip narrow, 4–5 mm. long, the segments setaceous; capsule ovoid, about 4 mm. long. Our western form has broader leaves than the type and if regarded as separate is the var. *nephrophylla* (Rydb.) Hult. (*L. nephrophylla* Rydb.) (*O. nephrophylla* Rydb.).

Woods, common, Pacific coastal districts of Alaska; circumboreal. (Fig. 351.)

10. PERAMIUM Salisb.

Herbs with creeping rootstocks and fleshy-fibrous roots; leaves basal, variegated, evergreen, strongly reticulated; flowers white or cream-colored, in one-sided racemes on scape-like, bracted stems; lateral sepals distinct, the upper united with the lateral petals; lip concave or saccate, roundish ovate with reflexed tip; anther with two pollinia attached to a small disc; inflorescence glandular. (Greek, referring to the pouch-like lip.) (*Goodyera* R. Br.)

Lip concave, the margins involute .... 1. *P. decipiens*
Lip saccate, the margins revolute .... 2. *P. repens*

1. *P. decipiens* (Hook.) Piper.   Menzies Rattlesnake Plantain
   P. menziesii (Lindl.) Morong.
   G. decipiens (Hook.) Hubbard.

Scape rather stout, 2–4 dm. tall, glandular-pubescent; leaves ovate-lanceolate, 4–8 cm. long, acute at both ends, usually whitish along the veins; perianth 7–9 mm. long; anther ovate, long-pointed.

Woods, southeastern Alaska—Que.—N. Hamp.—Minn.—Ariz.—Calif. (Fig. 352.)

2. *P. repens* (L.) Salisb. var. *ophioides* Fern.
                                                    Lesser Rattlesnake Plantain
   G. repens (L.) R. Br.

Appearing like a miniature of *P. decipiens*. Scapes 1–2 dm. tall; leaves ovate with whitish blotches, 10–25 mm. long, tapering into a sheathing petiole; perianth greenish-white, scarcely 4 mm. long; anther blunt; lip with a narrow recurved or spreading apex; column short.

Woods, central Alaska—Labr.—Newf.—N. Car.—N. Mex.—the whole species circumboreal. (Fig. 353.)

11. MALAXIS Soland.

Perennials with corms; leaves one to four, on lower part of stem; flowers small, in terminal spike-like racemes, whitish or greenish; sepals spreading, distinct; lip embracing the column; anther two-celled with 4 pollinia, without tail or glands. (Greek, in allusion to the soft tissue.) (*Microstylis* Nutt.) (*Acroanthes* Raf.)

Plants small, leaves two or four .... 1. *M. paludosa*
Plants larger, leaves one or two .... 2. *M. monophylla*

1. *M. paludosa* (L.) Sw.   Little Adder's Mouth

Stems 3–12 cm. tall; leaves ovate, 6–20 mm. long; flowers about 6 mm.

long; lip about 3.5 mm. long, narrow, slightly tapering toward the rounded apex; pedicels ascending; capsule 3-4 mm. long.

Muskeags, southeastern Alaska and scattered circumboreal stations. (Fig. 354.)

2. **M. monophylla** (L.) Sw. White Adder's Tongue
Microstylis monophyllos (L.) Lindl.

Stems glabrous, striate, 10-25 cm. tall; main leaf one, the blade 3-8 cm. long, 1-3 cm. wide, with usually a second leaf that may vary from scale-like to nearly as large as the main leaf; sepals and lip about 2 mm. long; lip 1.5 mm. wide; capsule 4-5 mm. long, 3-4 mm. wide.

Unalaska eastward along the coast; circumboreal. (Fig. 355.)

12. CALYPSO Salisb.

Low, scapose, one-flowered plants; corm superficial; leaf solitary, basal, flower terminal, showy; sepals and lateral petals similar, spreading or ascending, oblong-lanceolate, pinkish, lip large, saccate, spotted brown-purple, hairy within, with two short spurs near the apex; column winged, petal-like, with the anther just below the summit; pollinia two in each sac. (Greek god, Calypso, whose name signifies concealment.) *Cytherea* Salisb.

C. bulbosa (L.) Rchb. Calypso
Cytherea bulbosa (L.) House.

Scape 5-15 cm. tall; leaf ovate, 25-50 mm. long, 15-30 mm. wide, with a subcordate base; flowers variegated purple, pink, yellow; petals, sepals and lip 15-20 mm. long.

Mossy woods, rather rare except on some small islands, central Alaska south and east, nearly circumboreal. (Fig. 356.)

13. CORALLORRHIZA R. Br.

Brownish, purplish, or yellowish saprophytic plants; rootstocks coral-like masses from which the bracted, scape-like stems arise; flowers in terminal racemes; sepals nearly equal, the lateral ones united with the foot of the column and often forming a short spur, partly or wholly adnate to the top of the ovary; lip one to three ridged; anther terminal, lid-like, with four waxy pollinia. (Greek, meaning coral and root.)

Sepals and petals one-nerved .... 1. *C. trifida*
Sepals and petals three-nerved .... 2. *C. mertensiana*

1. **C. trifida** Chatelin. Early Coral-root
C. innata R. Br.
C. corallorrhiza (L.) Karst.

Stems slender, glabrous, 1-3 dm. tall with two or three sheathing bracts; flowers greenish-yellow or greenish-brown; sepals and petals about 5 mm. long; lip shorter than the petals, whitish, two-lobed; spur a small protuberance adnate to the top of the ovary; capsule reflexed, 8-12 mm. long.

Circumboreal; in our area from Kotzebue Sound south and east. (Fig. 357.)

2. *C. mertensiana* Bong.   Mertens Coral-root

Stems purple or brownish-purple, with two or three sheathing bracts, 2–5 dm. tall; flowers ten to twenty-five, sepals nearly 1 cm. long, the petals slightly shorter; lip with two sharp teeth at about the middle, crenulate, somewhat spotted, about same length as the sepals, narrowed at the base; column about 7 mm. long, spur free from the ovary at the apex; capsule 15–18 mm. long, narrowed at the base, reflexed.

Coniferous forest, southeastern Alaska—Mont.—Wyo.—Calif. (Fig. 358.)

PLATE XI

Fig.
192. *Rynchospora alba* (L.) Vahl. Inflorescence, scale and achene.
193. *Eleocharis acicularis* (L.) R. & S. Spike, achene and style.
194. *Eleocharis kamtschatica* (C. A. Mey.) Kom. Spike and achene.
195. *Eleocharis uniglumis* (Link.) Schult. Spike and achene.
196. *Eleocharis palustris* (L.) R. & S. Spike and achene.
197. *Eriophorum alpinum* L. Spike and achene.
198. *Eriophorum scheuchzeri* Hoppe. Achene.
199. *Eriophorum chamissonis* C. A. Mey. Achene.
200. *Eriophorum vaginatum* L. Achene and upper sheath.
201. *Eriophorum brachyantherum* Trautv. Achene and upper sheath.
202. *Eriophorum gracile* Koch. Achene.
203. *Eriophorum angustifolium* Roth. Achene.
204. *Scirpus caespitosus* L. var. *callosus* Bigel. Spike and achene.
205. *Scirpus americanus* Pers. Inflorescence, cross section of stem, scale and achene.
206. *Scirpus pacificus* Britt. Inflorescence, scale and achene.
207. *Scirpus rufus* (Huds.) Schrad. Inflorescence, spike and achene.
208. *Scirpus validus* Vahl. Cross section of stem, portion of inflorescence, achene and style.
209. *Scirpus microcarpus* Presl. Achene, portion of inflorescence and scale.
210. *Kobresia myosuroides* (Vill.) Fiori & Paol. Inflorescence and achene.
211. *Carex nardina* Fr. Achene and scale.
212. *Carex jacobi-peteri* Hult. Scale, achene and perigynium.
213. *Carex capitata* L. Achene, scale and perigynium.

FLORA OF ALASKA

PLATE XI

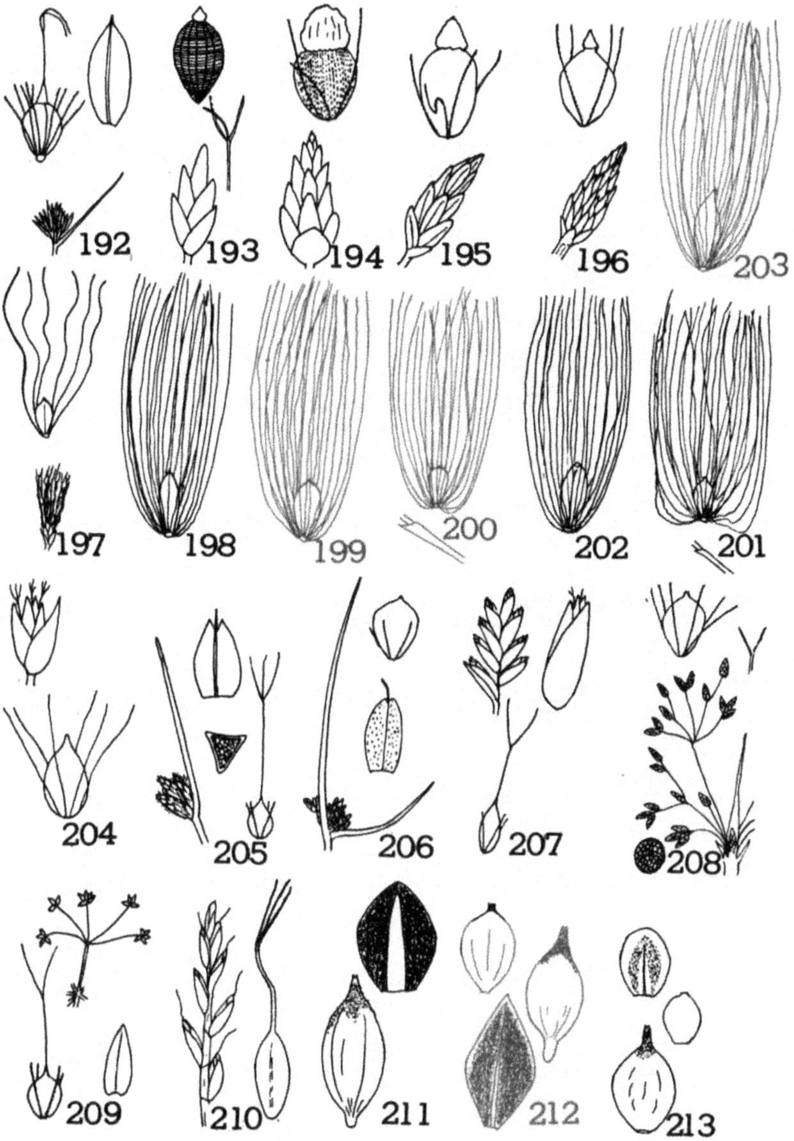

PLATE XII.

All these illustrations except fig. 216 and 218 show scale, perigynium and achene.

Fig.
214. Carex gynocrates Wormskj.
215. Carex scirpoidea Michx.
216. Carex anthoxanthea Presl.
217. Carex circinata C. A. Mey.
218. Carex leptalea Wahl. Top of spike, scale and perigynium.
219. Carex obtusata Lilj.
220. Carex pyrenaica Wahl.
221. Carex nigricans C. A. Mey.
222. Carex pauciflora Lightf.
223. Carex microglochin Wahl.
224. Carex maritima Gunner.
225. Carex chordorrhiza Ehrh.
226. Carex stipata Muhl.
227. Carex diandra Schk.
228. Carex athrostachya Olney.
229. Carex macloviana d'Urv. ssp. pachystachya (Cham.) Hult.
230. Carex praticola Rydb.
231. Carex crawfordii Fern.
232. Carex aenea Fern.
233. Carex lachenalii Schk.
234. Carex pribylovensis Macoun.
235. Carex glareosa Wahl.
236. Carex mackenziei Kretch.
237. Carex canescens L.
238. Carex brunnescens (Pers.) Poir.
239. Carex disperma Dewey.
240. Carex tenuiflora Wahl.
241. Carex loliacea L.
242. Carex stellulata Good.
243. Carex phyllomanica W. Boott.

FLORA OF ALASKA

PLATE XII.

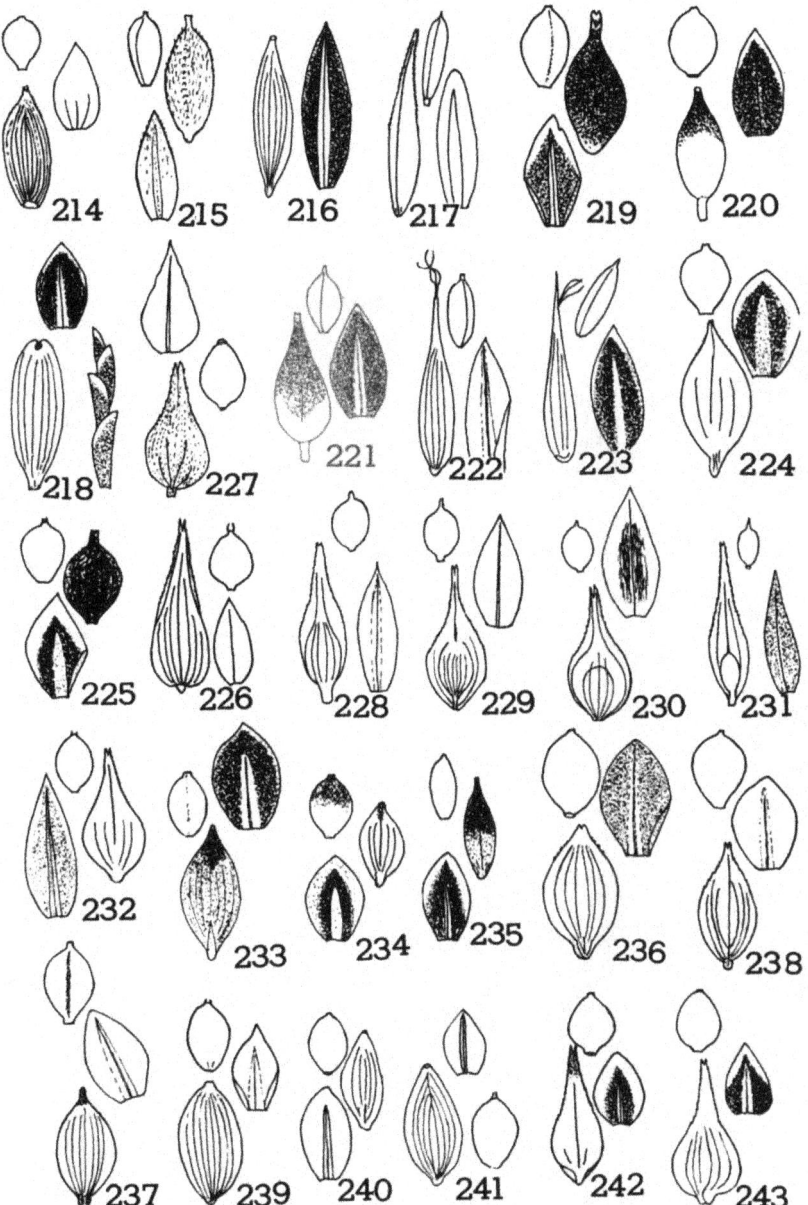

PLATE XIII

All illustrations of Carex show scale, perigynium and usually the achene.

Fig.
244. *Carex laeviculmis* Meinsh.
245. *Carex bicolor* All.
246. *Carex aurea* Nutt.
247. *Carex garberi* Fern. ssp. *bifaria* Fern.
248. *Carex bigelowii* Torr.
249. *Carex lugens* Holm.
250. *Carex kelloggii* W. Boott.
251. *Carex hindsii* C. B. Clarke.
252. *Carex aquatilis* Wahl.
253. *Carex sitchensis* Prescott.
254. *Carex subspathacea* Wormskj.
255. *Carex ramenskii* Kom.
256. *Carex lyngbyei* Hornem. ssp. *cryptocarpa* (C. A. Mey.) Hult.
257. *Carex norvegica* Retz. ssp. *inferalpina* (Wahl.) Hult.
258. *Carex buxbaumii* Wahl.
259. *Carex stylosa* C. A. Mey.
260. *Carex gmelini* Hook. & Arn.
261. *Carex atrata* L.
262. *Carex mertensii* Prescott.
263. *Carex macrochaeta* C. A. Mey.
264. *Carex montanensis* Bailey.
265. *Carex spectabilis* Dewey.
266. *Carex nesophila* Holm.
267. *Carex podocarpa* R. Br.
268. *Carex deflexa* Hornem.
269. *Carex rossii* Boott.
270. *Carex supina* Willd. ssp. *spaniocarpa* (Steud.) Hult.
271. *Carex concinna* R. Br.
272. *Carex glacialis* Mack.
273. *Carex rariflora* (Wahl.) J. E. Smith.

FLORA OF ALASKA

PLATE XIII

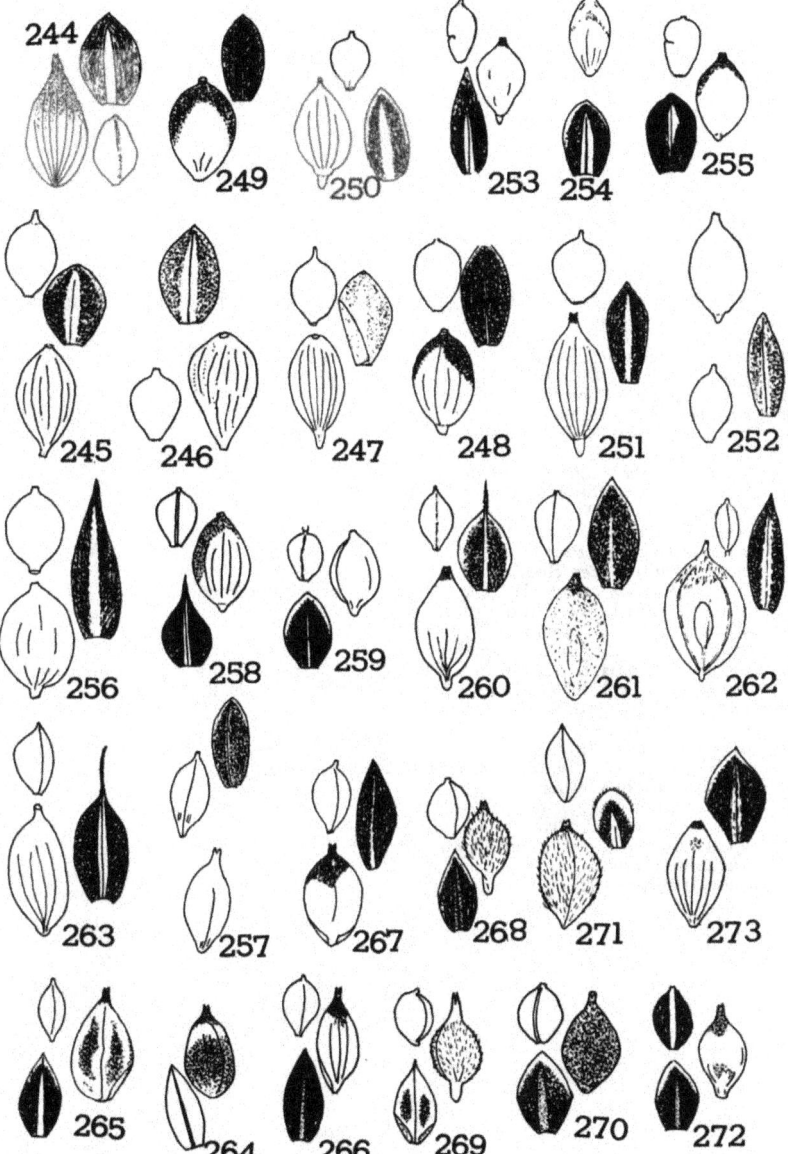

PLATE XIV.

Fig.
274. *Carex pluriflora* Hult. All the illustrations of Carex show a scale, perigynium and usually the achene.
275. *Carex limosa* L.
276. *Carex magellanica* Lam.
277. *Carex livida* (Wahl.) Willd.
278. *Carex vaginata* Tausch.
279. *Carex atrofusca* Schk.
280. *Carex misandra* R. Br.
281. *Carex capillaris* L.
282. *Carex viridula* Michx.
283. *Carex rostrata* Stokes.
284. *Carex rotundata* Wahl.
285. *Carex rhyncophysa* C. A. Mey.
286. *Carex physocarpa* Presl.
287. *Carex membranacea* Hook.
288. *Lysichitun americanum* Hult. & St. J. Leaf and inflorescence.
289. *Calla palustris* L. Leaf and fruiting spathe.
290. *Lemna trisulca* L. Group of fronds.
291. *Lemna minor* L. Floating frond.
292. *Juncus filiformis* L. All illustrations of Juncus show the capsule inclosed in the perianth with basal bracts when present and the seed.
293. *Juncus drummondii* E. Mey.
294. *Juncus effusus* L.
295. *Juncus arcticus* Willd.
296. *Juncus balticus* Willd. ssp. *sitchensis* (Engelm.) Hult.
297. *Juncus ensifolius* Wiks.
298. *Juncus bufonius* L.
299. *Juncus macer* S. F. Gray.
300. *Juncus stygius* L. ssp. *americanus* (Buch.) Hult.
301. *Juncus biglumis* L.
302. *Juncus triglumis* L.
303. *Juncus mertensianus* Bong.

FLORA OF ALASKA

PLATE XIV.

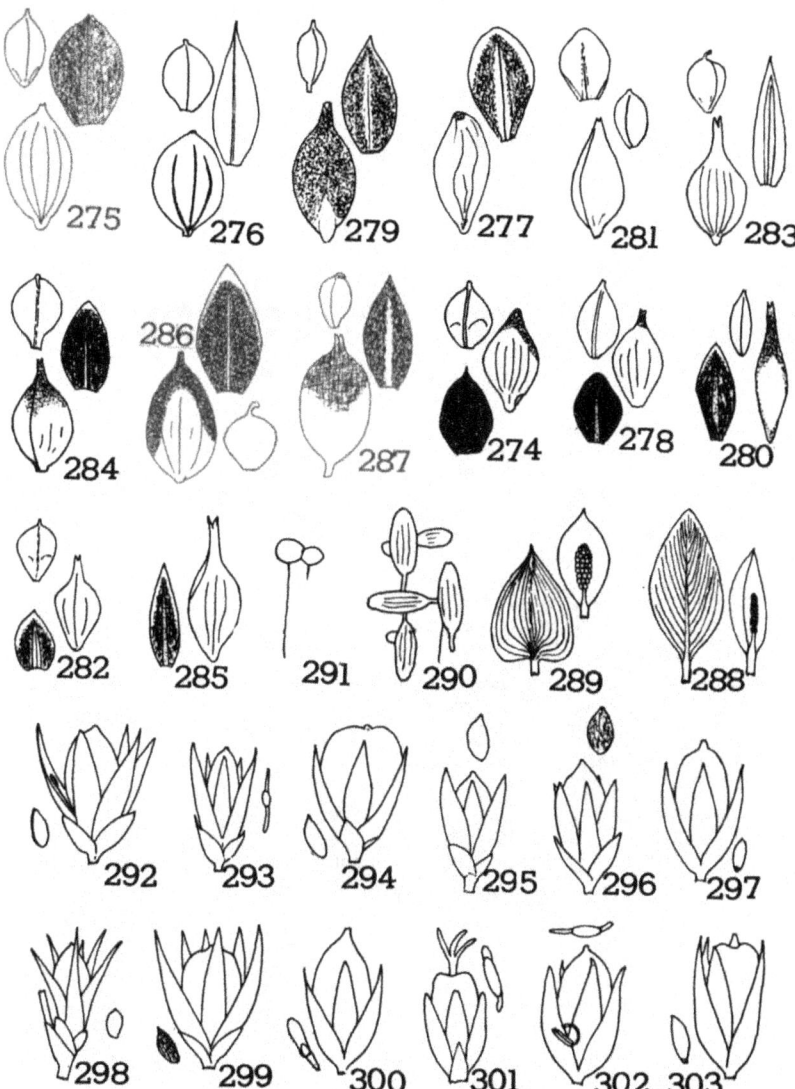

PLATE XV.

Fig.
304. *Juncus alpinus* Vill. ssp. *nodulosus* (Wahl.) Lindm. All illustrations of Juncus and Luzula show capsule inclosed in the perianth with bractlets and a seed.
305. *Juncus nodosus* L.
306. *Juncus falcatus* E. Mey. ssp. *sitchensis* (Buch.) Hult.
307. *Juncus castaneus* J. E. Smith.
308. *Luzula rufescens* Fisch.
309. *Luzula wahlenbergii* Rupr.
310. *Luzula parviflora* (Ehrh.) Desv.
311. *Luzula spicata* (L.) DC.
312. *Luzula arcuata* Wahl.
313. *Luzula hyperborea* R. Br.
314. *Luzula nivalis* (Laest.) Beurl.
315. *Luzula multiflora* (Retz.) Lej.
316. *Tofieldia occidentalis* S. Wats. Flower and leaf.
317. *Tofieldia pusilla* (Michx.) Pers. Flower and leaf.
318. *Tofieldia coccinea* Richards. Flower and leaf.
319. *Veratrum eschscholtzii* A. Gray. Leaves, flower and capsule.
320. *Zygadenus elegans* Pursh. Flower, petal and capsule.
321. *Allium sibiricum* L. Flower and leaf.
322. *Allium victoralis* L. ssp. *platyphyllum* Hult. Leaf.
323. *Lloydia serotina* (L.) Wats. Leaf, capsule and flower.
324. *Fritillaria camtschatcensis* (L.) Ker. Bulb, flower and leaf.
325. *Clintonia uniflora* (Schult.) Kunth. Flower, leaf and berry.
326. *Maianthemum dilitatum* (Wood) Nels. & Macb. Leaf and flower.
327. *Smilicina racemosa* (L.) Desf. Flower and leaf.
328. *Smilicina stellata* (L.) Desf. Flower and leaf.
329. *Kruhsea streptopoides* (Ledeb.) Kearney. Flower, leaf and petal.
330. *Streptopus amplexifolius* (L.) DC. Leaf, flower and petal.
331. *Streptopus roseus* Michx. ssp. *curvipes* (Vail) Hult. Leaf, petal and flower.

PLATE XV.

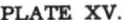

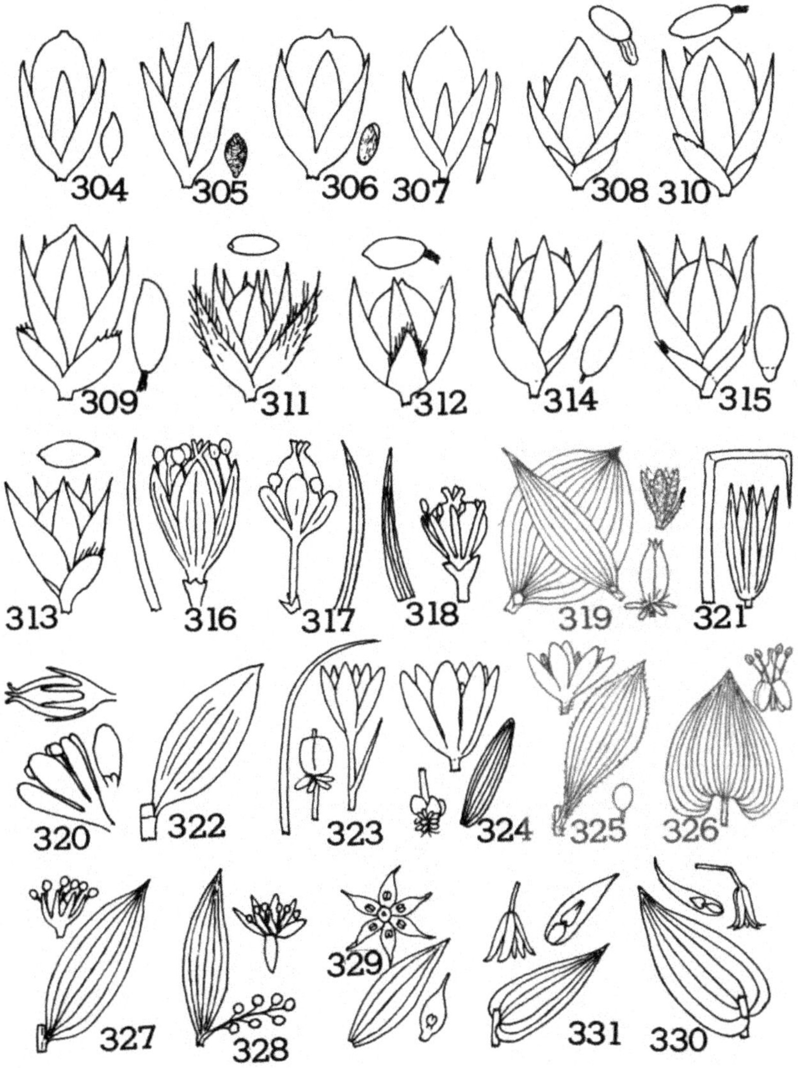
304
305
306
307
308
310
309
311
312
314
315
313
316
317
318
319
321
320
322
323
324
325
326
327
328
329
331
330

PLATE XVI.

Fig.
332. *Iris setosa* Pall. Flower and capsule.
333. *Sisyrinchium littorale* Greene. Flower and cluster of capsules with bracts.
334. *Cypripedium guttatum* Sw. Side view of flower.
335. *Cypripedium passerinum* Richards. Capsule.
336. *Orchis aristata* Fisch. Side view of flower.
337. *Orchis rotundifolia* Pursh. Front view of flower.
338. *Coeloglossum viride* (L.) Hartm. Flower and bract.
339. *Lysias orbiculata* (Pursh) Rydb. Side view of flower, petals hidden.
340. *Lysiella obtusata* (Pursh.) Rydb. Front view of flower.
341. *Limnorchis chorisiana* (Cham.) J. P. Anderson. Front view of flower and spur.
342. *Limnorchis convallariaefolia* (Fisch.) Rydb. Front view of flower and detached spur.
343. *Limnorchis stricta* (Lindl.) Rydb. Front view of flower with detached spur.
344. *Limnorchis hyperborea* (L.) Rydb. Flower, as above.
345. *Limnorchis dilitata* (Pursh.) Rydb. var. *leucostachys* (Lindl.) Ames. Front view of flower.
346. *Piperia unalaschensis* (Spreng.) Rydb. Flower.
347. *Spiranthes romanzoffiana* Cham. & Schleicht. Side view of flower.
348. *Listera borealis* Morong. Floral parts.
349. *Listera convallarioides* (Sw.) Torr. Flower with bract.
350. *Listera caurina* Piper. Flower with bract.
351. *Listera cordata* (L.) R. Br. Parts of flower in position.
352. *Peramium decipiens* (Hook.) Piper. Side view of flower.
353. *Peramium repens* (L.) Salisb. var. *ophioides* Fern. Side view of flower with one sepal removed.
354. *Malaxis paludosa* (L.) Sw. Front view of flower.
355. *Malaxis monophylla* (L.) Sw. Front view of flower.
356. *Calypso bulbosa* (L.) Rchb. Side view of flower.
357. *Corallorrhiza trifida* Chatelin. Flower and lip.
358. *Corallorrhiza mertensiana* Bong. Floral parts.

FLORA OF ALASKA

PLATE XVI.

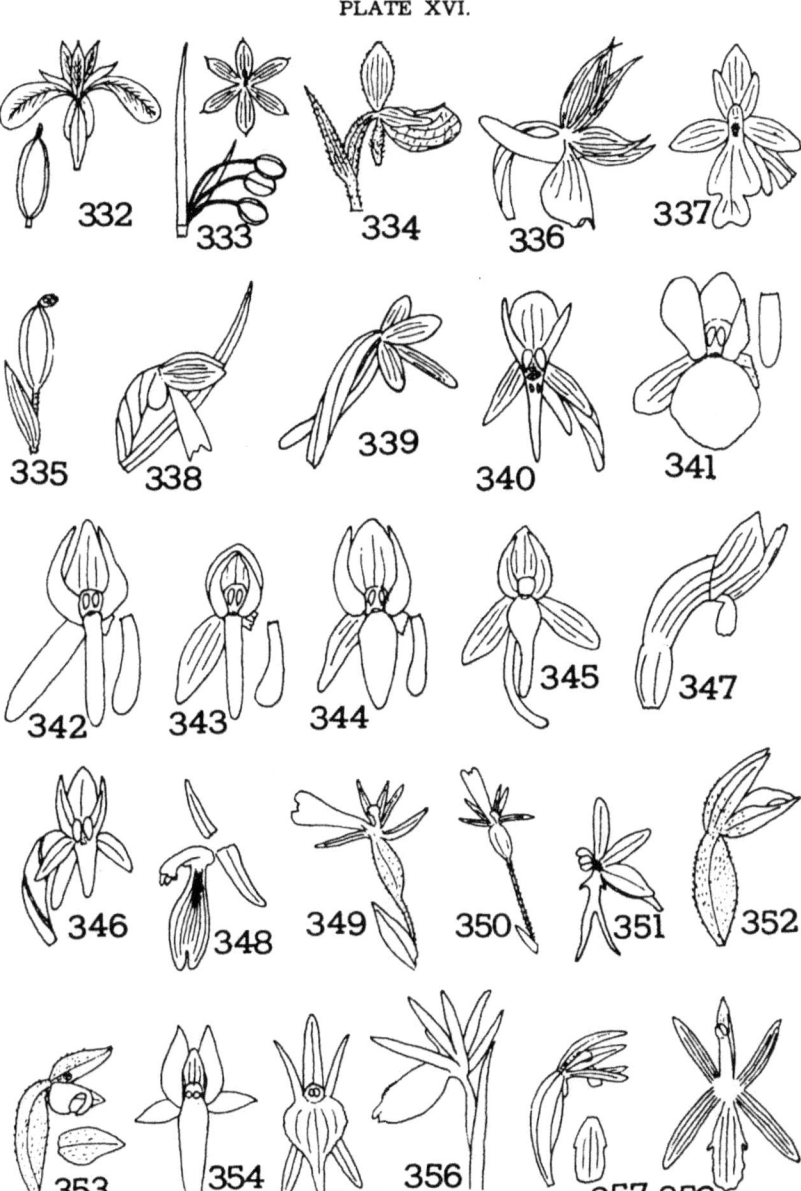

Subclass 2. *Dicotyledoneae*

1A. Woody plants; trees, shrubs or subshrubs.
  1B. Flowers without petals.
    1C. Flowers in aments.
      1D. Fruit a capsule, seed with a coma.................... 1. *Salicaceae*   P.184
      2D. Fruit a nutlet or drupaceous.
        1E. Staminate aments erect or ascending............... 2. *Myricaceae*   P.185
        2E. Staminate aments drooping........................ 3. *Betulaceae*   P.186
    2C. Flowers not in catkins.
      1D. Trailing shrub with small heather-like leaves.........27. *Empetraceae*   P.344
      2D. Upright shrubs with scaly leaves....................30. *Elaeagnaceae*   P.348

  2B. Flowers with petals.
    1C. Petals separate.
      1D. Ovary superior (see also Ledum in 38. *Ericaceae*).
        1E. Carpels usually 5 or more, sometimes enclosed in a
          fleshy receptacle. ....21. *Rosaceae*   P.293
        2E. Carpels 2, fruit winged...........................28. *Aceraceae*   P.345
      2D. Ovary inferior.
        1E. Fruit of papery or stony carpels enclosed in a
          fleshy pome. ....21. *Rosaceae*   P.293
        2E. Fruit a small-seeded berry.
          1F. Leaves palmately veined, carpels 2...............20. *Grossulariaceae* P.291
          2F. Leaves very small, carpels 4 *Oxycoccus* in.......39. *Vacciniaceae*   P.372
        3E. Fruit a drupe.
          1F. Petals 5, styles 2.................................33. *Araliaceae*   P.354
          2F. Petals 4, styles 1................................35. *Cornaceae*   P.361
    2C. Petals united.
      1D. Ovary superior.
        1E. Stamens inserted at the sinuses of the corolla........40. *Diapensiaceae*   P.374
        2E. Stamens inserted at the base of the corolla.........38. *Ericaceae*   P.365
      2D. Ovary inferior.
        1E. Flowers in small dense heads. *Artemisia* in......... 59. *Carduaceae*   P.466
        2E. Flowers not in heads.
          1F. Stamens 5 (or 4)................................54. *Caprifoliaceae*   P.431
          2F. Stamens 10. ....39. *Vacciniaceae*   P.372

2A. Herbaceous plants.
  1B. Flowers without petals.
    1C. Ovary superior.
      1D. Pistil of 1 or several and distinct carpels, each with solitary style and stigma.
        1E. Carpels solitary, flowers clustered.................. 4. *Urticaceae*   P.189
        2E. Carpels several or numerous....................... 12. *Ranunculaceae*   P.227
      2D. Pistil of 2 or more united carpels, stigmas or styles 2 or more, ovary 1-celled, 1-ovuled.
        See also *Sanguisorba* in .... 21. *Rosaceae*   P.293
        1E. Leaves with sheathing stipules................... 7. *Polygonaceae*   P.191
        2E. Leaves without sheathing stipules................. 8. *Chenopodiaceae*   P.199
      3D. Carpels 2 or more, ovary 1—several-celled, several—many-seeded.
        1E. Aquatic plants, styles 2. .... 26. *Callitrichaceae*   P.343
        2E. Plants not aquatic.
          1F. Carpels 2, stamens 2, 4 or 6. *Lepidium* in......... 16. *Brassicaceae*   P.245
          2F. Carpels 5, fleshy seashore plant. *Glaux* in........ 41. *Primulaceae*   P.374
    2C. Ovary inferior.
      1D. Parasitic on trees, without green leaves............... 5. *Loranthaceae*   P.189
      2D. Not parasitic, green leaves present.
        1E. Aquatic, or growing in wet places.................. 32. *Haloragidaceae*   P.352
        2E. Not aquatic.
          1F. Fruit a berry...................................... 6. *Santalaceae*   P.190
          2F. Fruit a capsule, low, small-leaved plants.
            Chrysosplenium in .... 19. *Saxifragaceae*   P.280
  2B. Petals present.
    1C. Petals distinct.
      1D. Carpels solitary or several and distinct or united only at the base.
        1E. Stamens inserted on main axis of flower.
          1F. Leaves peltate with glutinous covering. .... 11. *Cabombaceae*   P.227
          2F. Leaves not glutinous. .... 12. *Ranunculaceae*   P.227
        2E. Stamens inserted on an hypogynous disc.
          1F. Corolla irregular (bilateral)..................... 22. *Fabaceae*   P.311
          2F. Corolla regular (radial).
            1G. Stamens more than 10. .... 21. *Rosaceae*   P.293
            2G. Stamens 4–10.
              1H. Pistils usually of 2 carpels. .... 19. *Saxifragaceae*   P.280
              2H. Pistils of 4 or 5 carpels........................ 18. *Crassulaceae*   P.279
      2D. Carpels 2 or more and united.
        1E. Ovary superior.
          1F. Stamens numerous.
            1G. Calyx deciduous. .... 14. *Papaveraceae*   P.243
            2G. Calyx persistent. .... 13. *Nymphaeaceae*   P.242
          2F. Stamens not more than twice the number of petals.
            1G. Sepals 2.
              1H. Corolla regular. .... 9. *Portulaceae*   P.203
              2H. Corolla irregular. .... 15. *Fumariaceae*   P.244
            2G. Sepals 4 or 5.
              1H. Sepals and petals 4, stamens 6. .... 16. *Brassicaceae*   P.245
              2H. Stamens of same number or twice as many as sepals and petals.
                1I. Ovary 1-celled.
                  1J. Ovary 1-ovuled. .... 42. *Plumbaginaceae*   P.381
                  2J. Ovules more than 1.
                    1K. Placentae basal or central.............. 10. *Caryophyllaceae*   P.206
                    2K. Placentae parietal.
                      1L. Staminodia present. *Parnassia* in..... 19. *Saxifragaceae*   P.280
                    2L. Staminodia absent.

```
          1M. Stigmas 2-cleft, insectivorous plants
                with glandular-hispid leaves.......17. Droseraceae      P.268
          2M. Stigmas entire, corolla irregular....29. Violaceae        P.345
    2I. Ovary several-celled.
       1J. Stamens with wholly or partly united filaments.
          1K. Styles united around a central column,
                separating at maturity..............23. Geraniaceae     P.341
          2K. Filaments united at the base, each sinus
                with a staminodium. ...............24. Linaceae         P.343
       2J. Stamens with distinct filaments.
          1K. Anthers united, flowers irregular........25. Balsaminaceae  P.343
          2K. Anthers distinct, flowers regular.
             1L. Saprophytes without green leaves. ...37. Monotropaceae  P.364
             2L. Plants with green leaves and
                   rootstocks......................36. Pyrolaceae       P.362
2E. Ovary inferior.
   1F. Styles distinct.
      1G. Aquatic plants. ...............................32. Haloragidaceae  P.352
      2G. Not aquatic but often growing in wet places.
         1H. Fruit a berry-like drupe. ...................35. Cornaceae    P.361
         2H. Fruit dry, of 2 separating carpels............34. Ammiaceae   P.354
   2F. Styles united ....................................31. Onagraceae   P.348
2C. Petals more or less united.
   1D. Ovary superior.
      1E. Stamens free from the corolla.
         1F. Carpel 1, corolla irregular. .......................22. Fabaceae     P.311
         2F. Carpels 2 or more, united.
            1G. Filaments united. ...........................15. Fumariaceae  P.244
            2G. Filaments separate.
               1H. Saprophytes without green leaves. ..........37. Monotropaceae  P.364
               2H. Plants with green leaves, petals united
                     only at the base. .......................36. Pyrolaceae  P.362
      2E. Stamens adnate to the corolla.
         1F. Stamens opposite the lobes of the corolla.........41. Primulaceae  P.374
         2F. Stamens as many as the lobes of the corolla and
               alternate with them or fewer.
            1G. Corolla scarious. .............................52. Plantaginaceae  P.428
            2G. Corolla not scarious.
               1H. Carpels distinct except sometimes at the
                     apex. ....................................44. Apocynaceae  P.397
               2H. Carpels united.
                  1I. Ovary 1-celled with parietal placentae......43. Gentianaceae  P.393
                  2I. Ovary 2-4-celled or falsely 4-celled by
                        intrusion of placentae.
                     1J. Stamens 5.
                        1K. Carpels 3. ..........................45. Polemoniaceae  P.398
                        2K. Carpels 2.
                           1L. Fruit of 1-4 nutlets, ovary usually
                                 4-lobed. .......................47. Boraginaceae  P.402
                           2L. Fruit capsular.
                              1M. Corolla regular. .................46. Hydrophyllaceae  P.400
                              2M. Corolla irregular. ................49. Scrophulariaceae  P.412
                     2J. Stamens 4 and didymous or 2 or 1.
                        1K. Carpels ripening into 2 or 4 nutlets.....48. Lamiaceae  P.408
                        2K. Carpels ripening into a capsule.
                           1L. Placentae of ovary parietal, root
                                 parasites without chlorophyll.....51. Orobanchaceae  P.427
                           2L. Placentae of ovary axile.
```

    1M. Ovary usually 2-celled, land plants .49. *Scrophulariaceae* P. 412
    2M. Ovary usually 1-celled, bog plants. .50. *Lentibulariaceae* P. 426
  2D. Ovary inferior.
   1E. Stamens with filaments free from the corolla.........57. *Campanulaceae* P. 434
   2E. Stamens adnate to the corolla.
    1F. Ovary with 2 or more fertile cavities and 2-many ovules.
     1G. Stamens as many as the corolla lobes. ....53. *Rubiaceae*   P. 430
     2G. Stamens twice as many as the corolla lobes......55. *Adoxaceae*   P. 433
    2F. Ovary with 1 fertile cavity, calyx often modified.
     1G. Flowers not in heads.
      1H. Stamens 1-3. ....56. *Valerianaceae*   P. 433
      2H. Stamens 4 or 5. ....54. *Caprifoliaceae*   P. 431
     2G. Flowers in involucrate heads.
      1H. Corollas all ligulate (strap-shaped). . .58. *Cichoriaceae*   P. 447
      2H. Corollas tubular, only the ray-flowers with
        strap-shaped corollas. ....59. *Asteraceae*   P. 466

1. SALICACEAE (Willow Family)

  Dioecious trees and shrubs; leaves simple, alternate, stipitate; flowers in aments with solitary flowers in the axis of scale-like bracts; aments expanding before or with the leaves, the staminate ones often pendulous; stamens 1-many; pistils of 2-4 carpels, united to form a 1-celled ovary with 2-4 parietal placentae; fruit an ovoid, oblong or conic, 2-4-valved capsule with numerous minute seeds provided with a dense coma of white silky hairs.

Bractlets incised, stamens many............................ 1. *Populus*
Bractlets entire or denticulate, stamens few................. 2. *Salix*

1. POPULUS (Tourn.) L.

  Trees with soft wood; buds scaly and resinous; twigs terete or angled; leaves usually petioled, those on young, vigorous sprouts larger and more pointed than those on mature branches; both kinds of flowers in drooping aments; flowers from a cup-shaped disc subtended by a fringed bract; stamens 4-60, ovary sessile; stigmas 2-4; capsules 2-4-valved; coma long and copious. (The ancient Latin name.)

1A. Leaf-blades on mature branches usually less than 5 cm.
  long. .... 1. *P. tremuloides*
2A. Leaf-blades on mature branches usually more than 5 cm. long.
  1B. Pistils bicarpellary. .... 2. *P. tacamahacca*
  2B. Pistils tricarpellary. .... 3. *P. tricocarpa*

1. *P. tremuloides* Michx.   American Aspen. Quaking Aspen.

  A slender tree with light green or whitish bark; leaves glabrous with ciliolate margins when young, crenate-serrate with small incurved teeth, short-acuminate at apex, rounded to subcordate at the base, 25-50 mm. long and nearly as wide; petioles slender; fruiting aments up to 1 dm. long; capsule conical, narrow, warty. Often found in dense, almost pure stands, especially after forest fires.

  Most of Alaska—the Atlantic—northern Mexico in the mountains. Fig. 359.

2. *P. tacamahacca* Mill.   Balsam Poplar
P. balsamifera auct. not L.

Medium to large tree; bark of the branches light brown or gray; leaves ovate to ovate-lanceolate, shining above, pale beneath, acute or acuminate at the apex, cuneate or rounded at the base, crenulate, 6–10 cm. long or up to 20 cm. on young sprouts; fruiting aments 5–12 cm. long; capsule 2-valved, short-pedicelled. Rarely hybridizes with *P. tremuloides*. Reports of *P. candicans* from Alaska are based on forms of this species with wide leaves and light-colored bark.

Mostly in the interior in our area, Bering Sea—Labr.—N. Y.—Wyo.—Ore. Fig. 360.

3. *P. tricocarpa* Torr. & Gray.   Black Cottonwood

Our largest deciduous tree; branches pubescent; leaves broadly ovate to ovate-lanceolate, finely crenate-serrate, acute or acuminate at apex, cordate to rounded at the base, 6–12 cm. long, pale beneath; aments 5–12 cm. long, or in fruit up to 20 cm. long; capsules 3-valved.

Lowlands of the Pacific coast in our area, Alaska—Ida.—Calif. Fig. 361.

2. SALIX (Tourn.) L.

The text for the genus *Salix*, the willows, was to have been prepared by Dr. Carleton R. Ball, who is recognized as the leading authority on American willows. Up to the time of handing in the manuscript for this part of the Flora of Alaska and adjacent parts of Canada his treatment of the genus had not been received. It is hoped that it can be published later as a supplement.

2. MYRICACEAE L. (Bayberry Family)

Monoecious or dioecious shrubs or trees; leaves alternate, simple, usually coriaceous; flowers without perianth, borne in the axils of the bracts in erect or ascending aments; ovary 1-celled with a straight ovule and subtended by 2–8 bractlets; fruit a small oblong drupe or nut, its exocarp often waxy.

MYRICA L.

Our species is a deciduous shrub 5–15 dm. tall; leaves resinous-dotted; staminate aments oblong or cylindric, expanding with or before the leaves; stamens 4–8, with short filaments; pistillate aments ovoid or subglobose; ovary subtended by 2–4 bractlets; fruit resinous. (Ancient name of the Tamarisk.)

M. gale L.   Sweet Gale.

Leaves oblanceolate, obtuse and toothed at the apex, cuneate at the subsessile base, more or less puberulent beneath, 2–6 mm. long, 5–20 mm. broad; aments in fruit 6–10 mm. long, about 4 mm. thick; nutlets waxy-coated, of about same length as the 2 persistent bractlets which clasp it on each side and are adnate to the base. The form in eastern Asia and western America has leaves widest near the apex and under surface of the leaves more tomentose than the European form and if considered as distinct is var. *tomentosa* C.DC.

This species is circumboreal. Fig. 362.

3. BETULACEAE (Birch Family)

Monoecious trees and shrubs; leaves alternate, petioled, simple; flowers in aments, the staminate drooping; staminate flowers 1–3 in the axil of each bract, the calyx often wanting; stamens 2–10; pistillate flowers in ours without perianth, the 2 or 3 pistils at the base of each bract; fruit in ours a 1-celled, 1-seeded, usually winged nutlet.

Bracts of the fruiting aments thin, deciduous with the nutlet.. 1. *Betula*
Bracts of fruiting aments woody, persistent. . . . . 2. *Alnus*

1. BETULA (Tourn.) L.

Shrubs and trees with aromatic bark and scaly buds; leaves dentate or serrate; staminate flowers usually 3 in the axils of the bracts with a 4-toothed perianth; stamens divided, each fork bearing an anther-sac; pistillate bracts 3-lobed; fruit a compressed nutlet winged on both sides. The different species seem to hybridize freely and a large proportion of the birches in our region are probably hybrids. (The Latin name.)

1A. Low, spreading shrubs with rounded leaf-tips.
  1B. Leaves cuneate at the base, longer than wide. . . . . . 1. *B. glandulosa*
  2B. Leaves truncate or cordate at the base, often wider than
    long. . . . . 2. *B. nana exilis*
2A. Trees, leaf-tips acute.
  1B. Leaves ovate, double serrate. . . . . 3. *B. papyrifera occidentalis*
  2B. Leaves with truncate or cuneate base.
    1C. Leaves with prolonged apex. . . . . 5. *B. resinifera*
    2C. Leaves without prolonged apex. . . . . 4. *B. kenaica*

1. *B. glandulosa* Michx.   Glandular Scrub Birch

A shrub 5–15 dm. tall; twigs densely glandular and covered with a thin waxy layer; leaves 1–2 cm. long, longer than broad, the base toothless and cuneate, the apex rounded and crenate-dentate; petioles pubescent; fruiting aments 8–16 mm. long, 4–5 mm. thick, usually erect; bracts with a resiniferous hump on back, the central lobe not much longer than the divergent lateral ones; nutlet with very narrow wings.

Interior Alaska—Labr.—southern Greenl.—Maine—Colo.—Calif. Fig. 363.

2. *B. nana* L. ssp. *exilis* (Sukatch.) Hult.   Dwarf Alpine Birch
   B. glandulosa var. *sibirica* auct.

Resembles *B. glandulosa* but somewhat more dwarf; twigs less resiniferous and more pubescent; leaves reniform or orbicular, often broader than long; bracts of the fruiting aments without resiniferous hump on back; wings of the nutlets narrow but broader than in *B. glandulosa*.

The species is circumpolar, the ssp. nearly throughout Alaska, eastern Asia—Greenl. Fig. 364.

3. **B. papyrifera** Marsh. ssp. *occidentalis* (Hook.) Hult.
Western Paper Birch

A tree with white or brown exfoliating bark; young twigs pubescent and glandular, becoming smooth and orange-brown; leaves ovate, acute or acuminate; subcordate or subcuneate at the base, doubly serrate; petioles pubescent or puberulent; fruiting aments 25–40 mm. long, about 1 cm. thick; bracts with a long, narrow median lobe; wings wider than the nutlets.

Southeastern Alaska—northwestern Mont.—Wash., the typical form east of the Rocky Mts.—Newf.—Penn. Fig. 365.

4. **B. kenaica** W. H. Evans. Kenai Birch
   B. papyrifera var. kenaica (Evans) A. Henry.

A small- to medium-sized tree; bark exfoliating, grayish-white to dark brown; leaves ovate, acute to acuminate, broadly cuneate or rounded at the base, sharply and often doubly serrate, glandular-dotted beneath, usually more or less hairy on the upper surface; lobes of the bracts rounded, nearly equal in length; wings about as wide as the nutlets.

Central Alaska—Bering Sea—Alaska Penin. Fig. 366.

5. **B. resinifera** Britt. Alaska Birch
   B. alaskana Sarg. not Lesq.
   B. neoalaskana Sarg.

A forest tree of moderate size; bark exfoliating, white or rarely reddish or brownish, twigs brown, coated with a thin layer of wax; leaves ovate-rhombic, serrate, acute to long acuminate at the apex, sharply to widely cuneate at the base, 3–6 cm. long; fruiting aments 25–45 mm. long; bracts about 6 mm. long with ciliolate margins; wings of the nutlets as broad or broader than the body. The common White Birch of interior Alaska and Yukon.

Bering Sea—Mackenzie delta—Sask. Fig. 367.

The following hybrids have been recognized showing characters intermediate between the parent species and showing great variation.
Betula glandulosa × *nana exilis*
Betula glandulosa × *resinifera* (B. eastwoodae Sarg.). Figs. 368, 369.
Betula kenaica × *nana exilis* (B. hornei Butler). Fig. 370.
Betula kenaica × *resinifera*
Betula nana exilis × *resinifera* (B. beeniana A. Nels.).

2. ALNUS (Tourn.) L.

Shrubs or trees with astringent bark; leaves dentate or serrate; staminate flowers 3 in the axil of each bract in the pendulous aments, the perianth 3–5-parted; stamens 3–5, with simple filaments; pistillate aments erect, ovoid or ellipsoid in fruit, cone-like; pistillate flowers without perianth but with 1 or 2 minute bractlets. (Ancient Latin name.)

1A. Nutlets margined but without membranous wings. ...... 3. *A. incana*
2A. Nutlets with narrow wings. .... 4. *A. oregona*
3A. Nutlets with broad wings.
   1B. Peduncles pubescent. .... 1. *A. crispa*
   2B. Peduncles glandular but not pubescent. .... 2. *A. fruticosa*

1. *A. crispa* (Ait.) Pursh.   Green Alder
   A. alnobetula (Ehrh.) K. Koch.

A shrub 1–4 dm. tall; leaves oval or ovate, acute or obtuse at the apex, sharply and irregularly serrulate, glabrous above, usually more or less pubescent on the veins beneath, 4–8 cm. long; fruiting aments slender-peduncled, 10–15 mm. long, less than 1 cm. thick; nutlets elliptic, 2.5–3 mm. long; wings about as broad as the nutlet but variable and irregular.

Bering Sea eastward in our area, circumboreal. Fig. 371.

2. *A. fruticosa* Rupr.   Alaska Alder

A shrub or small tree, usually more or less decumbent and spreading; leaves broadly ovate, obtuse or short-acuminate, sharply and irregularly or doubly serrate, 6–12 cm. long; fruiting aments 12–20 mm. long, nearly 1 cm. thick; nutlets oval, about 3 mm. long. Related to *A. crispa* and by some regarded as only a variety or subspecies. Var. *sinuata* (Regel) Hult. (*A. sitchensis* Sarg.) is a more upright form that sometimes reaches tree size with trunk diameter of 15–20 cm.; the leaves are narrower and more sinuate.

Bering Str.—Alaska Range—Mont.—Ore. Figs. 372, 373.

3. *A. incana* (L.) Moench.   Mountain Alder
   A. tenuifolia Nutt.

A large shrub or small tree up to 10 m. tall and a trunk diameter of 22 cm.; leaves ovate or oval, shallowly lobed, acute or obtuse at the apex, rounded at the base, dentate with blunt teeth, 4–10 cm. long; fruiting aments 8–15 mm. long, less than 1 cm. thick; nutlets with a narrow border but without membranous wings, about 3 mm. long.

Western Alaska—Newf.—Penn.—N. Mex.—northern Calif., also Eu. & western Asia. Fig. 374.

4. *A. oregona* Nutt.   Red Alder. Oregon Alder
   A. rubra Bong. not Marsh.

A medium to large tree with gray bark; leaves ovate, rounded at the base, acute at the apex, doubly dentate with glandular, blunt teeth, tomentose beneath when young, 7–12 cm. long; fruiting aments 12–24 mm. long, about 1 cm. thick; nutlets ovate, about 3 mm. long, with narrow wings.

Yakutat Bay along the coast to northern Calif. Fig. 375.

4. URTICACEAE (Nettle Family)

Herbs; leaves simple, with stipules; flowers dioecious, monoecious, or polygamous, greenish, borne in axillary paniculate cymes; sepals 2-5, distinct or partly united; stamens 2-5, in pistillate flowers reduced to staminodia or lacking; pistil solitary, becoming a 1-seeded achene.

URTICA (Tourn.) L.

Ours are dioecious perennials; leaves opposite, toothed, 5-7-veined; flowers in spike-like, paniculate cymes; sepals 4, nearly distinct, in pistillate flowers the 2 outer smaller and spreading; staminate flowers with 4 stamens; stigmas sessile, tufted. (Latin, to burn, in allusion to the stinging hairs.)

Leaves wide, with cordate base. .... 1. *U. lyallii*
Leaves narrower, lanceolate to ovate. .... 2. *U. gracilis*

1. *U. lyallii* Wats.   Lyall Nettle

Stems 1-2 m. tall, sparingly bristly or nearly glabrous; leaves ovate, usually cordate at the base, more or less bristly above and on the veins beneath, coarsely and sharply serrate, acute or acuminate, 4-15 cm. long, 3-10 cm. wide; staminate flower clusters longer than the petioles but pistillate clusters often shorter than the petioles; sepals much shorter than the achene.

Near the coast, eastern Alaska—Ore. Fig. 376.

2. *U. gracilis* Ait.   Slender Nettle

Stems rather slender, 6-25 dm. tall; leaves sharply and deeply serrate, long-acuminate, narrowed or rounded at the base, 5-12 cm. long, 1-4 cm. wide; flower clusters slender, longer than the petioles, shorter than the leaves, hirsute; sepals nearly equaling the achene.

Western Alaska—Newf.—Conn.—N. Mex.—Ore. Fig. 377.

5. LORANTHACEAE (Mistletoe Family)

Evergreen shrubs or herbs parasitic on woody plants, nourished by means of specialized roots (haustoria) penetrating the tissues of the host plant; leaves in our plant reduced to opposite connate scales; flowers dioecious, regular, solitary or clustered, small, greenish; petals none; pistillate flowers with ovary adnate to the calyx tube; stamens 2-4; fruit a berry; seed solitary.

ARCEUTHOBIUM Marsch.-Bieb.

Small yellowish or greenish-brown fleshy plants with fragile, jointed, angled stems, and parasitic on coniferous plants; flowers solitary or a few in the axils of the scale-like leaves; calyx of staminate flowers 2-5-parted, usually bearing an equal number of stamens; berry fleshy, ovoid, more or less flattened. (Greek, meaning juniper, the original species being parasitic on Juniperus.)

A. tsugense (Rosend.) G. N. Jones   Hemlock Dwarf Mistletoe
Razoumofskya tsugensis Rosend.
R. douglasii var. *tsugensis* Piper.

Staminate plants much branched, 4–10 cm. tall; pistillate plants shorter and less branched; fruit 4–5 mm. long. This species is common on the Western Hemlock (*Tsuga heterophylla* (Raf.) Sarg.) around Juneau and Sitka and probably throughout southeastern Alaska but is usually high up in the trees and seldom noticed.

Alaska along the coast to Wash. Fig. 378.

6. SANTALACEAE (Sandalwood Family)

Herbs, shrubs, or trees; leaves entire, without stipules; flowers perfect, monecious, or dioecious, mostly greenish; calyx adnate to the base of the ovary or the disk, 4–5-lobed; petals none; stamens as many as the calyx-lobes and inserted near their bases or upon the lobes or annular disk; ovary 1-celled, ovules 2–4 but fruit a 1-seeded drupe or nut.

Flowers in terminal corymbose or paniculate cymes.......... 1. *Comandra*
Flowers on axillary 1–4-flowered peduncles................ 2. *Geocaulon*

1. COMANDRA Nutt.

Smooth perennial herbs usually more or less parasitic on the roots of other plants; leaves alternate, pinnately veined, nearly sessile; flowers perfect; calyx campanulate, 5-lobed, the tube with a 5-lobed disk; stamens 5, inserted in the lobes of the disk, attached to the calyx-lobes by tufts of hairs; fruit crowned by the persistent calyx. (Greek, referring to the hairy attachment of the anthers.)

C. pallida A. DC.   Pale Comandra

Stems slender, leafy, usually much branched, 15–45 cm. tall; leaves narrowly lanceolate or linear, or the lower oblong-elliptic, 15–35 mm. long; cymes clustered at the summit of the stems, the peduncles usually short; calyx purplish, about 4 mm. long, fruit ovoid, 6–8 $\times$ 4–5 mm.

Central Yukon—Man.—Texas—Ariz.

2. GEOCAULON Fern.

Creeping stems slender and cord-like; erect stems slender and simple; leaves alternate and short-petioled; flowers borne from the axils of the leaves in 1–4, but usually 3-flowered umbels, 1 or 2 of the flowers perfect, the others staminate; fruit a globose-oblong, edible drupe crowned by the ovate calyx-lobes. (Greek, referring to the subterranean stems.)

G. lividum (Rich.) Fern.   Northern Comandra
Comandra livida Rich.

Erect stems 1–3 dm. tall; leaves thin, oval, obtuse or rounded at the apex, 10–25 mm. long; peduncles 1–3; fruit reddish.

Common in interior Alaska, less so along the coast and extending to Gt. Slave L.—Labr.—N. Hamp.—Wash. Fig. 379.

7. POLYGONACEAE. Buckwheat Family

Herbs, or in warm climates sometimes woody plants; leaves usually entire, alternate, with stipules united to form a sheath; flowers small, regular, usually perfect; sepals 2–6, more or less united and often petaloid; corolla none; stamens 2–9; pistil of 2 or 3 carpels; ovary 1-celled; fruit a triangular or lenticular achene.

1A. Flower cluster subtended by an involucre. .... 1. *Koenigia*
2A. Flower cluster not involucrate.
  1B. Stigmas capitate. .... 4. *Polygonum*
  2B. Stigmas tufted.
    1C. Calyx 6-parted, style 3-parted. .... 2. *Rumex*
    2C. Calyx 4-parted, style 2-parted. .... 3. *Oxyria*

1. KOENIGIA L. (*Macounastrum* Small)

Small glabrous annual; stems slender, spreading or erect; leaves entire with funnelform, membranous sheaths; flowers minute, perfect, in terminal clusters subtended by a several-leaved involucre; calyx usually 3-parted, greenish-white, with equal valvate segments; stamens 2 or 4; achenes 3-angled. (Charles Dietrich Eberhard König 1774–1851, botanist.)

K. islandica L.   Koenigia
    M. islandicum (L.) Small.

Stems very slender, 5–15 cm. long, simple or forked; leaves obovate or oblong, 2–8 mm. long; involucre of 3–6 obovate leaves; flowers fascicled in the involucre and solitary or few in the axils of the upper leaves; calyx-segments ovate, obtuse; achenes about 1.5 mm. long, trigonous, the faces convex.

Wet places or in the edge of water, circumpolar. Fig. 380.

2. RUMEX L.

Mostly leafy-stemmed herbs with thick roots; leaves alternate or basal, often wavy or crisped; flowers green or reddish, perfect, dioecious, or polygamo-monoecious, borne in whorls; sepals 6, the 3 inner ones developing into entire, dentate, or fringed valves, one or all of which often bear a grain-like tubercle; stamens 6; ovary with 3 peltate, tufted styles; achene 3-angled. A very confusing group, many forms probably being hybrids. (The ancient Latin name.)

1A. Flowers mostly dioecious, basal leaves hastate or linear.
  1B. Inner sepals enlarging after flowering. .... 3. *R. acetosa*
  2B. Inner sepals not enlarging after flowering.
    1C. Flowers and stigmas large, leaves usually linear. ..... 1. *R. graminifolia*
    2C. Flowers and stigmas smaller, lower leaves hastate. ... 2. *R. acetosella*
2A. Flowers mostly perfect.
  1B. Valves deeply toothed or fringed.
    1C. Leaves large, cordate, broad. .... 4. *R. obtusifolius*
    2C. Leaves long and narrow. .... 5. *R. maritimus*
  2B. Valves entire or wavy-margined.
    1C. Stems erect.
      1D. One or more of the valves with tubercles. .... 6. *R. crispus*

2D. Valves without tubercles.
    1E. Valves broad, rounded, often broader than long.. 7. R. *domesticus*
    2E. Valves ovate or cordate, broadest near the base.
      1F. Leaves somewhat fleshy, those of the stem narrow. .... 8. R. *arcticus*
      2F. Leaves wavy or crisped, stem leaves broader.
        1G. Valves about 5 mm. long .... 9. R. *occidentalis*
        2G. Valves 7 mm. or more long. .... 10. R. *fenestratus*
  2C. Stems ascending or decumbent.
    1D. Valves 2.5–3 mm. long. .... 11..R. *sibiricus*
    2D. Valves 3–4 mm. long .... 12. R. *transitorius*

1. **R. graminifolius** Lamb.   Grass-leaved Sorrel

A rare species related to R. *acetosella* but distinguished by the very narrow and linear basal leaves and the much larger flowers and fruit.

Eastern Asia—Greenl.

2. **R. acetosella** L.   Sheep Sorrel

A glabrous, dioecious perennial with a creeping rootstock, 1–6 dm. tall; leaves narrowly hastate, some of the upper ones lanceolate or linear, 2.5–12 cm. long; flowers and achenes often reddish or purplish; achenes ovoid, triangular, minutely roughened, exceeding the persistent sepals, about 1.5 mm. long. Ssp. *angiocarpus* (Murb.) Murb. has the sepals adherent to the seed.

A common weed, native of Eurasia and widely naturalized. Fig. 381.

3. **R. acetosa** L.   Green Sorrel

Perennial; stem simple, grooved, 3–10 dm. tall; leaves ovate or oblong-ovate, usually with acute auricles at the base, crisped or erose on the margins, the basal few on long petioles, those of the upper part of the stem subsessile; panicle often reddish; pedicels nearly as long as the valves, jointed near the middle; valves cordate-orbicular, 3.5–5 mm. long; lower sepals reflexed. Most of the specimens collected in Alaska are the ssp. *alpestris* (Murb.) Murb. with ovate-triangular leaves and long, acute, rarely lacerate ocreae.

Bering Sea through central Alaska, circumboreal. Fig. 382.

4. **R. obtusifolius** L. ssp. *agrestis* (Fr.) Danser.   Bitter Dock

Stems stout, erect, conspicuously grooved, 6–12 dm. tall; lower leaves ovate-cordate, the margin wavy, long-petioled, 15–30 cm. long, the uppermost ovate-lanceolate; pedicels jointed below the middle; valves ovate, 4–5 mm. long, strongly reticulated, with a few spreading spiny teeth, one of the valves bearing a tubercle.

Introduced weed, native of Eurasia. Fig. 383.

5. **R. maritimus** L.   Golden Dock

Annual, pale green; stem with short pubescence, often diffusely branched; leaves narrow, papillate; flowers in dense whorls in leafy, compound racemes; valves with 1–3, usually 2, long bristle-like teeth on each margin, and bearing an oblong or lanceolate tubercle; achenes 1.5 mm. long, smooth and shining. Var. *fueginus* (Phil.) Dusen. (*R. persi-*

carioides Am. Auct.) has the median stem leaves slightly cordate or truncate at the base and more crisped. Also the fruit is darker in color.

Rare, the typical form has been collected at Dawson, the variety in Alaska—Anticosti—Penn.—Ill.—Calif., and in South America. The type form is Eurasiatic. Fig. 384.

6. R. *crispus* L.   Curled Dock

Stems erect, 5–10 dm. tall; leaves crisped and wavy-margined, oblong-lanceolate, 7–15 cm. long; inflorescence dense; pedicels longer than the valves, jointed at or below the middle, the joints conspicuous; valves cordate, 3–4 mm. long and wide, brown, with one valve or all 3 bearing a conspicuous, reddish, raised tubercle.

An introduced weed, native of Eurasia. Fig. 385.

7. R. *domesticus* Hartm.   Garden Dock

An upright perennial, 5–15 dm. tall; basal leaves broadly lanceolate, narrowed or rounded at the base, the margin wavy and somewhat crisped, up to 3 dm. long; panicle rather dense; pedicels jointed below the middle; valves round-reniform, usually broader than long, cordate, without tubercles, but one of them often showing a tendency toward a callosity at the base.

An introduced weed, native of Europe and western Asia. Fig. 386.

8. R. *arcticus* Trautv.   Arctic Dock

Stems erect, usually suffused reddish-purple, as low as 1 dm. tall in the high arctic to 1 m. further south; leaves cordate-lanceolate to linear-lanceolate, rather thick, not wavy, the margins sometimes finely crisped, the basal 6–25 cm. long; branches of the panicle few and simple; valves 4–8 mm. long, 3–4 mm. wide, usually reddish or brownish. Seems to hybridize with *R. fenestratus*. Var. *perlatus* Hult. Basal leaves elliptical, about 7 cm. long, 4–4.5 cm. wide.

A species of arctic-circumpolar distribution. Fig. 387.

9. R. *occidentalis* Wats.   Western Dock

Similar to *R. fenestratus* but less vigorous and with smaller fruits, the valves being about 5 mm. long and wide. Most reports of this species from Alaska refer to *R. fenestratus*.

Yukon—Que.—Maine—S. Dak.—N. Mex.—Calif.

10. R. *fenestratus* Greene.   Great Western Dock
    R. *occidentalis* Am. Auct. in part.

A vigorous grower, up to 2 m. tall; lower leaves cordate-ovate to cordate-lanceolate, up to 4 dm. long on petioles up to 6 dm. long, the margins wavy; pedicels longer than the valves, the articulation obscure; valves large, thin, translucent, 6–9 mm. wide, up to 10 mm. long, prominently reticulate-veined. Ssp. *puberulus* Hult. of southeastern Alaska has the stems, petioles, and lower surface of the leaves puberulent.

Alaska—Labr.—Que.—Mont.—Calif. Fig. 388.

11. R. sibiricus Hult. Siberian Dock

Resembles *R. transitorius* but has thick, narrow, grayish-green, smooth, not undulate leaves and smaller fruit, the valves 2.5–3 mm. long. Asia—Mackenzie district.

12. R. transitorius Rech. f. Beach Dock

Stems 3–6 dm. tall, usually decumbent at the base; leaves pale green, lanceolate, undulate or crisped, 4–15 cm. long, 1–3 cm. wide; inflorescence crowded, the branches erect or ascending; pedicels short, jointed near the base; valves ovate-lanceolate, acute, 3–4 mm. long, each with a prominent tubercle.

A plant of salt marshes, Alaska—Calif. Fig. 389.

3. OXYRIA Hill.

Somewhat fleshy, glabrous, alpine perennials with acid juice and rather fleshy taproot; leaves mostly basal, reniform to orbicular, cordate, long-petioled, palmately-veined; flowers perfect, small, green, in verticels arranged in panicled racemes; sepals 4, the outer larger than the inner; stamens 6, included; ovary 1-celled with 2-parted style; stigmas fimbriate, persistent on the wings of the calyx in fruit; achene ovate, lenticular. (Greek, sour, with reference to the acid leaves.)

O. digyna (L.) Hill. Mountain Sorrel

Stem erect, scapiform, 1–6 dm. tall; leaves 15–50 mm. wide, often undulate; racemes many-flowered; flowers slender-pedicelled; inner sepals erect, the outer somewhat reflexed in fruit; achenes broadly winged.

A circumboreal species found throughout our region. Fig. 390.

4. POLYGONUM (Tourn.) L.

Ours all herbs with alternate, entire or toothed leaves with sheathing stipules; flowers small, normally perfect, often spicate; sepals 4–6, united at the base and often colored; stamens 3–9; stigmas 2 or 3; achenes lenticular or triangular, enclosed by the persistent calyx. (Greek, many and knee, from the swollen joints of many species.)

```
1A. Stems twining, leaves cordate. ......................Subgenus Bilderdykia
2A. Stems not twining.
  1B. Stems unbranched, from a bulb-like caudex, inflores-
       cence a spike-like raceme. .....................Subgenus Bistorta
  2B. Stems branched.
    1C. Flowers paniculate or in axillary clusters, leaves
         ample. ........................................Subgenus Aconogonum
    2C. Flowers in spikes with very small bracts. ..........Subgenus Persicaria
    3C. Flowers in axillary clusters, or solitary, or spike-
         like with leafy bracts. .......................Subgenus Avicularia
                    Subgenus Bilderdykia
One species. ............................................... 1. P. convolvulus
                    Subgenus Bistorta
Racemes dense, without bulblets below the flowers. ........ 2. P. bistorta
Racemes less dense, usually with bulblets below the flowers... 3. P. viviparum
```

Subgenus *Aconogonum*
One species. .... 4. *P. alaskanum*
Subgenus *Persicaria*
1A. Plant usually floating, base of leaves ovate or cordate.. ... 5. *P. amphibium*
2A. Plant not floating, base of leaves cuneate.
  1B. Calyx and pedicels glandular.
    1C. Spikes dense, obtuse. .... 6. *P. scabrum*
    2C. Spikes narrow, acute.
      1D. Spikes interrupted, achenes dull. .... 7. *P. hydropiper*
      2D. Spikes not interrupted, achenes shining. .... 8. *P. nodosum*
  2B. Calyx and pedicels without glands.
    1C. Ocreae not fringed. .... 9. *P. pennsylvanicum*
    2C. Ocreae fringed with bristles.
      1D. Racemes slender, loosely-flowered. ....10. *P. hydropiperoides*
      2D. Racemes ovate, broad and compact. ....11. *P. persicaria*
Subgenus *Avicularia*
1A. Leaves acute.
  1B. Stems ascending, achenes smooth. ....18. *P. ramosissimum*
  2B. Stems prostrate, achenes dull. ....19. *P. neglectum*
2A. Leaves obtuse.
  1B. Achenes exerted.
    1C. Stems very slender, leaves small. ....12. *P. caurianum*
    2C. Stems coarser, leaves larger, maritime species. ....13. *P. fowleri*
  2B. Achenes included or only slightly exerted.
    1C. Flowers shorter than ocreae, plant much branched. ..14. *P. prolificum*
    2C. Flowers longer than the ocreae.
      1D. Stems prostrate. ....15. *P. buxiforme*
      2D. Stems ascending.
        1E. Leaves of the flowering branches much shorter
          than those of the stem. ....16. *P. heterophyllum*
        2E. Leaves of flowering branches like those of stem..17. *P. achoreum*

1. *P. convolvulus* L.   Black Bindweed
Bilderdykia convolvulus (L.) Dum.
Tiniaria convolvulus (L.) Webb. & Moq.

Stem climbing or trailing, 3–10 dm. long; leaves ovate-sagittate, acuminate, 2–6 cm. long; flowers greenish, 3.5–4 mm. long, in rather lax racemes 1–6 cm. long; three of the sepals keeled; pedicels slender, articulated, reflexed; achene triangular, black, minutely roughened.

Native of Eurasia but widely naturalized in temperate climates. Fig. 391.

2. *P. bistorta* L. ssp. *plumosum* (Small) Hult.   Mountain Meadow Bistort
Bistorta lilacina Greene.

Erect perennial, 5–50 cm. tall; leaves mostly basal, long-petioled, glabrous above, scabrous-puberulent beneath, 5–15 cm. long; stem leaves usually 2; spike terminal, 2–7 cm. long, more than 1 cm. thick, dense; perianth rose; stamens 8, exserted; achenes triangular, acuminate, 4–5 mm. long.

The species is circumboreal. Fig. 392.

3. *P. viviparum* L.   Alpine Bistort
Bistorta vivipara (L.) S. F. Gray.

A very variable species, some alpine forms occasionally less than 1 dm. tall, lowland forms up to 4 dm. or more tall; leaves ovate, lanceolate, or linear, the blades 1-15 cm. long, acute to subcordate at the base, acute or obtuse at the apex, reticulately veined and the midrib prominent; spikes 2-10 cm. long, less than 1 cm. thick, bulblet-bearing below and sometimes throughout; flowers white or light rose; stamens 8, exserted; achenes dark brown, granular, dull.

Wet soil, alpine-arctic to lowlands, circumboreal. Fig. 393.

4. *P. alaskanum* (Small) Wight.   Wild Rhubarb
P. alpinum alaskanum Small.
Aconogonum phytolaccaefolium Auct. in part.

Stem branched, erect or ascending, 8-18 dm. tall; leaves lanceolate, acute or acuminate at the apex, narrowed or truncate at the base, somewhat crisped, 6-20 cm. long, inflorescence showy; pedicels jointed near the base; calyx 3-4 mm. long; achenes 4 mm. long, light straw-colored, shining.

Interior Alaska from Bering Sea east and in Yukon. Fig. 394.

5.* *P. amphibium* L. ssp. *laevimarginatum* Hult.   Water Persicaria
P. coccineum Muhl.
P. natans (Michx.) Eaton.
Persicaria amphibia (L.) S. F. Gray.

An exceedingly variable species, the aquatic form with floating stems the leaves of which are smooth, glossy, tinged with red, oblong or elliptic; the amphibious form often with erect stems, lanceolate, acute leaves with stiff pubescence; spikes terminal, dense, 15-30 mm. long, more than 1 cm. thick; flowers rose; achenes lenticular, biconvex.

A circumboreal species, the subspecies in eastern Asia and across North America to Newf. and southward. Fig. 395.

6. *P. scabrum* Moench.   Tomentose Persicaria
P. tomentosa Schrank.

Annual, 1-5 dm. tall; leaves lanceolate, some of the lower ones retaining some flocculent tomentum on the under surface; spikes thickish, the lateral ones scarcely peduncled; flowers pale; achenes lenticular, the sides concave with a slight ridge through the center.

Sparingly introduced in our area, native of Eurasia.

7. *P. hydropiper* L.   Water Pepper

Annual; stems glabrous, simple to much branched, 2-6 dm. tall; leaves ovate-lanceolate to linear-lanceolate, acute at apex, narrowed into a short petiole at the base, papillose and punctate, very acrid, 2-9 cm. long; racemes 2-6 cm. long, interrupted and drooping; sepals greenish with pale or rose margins; achenes lenticular or 3-angled, granular and dull.

Sparingly introduced, native of Europe.

8. *P. nodosum* Pers.   Dock-leaved Persicaria
P. lapathifolium L. var. *nodosum* (Pers.) Weinm.

Annual, glabrous, 3–7 dm. tall; leaves lanceolate, punctate, ciliolate on the margins, cuneate at the base, 5–20 cm. long; racemes spike-like, panicled, 2–8 cm. long, erect or nodding; flowers greenish-white or tinted rose; achenes lenticular, broadly ovoid, about 2 mm. long and nearly as broad, shining, the faces concave.

A sparingly introduced weed native of Eurasia but widely distributed. Fig. 396.

9. *P. pennsylvanicum* L.   Pennsylvania Persicaria

Annual, glabrous below; stem simple or more usually branched; 3–8 dm. tall; leaves lanceolate, acuminate, petioled, ciliate on the margins, 4–20 cm. long; racemes panicled, oblong or cylindric, dense, the peduncles beset with stipitate glands; calyx deep pink or rose, 3–4 mm. long; achene orbicular, short-pointed, lenticular, about 3 mm. wide, smooth, shining.

An introduced weed, native of eastern U. S.

10. *P. hydropiperoides* Michx.   Mild Water Pepper

Perennial, glabrous or strigillose, 3–9 dm. tall; leaves oblong-lanceolate to linear-lanceolate, 5–15 cm. long, short-petioled, ciliate, pubescent with appressed hairs on the midrib beneath, racemes slender and interrupted, 3–8 cm. long; calyx white to rose; achenes 3-angled, ovoid or oblong, 2 mm. long, smooth and shining.

Rare, central Alaska—Que.—Fla.—Mex.

11. *P. persicaria* L.   Lady's Thumb
Persicaria maculosa S. F. Gray.

Annual, glabrous or nearly so, 2–6 dm. tall; leaves lanceolate or linear-lanceolate, punctate or roughened beneath, somewhat ciliate, 3–15 cm. long; ocreae cylindric with a fringed margin; spikes erect, 1–4 cm. long; achenes lenticular with convex sides, ovoid, about 2.5 mm. long and 2 mm. wide, rarely triangular.

An introduced weed, native of Eurasia. Fig. 397.

12. *P. caurianum* Robins.   Alaska Knotweed

Annual, usually more or less reddish; stems slender to very slender, prostrate or ascending, sparsely to profusely branched, 12–50 cm. long; leaves narrowly elliptical or oblong, 10–16 mm. long, 3–5 mm. wide, rounded at the apex, narrowed to a short petiole at the base; sepals rounded, the inner ones and often all of them with petaloid margins; achenes dark brown or black, minutely puncticulate, 2–3 mm. long, sometimes much longer than the calyx.

Northeastern Asia and northwestern America. Fig. 398.

13. *P. fowleri* Robins.   Fowler Knotweed

Perennial; stems ascending, decumbent, or prostrate, 2–6 dm. long; leaves all alike, oblong, oblanceolate, or elliptic-lanceolate, petioled, 1–3 cm. long, up to 1 cm. wide; sepals tipped and margined white, pink, or red, 2.5–3.5 mm. long in fruit and slightly shorter than the reddish-brown, acute achene.

Sea beaches, eastern Asia—Alaska—Wash. and Labr.—N. S. Fig. 399.

14. *P. prolificum* (Small) Robins.   Proliferous Knotweed

Annual; stems up to 5 dm. tall, much branched, strongly striate; leaves narrow, linear-oblong or linear, thick, dark green, 1–2 cm. long; perianth about 2 mm. long, pinkish; achenes brown, about 2 mm. long, abruptly contracted at the apex.

Probably introduced, Yukon—Mont.—Que.—Maine—Va.—Colo.

15. *P. buxiforme* Small.   Common Knotweed

Annual, stems decumbent or prostrate, diffusely branched, striate, 2–12 dm. long; leaves oblong, elliptic, or oblanceolate, usually obtuse, 5–25 mm. long, often crisped on the margin; flowers 2–6 in a cluster; sepals green with whitish or pinkish margins; achenes dark brown, somewhat roughened, 2–3 mm. long. Many reports of *P. aviculare* L. refer to this species.

Nome—Mayo—Ont.—Va.—Texas—Calif. Fig. 400.

16. *P. heterophyllum* Lindm.   Various-leaved Knotweed

Stems ascending, more or less branched, 3–9 dm. tall; lower leaves obovate or oblanceolate, 15–45 mm. long, 5–15 mm. wide; upper leaves reduced, narrower and acute; sepals whitish or pinkish at the tip, in fruit 3.5–4 mm. long, strongly reticulate-veined and enclosing the achene.

An introduced weed, native of Europe. Fig. 401.

17. *P. achoreum* Blake.

Annual; stems ascending, much branched, striate, glabrous, 15–40 cm. tall; leaves numerous, elliptic, oval, or obovate, rounded at the apex, 8–30 mm. long, 4–14 mm. wide; sepals in fruit 3.5–4 mm. long, the inner ones white- or pink-margined; achenes included, dull, about 2.5 mm. long.

Central Alaska—Que.—Vt.—Mo.—Kans.—Mont. Fig. 402.

18. *P. ramosissimum* Michx.   Bushy Knotweed

Annual, yellowish-green, glabrous; stems erect or ascending, usually much branched, 1–12 dm. tall; leaves lanceolate or linear-obling, short-petioled; 5–20 mm. long, acute at both ends; flowers short-pedicelled; sepals yellowish or with yellow margins, 2.5–3 mm. long; achenes black, shining, included or slightly protruding.

Southeastern Alaska—Minn.—Ill.—N. Mex.—Calif. Introduced in eastern U. S. and Canada.

19. *P. neglectum* Bess.

Annual or perennial; stems prostrate, diffusely branched, striate, 1–5 dm. long; leaves narrow, elliptic-lanceolate or linear, 6–18 mm. long;

flowers nearly sessile; sepals about 2 mm. long, the margins usually suffused with pink; achene reddish-brown, about 2.5 mm. long, definitely longer than the sepals.

An introduced weed, native of Europe. Fig. 403.

Fagopyrum esculentum Moench, the cultivated Buckwheat, sometimes persists for a few years after cultivation. It is an erect annual, 3–8 dm. tall; leaves hastate, 3–8 cm. long; sepals white or whitish; achenes about 5 mm. long, about twice as long as the calyx. It is a native of eastern Europe or western Asia.

8. CHENOPODIACEAE (Goosefoot Family)

Ours all more or less fleshy herbs, often white-mealy; leaves simple, in *Salicornia* reduced to mere ridges; flowers sessile in axillary or terminal clusters or in spikelets; calyx of 1–5 sepals, usually small; corolla none; stamens 1–5; pistil of 2–5 united carpels with 1-celled ovary and 2–5 styles; fruit a utricle with embryo curved around the endosperm.

1A. Leaves reduced to scales, stems fleshy, jointed.......... 6. *Salicornia*
2A. Leaves present, stems not jointed.
  1B. Leaves linear or subulate.
    1C. Calyx of 1 sepal. .... 4. *Corispermum*
    2C. Calyx 5-parted. .... 5. *Suaeda*
  2B. Leaves broader.
    1C. Sepals 1, stamen 1. .... 2. *Monolepis*
    2C. Calyx-lobes 3–5, stamens usually 5.
      1D. Flowers monoecious or dioecious. .... 3. *Atriplex*
      2D. Flowers perfect. .... 1. *Chenopodium*

1. CHENOPODIUM (Tourn.) L.

Ours all annual herbs; leaves alternate, mealy-coated or glandular; flowers very small, green, in axillary or terminal spikes or glomerules; sepals persistent, more or less enclosing the utricle; utricle 1-seeded, the embryo a complete ring. (Greek, goose and foot, from the shape of the leaves of some species.)

1A. Leaves triangular, cordate or hastate, sinuate-dentate or coarsely toothed.
  1B. Flowers in globose sessile heads, becoming berry-like
    in fruit. .... 1. *C. capitatum*
  2B. Flowers in loosely panicled racemes. .... 2. *C. gigantospermum*
2A. Leaves entire to sinuate-dentate, linear, oblong, or rhombic-ovate.
  1B. Plant decumbent. .... 3. *C. glaucum*
  2B. Stems usually erect.
    1C. Seeds covered with shallow, honeycomb-like pits on
      upper surface. .... 4. *C. berlandieri*
    2C. Seeds with radial furrows or nearly smooth.
      1D. Leaves linear, mostly entire. .... 5. *C. leptophyllum*
      2D. Leaves broader, mostly toothed. .... 6. *C. album*

1. **C. capitatum (L.) Achers.**   Strawberry Spinach
Blitum capitatum L.

Stems usually branched from the base, 2-5 dm. tall; leaves triangular-lanceolate, 3-7 cm. long, sinuate-toothed, the upper entire; flower heads becoming red, globular clusters, 7-14 mm. in diameter in fruit; seed compressed, ovate, acutely margined or keeled.

Central Alaska—N. S.—N. Jer.—Minn.—Colo.—Nev. Fig. 404.

2. **C. gigantospermum Aellen.**   Maple-leaved Goosefoot
C. hybridum Am. Auct.

Glabrous, bright green annual, sometimes mealy in the inflorescence; stems usually branched, 3-14 dm. tall; leaves with 1-4 large, triangular teeth on each side, the uppermost sometimes entire; flowers in large axillary or terminal panicles; calyx lobes not completely enclosing the fruit, often spreading as the fruit ripens; fruit flat, brownish-black, 1-2 mm. in diameter.

Dawson—Alta.—Maine—Va.—Okla.—N. Mex.—Calif.

3. **C. glaucum L. ssp. salinum (Standl.) Aellen.**   Oak-leaved Goosefoot

Low, succulent, spreading or prostrate; leaves green above, white-mealy beneath, 1-5 cm. long; flowers in small axillary clusters shorter than the leaves, or the upper panicled; calyx about 1 mm. broad, neither fleshy nor keeled in fruit, not entirely covering the utricle.

Manly Hot Springs—L. Athabasca—Man.—N. Mex.—Ariz. The main form is Eurasiatic. Fig. 405.

4. **C. berlandieri Moq. ssp. zschackei (Murr.) Zobel.**
                                                             Zschacke Goosefoot

Similar to *C. album*; stems erect, 3-9 dm. tall, branched, striate; leaves lanceolate, oblong, ovate, or somewhat rhombic, 15-40 mm. long, mucronulate, often with a few teeth; calyx densely farinose; utricle 0.8-1 mm. broad, puncticulate.

Collected a few times in Alaska, probably introduced. Ore.—Minn.—La.—Mex.—Calif.

5. **C. leptophyllum Nutt.**   Narrow-leaved Goosefoot

Annual; stems slender, striate or grooved, 2-7 dm. tall; leaves linear to linear-lanceolate, usually entire, 15-45 mm. long, farinose beneath; calyx densely farinose, completely enclosing the utricle; pericarp free; utricle about 1 mm. broad, nearly black, smooth and shining.

Introduced in our area, Yukon—Man.—Ill.—Mex.—Calif. Also adventive in eastern states, Argentina and Europe.

6. **C. album L.**   Lamb's Quarters

Stout and branched if not crowded, 3-20 dm. tall; leaves dentate, except the upper ones, 2-8 cm. long; spikes terminal and axillary, usually compound and often panicled; sepals keeled in fruit; usually enclosing the black, shining utricle.

A weed introduced in all temperate regions, native of Eurasia. Fig. 406.

2. MONOLEPIS Schrad.

Low branching annuals; leaves alternate; flowers perfect or polygamous, borne in small axillary clusters; calyx of a single herbaceous sepal; stamen 1, styles 2, slender; utricle vertical, flattened, the pericarp persistent; embryo a nearly complete ring. (Greek, one and scale, from the single sepal.)

M. nuttalliana (Schult.) Greene.   Nuttall Monolepis, Poverty Weed

Glabrous, or somewhat mealy when young, branched from near the base, 10–25 cm. tall; leaves hastate-lanceolate with 2 spreading lobes near the middle, short-petioled or the upper sessile and sometimes entire, 15–60 mm. long; pericarp minutely pitted.

Dry soil, central Alaska—N. W. Terr.—Minn.—Mo.—N. Mex.—Calif. Fig. 407.

3. ATRIPLEX (Tourn.) L.

Ours annual herbs with scurfy or mealy leaves; flowers monoecious or dioecious, borne in panicled spikes or congested axillary clusters; staminate flowers bractless, with 3–5 each of sepals and stamens; pistillate flowers usually without sepals but subtended by 2 more or less united bracts which enlarge in fruit; stigmas 2; utricle vertical; embryo a ring in the mealy endosperm. (From a Greek name of orache.)

1A. Pistillate flowers all alike, without calyx.
  1B. Leaves sessile. .... 1. *A. drymarioides*
  2B. At least the lower leaves petioled.
    1C. Bracts dentate, leaves with forward-pointing teeth. ... 2. *A. patula*
    2C. Bracts entire, leaves usually entire.
      1D. Fruiting bracts 6–20 mm. long. .... 3. *A. alaskensis*
      2D. Fruiting bracts 3–10 mm. long. .... 4. *A. gmelini*
2A. Pistillate flowers of 2 kinds, some with a 3–5-lobed
    calyx, others without perianth but with bracts. ...... 5. *A. hortensis*

1. *A. drymarioides* Standl.

Stems sparsely branched, erect or spreading, sparsely farinose, 6–10 cm. tall; lower leaves opposite, the upper alternate, sessile, cuneate-obovate to oblong, 9–17 mm. long, 4–8 mm. wide, rounded or obtuse at the apex, cuneate at the base, finely farinose; fruiting bracts usually on long, slender pedicels, 4–6 mm. long, usually narrower at the base than the utricle.

Pacific coast of Alaska.

2. *A. patula* L.   Spear Orache

Stems erect to procumbent, the branches 3–9 dm. long; lowest leaves opposite, the upper alternate; leaves lanceolate to rhombic-lanceolate, sometimes hastate, 25-80 mm. long, entire or sinuate-dentate, glabrous or farinose beneath; fruiting bracts 2-6 mm. long, often subhastate, acute or acutish, tuberculate, the margins usually denticulate.

Introduced, native of Eurasia.

3. *A. alaskensis* Wats. Alaska Saltweed

Profusely branched; stems 4-8 dm. tall; leaves lanceolate, petioled, entire or with a few teeth, 6-15 cm. long; fruiting bracts entire, attenuate at the apex, up to 10 mm. long and 8 mm. wide, reticulated, united only near the base; utricle minutely pitted.

Sandy beaches, Pacific coast of Alaska. Fig. 408.

4. *A. gmelini* C. A. Mey. Gmelin Saltweed

Stems simple to much branched, ascending, 1-5 dm. tall; leaves oblong, lanceolate, or linear, entire, sparingly toothed, or slightly 3-lobed near the base, 2-8 cm. long; fruiting bracts united only at the base, triangular-rhombic, their sides often tubercled, much smaller than in *A. alaskensis*.

Sea beaches, Japan—Kotzebue—northern California. Fig. 409.

5. *A. hortensis* L. Garden Orache. Sea Purslane

Stout, erect, 5-25 dm. tall, sparsely branched, the branches slender, ascending; lower leaves opposite, the upper alternate, broadly triangular or lance-oblong, 5-12, or even 20 cm. long, often hastately lobed, acute or obtuse at the apex, rounded, truncate, or subcordate at the base, sinuate-dentate to entire or undulate, farinose when young; fruiting bracts broadly oval or ovate, 5-18 mm. long, rounded to acute at apex, entire or denticulate.

Introduced at Fairbanks, native to central Asia.

4. CORISPERMUM (A. Juss.) L.

Annuals with narrow, entire, 1-nerved leaves; flowers small, perfect, bractless, produced in the axils of the modified upper leaves and forming terminal spikes; sepal broad; stamens 1-3; pericarp of the utricle adherent to the seed. (Greek, bug-seed.)

C. hyssopifolium L. Bug-seed

Usually pubescent, somewhat fleshy; stem striate, usually much branched, 12-50 cm. tall; lower leaves narrowly linear, sessile 15-50 mm. long; upper leaves ovate or lanceolate, acute or acuminate, usually imbricate; utricle ellipsoid, narrowly winged, the base of the styles persistent.

A circumboreal species found along the Yukon. Fig. 410.

5. SUAEDA Forsk.

Plants fleshy; leaves alternate, narrowly linear, thick, entire, sessile; flowers perfect or polygamous, solitary or clustered in the axils of the upper leaves; sepals 5, keeled or narrowly winged in fruit and enclosing the utricle; stamens 5; styles usually 2; seed separating from the pericarp. (Name Arabic.)

S. maritima (L.) Dumort. Low Sea-blite
Dondia maritima (L.) Druce.

A much branched, erect or decumbent annual, 6-20 cm. tall, somewhat glaucous; leaves 7-15 mm. long; sepals rounded or very obtusely keeled; seeds orbicular, slightly concave on one side, brownish-black, shining.

Sea beaches, Cook Inlet—southeastern Alaska and Atlantic coasts of America and Europe. Fig. 411.

6. SALICORNIA (Tourn.) L.

Fleshy glabrous herbs with opposite branches; leaves reduced to scales at the nodes; flowers perfect or polygamous in cylindrical terminal spikes, sunk into the internodes; calyx fleshy, the border truncate or 3-4 toothed; stamens 1 or 2, exserted; styles or stigmas 3; utricle enclosed in the spongy calyx. Plants growing in saline soil. (Greek, salt and horn, from the habitat and the horn-like branches.)

Plant annual. .... 1. *S. herbacea*
Plant perennial. .... 2. *S. pacifica*

1. *S. herbacea* L. Slender Glasswort
   S. europea Am. auct.

Stems usually upright and much branched, 5-15 cm. tall, often turning bright red; fruiting spikes slender, 1-3 cm. long, the apex acute; flowers 3 at each node, the middle one much higher than the lateral, but shorter than the internode.

Cook Inlet—Calif., Atlantic coast, Eurasia and Africa. Fig. 412.

2. *S. pacifica* Standl.

Stems usually more or less decumbent, 8-20 cm. long, with ascending branches, green or grayish; scales broad; fruiting spike 1-4 cm. long, about 4 mm. thick, blunt at the tip; flowers 3 at each node, on nearly the same level and about equaling the node.

Sea beaches, southeastern Alaska—Mexico. Fig. 413.

9. PORTULACEAE (Purslane Family)

Ours succulent herbs with perfect flowers; sepals usually 2; stamens opposite petals when of the same number; ovary superior, 1-celled, with central or basal placenta; styles usually 3, more or less united; fruit a 3-valved capsule; seeds few, usually black and shining, minutely roughened.

Petals 5, separate; stamens 5. .... 1. *Claytonia*
Petals more or less united, stamens 3. .... 2. *Montia*

1. CLAYTONIA L.

Mostly perennials; sepals 2, herbaceous; petals pink or white, usually showy; ovules 3-6; seeds compressed. The corms of *C. tuberosa* and the fleshy roots of *C. acutifolia* are eaten by the Eskimo. (John Clayton was an early American botanist.)

1A. Rootstock a subterranean corm. .... 1. *C. tuberosa*
2A. Rootstock a large, fleshy root.
  1B. Sepals 7 mm. or more long. .... 2. *C. acutifolia*
  2B. Sepals 5-6 mm. long. .... 3. *C. arctica*
3A. Roots fibrous.
  1B. Stems with 2 opposite leaves and sometimes a leaf-like bract.
    1C. Stems 1-flowered. .... 6. *C. scammaniana*
    2C. Stems few-several-flowered.
      1D. Stem leaves united into a cup. .... 7. *C. perfoliata*
      2D. Stem leaves not united.
        1E. Sepals 5-6 mm. long. .... 3. *C. arctica*
        2E. Sepals 3-4 mm. long.
          1F. Petals 6-9 mm. long. .... 4. *C. sibirica*
          2F. Petals 10-15 mm. long. .... 5. *C. sarmentosa*
  2B. Stems leafy, plants with stolons.
    1C. Leaves oblanceolate. .... 8. *C. chamissoi*
    2C. Leaves not oblanceolate, small.
      1D. Petals 7-8 mm. long .... 9. *C. parvifolia*
      2D. Petals 12-15 mm. long. ....10. *C. flagellaris*

1. *C. tuberosa* Pall.   Tuberous Spring Beauty

    Stems usually 1, occasionally more, 8-18 cm. tall, arising from a subterranean corm 1-2 cm. in diameter; basal leaves 1-few, arising directly from the corm, lanceolate to linear-lanceolate; stem leaves similar but sessile, 2-5 cm. long, 2-5 mm. wide; racemes 2-7-flowered; sepals 5-7 mm. long, obtuse; petals white, 9-12 mm. long; seeds black, orbicular, 2-2.5 mm. long.

    Eastern Asia—Yukon. Fig. 414.

2. *C. acutifolia* Pall.   Bering Sea Spring Beauty

    Stems usually several, 5-15 cm. tall, arising directly from the thick fleshy root; basal leaves narrowly lanceolate to linear, arising directly from the crown of the root, stem leaves similar but smaller; racemes 2-5-flowered; sepals 7-14 mm. long, petals usually white, rarely pink, 12-15 mm. long; seed rounded-oval, nearly 3 mm. in diameter. Our Alaskan form has narrower leaves and bracts than the type and has been described as ssp. *graminifolia* Hult.

    Eastern Asia—central Alaska. Fig. 415.

3. *C. arctica* Adams.   Arctic Spring Beauty

    Root somewhat fleshy; stems several, 6-15 cm. tall; basal leaves 3-7 cm. long, the blade spatulate, up to 1 cm. or more wide, decurrent on the petiole; stem leaves sessile, ovate, 1 cm. or more long; racemes 3-7-flowered; sepals somewhat unequal, 5-6 mm. long; petals white, 10-12 mm. long.

    Siberia and the Aleutian Islands. Fig. 416.

4. **C. sibirica L.** Siberian Spring Beauty
   C. alsinoides Sims.
   C. asarifolia Bong.
   Montia sibirica (L.) Howell.
   Limnia sibirica (L.) Haw.

   Stems few to many, ascending, 1–5 dm. tall; basal leaves long-petioled, ovate, lanceolate, or orbicular-lanceolate, 6–60 mm. wide, the petioles dilated at the base; stem leaves typically broadly ovate; racemes usually elongated, often bearing a small leaf; flowers varying in color from white to rose; capsule about as long as the sepals.

   Common in the coastal districts, Commander Islands—Mont.—Utah—Calif. Fig. 417.

5. **C. sarmentosa C. A. Mey.** Alaska Spring Beauty
   Montia sarmentosa (C. A. Mey.) Robins.
   Limnia sarmentosa (C. A. Mey.) Rydb.

   Stems spreading or ascending, 5–15 cm. long; basal leaves ovate, oblanceolate, or spatulate, narrowed into a petiole, the whole 2–9 dm. long, 3–15 mm. wide; racemes 2–6-flowered; sepals orbicular, 3–4 mm. long and about as wide; petals various shades of pink or even white, 9–15 mm. long; seeds black, about 2 mm. in diameter.

   Eastern Asia—Cape Lisburne—B. C. Fig. 418.

6. **C. scammaniana Hult.** Scamman Spring Beauty

   Stems several, usually 1-flowered, 4–9 cm. tall; basal leaves narrow, spatulate, 2–7 cm. long, 2–6 mm. wide; stem leaves ovate, less than 1 cm. long; sepals roundish-ovate, 4–7 mm. long; petals mostly bright rose, occasionally white, 10–15 mm. long.

   Central Alaska. Fig. 419.

7. **C. perfoliata Donn.** Small-flowered Spring Beauty
   C. parviflora Dougl.
   Montia parviflora (Dougl.) Howell.
   Limnia parviflora (Dougl.) Rydb.

   Annual; stems several, 5–30 cm. tall; basal leaves variable; stem leaves connate, forming a suborbicular disk 1–3 cm. wide; sepals less than 2.5 mm. long; petals white or pink, less than 5 mm. long; seed 1 mm. or more long.

   Introduced at Unalaska, B. C.—Ida.—Lower Calif. Fig. 420.

8. **C. chamissoi Esch.** Toad-lily
   Montia chamissonis (Esch.) Greene.
   Crunocallis chamissonis (Esch.) Rydb.

   Stems slender and weak but usually ascending, 6–30 cm. long, producing long filiform stolons; leaves opposite, oblanceolate, narrowed into a short petiole or sessile, 2–5 cm. long; flowers in axillary or terminal, 1–9-flowered racemes; sepals about 2 mm. long; petals white or pinkish, 6–10 mm. long.

   Aleutian Islands—central Alaska—Man.—Iowa—Calif. Fig. 421.

9. **C. parvifolia** Moc.   Small-leaved Spring Beauty
Montia parvifolia (Moc.) Greene.
Naiocrene parvifolia (Moc.) Rydb.

Perennial; stem weak, spreading or decumbent, 5–20 cm. long; leaves thick, crowded on the caudex and alternate on the stem and stolons, the basal with petioles up to 25 mm. long, those of the stem shorter and reduced in size, sometimes to mere bracts; flowers in few-flowered racemes; sepals roundish, 2–3 mm. long; petals pink.

Southeastern Alaska—Mont.—Calif.

10. **C. flagellaris** Bong.   Long-branched Spring Beauty
Montia flagellaris (Bong.) Robins.
Naiocrene flagellaris (Bong.) Heller.

Similar to *C. parvifolia*; rootstock more elongated, horizontal; flagelliform branches 2–4 dm. long, some of them flower-bearing at the end; leaves orbicular or broadly ovate; petals 11–14 cm. long.

Along the coast, southeastern Alaska—Ore. Fig. 422.

2. MONTIA (Micheli) L.

Small annual, glabrous herbs growing in water or wet situations; leaves opposite, fleshy, narrow; flowers minute, nodding, solitary or in short racemes; ovary 3-ovuled; styles 3, united below; seeds 1–3, compressed, suborbicular. (Guiseppe Monti was an Italian botanist.)

Ripe seed dark brown, reticulate-furrowed, shining, about
   1.5 mm. long. .... 1. *M. lamprosperma*
Ripe seed black, smaller, muricate-tuberculate. .... 2. *M. hallii*

1. **M. lamprosperma** Cham.   Blinks. Water Chickweed
M. fontana Auct.

Stems slender, much branched, not rooting at the nodes, seldom more than 8 cm. long when growing on soil but up to 25 cm. long when in water; leaves 1–2 cm. long, the lower petioled, the upper sessile, submerged leaves rather thin; flowers axillary or in small terminal racemes; sepals broad, about 1.5 mm. long.

Widely distributed in our territory, circumboreal. Fig. 423.

2. **M. hallii** (Gray) Greene.

Stems slender, branched, 5–15 cm. long, often rooting at the nodes; lower leaves petioled, spatulate, 5–10 mm. long, the petioles dilated at the base; middle and upper leaves sessile; racemes axillary and terminal, 3–10-flowered; sepals reniform, 1 mm. long; capsule slightly exceeding the sepals.

Kamchatka—Pribylof Islands—Nev.—Calif.

10. CARYOPHYLLACEAE (Pink Family)

Herbs, often with swollen nodes; leaves opposite, entire; flowers regular, usually perfect; sepals 4–5; petals of same number or wanting;

stamens twice the number of sepals or less; carpels 2-5, united into a 1-celled ovary with central or basal placenta; styles 2-5; fruit in ours a capsule opening by teeth or valves.

Sepals distinct. .... Subfamily *Alsineae*
Sepals united. .... Subfamily *Sileneae*
                        Subfamily *Alsineae*
1A. Capsule cylindric. .... 1. *Cerastium*
2A. Capsule ovoid or globose.
  1B. Stipules present, scarious.
    1C. Styles and capsule valves 5. .... 5. *Spergula*
    2C. Styles and capsule valves 3. .... 6. *Spergularia*
  2B. Stipules wanting.
    1C. Petals deeply 2-cleft or none. .... 2. *Stellaria*
    2C. Petals entire or emarginate.
      1D. Styles as many as the sepals and alternate with
          them. .... 4. *Sagina*
      2D. Styles fewer than the sepals. .... 3. *Arenaria*
                      Subfamily *Sileneae*
1A. Styles 5. .... 9. *Lychnis*
2A. Styles 3. .... 7. *Silene*
3A. Styles 2.
  1B. Calyx 5-nerved. .... 10. *Saponaria*
  2B. Calyx many-nerved. .... 8. *Dianthus*

1. CERASTIUM L.

Pubescent, often viscid annuals or perennials; leaves opposite; flowers in terminal dichotomous cymes; sepals usually 5; petals white, 2-cleft; stamens usually 10; styles usually 5; capsule cylindric, often curved, opening by usually 10 tooth-like valves; seeds rough. (Greek, horn, referring to the capsules.)

1A. Plant annual. .... 3. *C. glomeratum*
2A. Plant perennial.
  1B. Stem simple, erect. .... 1. *C. maximum*
  2B. Plants more or less caespitose.
    1C. Petals about same length as the sepals. .... 4. *C. caespitosum*
    2C. Petals markedly longer than the sepals.
      1D. Plants with sterile shoots in the axils of the upper
          leaves. .... 2. *C. arvense*
      2D. Plants without sterile shoots in the axils.
        1E. Petals 6-9 mm. long.
          1F. Leaves viscid-puberulent. .... 5. *C. beeringianum*
          2F. Leaves glabrescent with ciliate margins. ...... 6. *C. aleuticum*
        2E. Petals 9-14 mm. long.
          1F. Low growing, densely caespitose. .... 8. *C. arcticum*
          2F. Taller, loosely caespitose. .... 7. *C. fischerianum*

1. *C. maximum* L.   Great Chickweed

Stems simple, erect, finely puberulent, up to 6 dm. tall; leaves lanceolate to linear-lanceolate, long-acuminate, 5-10 cm. long, 4-12 mm. wide; inflorescence 1-5-flowered; sepals 8-10 mm. long; petals up to 2 cm. long; capsule 16-20 mm. long, 5-8 mm. wide, the teeth recurved; seeds flat, 2 mm. wide.

Woods, Yukon valley, Arctic coast and northern Eurasia. Fig. 424.

2. **C. arvense** L.   Field Chickweed

Stems caespitose, glandular-pubescent, 1–3 dm. tall; leaves narrowly lanceolate or oblanceolate, acute, 1–3 cm. long, 1–4 mm. wide, those at the base of the cyme shorter and wider; sepals 5–7 mm. long, petals about 1 cm. long, capsule scarcely exceeding the calyx.

Rocky places, circumboreal. Fig. 425.

3. **C. glomeratum** Thuill.   Mouse-ear Chickweed
    C. viscosum auct.

Stems tufted, viscid-pubescent, 1–3 dm. tall; leaves ovate to obovate, obtuse but often mucronate, 8–22 mm. long, 5–14 mm. wide; flowers usually more or less congested; sepals acute, about 4 mm. long; petals shorter than the sepals; capsule 6–8 mm. long, slender, on a short pedicel.

An introduced weed, native of Europe. Fig. 426.

4. **C. caespitosum** Gilib.   Larger Mouse-ear Chickweed
    C. vulgatum auct.

Stems viscid-pubescent, 1–4 dm. tall, leaves oblong, the upper becoming more or less lanceolate, obtuse, 1–3 cm. long, 3–8 mm. wide, villous; cymes leafy-bracted; sepals scarious-margined, often suffused with purple, 5–6 mm. long, about equaling the petals; capsule about 1 cm. long, slightly curved; pedicels 6–12 mm. long.

An introduced weed, native of Europe. Fig. 427.

5. **C. beeringianum** C. & S.   Beering Chickweed

Stems densely or loosely matted, spreading or ascending, glandular-pilose, 4–20 cm. long; leaves sometimes acute but mostly obtuse, 5–25 mm. long, more or less viscid-puberulent; cymes 1–4-flowered; sepals 3.5–8 mm. long, broadly lanceolate to oblong-ovate, the inner scarious-margined; capsules 8–12 mm. long.

Our commonest *Cerastium*, circumboreal. Fig. 428.

6. **C. aleuticum** Hult.   Aleutian Chickweed

About 5 cm. tall; leaves elliptic-lanceolate to obtuse-lanceolate, glabrous or with a few hairs on the surfaces, the margins strongly ciliate; sepals lanceolate, acute, pubescent, the margins scarious, 5–7 mm. long; petals about 9 mm. long. May be only a high alpine race of *C. beeringianum*.

Aleutian and Pribylof Islands.

7. **C. fischerianum** Sér.   Fischer Chickweed

Loosely matted; stems spreading or ascending, glandular-hispid, densely retrorsely hirsute below the nodes, 7–40 cm. long, the upper nodes usually elongated; leaves thick, lanceolate or ovate to lance-linear, usually acute, pilose on both surfaces, 1–4 cm. long, 3–16 mm. wide; cymes 3–27-flowered; sepals 4.5–9 mm. long, the margins hyaline, lanceolate to oblong, acute or acuminate.

Eastern Asia and Alaska. Fig. 429.

8. *C. arcticum* Lange. Arctic Chickweed

Plant densely tufted, stems viscid, pilose, 3–20 cm. long; leaves oval or elliptical, acute or obtuse, pilose, 5–25 mm. long; cymes 1–3-flowered; sepals ovate or ovate-lanceolate, scarious-margined, 4–8 mm. long; capsules 1.5–2 times as long as the sepals.

Eastern arctic Asia—Greenl.—northern Scandinavia.

2. STELLARIA L.

Tufted, weak, erect or spreading, annual or perennial herbs; leaves opposite; flowers usually in open cymes, sometimes solitary and axillary; sepals usually 5, rarely 4; petals white, deeply 2-cleft, or wanting; stamens 10 or fewer; styles 3, rarely 4 or 5; capsule globose to oblong, opening by twice as many valves as there are styles. (Latin, star, with reference to the star-shaped flower.)

```
1A. Flowers in the axis of scarious bracts or scarious-margined leaves.
  1B. Leaves linear-lanceolate, stems scabrous. ............. 5. S. longifolia
  2B. Leaves broader, stems smooth.
    1C. Sepals pubescent on back or ciliate on the margin. .. 6. S. laeta
    2C. Sepals glabrous or essentially so.
      1D. Sepals 5 mm. or more long. ...................... 2. S. alaskana
      2D. Sepals 3–4 mm. long. ........................... 7. S. longipes
2A. Flowers in the axils of green, not scarious-margined leaves.
  1B. Lower leaves long-petioled. ........................ 1. S. media
  2B. All leaves sessile.
    1C. Leaves lustrous, carinate. ......................... 6. S. laeta
    2C. Leaves not lustrous, flat.
      1D. Leaves ovate or ovate-lanceolate.
        1E. Leaves thin with translucent margins. .......... 4. S. crispa
        2E. Leaves thick, coriaceous, glaucous. ............ 3. S. ruscifolia
      2D. Leaves narrower.
        1E. Flowers axillary.
          1F. Sepals as long as the capsule. ................ 8. S. humifusa
          2F. Sepals shorter than the capsule. .............. 9. S. crassifolia
        2E. Flowers in terminal cymes.
          1F. Sepals 2–3 mm. long. .......................10. S. calycantha
          2F. Sepals 3–4 mm. long. .......................11. S. sitchana
```

1. *S. media* (L.) Cyril. Common Chickweed
Alsine media L.

A diffusely branching, decumbent or procumbent annual often rooting at the nodes; lower leaves cordate to ovate and petioled, 10–35 mm. long; upper leaves oval or ovate, becoming sessile at the inflorescence; inflorescence pubescent; sepals oblong-lanceolate, glandular-pubescent, about 5 mm. long; petals shorter than the sepals; capsule scarcely longer than the sepals.

Our most persistent weed, probably introduced, native of Europe. Fig. 430.

2. *S. alaskana* Hult. Alaska Starwort

Loosely tufted, glabrous and glaucous, 4–12 cm. tall; leaves crowded on the lower part of stem, lanceolate, acute or acuminate, 8–18 mm. long,

3–7 mm. wide; flowers 1 or 2; sepals narrowly triangular-lanceolate, prominently 3-nerved, acute, scarious-margined, 7–9 mm. long; petals scarcely equaling the sepals; capsule about as long as the sepals.

Central Alaska—Yukon—southeastern Alaska. Fig. 431.

3. *S. ruscifolia* Pall. ssp. *aleutica* Hult.   Ruscus-leaved Starwort

Stems loosely tufted, leafy, glaucous, 6–15 cm. tall; leaves lanceolate or ovate-lanceolate, acute, up to 18 mm. long; flowers long-peduncled, solitary, axillary but appearing terminal; sepals triangular-lanceolate, acute, scarious-margined, 5–7 mm. long; petals longer than the sepals, cleft half way.

Aleutians—Wiseman—southeastern Alaska, main species in eastern Asia. Fig. 432.

4. *S. crispa* C. & S.   Crisp Starwort
Alsine crispa (C. & S.) Holz.

Stems weak and decumbent, 1–4 dm. long; leaves ovate, acuminate, with crisp margins, 5–18 mm. long, nearly half as wide; flowers axillary; sepals lanceolate, acute, 3-nerved and with wide, scarious margins; petals minute or none; capsule longer than the calyx; seed brown, nearly smooth.

Woods, Aleutians—Wyo.—northern California. Fig. 433.

5. *S. longifolia* Muhl.   Long-leaved Starwort
Alsine longifolia (Muhl.) Britt.

Erect or ascending and diffusely branched, glabrous, the stem sharply 4-angled, 2–5 dm. long; leaves linear, sometimes ciliate near the base, 2–5 cm. long, 2–4 mm. wide, acute at both ends; inflorescence spreading; sepals lanceolate, acute about 3 mm. long, 3-nerved; petals slightly longer than the sepals; capsule exceeding the calyx.

Circumboreal. Fig. 434.

6. *S. laeta* Rich.   Shining Starwort
S. ciliatosepala Trautv.
S. laxmanni Fisch.
S. monantha Hult.
Alsine laeta (Rich.) Rydb.

Stems tufted, very leafy, 5–15 cm. tall; leaves lanceolate, sometimes glaucous, 8–18 mm. long, 2–4 mm. wide; flowers 1–few on rather long, erect peduncles; sepals lanceolate, about 4 mm. long; petals about 5 mm. long; capsule longer than the sepals. This group is very variable and several forms have been described as species. Perhaps these forms should be regarded as varieties.

Alpine and rocky places, probably circumboreal. Fig. 435.

7. *S. longipes* Goldie.   Long-stalked Starwort
Alsine longipes (Goldie) Cov.

Stems tufted, erect or ascending, simple or sparingly branched, 4-angled, 1–4 dm. tall; leaves linear-lanceolate, attenuate, rather firm and shining, 1–3 cm. long; flowers few to many in a terminal cyme; sepals 3–4 mm. long with scarious margins; petals exceeding the sepals; capsules about 5 mm. long, black and shining.

Circumboreal. Fig. 436.

8. *S. humifusa* Rottb.   Low Chickweed
Alsine humifusa Britt.

More or less fleshy; stems spreading or ascending, 5–25 cm. long; leaves ovate or oblong, 5–20 mm. long; flowers 1–few, axillary or terminal; sepals ovate-lanceolate, 4–5 mm. long; petals equaling or exceeding the sepals; capsule about as long as the sepals; seeds smooth.

Beaches, circumpolar. Fig. 437.

9. *S. crassifolia* Ehrh.   Fleshy Starwort
Alsine crassifolia (Ehrh.) Britt.

Stems weak, slender, diffuse, often growing in water, 5–25 cm. long; leaves small, 4–15 mm. long, 1.5–3 mm. wide; cymes terminal, few-flowered, or the flowers axillary and solitary; peduncles slender, sepals ovate-lanceolate, acuminate, exceeded by the petals and the capsule.

Widely distributed in our area, circumpolar. Fig. 438.

10. *S. calycantha* Bong.
Alsine calycantha (Bong.) Rydb.

Stems tufted, weak, 10–25 cm. tall; leaves ovate-lanceolate to linear-lanceolate, ciliolate at least in part, 5–25 mm. long, 2–6 mm. wide; cyme terminal, few-many-flowered; sepals lanceolate, acute, about 3 mm. long, longer than the petals, somewhat shorter than the capsule. Ssp. *interior* Hult. has roughened stem and smaller flowers, the sepals being about 1 mm. long.

The typical form occurs in the coastal districts, the ssp. in the interior of our area, circumboreal. Fig. 439.

11. *S. sitchana* Steud.   Sitka Starwort
S. borealis auct. in part.

Stems erect or ascending, sometimes weak and diffuse, 1–5 dm. long; leaves lanceolate or lance-linear, 1–5 cm. long, 3–8 mm. wide, often ciliolate at the base; cymes many-flowered; pedicels often reflexed in fruit; sepals ovate-lanceolate, acute, 4–5 mm. long, longer than the petals and about two-thirds as long as the capsule. Var. *bongardiana* (Fern.) Hult. has but few flowers which are axillary or terminal, the upper leaves but little reduced.

Wet soil, eastern Asia—Ida.—Calif. and in eastern America. Fig. 440.

3. ARENARIA L.

Annual or more often perennial herbs; stems usually tufted, erect or decumbent; leaves sessile, opposite or fascicled; flowers solitary in the axils or borne in cymes; sepals usually 5; petals 5, white, entire or slightly notched, or none; stamens 10, styles usually 3, many-ovuled. (Latin, sand, in allusion to the habitat of some of the species.)

1A. Leaves ovate, elliptical or lanceolate.
  1B. Leaves thin.
    1C. Plants 5-15 cm. tall. .... 3. *A. lateriflora*
    2C. Plants 2-6 cm. tall. .... 4. *A. humifusa*
  2B. Leaves thick.
    1C. Fleshy seashore plant. .... 1. *A. peploides*
    2C. Leaves less fleshy.
      1D. Flowers axillary. .... 2. *A. physodes*
      2D. Flowers terminal. .... 5. *A. dicranoides*
2A. Leaves very narrow.
  1B. Capsule opening with 6 teeth. .... 6. *A. capillaris*
  2B. Capsule opening with 3 teeth.
    1C. Stem and leaves glabrous.
      1D. Inflorescence 1-flowered. .... 7. *A. rossii*
      2D. Inflorescence branched. .... 8. *A. stricta*
    2C. Stems pubescent.
      1D. Sepals acute. .... 9. *A. rubella*
      2D. Sepals obtuse.
        1E. Leaves very acute. ....10. *A. laricifolia*
        2E. Leaves obtuse.
          1F. Leaves 3-nerved. ....14. *A. macrocarpa*
          2F. Leaves 1-nerved.
            1G. Seed smooth. ....11. *A. biflora*
            2G. Seed tuberculate.
              Sepals 3-4 mm. long. ....12. *A. obtusiloba*
              Sepals 5-8 mm. long. ....13. *A. arctica*

1. *A. peploides* L.   Sea-beach Sandwort
    Ammodenia peploides (L.) Rupr.
    Honckenya peploides (L.) Ehrh.

Stems glabrous, 1-6 dm. long, often much branched; leaves oblong to ovate, acute, clasping, 12-50 mm. long; flowers axillary or terminal; peduncles stout; sepals ovate, acute, 4-5 mm. long; petals greenish, about equaling the sepals; ovary 3-5-celled; capsule subglobose; seed smooth, obovoid. This species is represented in our area by two variants. The Pacific coast form is ssp. *major* (Hook.) Hult. which has longer stems, relatively narrower leaves and often several-flowered cymes as compared to the Arctic-Bering Sea form which is ssp. *latifolia* (Fenzl) Maguire.

The full species is circumpolar. Fig. 441.

2. *A. physodes* Fisch.   Merckia
    Merckia physodes (Fisch.) Fisch.

Stems trailing or decumbent, 1-3 dm. long, glandular-pubescent; leaves glabrous or nearly so, oval or ovate, 6-18 mm. long; sepals ovate, acute, 5-6 mm. long; petals white, about as long as the sepals; capsule 3-6-celled, about 6 mm. high and up to 1 cm. broad.

Wet places, mouth of Lena R.—northern Kamchatka—Mackenzie R. Fig. 442.

3. *A. lateriflora* L. Blunt-leaved Sandwort
    Moehringia lateriflora (L.) Fenzl.
    Stem slender, minutely pubescent, decumbent at base or ascending, 8–20 cm. tall; leaves oblong to ovate, obtuse or rounded at apex, ciliolate on margins and ribs beneath, 1–3 cm. long, 3–10 mm. wide; cymes 1–6-flowered; sepals ovate, 2–3 mm. long; petals obovate, 4–6 mm. long; capsule about 5 mm. long; seeds dark, appendaged.
    A widely distributed circumboreal species. Fig. 443.

4. *A. humifusa* Wahl. Low Sandwort
    Stems loosely to densely tufted, 2–8 cm. tall; leaves lanceolate or oblanceolate, papillose, 3–7 mm. long; flowers solitary, terminal, on puberulent peduncles 1–3 cm. long; sepals about 3.5 mm. long, exceeded by the capsule; seed brown, scarcely 1 mm. long.
    Seward Penin.—northern Finland. Fig. 444.

5. *A. dicranoides* (C. & S.) Hult. Matted Sandwort
    Cherleria dicranoides C. & S.
    Stellaria dicranoides (C. & S.) Seem.
    Stems glabrous, densely caespitose, forming small mats and only 1 or 2 cm. high; leaves imbricated, oblanceolate or obovate, 3–7 mm. long, 1–2 mm. wide, the old ones persisting; flowers solitary, terminal; peduncles 1–7 mm. long; sepals 2.5–4 mm. long; petals none; stamens borne on a prominent lobed disc; capsule nearly as long as the sepals; seed fully 1 mm. long, brown.
    Arctic-alpine, St. Lawrence Bay, Siberia—central Alaska. Fig. 445.

6. *A. capillaris* Poir. Beautiful Sandwort
    Caespitose, glabrous, branches of the caudex decumbent; stems usually erect, 8–20 cm. tall; leaves filiform with subulate tip, 2–7 cm. long, minutely ciliolate; cymes few-flowered; sepals 3.5–7 mm. long with scarious or colored margins and strong midvein; petals longer than the sepals; capsule as long as or longer than the sepals; seed black, about 1 mm. long.
    Central Asia—Yukon—?. Fig. 446.

7. *A. elegans* C. & S. Ross Sandwort
    A. rossii R. Br.
    Minuartia elegans (C. & S.) Schischkin.
    Stems densely tufted, 2–6 cm. tall, glabrous or nearly so; leaves linear, fleshy, 4–8 mm. long, 1-nerved; flowers usually solitary on rather long peduncles; sepals about 3 mm. long; petals and capsule about as long as the sepals; seed brown.
    Seward Penin.—Greenl.—Spitzbergen—Colo.—Ore. Fig. 447.

8. *A. stricta* (Sw.) Michx. Rock Sandwort
    A. dawsonensis Britt.
    Minuartia stricta (Sw.) Hiern.
    Stems slender, much branched from the base, 1–3 dm. tall; leaves

filiform or linear-subulate, 8–20 mm. long; cymes spreading; bracts lanceolate or subulate; sepals acute, 3-nerved, 3–4 mm. long; petals nearly as long as the sepals; capsule exceeding the sepals; seed dark brown, about 0.6 mm. long.

Circumpolar. Fig. 448.

9. *A. rubella* (Wahl.) Sm.
Minuartia rubella (Wahl.) Graebn.

Glandular-puberulent, branched from the base and spreading, 4–15 cm. tall; leaves linear-subulate, ascending, 3-nerved, 5–10 mm. long, less than 1 mm. wide; sepals lanceolate, acute, 3-nerved, scarcely 3 mm. long; petals about as long as the sepals; capsule slightly longer than the sepals; seed brownish-black.

Circumpolar. Fig. 449.

10. *A. laricifolia* (L.) Gray. Larch-leaved Sandwort
Minuartia laricifolia (L.) Schinz & Thell.
Alsinopsis laricifolia (L.) Heller.

Stems tufted, decumbent below, erect or ascending above, 8–18 cm. tall; leaves linear-filiform, ciliolate or glabrous, up to 15 mm. long; cymes 1–4 flowered; sepals oblong, 3-nerved, puberulent, 5–7 mm. long; petals about 1 cm. long; capsule slightly exceeding the calyx.

Western and central Alaska—Yukon—Ida.—Mont.—Wash. and central Europe. Fig. 450.

11. *A. biflora* (L.) Wats. Two-flowered Sandwort
Minuartia biflora (L.) Schinz & Thell.

Caespitose, 5–12 cm. tall; leaves flat, linear, 4–8 mm. long; sepals 3-nerved, about 4 mm. long; petals about same length as sepals; capsule exceeding the calyx.

Rare in our area, probably circumboreal.

12. *A. obtusiloba* (Rydb.) Fern. Alpine Sandwort
Alsinopsis obtusifolia Rydb.
Minuartia obtusifolia (Rydb.) House.

Densely caespitose, the lower part of stem clothed with old leaves, 1–6 cm. tall; leaves imbricate, linear, 4–8 mm. long, rather rigid, ciliolate on the margins; flowers usually solitary; sepals glandular-pubescent, 3-nerved, 3–4 mm. long; petals and capsules longer than the sepals.

Northern and central Alaska—Yukon—Alta.—Utah—N. Mex.

13. *A. arctica* Stev. Arctic Sandwort
Minuartia arctica (Stev.) Ascher. & Graebn.

Loosely to densely caespitose, 2–10 cm. tall; leaves linear, glabrous, the margins entire; flowers solitary; sepals obtuse, pubescent, 5–8 mm. long; petals 7–10 mm. long; capsule 8–10 mm. long. A very variable species, probably hybridizing with the next and other species.

Arctic-alpine, common in western Alaska, less so eastward to Yukon. Fig. 451.

14. *A. macrocarpa* Pursh.   Long-podded Sandwort
Minuartia macrocarpa (Pursh.) Ostenf.

More or less caespitose, 2–10 cm. tall; leaves linear, obtuse, with ciliate margins, 5–12 mm. long; flowers usually solitary; sepals 5–7 mm. long; petals 8–11 mm. long; capsules 10–15 mm. long, seed with long spines or tubercles.

Arctic-alpine, Nova Zemla—Siberia—Alaska. Fig. 452.

4. SAGINA L.

Low tufted or matted herbs; leaves opposite, filiform or subulate; flowers small, whitish, on more or less elongated pedicels; sepals 4 or 5, persistent; petals 4 or 5 or wanting; stamens as many as the sepals, fewer or twice as many; styles as many as the sepals; capsules dehiscent to the base, the valves opposite the sepals. (Ancient name of the spurry.)

```
1A. Annual, without basal rosettes. ........................ 1. S. occidentalis
2A. Perennials, with basal rosette of leaves.
  1B. Pedicel and calyx glandular. ........................ 2. S. litoralis
  2B. Pedicels and calyx glabrous.
    1C. Branches rooting at the nodes. .................... 3. S. linnaei
    2C. Caespitose, not rooting at the nodes.
      1D. Calyx 1.5–2 mm. long. ........................... 4. S. intermedia
      2D. Calyx about 3 mm. long. ........................ 5. S. crassicaulis
```

1. *S. occidentalis* Wats.   Western Pearlwort

Stems slender, more or less branched, decumbent or ascending, 3–10 cm. tall; leaves linear, acute; calyx rounded at the base, the sepals about 2 mm. long; petals when present shorter than the sepals; capsules about 3 mm. long.

Occasionally found introduced, native B. C.—Calif.

2. *S. litoralis* Hult.   Beach Pearlwort

Stems branched from the base, 5–10 cm. long; leaves glabrous, the basal filiform, about 15 mm. long; stem leaves subulate, 4–6 mm. long; peduncles 15–20 mm. long; sepals elliptic-ovate; petals shorter than the sepals; capsule acute, exceeding the sepals; seed with low papillae, about 0.6 mm. long.

Eastern Asia—southeastern Alaska.

3. *S. linnaei* Presl.   Arctic Pearlwort
S. saginoides (L.) Britt.

Stems decumbent, tufted, glabrous, 3–10 cm. long; leaves subulate, 5–15 mm. long; flowers usually solitary at the end of the stems; sepals oval, obtuse, 1.5–2 mm. long; petals scarcely as long as the sepals; capsules 3 mm. long; seed about 0.3 mm. long.

Circumboreal. Fig. 453.

4. *S. intermedia* Fenzl.   Snow Pearlwort
S. nivalis auct.

Stems densely caespitose, 1–5 cm. tall, 1–3-flowered; leaves crowded, subulate, 3–8 mm. long; sepals oval, rounded at the tip, purple-edged, scarcely 2 mm. long; petals short and narrow; capsules about 3 mm. long on pedicels 3–10 mm. long; seed about 0.5 mm. long.

Circumpolar. Fig. 454.

5. *S. crassicaulis* Wats.   Fleshy Pearlwort

Stems caespitose, glabrous, somewhat fleshy, branching, 3–10 cm. long; basal leaves linear, 1–2 cm. long; stem leaves shorter, connate; peduncles 1–4 cm. long; sepals oval; petals scarcely equaling the sepals; capsules about 4 mm. long; seed about 0.4 mm. long.

Along the coast, eastern Asia—Calif. Fig. 455.

5. SPERGULA L.

Annual branching herbs; leaves subulate or filiform, succulent, borne in whorls; flowers small, white, in terminal cymes; sepals, petals, styles and valves of the capsule each 5; stamens 5 or 10; seed compressed, narrowly winged. (Latin, to scatter.)

Spergula arvensis L.   Spurry

Slender, sparingly pubescent, 15–50 cm. tall; leaves linear-filiform, 2–5 cm. long; cymes loose, many-flowered; pedicels reflexed in fruit; sepals 3–4 mm. long; petals slightly exceeding the sepals; capsule ovoid, longer than the sepals; seed black.

An introduced weed, native of Europe. Fig. 456.

6. SPERGULARIA Presl

Low herbs; leaves somewhat succulent with scarious stipules and secondary leaves fascicled in their axils; sepals 5; petals 5, fewer or none; stamens 2–10; styles 3, capsule 3-valved. (Diminutive of *Spergula*.)

Seeds winged. .... 1. *S. canadensis*
Seeds not winged. .... 2. *S. rubra*

1. *S. canadensis* (Pers.) G. Don.   Canadian Sand Spurry
Tissa canadensis (Pers.) Britt.

Stems erect, spreading or decumbent, more or less pubescent, at least above, about 1 dm. tall; leaves linear-filiform, 1–4 cm. long; sepals ovate, 2.5–3.5 mm. long; petals pink or white, shorter than the sepals; capsule exceeding the calyx, more or less deflexed; seed brown, 1–1.4 mm. long, surrounded by an erose, membranous wing varying from a mere ridge to 0.5 mm. wide.

Sea beaches, Kodiak Isl.—Queen Charlotte Isls. and Labr.—N. Y. Fig. 457.

2. *S. rubra* (L.) Presl.   Purple Sand Spurry
Tissa rubra (L.) Britt.

Stems prostrate or decumbent, often forming dense mats, 6–25 cm. long; leaves linear, flat, fascicled, 6–12 mm. long; sepals acute, about 4 mm. long; petals bright pink, scarcely as long as the sepals; capsule sometimes exceeding the calyx; seed dark brown, sculptured, about 0.5 mm. long.

Introduced, native of Eurasia.

7. SILENE L.

Herbs with perfect flowers in terminal cymes or solitary; calyx with or more or less inflated tube, 10- or more-nerved; petals 5, in ours pink or white, with an appendaged crown, usually notched or cleft; stamens 10; styles usually 3; ovary sometimes incompletely 2–4-celled; capsule often stipitate, opening by usually 6 valves; seed tuberculate or echinate. (Greek, saliva, in allusion to the viscid secretion of some species.)

1A. Dwarf matted alpine perennial. .... 1. *S. acaulis*
2A. Taller plants.
  1B. Introduced annual weed. .... 5. *S. noctiflora*
  2B. Native perennials.
    1C. Calyx rose colored. .... 2. *S. repens*
    2C. Calyx green.
      1D. Calyx 8–12 mm. long. .... 4. *S. williamsii*
      2D. Calyx 5–7 mm. long. .... 3. *S. menziesii*

1. *S. acaulis* L.   Moss Campion. Moss Pink

Stems very densely caespitose in moss-like cushions; leaves crowded, linear, 5–15 mm. long, the margins glandular-ciliolate; flowers solitary at the end of the branches, pink or purplish, on short peduncles; calyx 5–6 mm. long; petals emarginate or 2-lobed.

Rocky places, circumboreal. Fig. 458.

2. *S. repens* Patin.   Pink Campion

Stems several, leafy, puberulent, more or less decumbent at the base, 10–25 cm. tall; leaves linear-lanceolate, finely pubescent to nearly glabrous; the margins ciliolate, 2–5 cm. long; calyx villous, 10–12 mm. long, the lobes rounded; petals rose-pink, much longer than the calyx, the blades bifid.

Interior Alaska—Yukon—Mont. and northern Europe. Fig. 459.

3. *S. menziesii* Hook.   Menzies Campion

Stems 1–4 dm. tall, usually much branched; leaves ovate-lanceolate, acute at both ends, more or less pubescent on both surfaces, 2–8 cm. long, 5–25 mm. wide; inflorescence a leafy-bracted cyme, calyx campanulate, the lobes often purplish; petals white, a little longer than the calyx; seed black, shining.

Kenai Penin.—Yukon—Man.—N. Mex.—Calif. Fig. 460.

4. *S. williamsii* Britt.   Williams Campion

Glandular-pubescent throughout, leafy, 1–4 dm. tall; leaves sessile, lanceolate to linear-lanceolate, 2–8 cm. long, 3–15 mm. wide; inflorescence dichotomous; petals white, forked, slightly or not at all exceeding the calyx; capsule as long as or slightly longer than the calyx; seed brown, tuberculate.

Central Alaska—Mackenzie R. Fig. 461.

5. *S. noctiflora* L.   Night-flowering Catchfly

A coarse, viscid-pubescent weed, 3–10 dm. tall; lowermost leaves obovate, narrowed in a petiole; upper leaves lanceolate and acute or acuminate, sessile, 4–10 cm. long; calyx at flowering tubular, becoming inflated in fruit, 2–3 cm. long with subulate teeth; petals white or pinkish, exceeding the calyx.

Native of Europe.

8. DIANTHUS L.

Mainly perennial plants with narrow leaves and terminal, usually solitary flowers; calyx tubular, 5-toothed, finely many-striate, with bracts at the base; petals 5, dentate or crenate, long-clawed; stamens 10; styles 2; pod 4-valved, seed flattened. (Greek, the flower of Jove (Zeus).)

D. repens Willd.   Northern Pink

Stems more or less decumbent, 5–15 cm. tall; leaves linear or linear-lanceolate, 2–4 cm. long, connate at the base; calyx somewhat inflated, 12–14 mm. long; petals pink or purplish, the spreading limb about 1 cm. long.

Rocky places, northern Eurasia—central Alaska.

9. LYCHNIS (Tourn.) L.

Ours perennials; calyx ovoid, more or less inflated, 5-toothed, 10-nerved; petals in ours usually inconspicuous, with small crown and 2-cleft blades; stamens 10, styles usually 5; capsule opening by twice as many valves as there are styles. (Greek, lamp, in allusion to the flame-colored flowers of some species.)

1A. Seeds 1.8 mm. or more in diameter.
  1B. Flowers 1, rarely 2, petals purplish. .... 1. *L. apetala*
  2B. Flowers 1–3, petals pale rose. .... 2. *L. macrosperma*
2A. Seeds less than 1.8 mm. in diameter.
  1B. Seeds small, wingless. .... 6. *L. dawsonii*
  2B. Seeds more or less winged.
    1C. Plants 3–5 dm. tall. .... 5. *L. taylorae*
    2C. Plants 10–25 cm. tall.
      1D. Petals white. .... 3. *L. furcata*
      2D. Petals reddish-violet. .... 4. *L. soczavianum*

1. *L. apetala* L.   Nodding Lychnis
Melandrium apetalum (L.) Fenzl.
Wahlbergella apetala (L.) Fries.

Stems solitary or a few together, glandular-pubescent, at least above; flowers usually solitary, nodding but becoming erect in fruit; calyx ellipsoid, much inflated, purple-veined, 12–15 mm. long with broad teeth; petals slightly longer than the calyx; seed brown with nearly circular wing, 1.8–2.4 mm. wide.

Alpine-arctic, circumpolar. Fig. 462.

2. *L. macrosperma* (Pors.) J. P. Anderson, n. comb.
                                                               Large-seeded Lychnis
Melandrium macrospermum A. E. Porsild in Rhodora 41 (1939) p. 225.

Stems few, densely pubescent, conspicuously flexuous, 10–30 cm. tall;

base leaves numerous, oblanceolate; inflorescence of 1-3 flowers; calyx about 15 mm. long, 10 mm. wide; petals barely exserted; seed dark brown with thick wings.

Bering Sea—Mt. McKinley Park.

3. *L. furcata* (Raf.) Fern. Arctic Lychnis
 L. affinis Am. auct.
 Melandrium furcatum (Raf.) Hult.

Stems tufted, glandular-pubescent, 5-30 cm. tall; leaves linear or narrowly oblanceolate, up to 3 cm. long; calyx ellipsoid, 8-12 mm. long, inflated in fruit; petals white, exserted; seed tuberculate-striate with irregular wings, 1-1.5 mm. wide.

Arctic-alpine, circumpolar.

4. *L. soczavianum* (Schischk.) J. P. Anderson n. comb.
 Melandrium soczavianum Schischk. in Journ. Soc. Bot. Russe 16 (1931) p. 83, et. fig. p. 84.

Resembles *L. furcata*; stems caespitose, erect or ascending, 7-20 cm. tall, 1-3-flowered; flowers usually nodding; calyx 10-14 mm. long.

Bering Sea region of Asia and Alaska.

5. *L. taylorae* Robins. Taylor Lychnis
 Melandrium taylorae (Robins.) Tolm.

More or less viscid-puberulent; basal leaves linear-oblanceolate, narrowed into a margined petiole; stem leaves sessile and clasping, 3-8 cm. long; flowers long-peduncled; petals exserted; capsule 10-15 mm. long, seed as in *L. furcata*.

Yenisei River—Mackenzie district.

6. *L. dawsonii* (Robins.) J. P. Anderson, n. comb. Dawson Lychnis
 L. triflora R. Br. var. *dawsonii* Robins. in Proc. Amer. Acad. 28 (1893) p. 149.
 Melandrium dawsonii (Robins.) Hult.

Stems 2-4 dm. tall; calyx scarcely inflated, about 1 cm. long, 5 mm. wide in fruit, densely pubescent; petals decidedly longer than the calyx; flowers axillary or glomerulate at the top.

Copper Center—Mackenzie district—B. C.

10. SAPONARIA L.

Caulescent herbs; leaves clasping, flowers slender-pedicelled in cymes; calyx inflated in fruit; stamens 10; styles 2; capsule 4-toothed. (Latin, soap, from the saponin in the stems.)

S. vaccaria L. Cow Herb
 Vaccaria segetalis (Neck.) Garcke.

An introduced weed, 3-10 dm. tall; leaves ovate-lanceolate, 3-8 cm. long; flowers long-pedicelled; calyx 5-winged; petals pale red.

Native of Eurasia.

Agrostemma githago L., the Corn Cockle, has been collected a few times in Alaska. Stems erect, simple or with a few branches, densely pubescent with appressed hairs, 3–9 dm. tall; leaves linear-lanceolate; flowers showy; calyx ovoid, its lobes linear, foliaceous, exceeding the petals; deciduous in fruit; seeds numerous, black.

PLATE XVII

Scale marked in millimeters.
FIG.
359. *Populus tremuloides* Michx. Leaf and capsule.
360. *Populus tacamahacca* Mill. Leaf and young capsule.
361. *Populus tricocarpa* T. & G. Leaf and dehisced capsule.
362. *Myrica gale* L. Leaf and drupe.
363. *Betula glandulosa* Michx. All drawings of *Betula* show leaf, scale and nutlet.
364. *Betula nana exilis* (Sukatch.) Hult.
365. *Betula papyrifera occidentalis* (Hook.) Hult.
366. *Betula kenaica* W. H. Evans.
367. *Betula resinifera* Britt.
368. *Betula glandulosa* × *resinifera* (*B. eastwoodae* Sarg.)
369. *Betula glandulosa* × *resinifera* another form.
370. *Betula kenaica* × *nana exilis* (*B. hornei* Butler)
371. *Alnus crispa* (Ait.) Pursh. Illustrations of *Alnus* show leaf and nutlet.
372. *Alnus fruticosa* Rupr.
373. *Alnus fruticosa* var. *sinuata* (Regel) Hult.
374. *Alnus incana* (L.) Moench.
375. *Alnus oregona* Nutt.
376. *Urtica lyallii* Wats. Leaf and fruit.
377. *Urtica gracilis* Ait. Leaf, flower, fruit, and utricle.
378. *Arceuthobium tsugense* (Rosend.) G. N. Jones. End of branch.
379. *Geocaulon lividum* (Rich.) Fern. Leaf, flower, fruit.
380. *Koenigia islandica* L. Node with leaf, fruit.
381. *Rumex acetosella* L. Leaves, fruit.
382. *Rumex acetosa* L. Leaves and fruit.
383. *Rumex obtusifolius agrestis* (Fr.) Danser. Leaf and fruit.
384. *Rumex maritimus* L. Leaf and fruit.
385. *Rumex crispus* L. Basal leaf, stem leaf, and fruit.
386. *Rumex domesticus* Hartm. Leaf and fruit.
387. *Rumex arcticus* Trautv. Leaves and fruit.
387a. *Rumex arcticus* Trautv. An extreme form.
388. *Rumex fenestratus* Greene. Leaf and fruit.
389. *Rumex transitorius* Rech. f. Leaf, fruit, and achene.
390. *Oxyria digyna* (L.) Hill. Fruit and leaf.

FLORA OF ALASKA 221

PLATE XVII

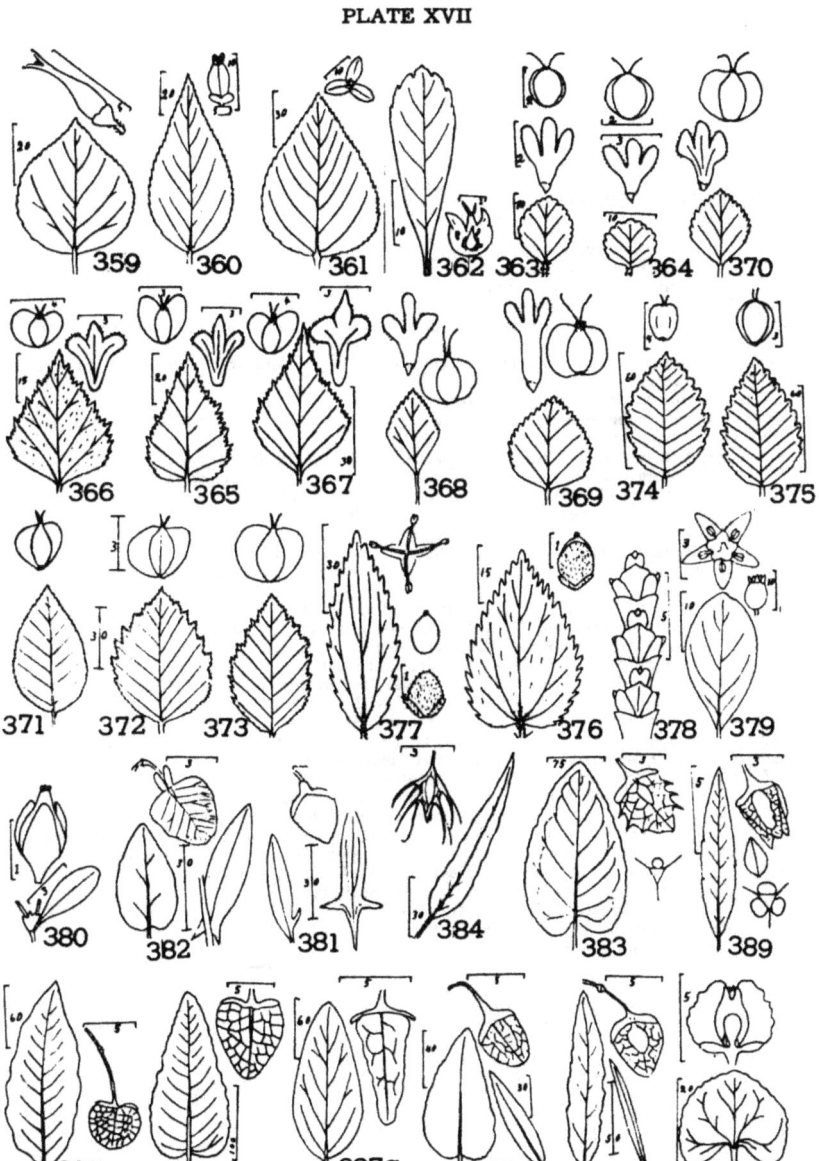

PLATE XVIII

Scale marked in millimeters.

FIG.
391. *Polygonum convolvulus* L. Leaf, fruit, and achene.
392. *Polygonum bistorta plumosum* (Small) Hult. Leaf, flower, and achene.
393. *Polygonum viviparum.* Leaves, flower, and bulblet.
394. *Polygonum alaskanum* (Small) Wight. Leaf, flower, and achene.
395. *Polygonum amphibium laevimarginatum* Hult. Leaf, flower, and fruit of aquatic form.
395a. *Polygonum amphibium laevimarginatum* Hult. Leaf and nodal sheaf of terrestrial form.
396. *Polygonum nodosum* Pers. Node with leaf and achene.
397. *Polygonum persicaria* L. Node with leaf and triangular and flat achenes.
398. *Polygonum caurianum* Robins. Leaf and fruit.
399. *Polygonum fowleri* Robins. Node with leaf, fruit, and achene.
400. *Polygonum buxiforme* Small. Node with leaf, fruit, and achene.
401. *Polygonum heterophyllum* Lindm. Leaf, fruit, and achene.
402. *Polygonum achoreum* Blake. Node with leaf, fruit, and achene.
403. *Polygonum neglectum* Bess. Leaf, fruit, and achene.
404. *Chenopodium capitatum* (L.) Achers. Leaf and utricle.
405. *Chenopodium glaucum salinum* (Stand.) Aellen. Leaf and top and side view of fruit.
406. *Chenopodium album* L. Leaf, flower, and utricle.
407. *Monolepis nuttalliana* (Schult.) Greene. Leaf, fruit, and sepal.
408. *Atriplex alaskensis* Wats. Leaf and fruit.
409. *Atriplex gmelini* C. A. Mey. Leaves and fruit.
410. *Corispermum hyssopifolium* L. Lower leaf, upper leaf, and utricle.
411. *Suaeda maritima* (L.) Dumort. Node with leaf, fruit, and utricle.
412. *Salicornia herbacea* L. Flowering spike, a portion enlarged.
413. *Salicornia pacifica* Standl. Flowering spike enlarged.
414. *Claytonia tuberosa* Pall. Tuber with leaf, calyx, and seed.
415. *Claytonia acutifolia* Pall. Leaf, petal, seed, and calyx.
416. *Claytonia arctica* Adams. Basal leaf, petal, calyx, and stem leaves.
417. *Claytonia sibirica* L. Basal leaf, stem leaves seed, petal, and sepal.
418. *Claytonia sarmentosa* C. A. Mey. Basal leaf, sepal, stem leaves, petal, and seed.
419. *Claytonia scammaniana* Hult. Basal leaf, petal, calyx, and stem leaves.
420. *Claytonia perfoliata* Donn. Basal leaf, seed, calyx, and stem leaves.
421. *Claytonia chamissoi* Esch. Leaf, petal, and sepal.
422. *Claytonia flagellaris* Bong. Calyx, basal leaf, and petal.
423. *Montia lamprosperma* Cham. Fruit, leaf, seed.
424. *Cerastium maximum* L. Leaf and fruit.

FLORA OF ALASKA

PLATE XVIII

PLATE XIX

Scale marked in millimeters.

FIG.
425. *Cerastium arvense* L. Node and fruit.
426. *Cerastium glomeratum* Thuill. Leaf and fruit.
427. *Cerastium caespitosum* Gilib. Leaf and fruit.
428. *Cerastium beeringianum* C. & S. Leaf and fruit.
429. *Cerastium fischerianum* Sér. Leaf and fruit.
430. *Stellaria media* (L.) Cyril. Leaf and fruit.
431. *Stellaria alaskana* Hult. Leaf and flower.
432. *Stellaria ruscifolia aleutica* Hult. Leaf and flower.
433. *Stellaria crispa* C. & S. Leaf and fruit.
434. *Stellaria longifolia* Muhl. Leaf and fruit.
435. *Stellaria laeta* Rich. Leaves and fruit.
436. *Stellaria longipes* Goldie. Leaf and fruit.
437. *Stellaria humifusa* Rottb. Leaf and fruit.
438. *Stellaria crassifolia* Ehrh. Leaf and fruit.
439. *Stellaria calycantha* Bong. Leaves and fruit.
440. *Stellaria sitchana* Steud. Leaves and fruit.
441. *Arenaria peploides major* (Hook.) Hult. Node, fruit, and seed.
442. *Arenaria physodes* Fisch. Node, fruit, and seed.
443. *Arenaria lateriflora* L. Leaf and fruit and seed.
444. *Arenaria humifusa* Wahl. Node, fruit, and seed.
445. *Arenaria dicranoides* (C. & S.) Hult. Top of flowering stem and seed.
446. *Arenaria capillaris* Poir. Capsule, seed, and leaf.
447. *Arenaria elegans* C. & S. Leaf, seed, and fruit.
448. *Arenaria stricta* (Sw.) Michx. Leaf, seed, and fruit.
449. *Arenaria rubella* (Wahl.) Sm. Fruit, seed, and leaf.
450. *Arenaria laricifolia* (L.) Gray. Flower and leaf.
451. *Arenaria arctica* Stev. Seed, leaf, and fruit.
452. *Arenaria macrocarpa* Pursh. Leaf, fruit, and seed.
453. *Sagina linnaei* Presl. Fruit, seed, and leaf.
454. *Sagina intermedia* Fenzl. Fruit, seed, and leaf.
455. *Sagina crassicaulis* Wats. Fruit, seed, and node.
456. *Spergula arvensis* L. Fruit, seed, and node.
457. *Spergularia canadensis* (Pers.) G. Don. Fruit, seed, and node.
458. *Silene acaulis* L. Leaf and calyx.
459. *Silene repens* Patin. Calyx and leaf.
460. *Silene menziesii* Hook. Fruit (the capsule dehisced) and leaf.
461. *Silene williamsii* Britt. Fruit and leaf.
462. *Lychnis apetala* L. Seed, fruit, and leaf.

PLATE XIX

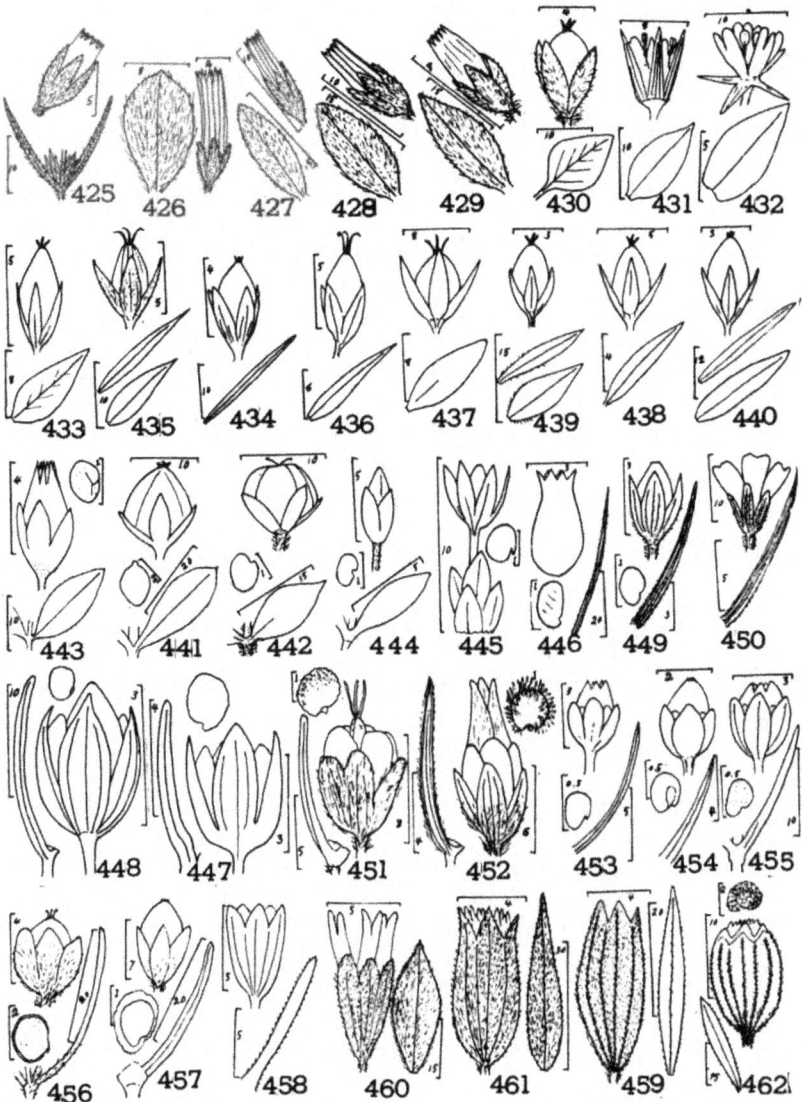

11. CABOMBACEAE (Water-shield Family)

Aquatic perennials; stems mucilage-coated; flowers solitary, axillary; sepals and petals usually 3; stamens 3–18; carpels 2–18, distinct; ovules 2 or 3; fruit indehiscent, septate; seeds 1–3, borne on the dorsal suture.

BRASENIA Schreb.

Stems slender, branching, covered with gelatinous matter as are also the petioles, peduncles and under surface of the leaves; leaves alternate, oval or elliptical, entire, floating, the petiole attached at center of under surface; flowers axillary, purple; sepals and petals each 3; stamens 12–18; carpels 4–18. (Name unexplained.)

B. schreberi Gmel.   Water-shield.

Leaves 4–8 cm. long, 3–5 cm. wide, tinged with purple, especially underneath; sepals and petals deep purple; stamens purple.

Southeastern Alaska, of scattered circumboreal distribution. Fig. 463.

12. RANUNCULACEAE (Crowfoot Family)

Ours all herbs; leaves alternate, without stipules but often with the base of the petioles clasping or sheathing the stem; sepals 3–15, green and caducaceous, or in some genera petaloid and persistent; petals as many as the sepals or wanting; stamens usually many; carpels few–many, rarely solitary, 1-celled, 1–many ovuled; fruit a berry or composed of achenes or follicles.

```
1A. Carpels 1-ovuled, fruit composed of achenes.
   1B. Petals usually present............................... 1. Ranunculus
   2B. Petals wanting, but sepals usually petal-like.
      1C. Flowers subtended by an involucre of leaf-like
          bracts, these sometimes remote from the calyx.... 2. Anemone
      2C. Flowers not subtended by an involucre........... 3. Thalictrum
2A. Carpels several-ovuled.
   1B. Fruit a berry ........................................ 7. Actaea
   2B. Fruit composed of follicles.
      1C. Flowers regular.
```

1D. Follicles decidedly stipitate.....................6. *Coptis*
   2D. Follicles short-stipitate or sessile.
      1E. Petals none, sepals petaloid.....................4. *Caltha*
      2E. Petals present.
         1F. Petals small and inconspicuous, sepals petaloid 5. *Trollius*
         2F. Petals showy, spurred.....................8. *Aquilegia*
2C. Flowers irregular.
   1D. Posterior sepal spurred..........................9. *Delphinium*
   2D. Posterior sepal forming a hood..................10. *Aconitum*

1. RANUNCULUS (Tourn.) L.

Mostly biennial or perennial plants with yellow or white, rarely reddish, flowers; leaves entire, lobed, divided, or dissected; sepals 5, deciduous; petals usually 5, occasionally more, each with a nectiferous gland and a scale at the base; carpels many, each developing into a flattened achene tipped with the style which forms a beak. (Latin, diminutive of frog, from the marshy habitat of many of the species.)

1A. Petals white or red.
   1B. Aquatic, achenes transversely ridged................27. *R. aquatilis*
   2B. Achenes not transversely ridged.
      1C. Larger leaves 3-lobed...........................28. *R. pallasii*
      2C. Lower leaves many-lobed......................26. *R. chamissonis*
2A. Petals yellow.
   1B. Achenes longitudinally ribbed......................23. *R. cymbalaria*
   2B. Achenes not longitudinally ribbed.
      1C. Plant scapose from filiform rootstock..............29. *R. lapponicus*
      2C. Plant not from filiform rootstock.
         1D. Petals 7-15.
            1E. Leaves deeply 3–5-lobed.....................24. *R. cooleyae*
            2E. Leaves entire or toothed.....................25. *R. kamchaticus*
         2D. Petals usually 5, sometimes more or less.
            1E. Leaves entire, narrow, stems creeping.........19. *R. flammula*
            2E. At least some of the leaves lobed, parted, or divided.
               1F. Palustrine or aquatic species.
                  1G. Leaves small, 3-lobed.....................20. *R. hyperboreus*
                  2G Leaves larger, 3–5-lobed.
                     1H. Plant usually erect with thick reniform
                         leaves ....21. *R. sceleratus*
                     2H. Plant floating or creeping, leaves
                         orbicular ....22. *R. gmelini*
               2F. Plants terrestrial but often growing in wet places.
                   (see also 21. *R. sceleratus*)
                  1G. Petals scarcely exceeding the sepals.
                     1H. Plants less than 1 dm. tall.
                        1I. Sepals glabrous ....17. *R. pygmaeus*
                        2I. Sepals copiously pubescent..............11. *R. verecundus*
                     2H. Plants more than 1 dm. tall.
                        1I. Stems glabrous or nearly so.
                           1J. Basal leaves crenate or somewhat
                               lobed ....9. *R. abortivus*
                           2J. Basal leaves divided and cleft........ 5. *R. bongardii* var.
                        2I. Stems pubescent.
                           1J. Beak of achenes triangular........... 7. *R. pennsylvanicus*
                           2J. Beak of achenes hooked.............. 5. *R. bongardii*
                  2G. Petals conspicuously longer than the sepals.
                     1H. Stems decumbent.
                        1I. Beak of achenes short.................. 2. *R. repens*
                        2I. Beak of achenes long................... 3. *R. septentrionalis*
                     2H. Stems erect or ascending.
                        1I. Sepals pubescent.
                           1J. Pubescence of sepals dark brown.

```
    1K. Receptacle glabrous..............15. R. nivalis
    2K. Receptacle brown-hispid..........16. R. sulphureus
   2J. Pubescence of calyx light-colored.
       (See also 10. R. eastwoodianus).
    1K. Plants 3–8 dm. tall.
     1L. Receptacle elongated in fruit,
         hairy ........................... 6. R. macounii
     2L. Receptacle but little elongated in
         fruit, glabrous.
      1M. Beak of achenes recurved
          about 1.5 mm. long............ 4. R. occidentalis & vars.
      2M. Beak of achenes short,
          slightly curved................ 1. R. acris
    2K. Plants less than 3 dm. tall.
     1L. Plants less than 1 dm. tall........14. R. grayi
     2L. Plants taller.
      1M. Achenes about 1.5 mm. long...13. R. eschscholtzii
      2M. Achenes with beak about
          2.5 mm. long..................12. R. pedatifidus
      3M. Achenes with beak about
          4 mm. long ................... 4. R. occidentalis
 2I. Sepals glabrous or nearly so.
     (See also 13. R. eschscholtzii).
  1J. Radical leaves lacking.................18. R. verticillatus
  2J. Radical leaves present.
    1K. Stems less than 3 dm. tall..........10. R. eastwoodianus
    2K. Stems 4–8 dm. tall................. 8. R. orthorhynchus
```

1. *R. acris* L.   Tall Buttercup

Stems erect, more or less pubescent, 3–9 dm. tall; lower leaves hairy, 3- to 5-divided to near the base, the divisions more or less cleft and divided into lanceolate lobes; petals bright yellow, 9–12 mm. long, twice the length of the hairy sepals; head of fruit globose; achenes with a short, curved beak. Var. *frigidus* Regel. Less vigorous than the type, radical leaves truncate at the base and palmately tripartate.

The typical form is introduced and native of Europe, the variety in east Asia and the Aleutians. Fig. 464.

2. *R. repens* L.   Creeping Buttercup.

More or less hairy, spreading by means of decumbent stems which root at the nodes; leaves ternate, the divisions petiolate, ternately cleft and toothed; petals ovate, about 8 mm. long, about twice the length of the sepals; fruiting heads globose; achenes margined, about 4 mm. long including the acute, slightly curved beak which is nearly 1 mm. long.

An introduced weed, native of Europe. Fig 465.

3. *R. septentrionalis* Poir.   Swamp Buttercup.

Plants subglabrous to hispid, branching, 2–6 dm. tall, some of the branches procumbent; leaves usually 3-divided, the divisions stalked; leaflets 3-lobed, -cleft, or -parted, and again toothed or lobed; sepals spreading or reflexed; petals bright yellow, twice as long as the sepals; achenes with long, strongly-margined, subulate beak. Our form has wider and shorter beak than the type and has been described as ssp. *pacifica* Hult.

N. Dak.—Labr.—Va.—Mo.—Texas, the subspecies known only from southeastern Alaska.

4. **R. occidentalis** Nutt. Western Buttercup.

Stems rather slender, 2–5 dm. tall, more or less hirsute or pilose; lower leaves pubescent, 2–5 cm. wide, 3-parted, the divisions cleft and toothed; upper leaves with linear divisions; petals about 1 cm. long; fruiting head globose; achenes about 2.5 x 2 mm. with a beak about 1.5 mm. long. Var. *brevistylis* Greene has a shorter beak and is somewhat more robust and nearly glabrous. ssp. *nelsoni* (DC.) Hult. grows up to 8 dm. tall; leaves deeply 3-parted, the central division again 3- to 9-lobed or toothed; achenes about 3 x 2.5 mm. with beak up to 2 mm. long. Ssp. *turneri* (Greene) J. P. Anderson, comb. nov. (*R. turneri* Greene, Pittonia 2: 296. 1892) hirsute, primary divisions of the radical leaves 3-lobed; the lateral ones bifid, all incisely cleft; flowers large; achene beak long and slender, recurved. Ssp. *insularis* Hult. is a dwarfer type with silky-gray pubescent leaves and the beak of the achenes short and broad.

Coastal districts of Alaska—Ida.—Wyo.—Calif.; the ssp. *nelsoni* in the southwestern Pacific Coast and Aleutians; ssp. *insularis* the middle and western Aleutians; ssp. *turneri* from Bering Sea—Mackenzie R. Fig. 466.

5. **R. bongardii** Greene Bongard Buttercup.

A rather stout weedy plant 4–8 dm. tall; leaves and petioles hispid, the blades of the basal leaves 3–9 cm. long, 4–14 cm. wide, deeply 3-lobed, the terminal lobe 3-cleft, the lateral ones 2- to 4-cleft, toothed; petals and sepals about equal, 3–5 mm. long; heads of fruit globose; achenes pubescent, at least when young, the body about 2 mm. long, with a hooked beak of the same length. Var. *tenellus* (Nutt.) Greene (*R. douglasii* Howell) is a nearly glabrous plant with smaller leaves, slightly larger achenes hirsute only on the edge or not at all and a proportionally shorter beak.

Aleutian Islands—Ida.—Calif. Hybridizes with *R. acris*. Fig. 467.

6. **R. macounii** Britt. Macoun Buttercup.

Stems rather stout, hirsute, 3–8 dm. tall; lower leaves 5–15 cm. wide, hirsute, ternate, the divisions more or less stalked, variously cleft and toothed; petals 5–8 mm. long; heads of achenes short-ovate to subglobose; achenes about 3 mm. long with wide-based beak 1–1.5 mm. long. There is a glabrous form, var. *oreganus* (Gray) Davis.

Central and southwestern Alaska—Ont.—Iowa—Ore. Fig. 468.

7. **R. pennsylvanicus** L. Bristly Buttercup.

An erect, branching, leafy, pilose-hispid plant 3–8 dm. tall; leaves ternate, the divisions, at least the central one, stalked, ternately cleft and sharply toothed; petals 2–4 mm. long, often shorter than the reflexed sepals; heads of achenes ovoid to cylindrical; achenes about 2.5 mm. long with a flat triangular beak scarcely 1 mm. long.

Introduced in our area, central Alaska—N. S.—Ga.—Colo.—B. C. Fig. 469.

8. **R. orthorhynchus** Hook. ssp. *alachensis* (Benson) Hult.
Straight-beaked Buttercup.

Stems in our form glabrous, 4–8 dm. tall; lower leaves pinnate with 3–5 leaflets; leaflets cleft and toothed, cuneate, the terminal one 3-lobed; petals 8–12 mm. long; fruiting head globose; achenes margined, 3–4 mm. long with a beak of about the same length.

The ssp. in southeastern Alaska only, typical form Vancouver Isl.—Wyo.—Calif. Fig. 470.

9. **R. abortivus** L. Smooth-leaved Crowfoot

Somewhat fleshy, glabrous or slightly pubescent, 2–6 dm. tall; radical leaves undivided, crenate, 1–5 cm. wide; stem leaves 3-cleft, the uppermost with linear or oblong divisions and sessile; petals 2–3 mm. long, shorter than the reflexed sepals; head of achenes 4–5 mm. long, 3–4 mm. wide; achenes about 1.5 mm. long with a very minute beak on the side near apex.

South central Alaska—Labr.—N. S.—Fla.—Ark.—Colo. Fig. 471.

10. **R. eastwoodianus** Benson. Eastwood Buttercup.

Plant nearly glabrous, stems erect, up to 3 dm. tall, striate; radical and lower stem leaves fan-shaped in outline, 25–30 mm. long, 2–3 cm. wide, divided and parted into about 7 linear divisions; peduncles thinly pilose, 2–7 cm. long in flower; sepals 5, yellowish-green, spreading, narrowly elliptic, 5 mm. long, thinly pilose dorsally; petals 5, cuneate-obovate, 9–10 mm. long, 5 mm. or more broad.

Known only from Nome and Skagway. Fig. 472.

11. **R. verecundus** Robins.

Stems 5–10 cm. tall; leaves reniform to suborbicular, glabrous, conspicuously cordate, 3-parted, the segments 3- to 5-lobed or deeply crenate; petioles 2–4 cm. long; peduncles glabrous, 3–7 cm. long; sepals pubescent, purplish on the back, 2–4 mm. long; petals obovate, about 5 mm. long; heads ovoid or short-cylindric; achenes 1.8 mm. long; obovoid, with a short, recurved beak.

Central Alaska—Alta.—Mont.—Ore.

12. **R. pedatifidus** Sm. Northern Buttercup.
R. *affinis* R. Br.

Stems 1–3 dm. tall, branched, sparingly silky or glabrate; basal leaves 2–4 cm. wide, the earliest 3-cleft and toothed, the rest divided into narrow, cleft segments, those of the stem sessile and with linear divisions; calyx and upper part of peduncle softly pubescent; petals longer than the sepals; achenes about 2.5 mm. long with a long, weak, recurved or twisted beak which is sometimes broken off.

Arctic-alpine, more or less circumpolar. Fig. 473.

13. **R. eschscholtzii** Schlecht. Eschscholtz Buttercup.

Stems nearly glabrous, 1–3 dm. tall; basal leaves 3- to 5-parted, the divisions again cleft, often ciliate, 1–3 cm. wide; upper leaves with 3–5 long, entire lobes; petals 6-8 mm. long, often retuse, sepals usually pubescent; head of achenes oblong; achenes about 1.5 mm. long, plump, with a slender curved beak less than half as long.

Alpine-arctic, eastern Asia—western Mont.—northern Ore. Fig. 474.

14. R. grayi Britt.   Gray Buttercup.
R. gelidus Karel & Kiril.

Stems 5-10 cm. tall; basal leaf-blades biternately or pedately divided and parted into oblong to spatulate lobes; sepals ovate, externally pubescent; petals about 5 mm. long; head of achenes globose; beak recurved, nearly as long as the achene.

Arctic coast of Yukon, central and southwestern Alaska, Rocky Mts., Alta.—Colo., and central Asia.

15. R. nivalis L.   Snow Buttercup.

Stems glabrous or minutely pubescent, 1-3 dm. tall; basal leaves usually only 1 or 2, reniform, 6-20 mm. wide, usually 3-cleft, some of the lobes with crenate teeth or secondary lobes; sepals densely pubescent with brown hairs; petals broadly ovate, about 1 cm. long; head of achenes ovoid to cylindric; achenes 1.5-2 mm. long, the rather weak beak about 1 mm. long.

Arctic-alpine, circumboreal. Fig. 475.

16. R. sulphureus Phipps.   Sulphur-colored Buttercup.

Stems 1-4 dm. tall, sparingly or not at all branched, glabrous below, hirsute above; some of the basal leaves merely deeply crenate but most of the leaves variously cut and divided, up to 5 cm. wide but usually much smaller; petals longer than the sepals, 8-10 mm. long; heads of achenes short-ovoid; achenes 2-2.5 mm. long with acute, recurved beaks up to 1.5 mm. long.

Circumpolar. Fig. 476.

17. R. pygmaeus Wahl.   Pygmy Buttercup.

Stems 3-8 or in fruit up to 15 cm. tall; leaf-blades reniform, variously lobed and divided, 6-12 mm. wide; peduncles pubescent, elongating in fruit; petals shorter than the sepals, 2-3 mm. long; head of achenes ovoid to nearly globose; achenes flattened but little, about 1.25 mm. long with a short hooked beak.

Arctic-alpine, circumpolar. Fig. 477.

18. R. verticillatus Eastw.   Verticillate-leaved Buttercup.

Stems slender, with few branches, glabrous at base, up to 4 dm. tall; basal leaves not known to occur; stem leaves divided to the base into 3-7 linear lobes giving the appearance of whorled leaves, the lobes 10-45 mm. long, about 3 mm. wide, minutely appressed-ciliate on the margins; peduncles finely pilose beneath the flower; sepals wooly-pubescent, boat-shaped, 5-6 mm. long; petals about 7 mm. long; head of achenes subglobose or short-ovoid; achenes pubescent, orbicular, about 2 mm. long with beak at least 1 mm. long.

Known only from Nome. Fig. 478.

19. R. flammula L.   Creeping Spearwort.

Stems reclining or stoloniferous, rooting at the lower nodes, branched near the base, usually glabrous but sometimes hirsute; leaves simple, entire or serrulate; head of achenes globose; achenes 1.4–1.7 mm. long, 1–1.2 mm. wide with a short stout beak. This species is represented in our area by 2 forms. Var. ovalis (Bigel.) Benson (R. unalaschensis Bess.). Stems 1–5 dm. long, leaf-blades much wider than the petioles, 1–5 cm. long, up to 8 mm. wide; petioles 2–12 cm. long; flowers larger than in the next. Var. filiformis (Michx.) Hook. (R. reptans L.). Stems 1–3 dm. long, leaves very narrow, the blade scarcely distinguishable from the petiole, 15–60 mm. long; cauline leaves in clusters at the nodes; petals 2–4 mm. long.

Wet soil, var. filiformis circumboreal, var. ovalis the Aleutians—Newf.—N. Y.—Minn. Fig. 479.

20. R. hyperboreus Rottb.   Arctic Buttercup.

Growing on mud or in shallow water; stem very slender, glabrous; leaves reniform, 6–15 mm. wide, palmately 3-lobed, or occasionally 4- to 5-lobed, the lobes of the immersed leaves very slender; petals shorter than the sepals, 2–3 mm. long; head of achenes subglobose; achenes 1–1.3 mm. long, the beak small.

Circumboreal. Fig. 480.

21. R. sceleratus L. ssp. multifidus (Nutt.) Hult.

Celery-leaved Crowfoot.

Plant somewhat fleshy, branching, glabrate, 15–50 cm. tall; basal leaves reniform, 3- to 5-lobed or -parted, 2–5 cm. wide, the segments round-lobed; upper leaves sessile with narrow lobes; sepals hairy, 3–4 mm. long; petals about same length as sepals; head of achenes oblong; achenes numerous, smooth, slightly more than 1 mm. long with a short beak.

Swampy soil, Nome—Minn.—N. Mex.—Calif., the type form Eurasiatic. Fig. 481.

22. R. gmelini DC.

This species is represented in our area by two varieties. Var. terrestris (Ledeb.) Benson (R. purshii Richards.). Plant palustrine or aquatic; stems reclining or floating, 1–4 dm. long, but little branched; leaves usually all cauline, pentagonal in outline, the submersed ones dissected into ribbon-like segments, the emersed ones 3- to 5-cleft, the divisions toothed or incised; petals 4–7 mm. long; achenes in a subglobose head, about 1.5 mm. long. Var. yukonensis (Britt.) Benson (R. yukonensis Britt.). Stems 5–20 cm. long, delicate, leaves usually 1 cm. or less wide, deeply cleft, flowers smaller.

Var. terrestris from central Alaska—Keewatin—N. S.—N. Mex.—Colo., var. yukonensis from Arctic coast—B. C. Fig. 482.

23. *R. cymbalaria* Pursh.   Seaside Crowfoot.
Halerpestis cymbalaria (Pursh.) Greene.

Low plants with runners; leaves mostly basal, glabrous, more or less fleshy, reniform to ovate, crenate-toothed or slightly lobed, the base cordate or truncate, 5–20 mm. long; scapes usually 1-flowered, but sometimes 2- to 7-flowered, 2–20 cm. tall; petals 3–5 mm. long; head of achenes usually ovoid, 4–14 mm. long; achenes nearly 2 mm. long, striate, with a small beak.

Wet or saline soil, western half of N. Am.—S. Am.—Asia. Fig. 483.

24. *R. cooleyae* Vasey & Rose.   Cooley Buttercup.
Arctoranthis cooleyae (Vasey & Rose) Greene.

Leaves all radical; petioles 4–10 cm. long; blades orbicular or reniform, the cordate base often nearly closed, usually 5-divided to near the base, the segments again cut and crenately toothed, 2–4 cm. wide; scape-like stems 5–25 cm. tall, usually 1-flowered but sometimes 2-flowered; petals 11–15, 5–8 mm. long; head of achenes globose; achenes about 3 mm. long, with 3 or 4 prominent ribs on each side, the back winged, borne on jointed pedicels; beak fully 1 mm. long, hooked.

St. Elias Range—northern B .C. Fig. 484.

25. *R. kamchaticus* DC.   Kamchatka Buttercup.
Oxygraphis glacialis (Fisch.) Bunge.

A glabrous, slightly fleshy perennial; leaves petioled, ovate, entire or toothed, cordate; flowers borne singly on naked scapes; sepals deciduous; petals 7–12.

An alpine plant of eastern Asia extending into the Aleutian and Shumagin Isls. and Seward Penin.

26. *R. chamissonis* Schlecht.   Chamisso Buttercup.
R. glacialis L. ssp. *chamissonis* (Schlecht.) Hult.

Stems 1–2 dm. tall, glabrous below, pubescent above with long, dark-brown hairs; basal leaves cut into 3 segments, these variously cut or lobed; stem leaves reduced; flowers solitary; sepals about 1 cm. long, densely wooly with long, dark-brown hairs; petals white, 10–16 cm. long; body of the achenes 2–2.5 mm. long with a very broad flat beak as long as the body.

Eastern Asia—Seward Penin. Fig. 485.

27. *R. aquatilis* L.   White Water-Crowfoot.

Stems submerged, branching, glabrous, and flaccid; submerged leaves 2–4 times ternately divided into filiform segments. This species is represented in our area by 3 varieties. Var. *capillaceus* DC. (*R. tricophyllus* Chaix.). Leaves all submerged; petals 5–8 mm. long; achenes in a globose head, about 1.5 mm. long, transversely ridged, the beak small. Var. *hispidulus* E. R. Drew (*R. grayanus* Freyn.). Some of the leaves floating, reniform, cut into 3 or 5 lobes. Var. *eradicatus* E. R. Drew (*R. confervoides* Fr.). Stems very slender, leaves all submersed, dissected into divisions 0.1 mm. wide; achenes about 1 mm. long.

Vars. *capillaceus* and *eradicatus* are circumboreal, var. *hispidulus* Aleutians—B. C.—western Mont.—Utah.—Calif. Fig. 486.

28. **R. pallasii** Schlecht. Pallas Buttercup.

Glabrous subaquatic perennial with thick rhizome; flowering branches with basal leaves usually deeply 3-cleft, other leaves entire and ovate; flowers usually 2; petals 6–10, 6–10 mm. long; head of achenes globose, up to 15 mm. wide; achenes thick, 5–6 mm. long with a short beak.

N. Bering Sea & Arctic Coasts—Que. Fig. 487.

29. **R. lapponicus** L. Lapland Buttercup.
Coptidium lapponicum (L.) Gand.

Scapose from slender running rootstocks usually in moss; leaves basal, glabrous, the blades reniform, ternate, the divisions crenate and usually incised, 2–5 cm. wide; scapes naked or with 1 leaf, 8–20 cm. tall; petals 4–5 mm. long; achenes in a globose head, nearly 5 mm. long, with slender hooked beak, the seed confined to lower half.

Circumboreal. Fig. 488.

2. ANEMONE (Tourn.) L.

Perennial herbs with basal leaves and scapose stems bearing a whorl of leaves which form an involucre often remote from the flower; leaves palmately divided or dissected; sepals usually 5, often more, petal-like; stamens and carpels numerous; achenes compressed, 1-seeded. (Greek, the wind.)

```
1A. Styles plumose, elongating in fruit. (Genus
     Pulsatilla Mill.) ...................................8. A. patens multifida
2A. Styles not plumose.
  1B. Achenes glabrous.
    1C. Flowers yellow .................................1. A. richardsonii
    2C. Flowers white ..................................2. A. narcissiflora
  2B. Achenes more or less densely villous.
    1C. Leaves trifoliate, segments not dissected............3. A. deltoidea
    2C. Leaves ternate, segments crenate and often cleft.......4. A. parviflora
    3C. Leaves 2–3-times ternate.
      1D. Plants 2–3 dm. tall ..............................5. A. multifida
      2D. Plants 5–18 cm. tall.
        1E. Sepals blue on both sides.........................6. A. multiceps
        2E. Sepals white, sometimes tinted blue on outside.....7. A. drummondii
```

1. **A. richardsonii** Hook. Yellow Anemone.

Basal leaves 1 or few, round-reniform, 3- to 5-parted, crenate with mucronate teeth, 25–60 mm. wide; stems pubescent, 1-flowered, 5–20 cm. tall; sepals yellow, 8–15 mm. long; achenes few, 4–5 mm. long with a slender, minutely hooked beak fully as long.

Wet places, eastern Asia—northern Alaska—western Greenland—Alta.—Aleutians. Fig. 489.

2. **A. narcissiflora** L. Narcissus-flowered Anemone.

Rootstocks thick, oblique; leaves more or less silky-villous, in age sometimes almost glabrous, 4–12 cm. wide, long-petioled, those of the involucre sessile; sepals white, sometimes tinged blue on the outside, 10–15 mm. long; achenes in a globose head, 5–8 mm. long, flat, broadly spatulate in outline. A variable species that has developed local races or subspecies, 4 of which occur in our area. Ssp. *villosissima* (DC.) Hult. Plant very villous; stems up to 6 dm. tall, several- to many-flowered; leaves round to reniform, quinate, the segments sessile. Range, Kuriles—Aleutians—Kodiak. Ssp. *alaskana* Hult. Stems up to 4 dm. tall, 1- to 5-flowered; leaves biternate, more or less pentagonal in outline, ultimate segments toothed. Range, along the coast, Alaska Penin.—Queen Charlotte Isls. Ssp. *interior* Hult. Stems 1-3 dm. tall, usually 1-flowered, but sometimes 2- or 3-flowered; leaves pentagonal in outline, the segments narrow; sepals much broader in the middle, white on outside; flowers large. Range, central Alaska—Alta. Ssp. *sibirica* (L.) Hult.. Stems 1-3 dm. tall, 1- to 5-flowered; leaves pentagonal in outline; sepals oval, often bluish on the outside. Range, Yenesei valley—western Alaska.

Other forms make the species circumboreal. Fig. 490.

3. *A. deltoidea* Hook. Columbia Wind-Flower.

Stems arising from very slender creeping rootstocks, 1-3 dm. tall; basal leaves usually solitary, trifoliate; leaflets ovate, dentate, 3-5 cm. long; involucral leaves 3, subsessile; sepals white; achenes glabrous above, short-hirsute toward the base.

Reported from Dease L., B. C.—B. C.—Calif.

4. *A. parviflora* Michx. Northern Anemone.

Basal leaves ternately divided, the cuneate parts more or less lobed and crenately toothed; scape more or less villous, 5-25 cm. tall, usually 1-flowered; sepals white, usually tinged with blue or rose on the back, 9-18 mm. long; heads of achenes nearly globular; achenes covered with long wool and tipped with a slender, fragile style.

Arctic-alpine, northeastern Asia—all of Alaska—Newf.—Colo. Fig. 491.

5. *A. multifida* Poir. Cut-leaved Anemone.
A. globosa Nutt.

Stems 2-6 dm. tall, 1- to 3-flowered, silky-villous; basal leaves 4-12 cm. broad, 2- to 3-ternate, pubescent, in age sparingly villous; sepals often more than 5, pubescent and tinged with blue or rose on outside; heads of achenes subglobose or ovoid; achenes densely wooly.

Meadows and hillsides; central Alaska—N. B.—Colo. Fig. 492.

6. *A. multiceps* (Greene) Standl. Alaska Blue Anemone.
Pulsatilla multiceps Greene.

Stems slender, usually less than 15 cm. tall, often very dwarf; leaves 10-35 mm. wide, ternately dissected into oblong-cuniform divisions; sepals blue or lavender, villous on the outside, 10-18 mm. long; body of achenes sparingly white-lanate.

Seward Penin.—Yukon. Fig. 493.

7. *A. drummondii* Wats. Drummond Anemone.

Stems usually 1-flowered, sometimes 2-flowered, 15-30 cm. tall in fruit; basal leaves 2-4 cm. wide, 2- to 3-ternate, glabrate or very sparingly pubescent, those of the involucre more villous, the divisions linear to cuneate-lanceolate; sepals 5-8, tinged with blue on the outside, 8-10 mm. long; styles prominently exerted, 2-4 mm. long; achenes densely wooly.

Bering Str.—arctic Yukon—Alta.—Ida.—Calif.

8. *A. patens* L. ssp. *multifida* (Pritzel) Zamels.

Pasque-flower. Wild Crocus.

A. patens var. *nuttalliana* (DC.) Gray.
A. patens var. *wolfgangiana* Koch.
Pulsatilla ludoviciana (Nutt.) Heller.

Silky-villous; stems 1-4 dm. tall, the involucral leaves sessile; leaves ternate and repeatedly divided into linear, acute lobes, becoming glabrate in age, at least on the upper surface, 5-10 cm. wide; sepals purple or violet, 25 mm. or more long; achenes with plumose styles about 3 cm. long.

Central Yukon valley—subarctic—Ill.—Texas—Wash. Fig. 494.

3. THALICTRUM L.

Erect perennials with ternately decompound leaves and small perfect dioecious, or polygamous flowers in panicles or racemes; sepals 4 or 5, usually greenish or greenish-white and deciduous; stamens numerous, the filaments often dilated; carpels few; fruit a head of ribbed achenes. (Greek name for some plant mentioned by Dioscorides.)

```
1A. Low growing alpine plant with scapose stems.................1. T. alpinum
2A. Taller plants with leafy stems.
  1B. Flowers dioecious ........................................4. T. occidentale
  2B. Flowers perfect.
    1C. Achenes strongly flattened............................2. T. sparsiflorum
    2C. Achenes subterete ....................................3. T. hultenii
```

1. *T. alpinum* L.   Arctic Meadow-rue

A glabrous alpine perennial, 6-25 cm. tall, with scaly rootstocks; leaves mostly basal, ternate-pinnate; leaflets less than 1 cm. long or wide, slightly lobed at the apex; flowers borne in a raceme on a scape-like stem; achenes few, strongly ribbed, about 2.25 mm. long with a short beak.

Circumpolar. Fig. 495.

2. *T. sparsiflorum* Turcz.   Few-flowered Meadow-rue.

Stems leafy, glabrous, 5-10 dm. tall; leaves mostly triternate; leaflets thin, rounded or cordate at the base, crenate, 8-18 mm. long; sepals whitish; filaments of the stamens enlarged and roughened above; achenes 6-15, straight-backed, sharp-beaked, with 3 or more ribs, 6-7 mm. long.

Western Siberia—Hudson Bay—Colo.—Calif. Fig. 496.

3. *T. hultenii* B. Boivin.

T. kemense E. Fries.
T. minus L. ssp. *kemense* (E. Fr.) Hult.

Stems erect, glabrous or glaucous, 3-8 dm. tall; leaves 2- to 3-ternate, the lower petioled; leaflets oval and narrowed at the base, 1- to 3-lobed at the apex; flowers in a loose panicle; anthers oblong, on slender filaments; achenes 8 or fewer, subsessile, obliquely ovate, about 6-grooved.

A species of northern Eurasia found in the eastern Aleutians. Fig. 497.

4. T. occidentale Gray.   Western Meadow-rue

Stems 5-10 dm. tall, glabrous and glaucous; leaves variable, 3- to 5-ternate; leaflets pale beneath, cuneate to cordate at the base, often broader than long, more or less deeply 3- to 8-cleft at the apex, 1-3 cm. long; panicle open; anthers slender, mucronate; achenes slightly compressed, the faces with 3 strong and often 1 or 2 secondary nerves, 6-8 mm. long.

Hyder, Alaska—Alta.—Wyo.—Calif. Fig. 498.

4. CALTHA (Rupp.) L.

Somewhat succulent perennials; leaves simple, mostly basal, flowers white or yellow; sepals 5 or more, petal-like; petals none; stamens numerous; carpels several or many, sessile, in fruit forming follicles with 2 rows of seeds along the ventral suture. (Latin name of the marigold.)

```
1A. Sepals yellow .................................................. 1. C. palustris
2A. Sepals white.
   1B. Aquatic, stems floating ...................................... 4. C. natans
   2B. Terrestrial but growing in wet places.
      1C. Leaves broader than long................................. 2. C. biflora
      2C. Leaves longer than broad................................. 3. C. leptosepala
```

1. C. palustris L.   Yellow Marsh Marigold.

Stems hollow, decumbent, often rooting at the lower nodes; basal leaves on long petioles, the blade cordate or reniform, with a deep, narrow sinus; follicles 3-12 or more, somewhat divergent, compressed. This is a circumboreal species represented in our area by 2 forms. Var. *arctica* (R. Br.) Huth. This group varies from plants approaching ssp. *asarifolia* to the extreme arctic form with small flowers and leaves about 1 cm. wide and follicles 4-5 mm. long. Ssp. *asarifolia* (DC.) Hult. (C. *asarifolia* DC.). Leaves 5-12 cm. wide, crenate; sepals 5-7, bright yellow, 15-20 mm. long with occasional double forms; follicles about 1 cm. long.

Var. *arctica*, Arctic and Bering Sea Coasts eastward through interior Alaska approaching the Pacific Coast in Kenai Penin., ssp. *asarifolia* along the coast, Aleutians—Ore. Fig. 499.

2. C. biflora DC.   Broad-leaved Marsh Marigold.

Leaves reniform, regularly and deeply crenate, up to 15 cm. wide; stems scape-like, with 1 leaf, usually 2-flowered, 1-4 dm. tall; sepals 6-10, white, 10-18 mm. long; follicles 3-10, about 15 mm. long, short-stipitate, erect, with beak 1-2 mm. long.

Wet woods, southeastern Alaska—Nev.—Calif. Fig. 500.

3. C. leptosepala DC.   Mountain Marigold.

Leaves oval with a narrow sinus at the base, crenate, 2-6 cm. wide, 3-8 cm. long; scape-like stems 1-4 dm. tall, 1- or 2-flowered, bearing a single leaf; sepals white, 10-18 mm. long; follicles several, 12-18 mm. long, erect, with curved beak 1 mm. long.

Wet alpine meadows, Pacific Coast districts of Alaska—Alta.—N. Mex.—Ore. Fig. 501.

4. *C. natans* Pall. Floating Marsh Marigold.

Stems floating or creeping and rooting at the nodes, 15-50 cm. long; leaves cordate-reniform, entire or crenate, 3-5 cm. wide, the upper leaves smaller; flowers white or pinkish; sepals 6-8 mm. long; follicles numerous, about 4 mm. long with a very short beak, in a globular head.

Northern Asia—Northwest Territories—Alta.—Minn. Fig. 502.

5. TROLLIUS L.

Erect or ascending perennials from thickened fibrous roots; leaves palmately divided or lobed; sepals 5 or more, petaloid; petals 5-many, small, linear, with nectiferous pit at the base of the blade; carpels 5 or more, becoming many-seeded follicles in fruit. (From an old German word meaning something round.)

T. *riederianus* Fisch. & Mey.

Stems less than 3 dm. tall, scape-like and 1-flowered; sepals 5, rarely more, yellow.

An Asiatic species found on Kiska Isl.

6. COPTIS Salisb.

Low, scapose perennials with yellow, spreading rootstocks; leaves compound; sepals 5-7, white or whitish; petals 5-7, small, filiform, enlarged and nectiferous at the apex or middle; stamens numerous; fruit composed of few to several stipitate follicles forming an umbell-like cluster. (Greek, to cut, in allusion to the leaves.)

Leaves trifoliate .... 1. *C. trifoliata*
Leaves ternate-pinnate .... 2. *C. asplenifolia*

1. *C. trifoliata* (L.) Salisb. Trifoliate Goldthread.

Leaves shining, evergreen; leaflets 3, ovate to obovate, with cuneate base, crenate or slightly lobed and with mucronate teeth, 10-25 mm. long; sepals white with yellow base; petals club-shaped with an orange-colored, enlarged nectiferous apex, shorter than the stamens; follicles 3-7, the stipe about equalling the body.

Bogs and swamps, eastern Asia—Greenl.—Newf.—Tenn.—Iowa—B. C. Fig. 503.

2. *C. asplenifolia* Salisb. Fern-leaved Goldthread.

Leaves shining, pinnately ternate into more or less incised and sharply toothed leaflets 6-20 mm. long; scapes 1- to 3-flowered, 1-3 dm. tall; sepals linear, greenish-white; petals enlarged near the middle; follicles 6-12, about 1 cm. long, slightly longer than the stipe.

Woods, central Alaska—southern B. C. Fig. 504.

7. ACTAEA L.

Erect perennials with thick rootstocks; leaves ternately decompound; flowers small, white, in terminal racemes; sepals 3-5, petal-like; petals

4–10, small; stamens numerous; pistil solitary, bicarpellary, with 2-lobed stigma; fruit a more or less poisonous berry. (Ancient name of the elder.)

Berry red ....1. A. *arguta*
Berry ivory-white ....2. A. *eburnea*

1. A. *arguta* Nutt.   Red Baneberry.

A. rubra (Ait.) Willd. ssp. *arguta* (Nutt.) Hult.

Stems glabrous or somewhat pubescent above, 6–10 dm. tall; leaves 2- to 3-ternate; leaflets thin, usually lobed and coarsely toothed; long acuminate, 3–10 cm. long; sepals with long claws and rhombic, acute blades; anthers white; berry red, globose to slightly elongated, 6–8 mm. long.

Woods, western Alaska—Nebr.—Calif. Fig. 505.

2. A. *eburnea* Rydb.   White Baneberry.

Similar to *A. arguta;* sepals orbicular and early deciduous; berry ivory-white, ellipsoid, 8–10 mm. long, attached somewhat obliquely. It is doubtful if this is more than a white-fruited form of *A. rubra* (Ait.) Willd.

Nearly same range in our area as the preceding.

8. AQUILEGIA (Tourn.) L.

Erect branching perennials; leaves ternately decompound; flowers perfect, regular; sepals 5, petaloid; petals 5, saccate and prolonged backward between the sepals into spurs; stamens numerous, the inner ones reduced to staminodia; carpels 5, developing into erect follicles in fruit; seeds many, smooth and shining. (Latin, Aquila, the eagle, on account of the spurs.)

Flowers blue .... 1. A. *brevistyla*
Flowers red and yellow .... 2. A. *formosa*

1. A. *brevistyla* Hook.   Small-flowered Columbine.

Stems slender, erect, pubescent above, usually branched, 15–45 cm. tall; leaves biternate, the leaflets nearly sessile; sepals blue, lanceolate, acute, about 15 mm. long; petals yellowish-white with short spurs; follicles pubescent, the beak short.

Central Alaska—Gt. Bear L.—L. Nipigon—S. Dak.—B. C. Fig. 506.

2. A. *formosa* Fisch.   Western Columbine.

A. columbiana Rydb.

Stems glabrous below, pubescent above, 4–10 dm. tall; leaves biternate, the leaflets round-ovate, deeply cleft and crenate; sepals and spurs red, the limb of the petals yellow; spurs 12–18 mm. long, shorter than the ovate-lanceolate sepals; follicles pubescent, the beak long.

Kenai Penin.—western Mont.—Utah.—Calif. Fig. 507.

9. DELPHINIUM (Tourn.) L.

Ours all erect perennials; leaves alternate, palmately lobed or divided; flowers irregular, blue or purple, borne in racemes; sepals 5, petaloid, the posterior one spurred; petals usually 4, two of them with spurs enclosed in the spur of the sepal; stamens numerous; carpels mostly 3, developing into many-seeded follicles. (Latin, dolphin, from some resemblance in the flower.)

1A. Plant 6–25 dm. tall .... 1. *D. glaucum*
2A. Plant less than 5 dm. tall.
  1B. Leaf-segments usually very narrow.................... 2. *D. brachycentrum*
  2B. Leaf-segments broader .... 3. *D. nutans*

1. *D. glaucum* Wats.   Glaucus Larkspur.
    D. alatum A. Nels.
    D. brownii Rydb.
    ?*D. hookeri* A. Nels.
    D. scopulorum Gray var. *glaucum* Gray.
    D. splendens G. N. Jones.

Stems stout, glabrate and usually glaucous; leaves pubescent, at least beneath, deeply cut into 5–7 variable divisions, these again cut into lanceolate, acute lobes; inflorescence 1–5 dm. long, often branched below; flowers blue or purple; spur 12 mm. or less long, longer than the sepals; follicles usually glabrous.
    Bering Str.—Gt. Slave L.—Wyo.—Calif. Fig. 508.

2. *D. brachycentrum* Ledeb.   Northern Dwarf Larkspur.
    D. blaisdellii Eastw.
    D. menziesii auct.
    D. ruthae A. Nels.

Pubescent throughout; leaves variable, deeply cut into narrow, gland-tipped segments; flowers dark blue or purple, pubescent except the light-colored upper petals; spur usually straight, 12–20 mm. long, slightly longer than the lateral sepals.
    East Asia—east central Alaska. Fig. 509.

3. *D. nutans* A. Nels.

Roots fascicled; stems 3–4 dm. tall, erect or ascending; pubescent; leaves pubescent on the petioles, the margins and the veins, the blade broader than long, parted into 3–5 divisions, each of which is 2- to 3-cleft and these again irregularly and deeply toothed; racemes few-flowered, with a few flowers in the axils of the upper leaves; follicles pubescent and obscurely glandular. This form may be a hybrid of *D. brachycentrum* x *D. glaucum*.
    Known from Mt. McKinley Park.

10. ACONITUM L.

Perennials with rootstocks or tubers; leaves palmately lobed or divided; flowers blue or purple, perfect, irregular, large and showy; sepals 5, the upper one forming a hood, petals 2–5, small, two of them hooded and concealed in the hooded sepal; stamens numerous; carpels 3–5, developing into many-seeded follicles. (Ancient Greek name.)

Hood boat-shaped... 1. *A. delphinifolium*
Hood helmet-shaped .... 2. *A. maximum*

1. **A. delphinifolium** DC. Delphinium-leaved Aconite.

Stems finely pubescent, 5–10 dm. tall, or in some forms as low as 1 dm.; leaves glabrate or ciliate on the margins and veins, divided to near the base into cuneate segments, these again cut into lanceolate lobes; racemes few-flowered; sepals pubescent, the lateral ones about 3 times as broad as the lower ones; hood 18–20 mm. long with a short beak. In addition to the typical form two subspecies occur. Ssp. *chamissonianum* (Rchb.) Hult. of the coast from the Aleutian and Pribylof Isls. eastward is relatively stouter with broader-lobed leaves. Ssp. *paradoxum* (Rchb.) Hult. (*A. nivatum* A. Nels.) of the Arctic and Bering Sea coasts—central Alaska is 1–3 dm. tall, 1- to few-flowered, the flowers relatively large.

East Asia—northern coast of Alaska—Alta.—B. C. Fig. 510.

2. **A. maximum** Pall. Kamchatka Aconite.
A. kamtschaticum Rchb.

Stem stout, erect, leafy, finely pubescent, 5–10 dm. tall; leaves pubescent, up to 14 cm. wide, deeply 3- to 5-lobed into cuneate divisions, these cut into acute, lanceolate lobes or teeth; racemes dense; flowers blue, the hood about 2 cm. long and wide.

East Asia—Aleutians—Alaska Penin. Fig. 511.

13. NYMPHACEAE (Water Lily Family)

Aquatic, acaulescent perennials; leaves large, leathery, floating, on long petioles arising from thick, horizontal rootstocks; flowers solitary, axillary, borne on long peduncles; sepals 4–12; petals usually numerous, often passing into staminodia; stamens numerous; pistil compound, of several more or less united carpels, the stigmas united into a disk; ovules numerous.

Flowers yellow .... 1. *Nuphar*
Flowers white .... 2. *Nymphaea*

1. NUPHAR (Sibth. & Smith)

Leaves cordate, large, the sinus deep; sepals 5–12, leathery, concave; petals 10–20, small and stamen-like, inserted with the petals under the ovary; stigmas forming a radiating disk. (From the Arabic.)

N. polysepalum Engelm. Yellow Pond Lily.
N. variegatum of reports.
Nymphaea polysepala (Engelm.) Greene.

Leaves oblong or ovate, 15–30 cm. long, 10–20 cm. wide, the sinus narrow or closed; sepals yellow or tinged with red; petals cuneate, 10–15 mm. long, half as wide, wider than the filaments; stigmatic rays 15–25; ovary contracted below the stigmatic disk.

Central and southwestern Alaska—Colo.—Calif. Fig. 512.

2. NYMPHAEA (Tourn.) L.

Plants with floating leaves and showy flowers; sepals 4; petals indefinite, gradually passing into the stamens; stamens numerous; stigmas 12- to 35-rayed; seeds numerous. (Greek, water-nymph.)

N. tetragona Georgi ssp. *leibergi* (Morong) Pors. White Water Lily.

Leaves ovate, 5–10 cm. long, 35–70 mm. wide, the sinus open, the veins sunk into the leaf-tissues below; flowers 3–6 cm. in diameter; sepals greenish; petals 6–10, white, scarcely as long as the sepals.

Southeastern and east central Alaska—Kewatin—Ont.—Ida. and in Eurasia, occurrence rare and scattered. Fig. 513.

14. PAPAVERACEAE (Poppy Family)

Annual or perennial plant with colored sap and acrid or narcotic properties; leaves alternate or mostly radical; flowers perfect, regular; sepals usually 2; petals usually 4, often more; stamens numerous; gynoecium of 2- to many united carpels; ovary 1-celled with parietal placentae; ovules numerous; fruit a capsule generally dehiscent by pores. Represented in our area by 1 genus.

PAPAVER (Tourn.) L.

Ours all perennials with milky sap; leaves all basal, lobed or dissected; flowers drooping in the bud, later erect; sepals usually 2, early deciduous; petals 4, rarely more; ovary with 3–20 internally projecting placentae, the stigma disk-like; capsule in ours pyriform, ovoid, or nearly cylindrical, opening by chinks near the summit; seeds numerous, with minute depressions. There are many local races, and besides the forms here recognized a number of varieties have been described. (Latin name of the poppy.)

```
1A. Leaves coriaceus, shining, simple or 3-lobed or
     -divided ................................................... 1. P. walpolei
2A. Leaves not coriaceus, nearly glabrous to densely
     hirsute, pinnately lobed or divided......................
  1B. Capsule 2-4 times at long as thick........................ 4. P. macounii
  2B. Capsule less than 2 times as long as thick.
    1C. Scapes more than 25 cm. long........................... 5. P. nudicaule
    2C. Scapes less than 25 cm. long.
      1D. Stigmas with long, narrow central projection........... 3. P. mcconnellii
      2D. Central projection of stigma, if present, short
           and thick.
        1E. Flowers white or rose.............................. 2. P. alboroseum
        2E. Flowers usually yellow............................. 6. P. radicatum
```

1. *P. walpolei* Pors. Walpole Poppy.

Densely caespitose; leaves crowded, short-petioled, 1–4 cm. long, including petiole, simple or 3-lobed, or -parted, glabrous or with a few stiff hairs; scape 5–10 cm. tall, erect, hirsute-strigose above; petals yellow or more often white, 10–18 mm. long; capsule obovoid-pyriform.

Teller—Goodnews Bay. Fig. 514.

2. *P. alboroseum* Hult.

Subacaulescent; leaves pinnatisect, the segments often 2- to 3-parted, the lobes mucronulate; scapes 6–15 cm. tall, bristly with light brown hairs; petals white or light rose, 6–10 mm. long; stigmatic lines 5 or 6.

An Asiatic species collected at Seward.

3. **P. mcconnellii** Hult. McConnell Poppy.

Leaves bipinnate, the ultimate divisions narrowly obovate-lanceolate to nearly linear; scapes about 15 cm. tall, pubescent with pale rigid hairs; flowers yellow, about 2 cm. in diameter; capsule obovoid, the stigmatic disk convex, with a very narrow beak about 1 mm. long.

District of Mackenzie near Yukon border.

4. **P. macounii** Greene. Macoun Poppy.

Leaves densely clustered on the short stem, somewhat hirsute-hispid but often nearly glabrous, ovate in outline, the pinnae oblong-lanceolate to nearly linear; scapes sparsely pubescent, up to 3 dm. long in fruit; petals 4, roundish-ovate, erose-dentate, up to 35 mm. long, yellow, fading greenish; capsule hispid, with 4–5 stigmatic lines.

Asia—Alaska—Mack. Fig. 515.

5. **P. nudicaule** L. Iceland Poppy.

Tufted; subacaulescent; leaves pinnate, some of the pinnae usually pinnatifid, the petiole with long, light-colored, spreading hairs; scapes 25–40 cm. tall, sparely hirsute; petals normally yellow, 18–35 mm. long; capsule obovoid, pubescent with stiff, ascending hairs.

Siberia—Yukon. Fig. 516.

6. **P. radicatum** Rottb. Arctic Poppy.

Leaves several to numerous, 2–10 cm. long, pinnately dissected, coarsely hirsute; scapes 6–25 cm. long, sparsely to densely hirsute; petals usually yellow, rarely white or tinged with red, 15–30 mm. long; capsule ovoid, hirsute, the stigmatic disk rather flat. Ssp. *alaskanum* (Hult.) J. P. Anderson, comb. nov. (*P. alaskanum* Hult. Fl. Aleut. Isls. [1937] p. 190). Distinguished by the very conspicuous, thickly packed light brown, strongly bristly sheaths covering the old stems.

Circumpolar, the ssp. in Aleutians and southwestern Alaska. Fig. 517.

15. FUMARIACEAE (Fumewort Family)

Herbs; leaves alternate, dissected; flowers in racemes or panicles, perfect, irregular; sepals 2, small and scale-like; petals 4, one or two of them spurred; stamens 6, diadelphous; pistil of 2 united carpels, the ovary 1-celled with 2 parietal placentae; fruit a capsule. Only one genus in our area.

CORYDALIS Vent.

Leaves alternate, bipinnately dissected; flowers in racemes; outer petals unlike, one of them spurred, the 2 inner petals narrow, keeled on the back; stamens 6, in 2 sets opposite the outer petals; fruit an elongated 2-valved capsule. (Greek, crested lark.)

```
1A. Corolla yellow ............................................ 1. C. aurea
2A. Corolla pink or purplish
  1B. Tall biennial ........................................ 2. C. sempervirens
  2B. Low arctic-alpine perennial .......................... 3. C. pauciflora
```

1. *C. aurea* Willd.   Golden Corydalis.
Capnoides aureum (Willd.) Kuntze.

Annual or biennial with a much branched glabrous and glaucous stem 1–5 dm. tall; leaves 2- to 3-pinnate and dissected into cuneate or oblong-ovate segments; corolla 12–15 mm. long, the spur one-third to one-fourth of its entire length; pod 2–3 cm. long, pendulous, strongly curved and torulose; seed shining, reticulated.

    Central Alaska—Gt. Bear L.—St. Lawrence R.—Texas.—Calif. Fig. 518.

2. *C. sempervirens* (L.) Pers.   Pink Corydalis.
Capnoides sempervirens (L.) Borkh.

Stems glabrous and glaucous, branched toward the top, 3–8 dm. tall; leaves 2- to 3-pinnatifid into obovate or cuneate divisions; corolla rose or purplish with yellow tip, 12–15 mm. long, the spur less than one-third the length of the body; pods ascending, 3–5 cm. long, slightly curved and torulose; seed shining, minutely reticulated.

    Southwestern and central Alaska—Newf.—Ga.—Minn.—Mont. Fig. 519.

3. *C. pauciflora* (Steph.) Pers.   Few-flowered Corydalis.

Stems almost scapose, glabrous, 7–20 cm. tall; leaves 2–5, borne at or near the base of the stem, with usually 3 stalked divisions, these again divided to near the base into 2–5 ovate or oblong lobes; flowers 2–6, bracted, corolla pink or purplish, 15–22 mm. long, the spur forming about one-half this length; pod 15–20 mm. long.

    An Asiatic species extending to the Forty Mile dist. and northern B. C. Fig. 520.

16. BRASSICACEAE (Mustard Family)

Mostly herbs with more or less acrid sap; leaves alternate; flowers perfect, regular or nearly so, borne in spikes or racemes; sepals 4, usually deciduous; petals 4, with spreading blades; stamens usually 6, four of them longer than the other two; carpels 2, united, the fruit usually 2-celled by a membranous partition with marginal placentae. The classification in this family depends largely on the mature fruit. If the fruit is much longer than broad it is known as a silique, if shorter as a silicle. In the seed the cotyledons are said to be accumbent if the radicle is turned to the edge of the cotyledons; incumbent if the radicle is turned to the back of one of the cotyledons; conduplicate if the cotyledons are curved around the radicle. The arrangement of the cotyledons is usually evident from the outline of the seed.

```
1A. Pod transversely 2-jointed.................................. 9. Cakile
2A. Pod not transversely 2-jointed.
  1B. Pods compressed contrary to the narrow partition.
    1C. Pubescence stellate .............................20. Smelowskia
    2C. Pubescence, if any, not stellate.
      1D. Cells 1-seeded ................................. 2. Lepidium
      2D. Cells several-seeded.
```

1E. Seeds 2-6 in each cell .... 3. *Thlaspi*
   2E. Seeds 10-12 in each cell ....16. *Capsella*
 2B. Pod not compressed contrary to the partition.
  1C. Pod 1-2.5 times longer than broad.
   1D. Pods indehiscent ....18. *Neslia*
   2D. Pods dehiscent.
    1E. Plant growing under water........................ 1. *Subularia*
    2E. Plant terrestrial.
     1F. Pods compressed parallel to the partition.
      1G. Pods ovate or oblong............................19. *Draba*
      2G. Pods orbicular ....24. *Alyssum*
     2F. Pods globose or ovoid, little or not compressed.
      1G. Pubescence stellate............................15. *Lesquerella*
      2G. Pubescence not stellate.
       1H. Flowers white .... 4. *Cochlearia*
       2H. Flowers yellow.
        1I. Pod margined, pyriform.....................17. *Camelina*
        2I. Pod not margined, globose—oblong...........13. *Rorippa*
  2C. Pods much longer than broad.
   1D. Pods flat.
    1E. Valves nerveless or nearly so.
     1F. Pods short, 15 mm. long or less....................19. *Draba*
     2F. Pods long.
      1G. Valves opening elastically, seeds wingless.......14. *Cardamine*
      2G. Valves not opening elastically, seed
           often winged................................21. *Arabis*
    2E. Valves nerved.
     1F. Pods torulose.
      1G. Pods narrow, 2 mm. wide or less..............25. *Braya*
      2G. Pods 4-6 mm. wide..........................26. *Parrya*
     2F. Pods not torulose.
      1G. Pods 2 cm. long or more......................21. *Arabis*
      2G. Pods 15 mm. long or less.
       1H. Septum entire ....19. *Draba*
       2H. Septum perforated or rudimentary.
        1I. Leaves toothed, lobed, or pinnatifid..........22. *Ermania*
        2I. Leaves entire.
         1J. Plant very low growing.................... 5. *Aphragmus*
         2J. Stems 1-3 dm. tall........................ 6. *Eutrema*
   2D. Pods not compressed.
    1E. Pods indehiscent ....11. *Raphanus*
    2E. Pods dehiscent.
     1F. Pods with a stout beak ....10. *Brassica*
     2F. Pods beakless, merely tipped with the style.
      1G. Pods terete.
       1H. Valves of the short pod nerveless..............13. *Rorippa*
       2H. Valves of the long pod nerved.
        1I. Pubescence of short hairs or none .... 7. *Sisymbrium*
        2I. Pubescence of forked hairs .... 8. *Descurainia*
      2G. Pods 4-angled by strong midribs.
       1H. Leaves lyrate-pinnatifid ....12. *Barbarea*
       2H. Leaves not lyrate-pinnatifid.
        1I. Leaves entire, pubescence appressed ....23. *Erysimum*
        2I. Leaves runcinate-pinnatifid .... 7. *Sisymbrium*

1. SUBULARIA L.

Small perennial aquatic herb; leaves basal, subulate; flowers small, white, borne in few-flowered racemes; pod ovoid or subglobose, short-stipitate, the valves 1-nerved; seeds few, in 2 rows in each cell; cotyledons incumbent. (Latin, an awl, from the shape of the leaves.)

S. aquatica L.   Awlwort.

Growing in shallow water; leaves tufted, erect or ascending, 12-50 mm. long; scapes 1-5 mm. long; pod obovoid, 2-3 mm. long.

Circumboreal. Fig. 521.

2. LEPIDIUM (Tourn.) L.

Ours annual or biennial introduced weeds; leaves entire, toothed or lobed; flowers small, perfect, borne in racemes; pod orbicular, notched at the apex, wing-margined; seed solitary in each cell; cotyledons incumbent. (Greek, a little scale, from the shape of the pod.)

Petals wanting .... 1. *L. densiflorum*
Petals present .... 2. *L. virginicum*

1. *L. densiflorum* Schrad. Common Pepper-grass.

Stems minutely puberulent, branched above, 1–5 dm. tall; basal leaves usually lacking at flowering time; stem leaves oblanceolate, entire or dentate with sharp teeth, puberulent; petals rudimentary or none; stamens 2 or 4; pod about 3.5 mm. long, 3 mm. wide.

Widely introduced weed, B. C.—Maine—Va.—Texas—Nev. Fig. 522.

2. *L. virginicum* L. Wild Pepper-grass.

Similar to *L. densiflorum* but the stem glabrate, the flowers with white petals about twice as long as the sepals and longer and more divergent pedicels.

Sparingly introduced, native of eastern N. America.

Lepidium sativum L., the cultivated pepper-grass, has been collected at Dawson. It is probably not established as a part of our flora.

Iberis amara L., the garden candytuft, self-sown may persist for a few years. It is an annual with white flowers and nearly orbicular, winged pods 5–8 mm. in diameter. Native of Europe.

3. THLASPI (Tourn.) L.

Erect annual or perennial herbs; basal leaves entire or toothed, the stem leaves clasping the stem; flowers racemose, perfect, small, white or purplish; pods very flat, cuneate or orbicular, crested or winged. (Greek, to flatten, referring to the pod.)

Introduced weed .... 1. *T. arvense*
Native perennial .... 2. *T. arcticum*

1. *T. arvense* L. Field Penny Cress.

Branched annual, 15–80 cm. tall; basal leaves oblanceolate, early deciduous, entire or sparingly toothed; upper leaves oblong-lanceolate, auricled and clasping at the base; pod broadly winged, nearly circular in outline with a notch at the apex, 12–18 mm. long, 10–15 mm. broad.

A widely introduced weed, native of Europe and Asia. Fig. 523.

2. *T. arcticum* Pors. Arctic Penny Cress.

Perennial with a many-branched caudex; basal leaves spatulate, subglaucous, fleshy, glabrous, entire, 10–25 mm. long, 5–8 mm. wide, the midrib prominent; stems at anthesis 3–5 cm. long, in fruit up to 18 cm. tall; stem leaves 3–5, linear, sessile; petals white, about 4.5 mm. long; pods 6–7 mm. long, 2–2.5 mm. wide, cuneate-clavate, the valves keeled; style slender, about 1 mm. long; septum incomplete; seeds 1.5–2 x 0.8–1 mm.

Arctic Coast—Yukon & southeastern Alaska.

4. COCHLEARIA L.

Low, glabrous, maritime herbs; leaves simple, succulent; flowers racemose, small, white; pods subglobose to oblong, inflated, the valves with strong midvein and often more or less reticulated; seed in 2 rows in each cell; cotyledons accumbent. (Greek, spoon, from the shape of the leaf.)

C. officinalis L.　　　　　　　　　　　　　　　　　Scurvy Grass.

Diffusely branched annual or perennial; lower leaves petioled, ovate, the stem leaves with cuneate or truncate base, dentate or entire, 6–20 mm. long; upper leaves sessile; pods 4–7 mm. long, 3–4 mm. wide. A very variable circumpolar species passing into many local races. Represented in our area mostly by 2 forms. Ssp. *arctica* (Schlecht.) Hult. has pods 1.5–2 times as long as broad. Ssp. *oblongifolia* (DC.) Hult. has pods 1–1.5 times as long as broad.

Ssp. *artica* is found along the Arctic and northern Bering Sea Coasts. Ssp. *oblongifolia* along the Pacific and most of the Bering Sea Coasts. The range of the two forms overlaps. Fig. 524.

5. APHRAGMUS Andrz.

Pods lanceolate, compressed; valves plain, marked with a median line; septum none, style very short; stigmas capitate; seeds oval, suspended from the upper part of the placentae. (Latin, referring to the lack of septum.)

A. eschscholtzianus Andrz.

A small plant with the appearance of *Cardamine bellidifolia;* leaves long-petioled, entire, obtuse or rounded; stems 12–50 mm. long, naked below, but with an involucre of 2–4 foliaceous bracts; flowers white. small; pods 8–12 mm. long, about 3 mm. broad, 4- to 10-seeded; seed long. adhering to the placentae after falling of the valves.

Seward Penin.—Aleutian and Shumagin Islands.

6. EUTREMA R. Br.

Sepals short, ovate; petals exserted, entire, obovate, short-clawed; stamens free and unappendaged; anthers short, ovate; styles short or almost none, stigma small, simple; pod oblong-lanceolate to linear, flattened parallel to the septum, narrowed at each end, the valves 1-nerved and slightly keeled; septum very incomplete or almost wanting. (Greek, well and opening, referring to the incomplete septum.)

E. edwardsii R. Br.

Glabrous, root thick, fleshy; stems 1–several, 3–30 cm. tall; leaves entire, ovate, the lower petioled, the upper sessile or nearly so; flowers small, white or pale purple, densely crowded, the fruiting racemes elongated; pods sharply pointed, 10–18 mm. long.

Alaska Range—Bering Sea northward, nearly circumpolar. Fig. 525.

7. SISYMBRIUM L.

Annual or biennial herbs; leaves alternate, pinnately lobed; flowers in racemes, perfect; petals small, in ours yellow; pods narrowly linear, terete or nearly so; stigmas 2-lobed, seeds oblong, not winged; cotyledons incumbent. (Ancient Greek name of some cruciferous plant.)

Pod short, appressed .... 1. *S. officinale*
Pod long, spreading .... 2. *S. altissimum*

1. *S. officinale* (L.) Scop.   Hedge Mustard.
Erysimum officinale L.

Stems branching, more or less hirsute, 3-8 dm. tall; leaves hirsute, pinnatifid to various degrees, the divisions more or less toothed; pods 15-20 mm. long, on very short, stout pedicels and closely appressed to the stem, somewhat torulose, the valves with a strong, prominent midvein.

A sparingly introduced weed, native of Europe. Fig. 526.

2. *S. altissimum* L.   Tumble Mustard.
Norta altissima (L.) Britt.

Stem branching, 4-10 dm. tall, rather sparingly ciliate; lower leaves pinnatifid about half way to the midrib; upper leaves pinnate into very narrow divisions; pods 6-10 cm. long, about 1 mm. wide, at first ascending, later widely divergent on stout pedicels 1 cm. or less long.

Widely introduced weed, native of Europe. Fig. 527.

8. DESCURAINIA Webb. & Berthel.

Annual or biennial herbs, pubescent with short, branched hairs; leaves twice pinnatifid or finely dissected; flowers small, yellow, in terminal racemes much elongated in fruit; pods linear, slender-pedicelled, the valves 1-nerved; styles short; seeds in 1 or 2 rows in each cell; cotyledons incumbent. (Francis Descurain was a friend of the botanist Jussieu.)

1A. All the leaves 2- to 3-pinnate .... 1. *D. sophia*
2A. Upper leaves simply pinnate.
  1B. Seed usually in 2 rows .... 4. *D. pinnata filipes*
  2B. Seed usually in 1 row.
    1C. Pods 12-25 mm. long, spreading .... 2. *D. sophioides*
    2C. Pods 7-12 mm. long, erect .... 3. *D. richardsonii*

1. *D. sophia* (L.) Webb.   Tansy Mustard.
Sophia sophia (L.) Britt.
Sisymbrium sophia L.

Annual; stems branched, 25-40 cm. tall; leaves canescent to glabrate, the ultimate segments linear to linear-oblong; sepals 2 mm. long; petals 1.5 mm. long; pods erect or ascending, mostly curved, 15-25 mm. long, on ascending pedicels 9-11 mm. long.

Introduced weed, native of Europe. Fig. 528.

2. *D. sophioides* (Fisch.) Schulz.   Northern Tansy Mustard.
Sophia sophioides (Fisch.) Heller.

Stems slightly puberulent, often glandular, simple or branched above, 3-9 dm. tall; leaves nearly glabrous, pinnate or bipinnate, the ultimate

divisions quite variable; pods 12–30 mm. long, 1 mm. wide, more or less curved, spreading, on slender pedicels 3–7 mm. in length.
Northern Asia—Alaska—Hudson Bay. Fig. 529.

3. *D. richardsonii* (Sweet) Schulz.   Mountain Tansy Mustard.

Biennial; stems finely canescent, branched, 3–10 dm. tall; leaves finely pubescent, the lower bipinnate with lanceolate to linear divisions; pods linear, glabrous, 7–12 mm. long, 1 mm. wide, erect on closely ascending pedicels 3–6 mm. long.
Alaska Range—Gt. Slave L.—Gt. Lakes—S. Dak.—northern Mex.—Ariz. Fig. 530.

4. *D. pinnata* (Walt.) Britt. ssp. *filipes* (Gray) Detling.
    Western Tansy Mustard.

Annual; stems 10–65 cm. tall, simple or short-branching; leaves dark green, glabrous to finely puberulent, pinnate, the leaflets sometimes pinnatifid, the terminal segment typically greatly elongated; pods linear to clavate, 10–15 mm. long, on slender spreading pedicels usually longer than the pods; seed biseriate but often crowded into 1 row.
Reported from Telegraph Creek, B. C.; from there it ranges southward.

9. CAKILE Gaertn.

Glabrous, fleshy, diffuse or ascending, branching seashore annuals; flowers purplish; pods flattened or ridged, 2-jointed, the joints indehiscent, 1-celled, 1-seeded, the lower joint usually not developing; cotyledons accumbent. (Old Arabic name.)

C. edentula (Bigel.) Hook.   Sea Rocket.

Much branched from a deep root, 3 dm. tall or less; leaves oblanceolate, sinuate-dentate or lobed, narrowed into a winged petiole, 3–8 cm. long; upper joint of the pod ovoid, ridged, about 1 cm. long, 5 mm. wide, 4 mm. thick. Consists of three races, our form being ssp. *californica* (Heller) Hult.
Ssp. *californica*, Kodiak Isl. & B. C.—Calif., ssp. *lacustris*, the Gt. Lakes region, the typical form, Iceland—Labr.—Fla. Fig. 531.

10. BRASSICA L.

Caulescent annual, biennial or perennial plants; leaves entire or pinnatifid; flowers in elongated racemes, perfect, yellow; pods elongate, linear, terete or 4-angled, with an elongate beak, the valves convex, 1- to 3-nerved; seed in 1 row in each cell, subglobose; cotyledons conduplicate. All species of Brassica are introduced plants about gardens, fields, and roadsides. (Latin name for the cabbage.)

1A. Upper stem leaves cordate-clasping .... 1. *B. campestris*
2A. Upper stem leaves short-petioled or simply sessile.
  1B. Pod knotty, the beak at least one-third of its length .... 3. *B. arvensis*
  2B. Beak less than one-third the length of the pod .... 2. *B. juncea*

1. *B. campestris* L. Rape. Rutabaga. Turnip.
   B. napus L.

Usually biennial; 3–10 dm. tall, glabrous and more or less glaucous, or pubescent below; lower leaves lyrate-pinnatifid; upper leaves lanceolate with cordate-clasping base; pod 4–7 cm. long, about 3 mm. wide; beak 8–14 mm. long. Consists of many races.
Native of Europe. Fig. 532.

2. *B. juncea* (L.) Cosson. Indian Mustard.

Stems erect, 3–12 dm. tall; glabrous or slightly pubescent, somewhat glaucous; lower leaves lyrate-pinnatifid and dentate, long-petioled; upper leaves sessile, lanceolate to linear and entire; pods 35–50 mm. long, about 2.5 mm. wide; beak 1 cm. or less in length.
Native of Asia. Fig. 533.

3. *B. arvensis* (L.) Ktze. Charlock. Wild Mustard.
   Sinapsis arvensis L.

Stems hispid below, 5–10 dm. tall; leaves lyrate or pinnatifid, the upper sessile and often merely toothed; pods glabrous, ascending, 3–4 cm. long, about 3 mm. wide, the beak one-third to two-fifths the length of the pod.
Native of Europe. Fig. 534.

11. RAPHANUS (Tourn.) L.

Erect, branching, annual or perennial herbs; leaves lyrate; flowers rather showy; pods elongated, linear, fleshy or corky, constricted or continuous with spongy tissue between the seeds, indehiscent, tapering to a long conical beak; seeds subglobose; cotyledons conduplicate. It is doubtful if these species can be considered as established in our area. (Greek, quick-appearing, from the rapid germination of the seed.)

Pods longitudinally grooved .... 1. *R. raphanistrum*
Pods not longitudinally grooved .... 2. *R. sativus*

1. *R. raphanistrum* L. Jointed Charlock.

Root slender; stem branching freely, 3–8 dm. tall; lower leaves lyrate-pinnatifid; petals yellow or purplish, fading to white, 15–20 mm. long; pod nearly cylindric when green, deeply constricted between the seeds when dry; beak 1–2 cm. long.
Introduced near Fairbanks, native of Europe and northern Asia.

2. *R. sativus* L. Garden Radish.

Root more or less fleshy; stems branched, 3–5 dm. tall; lower leaves lyrate-pinnatifid; petals variable in color but purple-veined; beak of the pod often equaling the seed-bearing part.
Occasionally spreads from cultivation, native of Europe.

12. BARBAREA R. Br.

Ours a biennial herb; stems angled; leaves lyrate-pinnatifid; flowers in racemes or panicles, perfect, yellow; pod linear, more or less 4-angled, the valves keeled or ribbed; style short; stigmas 2-lobed; seeds in 1 row in each cell, marginless; cotyledons accumbent. (Dedicated to St. Barbara.)

B. *orthoceras* Ledeb.   Winter Cress. Yellow Rocket.

Stems glabrous, often purple-tinged, 2-8 dm. tall, erect, simple or variously branched; leaves with a large terminal lobe and 1-4 pairs of small segments below, the lower leaves petioled, the upper with sagittate and clasping base; pods 20-45 mm. long; seeds ovate, reticulate, about 2 mm. long. Reports of *B. stricta* and *B. americana* from Alaska refer to this species.

Siberia—Labr.—N. Hamp.—Colo.—Ariz.—Calif.—Mongolia. Fig. 535.

13. RORIPPA Scop.

Annuals or perennials; leaves alternate, pinnately dissected or lobed; flowers white or yellow, perfect, borne in terminal or axillary racemes; sepals spreading; pods from subglobose to short cylindric; seeds in 2 rows in each cell; cotyledons accumbent. (Name unexplained.)

1A. Flowers white .... 3. *R. nasturtium-aquaticum*
2A. Flowers yellow.
   1B. Pods 2-valved .... 1. *R. palustris*
   2B. Pods 4-valved .... 2. *R. barbareaefolia*

1. *R. palustris* (L.) Bess.   Marsh Yellow-cress.
   R. clavata Rydb.
   R. williamsii Britt.
   Radicula palustris (L.) Moench.

Annual or biennial; stems branched, 3-10 dm. long; leaves lyrate-pinnatifid with toothed leaflets occasionally only dentate or pinnately lobed; pod 7-8 mm. long, about 2 mm. thick, often slightly curved; style 1 mm. or less long; pedicels 4-10 mm. long, spreading or divergent; seeds numerous, nearly 1 mm. long, light brown finely reticulate. Var. *hispida* (Desv.) Rydb., plant more or less hispid with spreading hairs; pods 1.5-2 times as long as thick.

A circumboreal species reaching Mexico. Fig. 536.

2. *R. barbareaefolia* (DC.) Pors.   Round-podded Yellow-cress.

An erect, more or less hirsute biennial, branched above, 5-10 dm. tall; leaves lyrate-pinnatifid with toothed divisions, 5-12 cm. long; petals about 2 mm. long; pods subglobose, about 4 mm. long and 3 mm. wide, appearing to be 4-valved; styles stout, scarcely 1 mm. long; pedicels ascending or spreading, 4-10 mm. long; seeds small, about 0.6 x 0.4 mm., reddish-brown, not reticulate.

Eastern Siberia—Yukon, common in interior Alaska. Fig. 537.

3. *R. nasturtium-aquaticum* (L.) Hayak.   Water-cress.

Stems glabrous, floating, creeping or ascending, rooting at the nodes; leaves of 3-9 segments, the terminal the largest and nearly orbicular;

racemes elongating in fruit; petals 3-4 mm. long; pods 1-3 cm. long, about 2 mm. wide, more or less spreading.

Established at Manly (Tanana) Hot Springs, native of Eurasia. Fig. 538.

14. CARDAMINE L.

Nearly glabrous annual or perennial herbs; leaves entire or more often pinnate; flowers usually small, perfect, white or purplish; pods linear, flattened, the valves nerveless or only faintly nerved, opening elastically from the base; seeds in a single row in each cell, not margined or winged; cotyledons accumbent. (Greek name of a cress.)

```
1A. Alpine plant with small, simple, entire leaves ............... 1. C. bellidifolia
2A. Leaves mostly trifoliate ................................. 2. C. angulata
3A. Leaves pinnate or digitate.
  1B. Flowers small, 4 mm. or less across.
    1C. Stem leaves few ....................................... 7. C. umbellata
    2C. Stems leafy.
      1D. Rootstocks fibrillose ............................... 8. C. regeliana
      2D. Rootstocks not fibrillose ........................... 9. C. pennsylvanica
  2B. Flowers larger.
    1C. Stems pubescent in the upper part .................... 3. C. purpurea
    2C. Stems glabrous throughout.
      1D. Stem leaves with several pairs of linear leaflets ....... 4. C. pratensis
      2D. Stem leaves with 1 or 2 pairs of leaflets or digitately 3- to 7-parted
        1E. All leaflets linear or lanceolate-linear ............. 5. C. richardsonii
        2E. At least the basal leaves with broad leaflets ........ 6. C. microphylla
```

1. *C. bellidifolia* L.   Alpine Cress.

A dwarf, tufted perennial growing 4-15 cm. tall; leaves oval or ovate, entire, the blades 4-12 mm. long; petals white, 3-4 mm. long; pods erect, 15-25 mm. long, about 1.5 mm. wide, the pedicels 4-10 mm. long. A race with some of the leaves slightly lobed occurs at Seward.

Arctic-alpine, circumpolar. Fig. 539.

2. *C. angulata* Hook.   Seaside Bitter-cress.

A perennial with stolons; stems 3-7 dm. tall, new growth hirsute-pubescent, old growth glabrate; leaves usually 3-foliate, terminal leaflet of base leaves rotund, the lateral ovate; leaflets of stem leaves rhombic-ovate, all coarsely toothed or lobed with rounded teeth, mucronate with vein-endings; petals white, 8-12 mm. long; pod about 2 cm. long, nearly 2 mm. wide.

Near the coast, southeastern Alaska—Ore. Fig. 540.

3. *C. purpurea* C. & S.   Purplish Bitter-cress.

An alpine perennial; leaves mostly basal, more or less hirsute; leaflets 3-7, mostly 5, the terminal one reniform, often slightly lobed, 5-10 mm. wide, the lateral ones inclined to be orbicular, 2-6 mm. wide; flowering stems hirsute and with 1-3 leaves, 5-15 cm. tall; flowers several, almost umbellate; petals white to rose-purple or violet-purple, 5-6 mm. long; pods erect, tapering at the apex, 10-25 mm. long, about 2 mm. broad.

Extreme northeastern Asia—Arctic Coast—Yukon—Alaskan Range. Fig. 541.

4. *C. pratensis* L.   Cuckoo Flower.

An erect perennial from a short rootstock, 1-4 dm. tall; leaflets 5-15, variable, those on the basal leaves broad, sometimes toothed, those of the stem leaves narrow and linear; flowers showy, white, pink, or purple; petals 7-11 mm. long; pods 2-3 cm. long, 2 mm. wide.

Wet places, widespread, circumboreal. Fig. 542.

5. *C. richardsonii* Hult.   Richardson Bitter-cress.
C. digitata Rich.

Stems 6-20 cm. tall, usually simple, glabrous, 2- to 5-leaved, purplish at the base; lower leaves consisting of 3-7 narrow, acute leaflets 12-40 mm. long, mucronulate at the apex; pedicels 6-12 mm. long at anthesis, up to 15 mm. long in fruit; sepals about 3 mm. long, ovate, often tinted red near the apex and with hyaline margins; petals white, 6-8 mm. long; pods 2-3 cm. long, about 1.5 mm. wide, narrowed at both ends.

Northeastern Siberia—Alaska—Hudson Bay. Fig. 543.

6. *C. microphylla* Adams.   Small-leaved Bitter-cress.
C. blaisdellii Eastw.

Rootstocks horizontal, slender, glabrous; stems erect or ascending, glabrous, 6-20 cm. tall; lower leaves with 3-5 leaflets, the terminal one broad and more or less 3-lobed, the lateral ones usually 2- or 3-lobed; inflorescence corymbose, lengthening into a raceme; pedicels flattened, becoming 2 cm. long; sepals oblong, yellow with lighter margin, obscurely 3-nerved, 3-3.5 mm. long; petals white, broadly spatulate; pods slender, 2-4 cm. long, the beak long.

Lena R., Siberia—west central Alaska. Fig. 544.

7. *C. umbellata* Greene.   Umbel-flowered Bitter-cress.

Perennial though often with the appearance of being annual; glabrous or nearly so, 1-5 dm. tall; leaflets 3-9, usually 7 or 5, varying greatly, those of the lower leaves broad and often toothed or lobed, those of the upper leaves narrow and usually entire; inflorescence a raceme, often shortened to resemble an umbel; petals white, 3-4 mm. long; pods erect, 2-3 cm. long, 1-1.5 mm. wide, with a beak less than 1 mm. long.

East Asia—Yukon—Alta.—Colo.—Ore. Fig. 545.

8. *C. regeliana* Miq.   Regel Bitter-cress.

Rootstock short and fibrillose; stems simple or branched, erect, up to 5 dm. tall; leaves somewhat fleshy, the basal and lower cauline 4-7 cm. long, the upper 45-95 mm. long; terminal leaflets large, irregularly lobed; racemes many-flowered, the flowers 3.5-6 mm. long; apex of sepals purplish or blackish; pods 20-25 mm. long; seeds 0.8-1 x 0.6-0.75 mm.

An eastern Asiatic species found on Attu Isl. and near Ketchikan.

9. *C. pennsylvanica* Muhl.   Pennsylvania Bitter-cress.

Stems glabrous or sparingly pubescent below, freely branching, 15-50 cm. tall; lower leaves 5-12 cm. long, the terminal segments ovate or obovate, sometimes lobed, the lateral segments 3-5 pairs, oblong, some of them often petiolulate; flowers small, white; pods 12-25 mm. long, 1 mm. wide, on slender spreading pedicels.

Near Ketchikan, probably introduced but occurring throughout most of temperate N. America.

15. LESQUERELLA Wats.

Ours a low perennial with stellate pubescence; flowers yellow; leaves simple; petals entire; pods globose or oblong, inflated, the valves nerveless; septum translucent; seeds several to many in each cell of the pod, flattened. (Lesquereux was a Swiss and American botanist.)

L. arctica (Wormskj.) Wats.   Arctic Bladder-pod.

Tufted; densely stellate-pubescent; stems 3-12 cm. tall, usually simple; leaves spatulate or oblanceolate, 25 mm. or less long, entire; pods 4-6 mm. long, with a narrow style 1-2 mm. long. Var. *scammanae* Rollins is taller and more robust; leaves including petioles may reach a length of 7 cm. and the style may be 2-3 mm. long.

Northern Asia—northern and central Alaska—Greenl.—Newf.—B. C. Fig. 546.

16. CAPSELLA Medic.

Erect, branching annuals, glabrate above, pubescent with both simple and branched hairs below; leaves largely clustered at the base, entire, lobed, or pinnatifid; pods flattened contrary to the narrow partition, triangular obcordate, the valves boat-shaped and keeled; cotyledons accumbent. (Latin, little box, from the shape of the pod.)

Pods with convex or straight sides .... 1. *C. bursa-pastoris*
Pods with concave sides .... 2. *C. rubella*

1. *C. bursa-pastoris* (L.) Medic.   Shepherd's Purse.
Bursa bursa-pastoris (L.) Britt.

Summer or winter annual, often forming a rosette over winter, 1-6 dm. tall; lower leaves usually lyrate-pinnatifid, lobed or dentate; stem leaves few ,lanceolate and usually sagittate at the base; flowers white, the petals decidedly longer than the sepals; pods triangular, 6-8 mm. long; pedicels spreading.

Widely introduced weed, native of Europe. Fig. 547.

2. *C. rubella* Reuter.

Similar to *C. bursa-pastoris*; pods larger with distinctly concave sides, often suffused with reddish-purple; petals scarcely longer than the sepals.

Kodiak Isl.—Unalaska—Nome, native of the Mediterranean region. Fig. 548.

17. CAMELINA Crantz.

Erect annual herbs; flowers small, yellowish, in terminal racemes; pods ovoid or pear-shaped, slightly flattened; valves strongly convex; 1-nerved; seeds in 2 rows, oblong, marginless. (Greek, low flax.)

C. sativa (L.) Crantz. False Flax.

Glabrous or nearly so; stems simple or branched above; 3-9 cm. tall; basal leaves petioled, 5-8 cm. long, lanceolate, toothed or entire; upper leaves smaller, sessile with a clasping sagittate base; racemes many-flowered; pods margined, 6-8 mm. long.

Introduced with grain, native of Europe.

18. NESLIA Desv.

Erect, leafy-stemmed annuals with branching pubescence; leaves sessile, entire; flowers racemose, yellow; pods small, globose, reticulated, indehiscent, usually 1-seeded by obliteration of the partition; style elongate; stigma simple; cotyledons incumbent. (J. A. N. de Nesle was a French botanist.)

N. paniculata (L.) Desv. Ball Mustard.

Stems slender, branched above, 2-9 dm. tall, rough-hispid; leaves lanceolate, 2-6 cm. long, the upper with sagittate-clasping base; racemes much elongated in fruit; pod depressed globose, strongly reticulated, about 2 mm. long, nearly 3 mm. wide; pedicels slender, 6-12 mm. long.

Introduced with grain, native of Europe. Fig. 549.

19. DRABA L.

Low, tufted, annual or more often perennial herbs; leaves simple, usually with stellate or forked pubescence; flowers yellow or white, perfect, borne in racemes; pods elliptic, ovate or linear, flat, the valves dehiscent, usually nerveless; seeds in 2 rows, usually wingless; cotyledons accumbent. (Greek name for some member of this family.) A very complicated group of plants about which there has been much confusion. No two writers on the genus agree on the species. The same name has been used for different forms and the same form has been reported under several names. Some of the forms are known only from a very few collections and more material is needed to obtain a clearer understanding of the group. Some of the forms here reported should perhaps be reduced to varieties or subspecies, but until further studies are made it is thought best to keep the forms separate.

```
1A. Plants annual ................................................ 1. D. nemorosa
2A. Plants perennial.
   1B. Leaves glabrous or slightly ciliated (see also 12. D.
         fladnizensis) ............................................ 4. D. crassifolia
   2B. Leaves pubescent on either or both sides.
      1C. Plant scapose, rarely more than 1 dm. tall.
         1D. Pubescence stellate, or only sparsely ciliate near the base.
            1E. Leaves carinate, narrow ........................ 5. D. oligosperma
            2E. Leaves flat, broader.
               1F. Flowers white ................................ 6. D. nivalis
               2F. Flowers yellow.
                  1G. Pods glabrous.
                     1H. Stigmas subsessile .................... 7. D. caesia
                     2H. Stigmas 0.5-1 mm. long ............... 8. D. chamissonis
                  2G. Pods pubescent.
                     1H. Sepals 1.5-2 mm. long................. 9. D. exalata
                     2H. Sepals 2.5-3 mm. long ................10. D ruaxes
```

2D. Pubescence of simple or forked hairs, sometimes mixed with stellate hairs.
  1E. Scapes very short, flowers yellow.
    1F. Pods pyriform, glabrous .... 2. *D. aleutica*
    2F. Pods spherical or oblong, pubescent .... 3. *D. densifolia*
  2E. Scapes longer, pods usually longer.
    1F. Scapes and pedicels glabrous, leaves narrow.
      1G. Plants stout, leaves carinate, flowers
        yellow .... 11. *D. pilosa*
      2G. Plants more delicate, flowers white.
        1H Leaves ciliate with exclusively simple
          hairs .... 12. *D. fladnizensis*
        2H. Leaves with mixed simple and forking
          hairs .... 13. *D. lactea*
    2F. Scapes pubescent.
      1G. Without stellate hairs, flowers yellow.
        1H. Densely tufted, pods large.................. 14. *D. macrocarpa*
        2H. Not densely tufted, pods smaller .... 15. *D. alpina*
      2G. Pubescence mixed with short stellate hairs.
        1H. Petals large, styles long .... 16. *D. eschscholtzii*
        2H. Petals smaller, styles short .... 17. *D. pseudopilosa*
2C. Stems normally with 1-many leaves.
  1D. Basal leaves 8–16 cm. long .... 27. *D. hyperborea*
  2D. Basal leaves much smaller.
    1E. High growing plants usually with several stem leaves.
      1F. Flowers yellow .... 26. *D. aurea*
      2F. Flowers white.
        1G. Plant up to 6 dm. tall with 6–15 leaves .... 25. *D. maxima*
        2G. Plant smaller with fewer leaves .... 24. *D. borealis*
    2E. Lower growing plants with fewer stem leaves.
      1F. Pods glabrous.
        1G. Pods long, narrow, stigmas sessile .... 18. *D. stenoloba*
        2G. Pods shorter, style evident.
          1H. Pedicels longer than the pods .... 19. *D. longipes*
          2H. Pedicels usually not longer than the pods.
            1I. Usually only 1 stem leaf .... 20. *D. kamtschatica*
            2I. Usually 3 or more stem leaves.............. 21. *D. glabella*
      2F. Pods pubescent.
        1G. Pods short-pediceled, appressed to the
          stem .... 22. *D. lanceolata*
        2G. Pods long-pedicelled, divaricate .... 23. *D. cinerea*

1. *D. nemorosa* L.   Wood Whitlow-grass.
    D. lutea Gilib.

    Winter annual, 5–25 cm. tall; leaves 1–3 cm. long, mostly basal or on lower part of stem, with branched and simple hairs; racemes lax, elongate, 10- to 40-flowered; pedicels 1–5 times as long as the pods, spreading or ascending; sepals about 1.5 mm. long; petals light yellow, about 2 mm. long; pods averaging about 8 mm. long. Our form is var. *leiocarpa* Lindb. with glabrous pods.
    May be introduced in our area, has an interrupted circumpolar distribution. Fig. 550.

2. *D. aleutica* E. Ech.   Aleutian Draba.

    A very low, diffuse, matted plant that seldom rises more than a centimeter or two above the surface of the habitat; leaves all basal, 5–10 x 2–4 mm., persistent, ciliate with long simple or occasionally forked hairs; scapes very short; sepals about 2 mm. long; petals yellowish, 2–3 mm. long; pods obovate-obcordate, 4–5 x 3–4 mm., inflated, glabrous, 4-seeded.
    Aleutian & Pribylof Isls.—?Seward. Fig. 551.

3. **D. densifolia** Nutt.

Low, caespitose plant; leaves densely crowded, 2–9 x 0.5–3 mm., the midvein prominent below, ciliate with stiff, straight cilia 0.5–1 mm. long, glabrous on the surfaces or with a few hairs on lower surface; scapes leafless, 1–3 cm. tall, glabrous to hirsute; racemes 3- to 15-flowered; sepals 2–3 mm. long; petals yellow, 2–6 mm. long, pods ovate or orbicular, 2–7 x 2–4 mm., more or less pubescent; styles 0.5–1 mm. long; seeds about 2 mm. long.

Rare, eastern Asia—Alaska—Wyo.—Calif. Fig. 552.

4. **D. crassifolia** Grah.

Stems 2–15 cm. long, usually scapiform, glabrous or nearly so; leaves numerous, narrowly oblanceolate, 5–15 mm. long, ciliate, usually entire; sepals oval, about 1 mm. long, glabrous to pilose; petals yellow, fading to white, 2–2.5 mm. long; pods glabrate, tapering to both ends, 5–12 mm. long; style almost lacking.

Rare, central Alaska—Greenl.—Rocky Mts.

5. **D. oligosperma** Hook.

D. inserta Pors. not Payson.

Caespitose matted perennial; leaves imbricate, 3–11 x 0.75–1.75 mm., the median vein prominent, the lower surface and sometimes the upper covered with appressed, pectinately branched hairs, the margins often ciliate; scapes 1–10 cm. tall, glabrous except sometimes near the base; racemes 3- to 15-flowered, often more than half the total height of the stem; sepals 2–2.25 mm. long; petals yellow, 3–4.5 mm. long; pods oval or ovate-lanceolate, 2.5–7 x 2–4 mm., with short, stiff, simple or branched hairs; seeds 1.4–1.8 mm. long.

Central Alaska—Gt. Bear L.—Colo.—Calif.

6. **D. nivalis** Lilj.

Caespitose with slender prostrate or ascending branches or the caudices ending in rosettes; leaves 5–15 x 1–5 mm., densely and finely pubescent with stellate hairs; flowering stems naked or with 1–4 denticulate leaves, 3–20 cm. tall, glabrous or finely stellate-pannose; sepals about 2 mm. long; petals white, 2.5–3 mm. long. In the typical form the pods are mostly 4–8 x 1.5–2 mm.; the stems are without a leaf and less than 1 dm. tall. In var. *denudata* (Schulz) C. L. Hitchc. the stems are up to 2 dm. tall with 1 or 2 dentate leaves and the pods 12–20 mm. long.

Circumboreal, the variety from Prince William Sound—Juneau. Fig. 553.

7. **D. caesia** Adams.

Closely related to *D. nivalis;* sepals oblong, obtuse, pubescent; petals yellow; pedicels and pods glabrous, lanceolate; stigmas subsessile.

Rare, Lena R., Siberia—Seward Penin.—Mackenzie R.

8. **D. chamissonis** G. Don.

Densely caespitose; leaves elliptic-oblanceolate, 4–8 x 1.5–2.5 mm., cinereous with fine stellate pubescence and simple cilia near the base, the midrib prominent and marcescent; scapes usually leafless, 4–8 cm. tall, sparingly pannose-stellate; racemes 5- to 20-flowered; lower pedicels 5–12 mm. long; sepals about 2.5 mm. long; petals yellow, 4–5 mm. long; pods elliptic-ovate, 4–8 mm. long, 2.5–4 mm. wide, glabrous, the valves reticulate-veined; styles 0.5–1 mm. long; seeds about 1 mm. long.

Cape Thompson—Teller—Elim. Fig. 554.

9. *D. exaltata* E. Eck.

Densely caespitose, the old leaves persistent; leaves in a congested rosette, obovate, short-petioled, the apex rounded, 4–6 mm. long, 2.5–3 mm. wide, pubescent with soft stellate hairs; scapes naked; sepals ovate, 1.5–2 mm. long, 0.75 mm. wide, pilose on the back; petals yellow, emarginate, 3–4 mm. long; pods ovate–oblong, acuminate, 3–5 x 2–3.25 mm.; styles about 0.5 mm. long; seed about 1 mm. long.

Very rare, Seward Penin. and Arctic Coast.

10. *D. ruaxes* Payson & St. J.

D. ventosa Gray var. *ruaxes* (Payson & St. J.) C. L. Hitch.

Low plant with branched caudex; leaves oblanceolate to nearly elliptic or ovate, entire, 5–18 mm. long, 2–4 mm. wide, densely pubescent with mostly 4- to many-forked and some simple or forked hairs; marcescent, the old midribs persisting several years; stems scapose, 2–5 cm. long, densely pubescent with mostly simple or forked hairs; sepals 2–2.5 mm. long, soft pilose; petals yellow, 4–5 mm. long; pods oval to ovate, 5–8 x 3–4 mm., the valves thick and firm, densely pubescent; styles about 0.7 mm. long; seed 1.5–2 mm. long.

A specimen from Mt. Crillon was doubtfully placed here, otherwise known only from one collection in British Columbia and one in Washington.

11. *D. pilosa* Adams.

Caudices densely covered with the marcescent leaf-bases; leaves 5–15 mm. long, oblong-linear, rigid, slightly fleshy, the midrib very prominent, partly glabrous, strongly ciliate with simple or forked hairs below and on the margins; stems scapose, glabrous; flowers yellow; pods glabrous, ovate, in a capitate cluster.

Central Siberia—Alaska—Mackenzie Delta.

12. *D. fladnizensis* Wulfen.

Caespitose, almost acaulescent perennial; leaves 5–10 mm. long, 1–2 mm. wide, ciliate with long simple hairs and pubescent with once- or twice-forked hairs or ciliate only; scapes 2–8 cm. tall, glabrous or pubescent near the base, leafless or with 1 or 2 small leaves; sepals 1–2 mm. long; petals white, 2–3 mm. long; pods 3–6 x 1.5–2 mm., usually glabrous; styles nearly lacking.

Some Alaska collections have doubtfully been referred to this species, distribution interrupted circumpolar.

13. *D. lactea* Adams.

Loosely pulvinate; leaves 5–15 x 1–4 mm., the midrib prominent, stiffly ciliate, the lower surface with more or less many-branched hairs; scapes leafless, 1–10 cm. tall, glabrous or pubescent below; racemes 3- to 5-flowered; pedicels usually short; sepals about 2 mm. long; petals white, about 4 mm. long; pods 4–10 x 2–3 mm., glabrous; styles 0.5–1 mm. long; seeds 1–1.5 mm. long.

Circumpolar and usually arctic. Fig. 555.

14. *D. macrocarpa* Adams.

Related to *D. alpina;* leaves densely tufted, those of previous years remaining long on the stems, pubescent with mostly simple hairs on the upper surface, mixed hairs on the lower surface; sepals pilose; pods larger than those of *D. alpina*.

An arctic species, Nova Zemla.—Greenl.—Nome.

15. *D. alpina* L.

Caespitose, with thick cushions of marcescent leaves; leaves all basal, rarely 1 cauline, 5–20 x 1.5–4 mm., conspicuously long-ciliate; the surfaces with long simple or once or twice forked hairs; scapes 3–10 cm. tall, pubescent; racemes 4- to 20-flowered; pedicels 3–10 mm. long; sepals 2–3.5 mm. long; petals yellow, about 5 mm. long; pods 5–9 x 2–4 mm., glabrous or hispidulous; styles 0.3–0.7 mm. long; seeds about 1.5 mm. long.

More or less circumpolar. Fig. 556.

16. *D. eschscholtzii* Pohle.

Related to *D. alpina;* leaves long and narrow, sparsely ciliate on the margins, the surfaces pubescent with short, simple, forked, or branched hairs, dense on the under side; petals white, emarginate; pods long and glabrous, usually longer than the pedicels; styles 1–1.75 mm. long.

East Asia—Lake Bennett.

17. *D. pseudopilosa* Pohle.

Resembling *D. lactea;* leaves, at least on the lower side pubescent with short, stellulate hairs often mixed with simple or forked hairs, ciliated with simple hairs; scapes pubescent; pedicels glabrous.

Northeastern Asia—Islands of Bering Sea—Arctic Coast of Alaska.

18. *D. stenoloba* Ledeb.

D. macouniana Rydb. not *D. macounii* Schulz.

Leaves mostly in a basal rosette, 10–40 x 3–8 mm., usually denticulate, hispidulous with simple or forked hairs; stems simple or branched, 5–30 cm. tall, with 1–7 leaves, sparingly strigose to stellate below, glabrous above; sepals 1–2.25 mm. long, pilose; petals yellow, 2–4.5 mm. long, often fading to white; pods acute, 8–12 x 1.5–2.3 mm.; styles nearly lacking; seeds 1 mm. or less long.

Unalaska—Alaska Range—Rocky Mts.—Calif. Fig. 557.

19. *D. longipes* Raup.

A diffuse plant with small rosettes; rosette leaves oblanceolate to obovate-oblanceolate, entire or with a few small teeth, 5–25 mm. long, 1.5–10 mm. wide, usually with 1- or 2-forked cilia and short-stalked or sessile or appressed 4-rayed trichomes; cauline leaves 1–3, rarely none, sessile, usually dentate with a few teeth; stems 5–20 cm. tall, pubescent; pedicels slender, 3–15 mm. long; pods broadly lanceolate to linear-lanceolate, 3–15 x 1–2.5 mm., glabrous or nearly so; styles 0.5–1 mm. long; seeds about 1 mm. long.

Bering Sea—northern B. C.—Fig. 558.

20. *D. kamtschatica* (Ledeb.) N. Busch.
D. nivalis var. *kamtschatica* Pohle.

Caespitose perennial with many slender branches ending in rosettes; leaves mostly basal, linear to oblanceolate or obovate, 5–15 x 1–5 mm., densely and finely stellate-pannose and canescent; stems slender, with 1–4 leaves, up to 12 cm. tall; sepals about 2 mm. long, glabrous to stellate-pilose; petals white, 2–2.5 mm. long; pod elliptic to oblong-lanceolate, 6–12 $\times$ 1–2 mm., contorted; seeds about 0.75 mm. long.

Siberia—Alaska—Vancouver Isl.

21. *D. glabella* Pursh.
D. hirta auct.

Loosely branched perennial with slender caudices; basal leaves 1–4 cm. long, 2–10 mm. wide, entire or remotely denticulate, the blades passing into slender petioles, pubescent with pectinately branched hairs; cauline leaves 1–10, sessile; stems 1–4 dm. tall, sparsely and finely pectinate-stellate; sepals 2–3 mm. long; petals white, 4–5 mm. long; pods lanceolate to ovate-lanceolate, 5–15 x 1.5–3 mm., glabrous or nearly so; seeds about 1 mm. long.

Our commonest *Draba*, circumboreal. Fig. 559.

22. *D. lanceolata* Royle.

Leaves of basal rosette 10–30 x 3–8 mm., mostly oblanceolate; stem leaves lanceolate, 5–25 mm. long, grayish with soft stellate or branched hairs, the basal often with a few cilia; stems several, 5–25 cm. tall, with soft simple and branched hairs up to 1 mm. long; racemes simple or compound, 10- to 50-flowered; pedicels ascending or appressed; sepals about 2 mm. long; petals white, emarginate, 3–5 mm. long; pods 4–12 x 1.5–3 mm., soft pubescent with short, simple, or branched hairs; seeds 0.6–1 mm. long.

Central Alaska, distribution interrupted circumboreal. Fig. 560.

23. *D. cinerea* Adams.

Caespitose, the caudices ending in rosettes 2–10 cm. wide; stems 1–4 dm. tall, usually bearing a few leaves, basal leaves 6–25 mm. long, 2–8 mm. wide, densely pannose; sepals about 2 mm. long, densely pubescent; petals about 3.5 mm. long; pods pubescent with appressed 4- to 6-rayed hairs; styles 0.5–0.8 mm. long.

Of erratic circumpolar distribution. Fig. 561.

24. **D. borealis** DC.
   D. unalaschensis DC.

   Stems 1-many, often decumbent at the base, erect, pubescent, 5-30 cm. tall; basal leaves obovate or oblanceolate, 1-3 cm. long, 2-18 mm. wide, entire or dentate, the pubescence of simple, forked, and 4- to 6-rayed stellate hairs; stem leaves 3-15, ovate or obovate, broader and more dentate than the basal ones; racemes many-flowered; sepals about 3 mm. long; petals white, about 5 mm. long; pods lanceolate, 8-12 x 2.5 mm., more or less pubescent, plane or more often contorted.

   Most of Alaska, Yukon, and northern B. C. Fig. 562.

25. **D. maxima** Hult.

   Probably biennial, pubescent throughout; stems usually several, 1-6 dm. tall, 6- to 15-leaved, pilose with simple and forked hairs; basal leaves oblanceolate, up to 45 mm. long, attenuate at the base, sometimes into short winged petioles; stem leaves sharply toothed, obovate to short-lanceolate, sessile; inflorescence at first congested, in fruit elongated; sepals 2.5-3 mm. long; petals white, 4-5 mm. long; pods lanceolate to ovate-lanceolate, 10-15 mm. long. This form was formerly included in *D. borealis*. It is our tallest species of *Draba*.

   Along the coast, Kodiak Isl.—southeastern Alaska. Fig. 563.

26. **D. aurea** Vahl.

   A variable species with a simple or branched caudex; leaves of the basal rosettes oblanceolate to spatulate, 1-5 cm. long, 2-15 mm. wide, mostly entire; stem leaves 3-30, entire or denticulate, more or less canescent with cruciform, branched or simple hairs; stems 1-several, 1-5 dm. tall, with some of the simple hairs quite long; racemes simple or compound, 5- to 50-flowered; pedicels 3-20 mm. long; sepals 2-3.5 mm. long; petals yellow, 4.5-6 mm. long; pods lanceolate to oblong-lanceolate, pubescent, usually contorted, seeds about 1 mm. long.

   Southwestern & central Alaska—Gt. Bear L.—Greenl.—S. Dak.—Ariz. Fig. 564.

27. **D. hyperborea** (L.) Desv.
   Nesodraba grandis (Langsd.) Greene.

   Loosely branched perennial from a thick rootstock, 10-35 cm. tall; pubescent with simple or forked hairs; basal leaves up to 17 cm. long including the winged petioles of about equal length with the blade, 5-35 mm. wide, pubescent, remotely dentate, the teeth often long, cauline leaves smaller, short-petioled to sessile; sepals 3-5 mm. long; petals yellow, about 5 mm. long; pods 8-25 mm. long, 3-8 mm. wide, glabrous; seeds about 1.5 mm. long.

   East Asia—Alaska—Queen Charlotte Isls. Fig. 565.

20. SMELOWSKIA C. A. Mey.

Low caespitose perennials canescent with fine stellate hairs; leaves pinnatifid or some or all of them entire; flowers small, white, yellowish or tinged with purple; petals obovate, exerted; pods obovate to lanceolate, the valves strongly keeled; stigmas nearly sessile; cotyledons incumbent. (T. Smelowski was a Russian botanist.)

S. calycina C. A. Mey. ssp. *integrifolia* (Seem.) Hult.

Densely caespitose from a branched caudex which is covered with the remains of old leaves; leaves sometimes pinnatifid but more usually entire, the entire ones oblanceolate to oblong-linear, densely stellate-pubescent with longer simple hairs at base; stems 5–15 cm. tall; pods lanceolate to oblanceolate, attenuate at both ends, 5–10 mm. long; seeds few, about 2 mm. long.

East Asia & B. C.—Alta.—Mont.—Colo.—Ore., the ssp. in east Asia—central Alaska. Fig. 566.

21. ARABIS L.

Biennial or perennial herbs; leaves alternate, mostly toothed; flowers perfect, white or purple, borne in terminal or axillary racemes; pubescence when present of simple or branched hairs; pods linear, flat to nearly orbicular in cross section, the valves usually nerved or veiny; seeds winged, margined ,or wingless; cotyledons usually accumbent. Name for Arabia, where many of the species grow.)

```
1A. Cauline leaves not auriculate and clasping at the base.
  1B. Cauline leaves attenuate at the base .................. 1. A. lyrata kamchatica
  2B. Cauline leaves merely sessile ....................... 2. A. arenicola
2A. Cauline leaves auriculate-clasping at the base.
  1B. Pedicels and pods reflexed ......................... 3. A. holboellii
  2B. Pedicels spreading.
    1C. Valves rounded ................................. 7. A. hookeri
    2C. Valves flat .................................... 4. A. divaricarpa
  3B. Pedicels erect or closely ascending.
    1C Seeds in 2 rows in each cell ..................... 5. A. drummondii
    2C. Seeds in 1 row in each cell.
      1D. Plant hirsute ............................... 6. A. hirsuta
      2D. Plant glaucous above ........................ 8. A. glabra
```

1. *A. lyrata* L. ssp. *kamchatica* (Fisch.) Hult.   Kamchatka Rock-cress.
A. ambigua DC.

Stems tufted, glabrous or nearly so, 1–3 dm. tall; basal leaves lyrately lobed, 15–40 mm. long; stem leaves spatulate to linear, usually entire but sometimes toothed, 1–3 cm. long; petals white, 4–8 mm. long; pedicels in fruit ascending, less than 1 cm. long; pods erect or nearly so, 2–3 cm. long, about 1 mm. wide.

Typical form in eastern states, the subspecies in east Asia—central Alaska—Sask.—Wash. Fig. 567.

2. *A. arenicola* (Rich.) Gelert.   Arctic Rock-cress.

Perennial from a slender root, somewhat pubescent, at least below, or sometimes entirely glabrous; stems 1-several, ascending, more or less flexuous, 7–25 cm. long; leaves chiefly basal, spatulate or oblong, entire or with 1 or 2 teeth on each side, the lower petioled, the upper sessile; stem leaves

2 or 3; flowers white or purplish; pods linear, flat, 15–25 mm. long, about 1.5 mm. wide.

Little Susitna valley near Matauska, also reported from Golovin and St. Michael, Ellesmereland—Greenl.—Labr.—L. Athabasca. Fig. 568.

3. *A. holboelli* Hornem.   Holboell Rock-cress.

Biennial or perennial, pubescent throughout or nearly glabrous, usually branched above, 2–8 dm. tall; basal leaves oblanceolate, densely pubescent, 1–5 cm. long; stem leaves lanceolate to auriculate and clasping at the base; sepals 3–4 mm. long, scarious-margined; petals 6–8 mm. long, white or pink; pedicels 6–16 mm. long, strictly reflexed or loosely descending; pods 3–6 cm. long, 1–2.5 mm. wide; seed narrowly winged all around, about 1 mm. broad. Var. *retrofracta* (Grah.) Rydb. is our more common form. The pods are usually adpressed to the stem, straight or nearly so, 35–80 mm. long.

Central Alaska—Greenl.—Alta.—Wash. Fig. 569.

4. *A. divaricarpa* A. Nels.   Spreading-pod Rock-cress

Stems 1–few from a biennial root, simple or branched above, pubescent below with appressed, branched hairs or sometimes glabrous throughout, 3–9 dm. tall; basal leaves oblanceolate to spatulate, pubescent with 3- to several-rayed hairs, 2–6 cm. long, 4–8 mm. wide, stem leaves narrowly oblong to lanceolate, the upper glabrous; sepals scarious-margined, 3–5 mm. long; petals pink to purplish, 6–10 mm. long; pods straight or curved, 2–8 cm. long, on spreading or decending pedicels 6–12 mm. long; seed about 1 mm. wide.

Central Alaska—Que.—N.Y.—Calif. Fig. 570.

5. *A. drummondii* Gray.   Drummond Rock-cress

Stems 1–3 from a simple caudex, simple or branched above, glabrous or somewhat pubescent at the base; basal leaves oblanceolate, usually entire, narrowing into a petiole, 2–8 cm. long; cauline leaves sessile, acute, crowded toward the base; petals white or pinkish, 7–10 mm. long; pedicels 1–2 cm. long; pods erect, often strict, straight, glabrous, 4–10 cm. long, 1.5–3 mm. wide; seed prominently winged on one end and sides; 1.5–2 mm. long, 1 mm. wide.

South central Alaska—Yukon—Labr.—Newf.—Del.—Calif. Fig. 571.

6. *A. hirsuta* (L.) Scop.   Hairy Rock-cress.

Stems usually simple, hirsute below, less so above, 2–6 dm. tall; basal leaves spatulate or oblanceolate, sinuately toothed, 2–7 cm. long; stem leaves lanceolate, cordate-clasping, 1–5 cm. long; petals white or tinged purple; pedicels 6–12 mm. long; pods 4–8 cm. long, 1 mm. or more wide, usually erect. Represented in our area by 2 subspecies. Ssp. *pycnocarpa* (Hopkins) Hult. Petals 3–5 mm. long; pods strictly erect. Central Alaska—Que.—Ga.—Calif. Ssp. *eschscholtziana* (Andriz) Hult. (*A. rupestris* Nutt.). Petals 5–9 mm. long; pods sometimes somewhat divergent;

upper part of stem hirsute. Along the Pacific Coast, Aleutian Isls.—Ore. Fig. 572.

The species is circumboreal.

7. A. hookeri Lange.   Hooker Rock-cress.
Arabidopsis mollis (Hook.) Schulz.

Stems several from a biennial, often branching rootstock, diffuse or ascending, up to 5 dm. tall. hirsute below; leaves oblanceolate, sinuate-dentate, acute, up to 5 cm. long; flowers small, white, sepals and pedicels hairy; pods 25–40 mm. long, ascending or occasionally spreading on spreading pedicels 6–12 mm. long; seeds minute, oblong.

Central Alaska—mouth of Mackenzie R.—western Greenland—Gt. Bear L. and probably farther south. Fig. 573.

8. A. glabra (L.) Bernh.   Tower Mustard.
Turritis glabra L.

Stems one or a few from a taproot, simple or branching above, pubescent below, 4–12 dm. tall; basal leaves spatulate to ovate, denticulate to pinnately parted, coarsely pubescent to nearly glabrous, the cauline sessile; flowers small, yellowish-white; pods strictly erect, only slightly flattened, glabrous, 4–10 cm. long, slightly more than 1 mm. wide; seed averaging 1 mm. long by 0.5 mm. wide.

A circumboreal species rather rare in Alaska. Fig. 574.

22. ERMANIA Cham.

Low, alpine, tomentose perennials; leaves small, more or less lobed; sepals persistent under the mature fruit; styles short with capitate stigmas; pods oblanceolate, the partition perforated or almost lacking, the valves strongly nerved.

Basal leaves with 3–5 crenate teeth or lobes .... 1. E. parryoides
Basal leaves deeply 7- to 9–lobed .... 2. E. borealis

1. E. parryoides Cham.

Leaves small, broad, usually 3-lobed; flowers yellowish-white; pods oblanceolate to oblong, not inflated.

A species of eastern Asia collected on rock slides of the Alaska Range.

2. E. borealis (Greene) Hult.

Basal leaves 10–15 mm. long, deeply cleft into 7–9 lobes; stems branched, racemosely floriferous throughout; flowers purple; pods obovate to broadly lanceolate, often oblique, irregularly inflated.

Known only from the Alaska-Yukon boundary north of the Yukon River and Mt. McKinley Park.

23. ERYSIMUM L.

Annual, biennial or perennial leafy-stemmed plants with appressed, forked hairs; flowers perfect, borne in terminal racemes; outer 2 sepals gibbous at base; petals in ours yellow or purple; pods elongate-linear, 4-angled or with a strong midrib; seeds in 1 row in each cell, numerous.

1A. Petals 4-5 mm. long .... 1. *E. cheiranthoides*
2A. Petals 6-10 mm. long .... 2. *E. inconspicuum*
3A. Petals 12-20 mm. long.
  1B. Petals yellow .... 3. *E. angustatum*
  2B. Petals purple. .... 4. *E. pallasii*

1. *E. cheiranthoides* L.   Wormseed Mustard.
Cheirinia cheiranthoides (L.) Link.

Stems minutely strigose-pubescent, 3-10 dm. tall; leaves lanceolate, entire or denticulate, 2-10 cm. long; pods finely pubescent, 2-3 cm. long, 1-1.5 mm. wide, erect or ascending on more or less spreading pedicels.

Moist soil, circumboreal. Fig. 575.

2. *E. inconspicuum* (Wats.) MacM.   Small-flowered Prairie-rocket.

Perennial; the whole plant sparsely cinereous and scabrous with mostly 2-pointed hairs; stems 3-10 dm. tall; leaves linear to oblanceolate, 25-75 mm. long, entire or with a few teeth; petals yellow; pedicels stout, 4-6 mm. long; pods 2-5 cm. long, about 1.5 mm. wide; styles short and thick.

Central and northern Alaska—lower MacKenzie R.—Ont.—Colo.—Nev.—B. C. Fig. 576.

3. *E. angustatum* Rydb.   Narrow-leaved Wallflower.

More or less caespitose perennial; stems 1-2 dm. tall; sparingly grayish-strigose; leaves very narrowly lanceolate-linear or linear, 4-7 cm. long, 1-2 mm. wide, grayish-strigose; sepals linear, obtuse, about 8 mm. long, the alternate ones deeply saccate at the base; petals lemon yellow, about 14 mm. long; pods 5-8 cm. long, 1.5 mm. wide on ascending pedicels 5-8 mm. long, with a distinct beak 3-5 mm. long, somewhat constricted between the seeds.

Known only from the region around Dawson.

4. *E. pallasii* (Pursh.) Fern.   Pallas Wallflower.

Dwarf biennial or perennial; leaves crowded at the base, linear or lanceolate-linear, entire or with a few teeth, pubescent with appressed, 2-pointed, white hairs; inflorescence very dense at anthesis; sepals oblong, saccate at the base, purple; petals purple, 10-18 mm. long; pods pubescent, 3-8 cm. long.

Seward Penin. & northern Alaska, interrupted circumpolar. Fig. 577.

24. ALYSSUM (Tourn.) L.

Low, branching, stellate-pubescent herbs; flowers yellow or whitish; sepals short, ovate or oblong, more or less spreading; petals entire; stamens with filaments more or less dilated at the base and toothed; pod with convex valve. (Greek, curing madness.)

A. americanum Green. American Alyssum.

Stems decumbent, 7–20 cm. long, leafy to the inflorescence; leaves spatulate, pale above, white beneath, entire, 6–12 mm. long, rounded at the apex; pedicels divaricate; petals with rounded, narrowly notched blade and slender claw; pods broadly ovate, about 4 mm. long with slender persistent styles, the cells 2-seeded.

Upper Koyukuk valley—Yukon.

25. BRAYA Stern. & Hoppe

Perennials with stout root, caespitose at the base; leaves mostly tufted at the base of the stems; flowers white or purplish; sepals short, ovate, equal at the base; styles short; stigmas more or less 2-lobed; pods subterete or somewhat flattened, the valves faintly 1-nerved. (Count F. G. deBray, botanist and French ambassador to Bavaria.)

```
1A. Pods about 1 mm. thick, 18–30 mm. long.................... 1. B. humilis
2A. Pods thicker and shorter.
  1B. Pods lanceolate, widest near the base .................... 2. B. henryae
  2B. Pods oblong, widest near the middle.
    1C. Leaves spatulate, glabrate ........................... 3. B. purpurescens
    2C. Leaves linear-lanceolate, pilose ...................... 4. B. pilosa
```

1. *B. humilis* Robins. Northern Rock-cress.
Arabidopsis richardsonii Rydb.

Stems branched and decumbent at the base, 1–3 dm. tall, pubescent with branched hairs; basal leaves spatulate, rather thick, often coarsely toothed, 1–3 cm. long; cauline leaves rather remote and small; flowers small, white, or purplish; pods linear, pubescent, torulose, 1 mm. wide.

Central Asia—Alaska—Victoria Land—Greenl.—Vermont—B. C. Fig. 578.

2. *B. henryae* Raup.

Stems scapose, 6–10 cm. tall, loosely pubescent with 2-branched hairs; leaves narrowly spatulate, gradually narrowed into petioles, glabrous, ciliate on the margins; inflorescence capitate in flower, 2–5 cm. long in fruit; sepals 3–3.5 mm. long, ovate; petals 5 mm. long, white, purplish at the base; pods 8–12 mm. long, 1–2 mm. wide at the base, pubescent; fruiting pedicels 2–3 mm. long; styles 1–1.6 mm. long, the lobes of the stigma spreading.

Chuckh Penin., Asia—Seward Penin.—also northeastern B. C.

3. *B. purpurescens* (R. Br.) Bunge.

Leaves fleshy, spatulate, usually entire, glabrate or ciliate toward the base, arising directly from the caudex; stems 1-several, 1 dm. or less tall, pubescent; sepals purplish, 2 mm. long; petals white or purplish; pods oblong, somewhat pubescent, 8–10 mm. long.

Alpine-arctic, not common, circumpolar. Fig. 579.

4. **B. pilosa** Hook.

Much as in *B. purpurescens;* leaves linear-lanceolate, pilose on both surfaces and on the margins, chiefly with simple hairs; flowers fragrant, appearing early; petals up to 7 mm. long.

Teller—Hudson Str.

26. PARRYA R. Br.

Perennials with thick, often branched caudices; flowers perfect, borne in racemes; sepals oblong, the lateral ones gibbous at the base; petals pink or purple, clawed, the blade broad; anthers included, sagittate at the base; pods flat, the valves nerved; stigmas 2-lobed; seed margined or winged; cotyledons accumbent. (Capt. W. E. Parry was an arctic explorer.)

P. nudicaulis (L.) Regel.

Leaves all basal, usually with a few teeth, hispidulous to glabrate, oblanceolate in outline, tapering into a petiole, 5–10 cm. long, including the petiole; scapes 1–3 dm. tall, glandular-hispidulous; petals white to rose-purple, about 15 mm. long; pedicels 1–5 cm. long, ascending or divergent; pods erect, glandular-hispidulous, the margins wavy, 2–5 cm. long, 4–7 mm. wide, acute at both ends. The form in the interior differs from that on the coast in having leaves with fewer teeth, narrower pods and longer styles. It has been described as ssp. *interior* Hult. A very large flowered form from central Alaska is the variety *grandiflora* Hult.

The species is circumpolar, ranging in Alaska south to the Aleutians and Shumagin Islands. Fig. 580.

17. DROSERACEAE (Sundew Family)

Perennial or biennial herbs, mostly with basal leaves bearing stout, sensitive hairs from which is secreted a viscid fluid in which small insects become entangled and are digested; sepals, petals, and stamens each 4–8; ovary 1-celled, with 2–5 parietal placentae.

DROSERA L.

Scapose bog plants with basal leaves; flowers regular, perfect, borne in secund racemes; sepals, petals, and stamens usually 5 each in our species; pistils 3; capsule 3-valved, many-seeded, loculicidally dehiscent. (Greek, dewy, from the appearance of the leaves.)

Leaf blades nearly round, as broad or broader than long .... 1. *D. rotundifolia*
Leaf blades elongate, more than twice as long as broad .... 2. *D. anglica*

1. *D. rotundifolia* L. Round-leaved Sundew.

Leaves 5–10 mm. wide, narrowing abruptly into petioles, the large, spreading, reddish hairs with a drop of secretion at the end; scapes glabrous or nearly so, 6–20 cm. tall; sepals about 3 mm. long; petals white, about 4 mm. long; capsule erect, 5–6 mm. long; seeds fusiform, smooth, pointed at both ends.

Growing in bogs, circumboreal, south to Fla. & Calif. Fig. 581.

2. **D. anglica** Huds. Long-leaved Sundew.
   D. longifolia of Am. manuals.

Leaves spatulate or oblanceolate, 15–25 mm. long, 3–4 mm. wide, tapering gradually into an almost glabrous petiole; scapes 6–18 cm. tall; flowers fewer and slightly larger than in *D. rotundifolia;* seed obtuse at both ends.

In bogs, circumboreal, south to Newf. and Calif. Fig. 582.

PLATE XX

Scale marked in millimeters

FIG.
463. *Brasenia screberi* Gmel. Flower, anther, leaf, and follicle.
464. *Ranunculus acris* L. Achene and leaf.
465. *Ranunculus repens* L. Achene and leaf.
466. *Ranunculus occidentalis* Nutt. Achene and leaf.
467. *Ranunculus bongardii* Greene. Achene and leaf.
468. *Ranunculus macounii* Britt. Achene and leaf.
469. *Ranunculus pennsylvanicus* L. Achene and leaf.
470. *Ranunculus orthorhynchus alaschensis* (Benson) Hult. Achene and leaf.
471. *Ranunculus abortivus* L. Basal leaf, stem leaf, and achene.
472. *Ranunculus eastwoodianus* Benson. Petal and leaf.
473. *Ranunculus pedatifidus* Sm. Achene and leaf.
474. *Ranunculus eschscholtzii* Schlecht. Achene and leaf.
475. *Ranunculus nivalis* L. Achene and leaf.
476. *Ranunculus sulphureus* Phipps. Achene and basal leaf.
477. *Ranunculus pygmaeus* Wahl. Achene and leaf.
478. *Ranunculus verticillatus* Eastw. Achene and leaf.
479. *Ranunculus flammula* L. Leaves and achene.
480. *Ranunculus hyperboreus* Rottb. Achene and leaf.
481. *Ranunculus sceleratus multifidus* (Nutt) Hult. Leaves and achene.
482. *Ranunculus gmelini* var. *terrestris* (Ledeb.) Benson. Leaves and achene.
483. *Ranunculus cymbalaria* Pursh. Achene and leaf.
484. *Ranunculus cooleyae* Vasey & Rose. Achene and leaf.
485. *Ranunculus chamissonis* Schlecht. Achene and leaf.
486. *Ranunculus aquatilis* var. *capillaceous* DC. Achene and leaf.
487. *Ranunculus pallasii* Schlecht. Achene and leaf.
488. *Ranunculus lapponicus* L. Achene and leaf
489. *Anemone richardsonii* Hook. Achene and leaf.
490. *Anemone narcissiflora alaskana* Hult. Achene and leaf.
491. *Anemone parviflora* Michx Achene and leaf.
492. *Anemone multifida* Poir. Achene and leaf.

FLORA OF ALASKA

PLATE XX

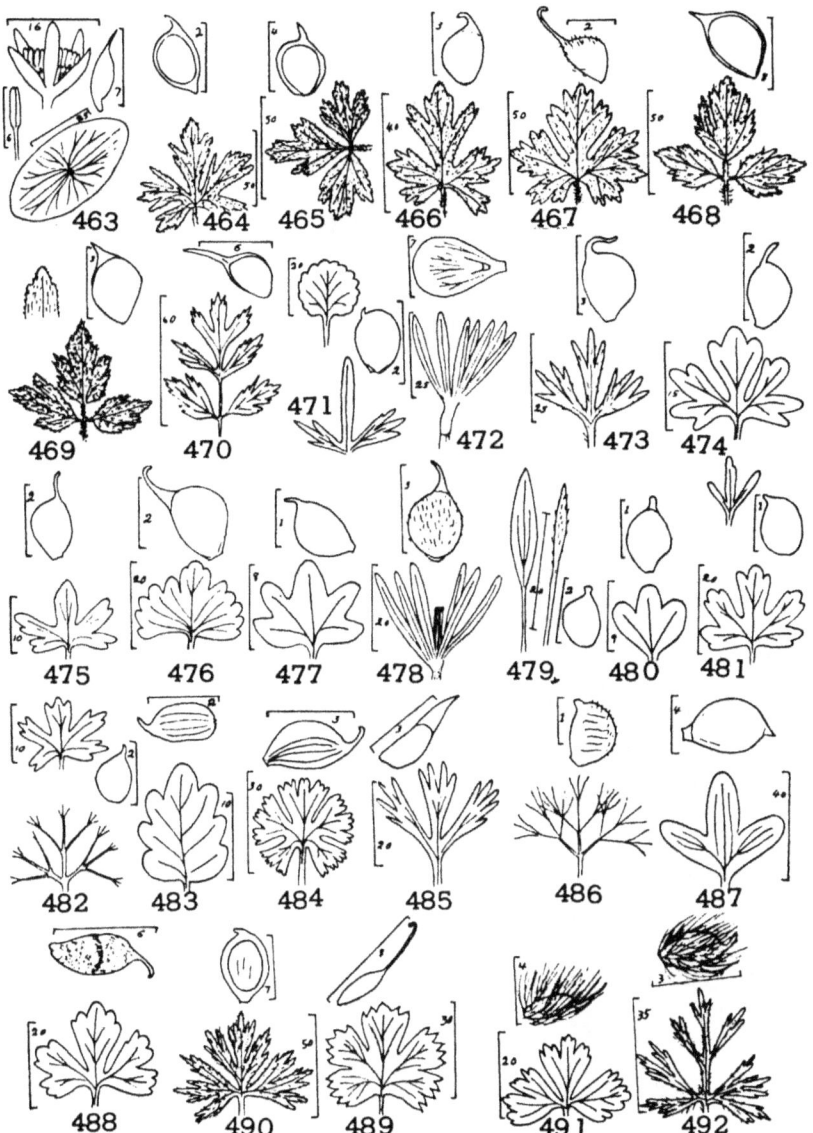

PLATE XXI

Scale marked in millimeters

Fig.
493. *Anemone multiceps* (Greene) Standl. Flower and leaf.
494. *Anemone patens multifida* (Pritzel) Zamels. Achene and leaf.
495. *Thalictrum alpinum* L. Leaf, achene, and anther.
496. *Thalictrum sparsiflorum* Turcz. Stamen, portion of leaf, and achene.
497. *Thalictrum hultenii* B. Boivin. Portion of leaf and anther.
498. *Thalictrum occidentale* Gray. Portion of leaf, anther, and achene.
499. *Caltha palustris asarifolia* (DC.) Hult. Follicle and leaf.
500. *Caltha biflora* DC. Follicle and leaf.
501. *Caltha leptosepala* DC. Follicle and leaf.
502. *Caltha natans* Pall. Follicle and leaf.
503. *Coptis trifoliata* (L.) Salisb. Follicle, leaf, petal, and sepal.
504. *Coptis asplenifolia* Salisb. Follicle, leaf, petal, and sepal.
505. *Actaea arguta* Nutt. Sepal, part of leaf, and berry.
506. *Aquilegia brevistyla* Hook. Flower, leaflets, and follicle.
507. *Aquilegia formosa* Fisch. Flower, leaflets, and follicle.
508. *Delphinium glaucum* Wats. Flower and leaf.
509. *Delphinium brachycentrum* Ledeb. Leaf.
510. *Aconitum delphinifolium* DC. Hood and leaves.
511. *Aconitum maximum* Pall. Hood and leaf.
512. *Nuphar polysepalum* Engelm. Leaf and section of fruit.
513. *Nymphaea tetragona leibergi* (Moreng) Pors. Flower and leaf.
514. *Papaver walpolei* Pors. Capsule and leaves.
515. *Papaver macounii* Greene. Capsule and leaf.
516. *Papaver nudicaule* L. Leaf and capsule.
517. **Papaver radicatum Rottb. Leaf and capsule.**
518. *Corydalis aurea* Willd. Portion of leaf, flower, and fruit.
519. *Corydalis sempervirens* (L.) Pers. Part of leaf, flower, and fruit.
520. *Corydalis pauciflora* (Steph.) Pers. Fruit, leaf and flower.

FLORA OF ALASKA

PLATE XXI

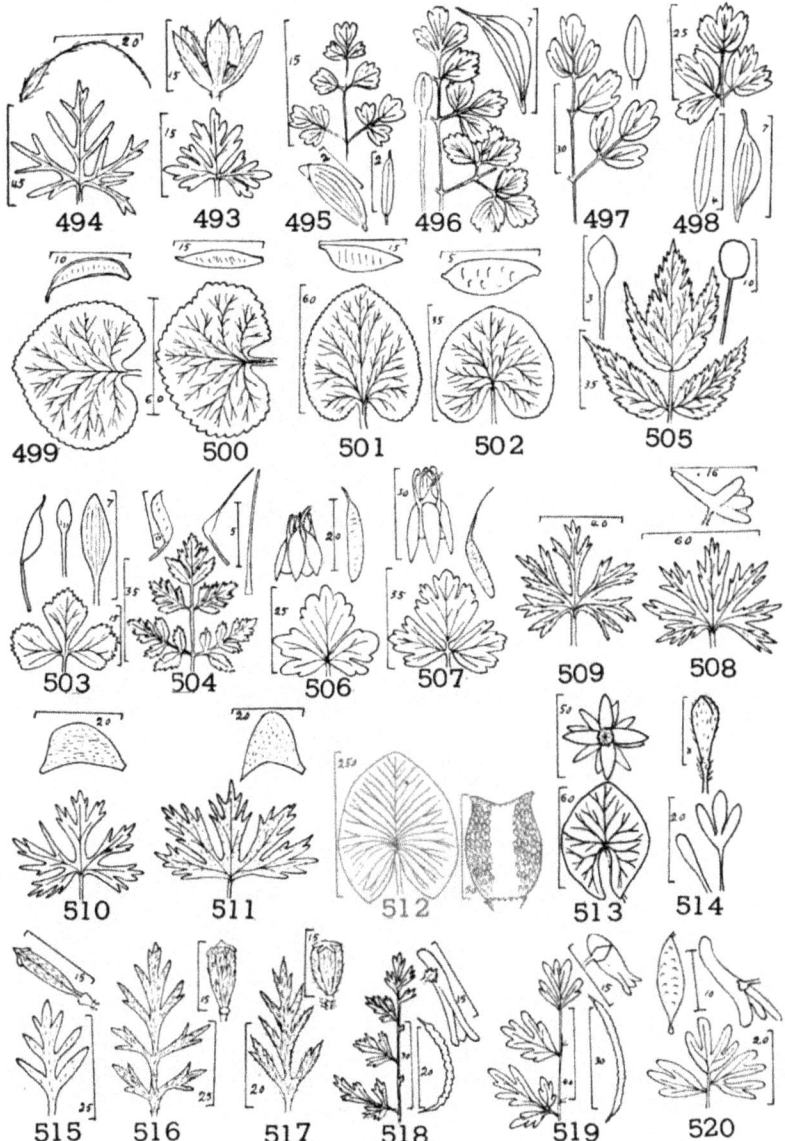

PLATE XXII

Scale marked in millimeters

Fig.
521. *Subularia aquatica* L. Leaf and capsule.
522. *Lepidium densiflorum* Schrad. Leaves and capsule.
523. *Thalaspi arvense* L. Capsule and leaves.
524. *Cochlearia officinalis* L. Leaves and A capsule of ssp. *arctica*, B. of ssp. *oblingifolia*.
525. *Eutrema edwardsii* R.Br. Leaf, seed, and capsule.
526. *Sisymbrium officialale* (L.) Scop. Leaf, seed, and capsule.
527. *Sisymbrium altissimum* L. Leaves and parts of capsule.
528. *Descurainia sophia* (L.) Webb. Leaf and capsule.
529. *Descurainia sophioides* (Fisch.) Schulz. Leaf and capsule.
530. *Descurainia richardsonii* (Sweet) Schulz. Leaf and capsule.
531. *Cakile edentula* (Bigel.) Hook. Leaf and capsule.
532. *Brassica campestris* L. Stem leaf and capsule.
533. *Brassica juncea* (L.) Cosson. Stem leaf and capsule.
534. *Brassica arvensis* (L.) Ktze. Capsule.
535. *Barbarea orthoceras* Ledeb. Seed, leaf, and capsule.
536. *Rorippa palustris* (L.) Bess. Leaf, seed, and capsule.
537. *Rorippa barbareaefolia* (DC.) Pors. Capsule and seed.
538. *Rorippa nasturtium-aquaticum* (L.) Hayak. Leaf, seed, and capsule.
539. *Cardamine bellidifolia* L. Leaf and capsule.
540. *Cardamine angulata* Hook. Leaf.
541. *Cardamine purpurea* C. & S. Stem leaf and basal leaf.
542. *Cardamine pratensis* L. Basal leaf and stem leaf.
543. *Cardamine richardsonii* Hult. Basal and stem leaves.
544. *Cardamine microphylla* Adams. Leaf.
545. *Cardamine umbellata* Greene. Leaves.
546. *Lesquerella arctica* (Wormskj.) Wats. Leaves and capsule.
547. *Capsella bursa-pastoris* (L.) Medic. Capsule and leaf.
548. *Capsella rubella* Reuter. Capsule and leaf.
549. *Neslia paniculata* (L.) Desv. Leaf and capsule.
550. *Draba nemorosa* L. Capsule and leaf.
551. *Draba aleutica* E.Ech. Capsule and leaf.
552. *Draba densifolia* Nutt. Capsule and leaf.
553. *Draba nivalis* Lilj. Capsule and leaf.
553A. *Draba nivalis* var. *denudata* (Schulz) C. L. Hitchc. Capsule.
554. *Draba chamissonis* D. Don. Capsule and leaf.
555. *Draba lactea* Adams. Capsule and leaf.
556. *Draba alpina* L. Capsule and leaf.
557. *Draba stenoloba* Ledeb. Capsule and leaf.
558. *Draba longipes* Raup. Capsule and leaf.

FLORA OF ALASKA

PLATE XXII

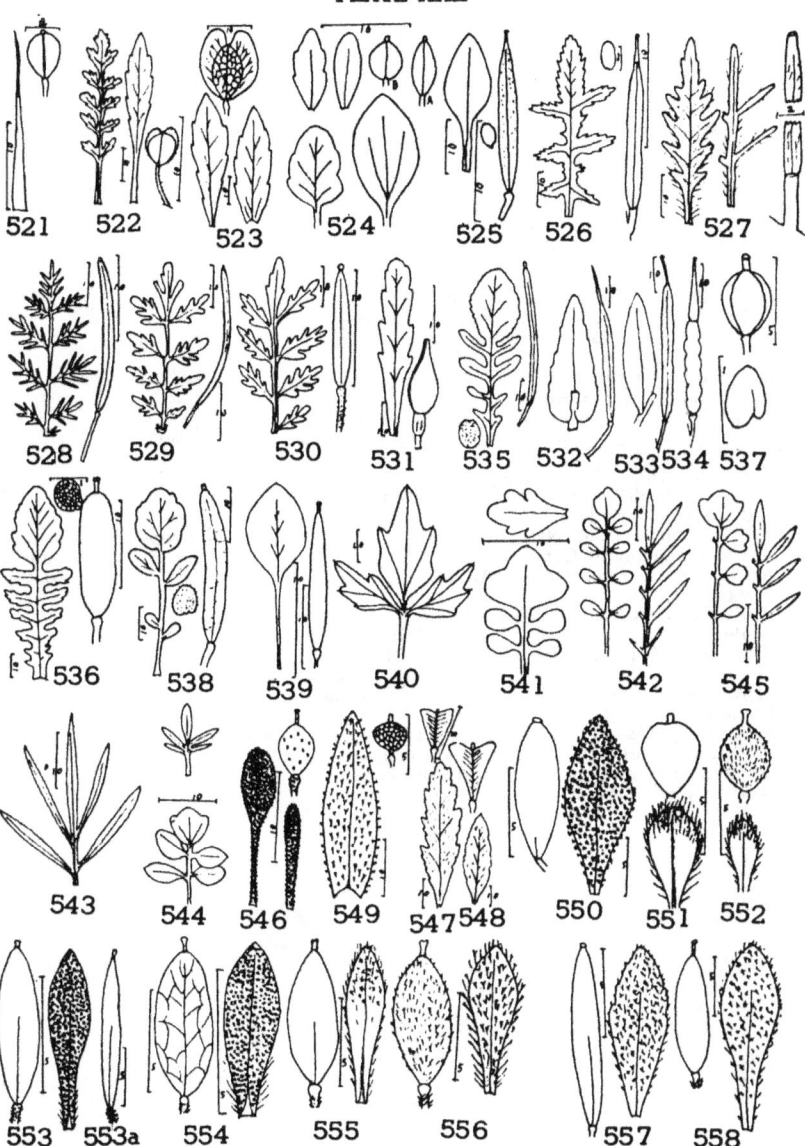

PLATE XXIII

Scale marked in millimeters.

FIG.
559. *Draba glabella* Pursh. Capsule and leaf.
560. *Draba lanceolata* Royle. Capsule and leaf.
561. *Draba cinerea* Adams. Capsule and leaf.
562. *Draba borealis* DC. Capsule and leaf.
563. *Draba maxima* Hult. Capsule and leaf.
564. *Draba aurea* Vahl. Capsule and leaf.
565. *Draba hyperborea* (L.) Desv. Capsule and leaf.
566. *Smelowskia calycina integrifolia* (Seem.) Hult. Leaf and capsule.
567. *Arabis lyrata kamchatica* (Fisch.) Hult. Leaves, seed and capsule.
568. *Arabis arnicola* (Rich.) Gelert. Seed, leaves and capsule.
569. *Arabis holboellii* var. *retrofracta* (Grah.) Rydb. Basal and stem leaves, seed and parts of capsule.
570. *Arabis divaricarpa* A. Nels. Basal leaf, stem leaf and parts of capsule.
571. *Arabis drummondii* Gray. Basal and stem leaves, seed and parts of capsule.
572. *Arabis hirsuta eschscholtzii* (Andriz) Hult. Basal and and stem leaves, seed and parts of capsule.
573. *Arabis hookeri* Lange. Basal and stem leaves, seed, and capsule.
574. *Arabis glabra* (L.) Bernh. Basal and stem leaves, seed and parts of capsule.
575. *Erysimum cheiranthoides* (L.) Link. Leaf, seed and capsule.
576. *Erysimum inconspicuum* (Wats.) MacM. Leaf and capsule.
577. *Erysimum pallasii* (Pursh) Fern. Leaf, capsule and part of capsule.
578. *Braya humilis* Robins. Leaf and capsule.
579. *Braya purpurescens* (R.Br.) Bunge. Leaf and capsule.
580. *Parrya nudicaulis* (L.) Regel. Leaf, seed and capsule.
581. *Drosera rotundifolia* L. Leaf and flower.
582. *Drosera anglica* Huds. Leaf.

FLORA OF ALASKA

PLATE XXIII

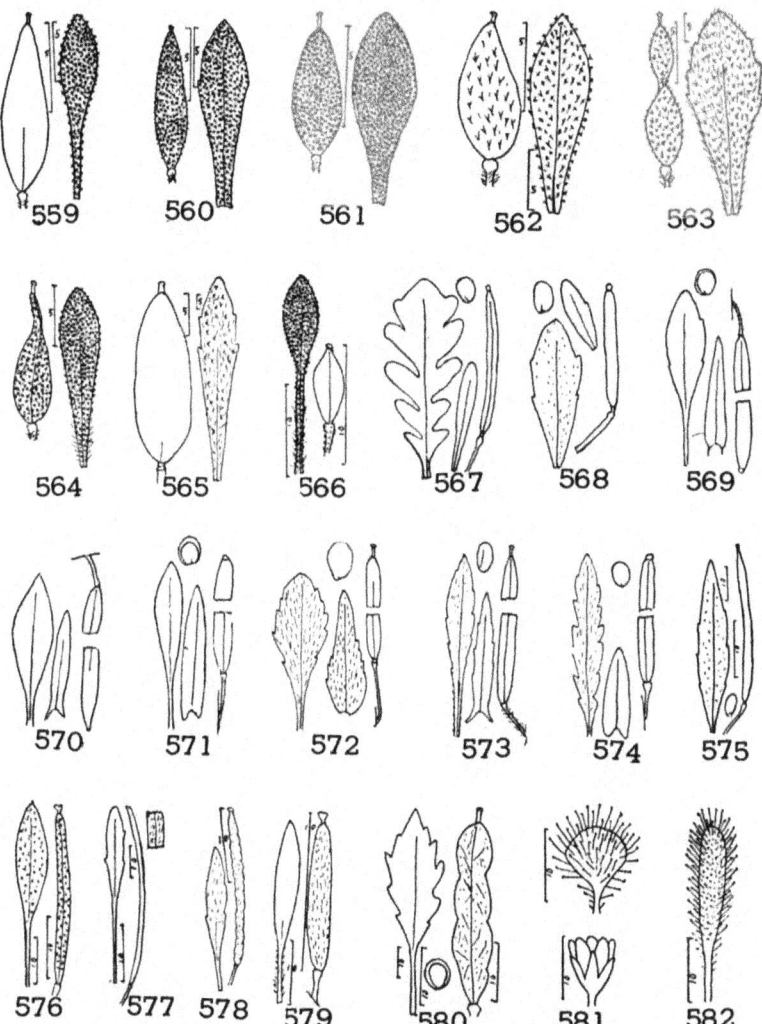

18. CRASSULACEAE (Stone-crop Family)

Mostly fleshy or succulent herbs; flowers regular, borne in cymes; sepals, petals, and carpels each 4 or 5, with stamens of the same number or twice as many; carpels distinct or nearly so with a small scale at the base of each; fruit composed of dry, dehiscent follicles.

SEDUM L.

Fleshy herbs with flowers borne in terminal, often one-sided cymes; leaves alternate, often imbricate; sepals distinct or somewhat united; stamens 8–10, the alternate ones usually attached to the petals; carpels 4 or 5, distinct or united at the base. (Latin, to sit, in allusion to the habit of the plants).

1A. Petals united below .... 1. *S. oregonum*
2A. Petals distinct.
  1B. Flowers polygamous or dioecious, leaves broad .... 2. *S. roseum*
  2B. Flowers perfect, leaves terete .... 3. *S. stenopetalum*

1. *S. oregonum* Nutt.   Oregon Sedum
   Gormania oregona (Nutt.) Britt.
   Rootstock rather slender, creeping; stems erect or ascending, often curved, 6–15 cm. tall; leaves spatulate-cuneate, glabrous, 8–20 mm. long; cymes rather congested, with a leafy involucre; calyx lobes lanceolate, about 4 mm. long; petals narrowly lanceolate, acuminate, 10–12 mm. long, united about one fourth their length, yellow often tinged rose.
   Southeastern Alaska—northern California. Fig. 583.

2. *S. roseum* (L.) Scop.   Roseroot. Rosewort
   Rhodiola rosea L.
   Rootstock thick, fleshy or woody, rose-scented; stems leafy, somewhat glaucous, 1–3 dm. tall; leaves oblanceolate or obovate, entire or dentate, 1–4 cm. long, the lower ones smaller; petals in the type form yellow, in the ssp. *integrifolium* (Raf.) Hult. (*Rhodiola integrifolia* Raf.) dark reddish purple; follicles erect with widely divergent tips. Var.

frigidum (Rydb.) Hult. (*Rhodiola alaskana* Rose) averages taller, leaves sharply toothed in upper third, drying thin.

The type form has been collected at Nome, ssp. *integrifolia* is common in most of our area, var. *frigidum* occurs in the Pacific Coast and Bering Sea regions, the species being circumboreal. Fig. 584.

3. *S. stenopetalum* Pursh. Narrow-petaled Sedum

Tufted perennial; rootstock slender, branching; stems 7–18 cm. tall; leaves linear, sessile, 5–15 mm. long, imbricate on the sterile shoots; cyme 3- to 7-forked; pedicels short; calyx lobes lanceolate; petals yellow, narrowly lanceolate, acuminate, 6–7 mm. long; follicles about 4 mm. long, the tips divergent.

S. Yukon—Sask.—Nebr.—northeastern Calif. Fig. 585.

19. SAXIFRAGACEAE (Saxifrage Family)

Ours all herbs, mostly perennial; leaves alternate, rarely opposite, or often all basal, usually without stipules; flowers in ours perfect and regular or nearly so; hypanthium often more or less adnate to the ovary; sepals and petals 5 or rarely 4 or the petals wanting; stamens as many or twice as many as the sepals except in *Tolmiea* which has only 3; carpels 1— several, usually 2, distinct or united; fruit a capsule or composed of follicles.

```
1A. Staminoidea present between the stamens .............. 1. Parnassia
2A. Staminoidea not present.
   1B. Petals none, low herbs ........................... 2. Chrysosplenium
   2B. Petals usually present.
      1C. Petals fringed or laciniate-lobed.
         1D. Calyx flat at base ......................... 3. Mitella
         2D. Calyx cup-shaped at base ................... 4. Tellima
      2C. Petals entire.
         1D. Stamens 3 ................................. 5. Tolmiea
         2D. Stamens 5.
            1E. Capsule 1-celled, leaves mostly basal ........... 6. Heuchera
            2E. Capsule 2-celled, stem leafy ................... 7. Boykinia
         3D. Stamens 8 or 10.
            1E. Carpels unequal............................ 8. Tiarella
            2E. Carpels equal.
               1F. Leaves leathery, carpels nearly distinct........ 9. Leptarrhena
               2F. Leaves not leathery, carpels
                   more or less united...................... 10. Saxifraga
```

1. PARNASSIA L.

Glabrous, scapose perennials; leaves basal, petioled, entire; flowers perfect, solitary, terminal, white or yellowish; scapes usually bearing 1 leaf; sepals and petals each 5; stamens 5, alternating with the petals and with 5 clusters of gland-tipped staminoidea; carpels 3 or 4, united; ovary with 3 or 4 parietal placentae; fruit a 1-celled loculicidal capsule. (Name from Mt. Parnassus in Greece.)

```
1A. Petals fimbriate on the sides.......................... 1. P. fimbriata
2A. Petals entire.
   1B. Petals scarcely equaling the sepals, 3-veined.......... 4. P. kotzebuei
```

2B. Petals longer than the sepals, 5- to 9-nerved.
  1C. Petals nearly twice as long as the sepals,
      staminoidea, 8–15 in each fascicle.................. 2. *P. palustris*
  2C. Petals only slightly exceeding the sepals,
      staminoidea 7–9 in each fascicle................... 3. *P. montanensis*

1. *P. fimbriata* Konig.   Fringed Grass-of-Parnassus

Leaf-blades reniform to cordate, 2–4 cm. wide; scapes 2–5 dm. tall, with 1 sessile leaf above the middle; sepals elliptical, obtuse, about 5 mm. long; petals nearly twice as long as the sepals, obovate; staminoidea 5–9 in each fascicle.

Central Pacific coast of Alaska—Yukon—Utah—N. Mex.—Calif. Fig. 586.

2. *P. palustris* L.   Northern Grass-of-Parnassus

Leaves cordate, 1–3 cm. wide; scapes 1–5 dm. tall, bearing a cordate-clasping leaf below the middle; sepals ovate-lanceolate, strongly veined, 5–7 mm. long; petals oval, 8–12 mm. long; capsule ovoid, about 1 cm. long. The inland race has more deltoid stem leaves, narrower sepals and broader-clawed staminoidea than the type form and has been separated as var. *neogaea* Fern.

Common in wet places, circumboreal. Fig. 587.

3. *P. montanensis* Fern. & Rydb.   Montana Grass-of-Parnassus

Leaves ovate with subcordate or rounded base, 10–20 × 8–18 mm.; scapes about 2 dm. tall, the leaf ample, ovate, borne below the middle; sepals acute, 7- to 9-veined, 7–9 mm. long; capsule round-ovoid, about 1 cm. long.

Yukon—Sask.—Mont.

4. *P. kotzebuei* C. & S.   Kotzebue Grass-of-Parnassus

Leaves ovate, narrowed, truncate or subcordate at the base, 1–2 cm. long; scapes naked or with a leaf near the base, 6–15 cm. tall; sepals oblong-lanceolate, 5–6 mm. long, about the same length as the petals; staminoidea 3–5 in each fascicle.

East Asia—Coronation Gulf—Labr.—Greenl.—Newf.—Wyo. Fig. 588.

P. parviflora DC. has been reported from Alaska but the reports need confirmation. It has leaves with acute bases, the stem-leaf at or a little below the middle, petals about the same length as the sepals, and staminoidea 5–7 in each fascicle.

2. CHRYSOSPLENIUM (Tourn.) L.

Low, glabrous, somewhat succulent herbs usually growing in very wet places; leaves petioled, crenate; flowers axillary or terminal; hypanthium adnate to the lower portion of the ovary; sepals usually 4; petals none; capsule 1-celled with 2 parietal placentae, many seeded; seed smooth, shining. (Greek, golden spleen, from reputed medicinal virtues.)

Stamens 4.. 1. *C. tetrandrum*
Stamens 8.. 2. *C. wrightii*

1. **C. tetrandrum** Th. Fries.   Northern Water Carpet

A stoloniferous perennial; stems 3–15 cm. tall, bearing several leaves; leaf blades reniform or orbicular with 3–5, rarely 7 rounded teeth, truncate to cordate at the base, 4–12 mm. wide; sepals usually 4; stamens opposite the sepals; seeds several, brownish red.

Wet places, circumpolar. Fig. 589.

2. **C. wrightii** Franch. & Sav.   Bering Sea Water Carpet
    C. beringianum Rose

Perennial with a rather thick, scaly rootstock; leaves thick, coriaceous, 3- to 7-lobed with rounded divisions, the petioles usually with brownish hairs; flowering stem short and stout, almost leafless except at apex, many-flowered, the flowers clustered; sepals short and broad, rounded.

E. Asia—Yukon—Aleutian Isls. Fig. 590.

3. MITELLA (Tourn.) L.

Perennials; leaves cordate, orbicular, or ovate, clustered on a scaly rootstock; stems scape-like, naked or with a few leaves; inflorescence a simple raceme; hypanthium saucer-shaped, adnate to the ovary; flowers white or greenish; petals 5, pectinately pinnatifid; filaments short; ovary 1-celled with 2 parietal or almost basal placentae; styles 2, very short. (Diminutive of Mitra, a cap.)

Stamens 5... 1. *M. pentandra*
Stamens 10.. 2. *M. nuda*

1. **M. pentandra** Hook.   Alpine Mitrewort
    Pectianthia pentandra (Hook.) Rydb.

Leaves cordate, crenate, indistinctly lobed, 2–5 cm. wide; flowering stem naked or with 1 small leaf, 1–3 dm. tall, hirsute, glandular above; racemes lax, the flowers often in pairs; sepals broadly triangular; petals cut into 5–9 capillary pinnae; stamens with very short incurved filaments and reniform anthers.

Wet alpine meadows, Kodiak—southeastern Alaska—Colo.—northern Calif. Fig. 591.

2. **M. nuda** L.   Stoloniferous Mitrewort

Stoloniferous; flowering stems usually naked, pubescent, 5–20 cm. tall; leaves reniform-orbicular, cordate at the base, crenate or doubly crenate, 12–40 mm. wide, pubescent with scattered hairs; flowers few, greenish; petals pinnately divided into filiform segments; filaments subulate, more than half as long as the sepals.

Southern Yukon—Newf.—N. S.—Penn. Fig. 592.

4. TELLIMA R. Br.

Hirsute perennial; rootstock thick and scaly; leaves palmately lobed, parted or divided; flowers in an elongated raceme on a scape-like stem; sepals ovate, erect; petals white, purplish, or yellowish, spreading or reflexed, pinnately laciniate; stamens 10, included; carpels 2, ovary 1-celled with 2 many-ovuled parietal placentae; capsule 2-valved, adherent to the base of the hypanthium. (An anagram of Mitella.)

T. grandiflora (Pursh) Dougl.   Fringe Cup

Leaves cordate or reniform, sparingly hirsute on both sides, shallowly lobed, dentate, 4–10 cm. wide; flowering stems 3–10 dm. long, hirsute with long hairs, bearing 2 or 3 leaves; inflorescence 1–3 dm. long, glandular; hypanthium cup-shaped, about 8 mm. long.

Unimak Isl.—southeastern Alaska—Selkirk Mts.—northern Calif. Fig. 593.

5. TOLMIEA Torr. & Gray

Perennial with a scaly caudex; leaves many, mostly basal, with stipules; flowers borne in long terminal racemes; sepals united into a long tube split on one side; petals filiform; ovary 1-celled, stipitate, with 2 equal carpels and parietal placentae. (Dr. W. F. Tolmie was a collector and surgeon of the Hudson Bay Co.)

Tolmiea menziesii (Pursh) Torr. & Gray   Youth-on-Age
Leptaxis menziesii (Pursh) Raf.

Basal leaves cordate, acute, cuspidate-toothed, hirsute, ciliate, 2–12 cm. wide, on long petioles; stems up to 1 m. tall with a few–several leaves; flowers on slender pedicels subtended by small fimbriate bracts; petals capillary, brown, exerted from the sinuses between the sepals; fruit protruding through the slit on the lower side of the hypanthium. Propagates vegetatively by new plants forming in the sinuses of the leaves.

Southeastern Alaska—Calif. Fig. 594.

6. HEUCHERA L.

Perennials with thick, scaly rootstocks; leaves mostly radical, long-petioled; stems scape-like, bearing racemes or panicles of small whitish or purplish flowers; hypanthium adherent to lower portion of ovary, often oblique; sepals 5, often unequal; petals 5, small; ovary 1-celled, with 2 parietal, many-ovuled placentae; styles 2, slender. (Johann Heinrich von Heucher was a German botanist.)

H. glabra Willd.   Alpine Heuchera

Basal leaves cordate, 5- to 7-lobed, thin and shining, doubly serrate, 4–12 cm. long; flowering stems 2–6 dm. tall, 1- to 3-leaved; panicle lax; sepals ovate, scarcely 1 mm. long; petals ovate, clawed, about twice as long as the sepals.

Bering Sea—central Alaska—Selkirk Mts.—Ore. Fig. 595.

7. BOYKINIA Nutt.

Glandular-pubescent perennials with thick, scaly rootstocks; leaves alternate, petioled; flowers in terminal panicles; hypanthium adnate to lower half of ovary; sepals 5, lanceolate or ovate-lanceolate, petals 5, whitish; filaments short; ovary and capsule 2-celled with axial placentae; seed numerous, shining, punctate. (Dr. Boykin was a physician of Georgia.)

B. *richardsonii* (Hook.) Gray   Richardson Saxifrage
Therefon richardsonii (Hook.) O.Kze.

Plant with large glands raised on thick pedicels, 3–10 dm. tall; leaves mostly basal with long petioles and blades reniform to orbicular in outline, shallowly lobed and doubly toothed, the margins with prominent glands, deeply cordate at the base, 5–15 cm. wide; stem leaves reduced; hypanthium campanulate, about 5 mm. long; sepals triangular-ovate, 4–5 mm. long; petals about 1 cm. long.

Bering Str.—Arctic coast—Yukon—central Alaska. Fig. 596.

8. TIARELLA L.

Perennials with scaly rootstocks; leaves mostly basal, petioled, with small stipules adnate to the base; stems erect, the flowers small, white; hypanthium short-campanulate, nearly free from the ovary; sepals 5, ovate or lanceolate, petals 5, clawed or filiform; stamens exerted; carpels 2, very uneven in fruit, membranous; seed few, smooth. (Diminutive of tiara, from the form of the capsule).

Leaves trifoliate... 1. *T. trifoliata*
Leaves not divided... 2. *T. unifoliata*

1. *T. trifoliata* L.   Trifoliate Foamflower

Leaves and upper part of petiole hirsute; leaflets ovate to rhomboid, slightly lobed and with mucronate teeth, 2–9 cm. long; flowering stems 15–50 cm. tall, 1- to 3-leaved, glabrate below, glandular-pubescent above; inflorescence a narrow panicle; sepals whitish, scarcely 2 mm. long; petals very narrow; valves of capsule in fruit 4–5 and 7–9 mm. long.

Unga Isl.—southeastern Alaska—Ore. Fig. 597.

2. *T. unifoliata* Hook.   Unifoliate Foamflower

Similar to *T. trifoliata* but the leaves broadly cordate, 3- to 5-lobed, 4–10 cm. wide; lower carpel of fruit twice as wide as the upper one.

Southeastern Alaska—western Alta.—western Mont.—Calif. Fig. 598.

9. LEPTARRHENA R. Br.

Perennial with horizontal rootstock; leaves thick, leathery, crowded at the base of the scape; flowers small in a terminal panicle; hypanthium flattened; sepals 5, erect; petals 5, white, persistent; filaments subulate; carpels 2, united at the base, the tips slightly divergent in fruit. (Greek, delicate and male, probably referring to the slender stamens.)

L. *pyrolifolia* (D. Don) Ser.   Leather-leaf Saxifrage

Leaves ovate to obovate, glabrous, deep green and shining above, pale beneath, obtuse, serrate, narrowed into a short petiole, the blade 3–12 cm. long; scape with 2 reduced and clasping leaves, glabrous below, glandular-pubescent above; sepals ovate, about 1.5 mm. long; petals narrow, longer than the sepals; follicles 6–8 mm. long.

Wet places, Aleutians—southeastern Alaska—western Mont.—Wash. Fig. 599.

10. SAXIFRAGA (Tourn.) L.

Perennials with perfect flowers; hypanthium free or adnate to the base of the usually 2-celled ovary; sepals and petals each 5; stamens 10; styles short; ovules numerous on axial placentae; Capsule 2-beaked (except in one or two species), many seeded; seed small. (Latin, rock and to break, referring to the habitat of many of the species.)

1A. Leaves opposite, plants matted..........................29. *S. oppositifolia*
2A. Leaves alternate or basal.
  1B. Leaves entire, not toothed.
    1C. Flowers not or very slightly rising above the leaves.
      1D. Margins of the leaves ciliate......................26. *S. eschscholtzii*
      2D. Leaves glabrous, not ciliate......................19. *S. aleutica*
    2C. Flowers on elongated stems.
      1D. Stems low, scapose, with 1 or 2 leaves.
        1E. Flowers yellow................................20. *S. serpyllifolia*
        2E. Flowers white.................................23. *S. tolmiei*
      2D. Stems taller with several leaves.
        1E. Leaves glabrous, not ciliated....................21. *S. hirculus*
        2E. Leaves ciliate or pubescent.
          1F. Plants with long flagelliform stolons..........22. *S. flagellaris*
          2F. Plants without stolons.
            1G. Leaves glabrous with ciliated margins.......24. *S. bronchialis*
            2G. Leaves glandular-pubescent.
              1H. Stems 2–8 cm. tall........................ 5. *S. adscendens*
              2H. Stems 2–4 dm. tall........................16. *S. integrifolia*
  2B. Leaves toothed or lobed.
    1C. Basal leaves orbicular or reniform as broad as long (see also 13. *S. lyallii*).
      1D. Flowering stems scape-like.
        1E. Leaves with 3-toothed lobes....................28. *S. mertensiana*
        2E. Leaves simply toothed.
          1F. Flowers in a narrow spike-like panicle.......12. *S. spicata*
          2F. Flowers in a head-like or corymb-like panicle.
            1G. Leaves small, 10–15 mm. wide...............27. *S. nudicaulis*
            2G. Leaves larger............................14. *S. punctata*
      2D. Flowering stems leafy.
        1E. Plants with bulblets, only terminal flower developing .... 3. *S. cernua*
        2E. Lateral flowers developed.
          1F. Petals about 1 cm. long...................... 4. *S. radiata*
          2F. Petals shorter.
            1G. Plants stout, leaves 5- to 8-lobed........... 1. *S. bracteata*
            2G. Plants slender, leaves 3- to 5-lobed......... 2. *S. rivularis*
    2C. Basal leaves longer than broad (except sometimes in *S. lyallii*).
      1D. Leaves 3- to 5-lobed............................. 6. *S. caespitosa*
      2D. Leaves not lobed.
        1E. Basal leaves cuneate-oblong or cuneate-oblanceolate.
          1F. Leaves stiff with 3 acute teeth at apex........25. *S. tricuspidata*
          2F. Leaves not stiff with 3 rounded teeth at apex.. 5. *S. adscendens*
          3F. Leaves with several teeth, bulblets usually present.
            1G. Inflorescence with long ascending branches..18. *S. ferruginea*
            2G. Inflorescence with short, rigid branches.....17. *S. foliolosa*
        2E. Basal leaves flabellate or cuneate-obovate.
          1F. Filaments clavate, broadest at middle.........13. *S. lyallii*
          2F. Filaments subulate.
            1G. Branches of inflorescence short and thick... 15. *S. unalaschensis*

  2G. Branches of inflorescence longer and
    thinner.................................... 10. *S. davurica*
 3E. Basal leaves ovate or oval.
  1F. Flowers in a spike-like panicle............... 7. *S. hieracifolia*
  2F. Flowers paniculate or in terminal cluster.
   1G. Leaves glabrous on both surfaces........... 8. *S. nivalis*
   2G. Leaves with reddish-brown pubescence
    on lower surface......................... 9. *S. rufidula*
   3G. Leaves pubescent on both surfaces......... 11. *S. reflexa*

1. *S. bracteata* D. Don.   Bracted Saxifrage

  Stems often tufted, pubescent, at least above, 3–20 cm. tall; leaf blades reniform or orbicular, 1–4 cm. wide, mostly 3- to 7-lobed, those of the upper part of stem often 3-lobed and nearly sessile, the basal on long petioles, cuneate to cordate at the base with bulblets at the base of the petiole; inflorescence rather congested; hypanthium 3–4 mm. long; sepals ovate, 3–4 mm. long; petals 5–6 mm. long; fruit 7–8 mm. long.

  East Asia—Bering Str. district—Kodiak Isl. Fig. 600.

2. *S. rivularis* L.   Alpine Brook Saxifrage

  Stems usually tufted, 1- to 3-flowered, glabrous or finely glandular-pubescent, 3–9 cm. tall; leaves fan-shaped or reniform, 3- to 5-lobed, those of the stem sometimes entire, 3–10 mm. wide; sepals ovate, about 2 mm. long, obtuse; petals white or purplish, nearly twice as long as the sepals; tips of the fruiting carpels widely divergent.

  Wet alpine situations, circumpolar. Fig. 601.

3. *S. cernua* L.   Nodding Saxifrage

  Stems slender, ascending, pubescent, 8–25 cm. tall, with bulblets at the base; basal and lower stem leaves petioled, reniform, 5- to 7-lobed, 6–25 mm. wide; upper stem leaves sessile, 3-lobed or entire, bearing bulblets in the axils; flower often nodding; sepals about 3 mm. long; petals 6–9 mm. long; fruit seldom developing.

  Alpine and circumpolar. Fig. 602.

4. *S. radiata* Small.

  Stems more or less glandular-pubescent, 7–20 cm. tall; basal and lower stem leaves reniform or orbicular-flabelliform, petioled, 5- to 7-lobed, 10–22 mm. wide; uppermost stem leaves simple; flowers 2–7, none replaced by bulblets; sepals 2–3 mm. long; petals 8–13 mm. long; fruiting carpels 7–8 mm. long.

  Bering Sea region of Asia and Alaska—Herschel Isl.—central Yukon. Fig. 603.

5. *S. adscendens* L. ssp. *oregonensis* (Raf.) Bacigalupi.
                Wedge-leaved Saxifrage

  Plants tufted, glandular-pubescent, 2–8 cm. tall; basal leaves imbricated, pubescent, oblong-spatulate, entire or with 3 rounded teeth at apex, 5–15 mm. long; stem leaves often purplish; sepals about 2 mm. long; petals white, 3–5 mm. long.

  Rare in Alaska—B.C.—Ore. Fig. 604.

6. *S. caespitosa* L. ssp. *sileneflora* (Sternb.) Hult.   Tufted Saxifrage
Muscaria sileneflora (Sternb.) Small.

Densely tufted, glandular-pubescent, with leaves crowded on the caudices; leaves 8–18 mm. long, fan-shaped, 3- to 5-lobed at apex, the lobes lanceolate to linear; scapes 5–15 cm. tall, 1- to 3-flowered, bearing 2 or 3 reduced leaves; hypanthium campanulate; sepals ovate, 2–3 mm. long; petals white, 4–6 mm. long; fruit 7–10 mm. long.

This species is circumboreal. Fig. 605.

7. *S. hieracifolia* Wallst. & Kit.   Hawkweed-leaved Saxifrage

Leaves ovate, narrowed into margined petioles, usually acute, the margins ciliolate, toothed, 3–7 cm. long; scapes 1–5 dm. tall, glandular-pubescent; inflorescence resembling a bracted, interrûpted spike, the flowers densely gregarious in the axils of the bracts; sepals triangular-ovate, 2–3 mm. long; petals purple, narrow, about as long as the sepals; fruit purplish, 5–6 mm. long. Var. *rufopilosa* Hult. has reddish-brown hairs on the under surface of the leaves.

Distribution interrupted circumpolar. Fig. 606.

8. *S. nivalis* L.   Alpine Saxifrage

Leaves ovate, crenate-serrate, cuneate at the base, rounded at the apex, 1–4 cm. long, purplish beneath; scapes 4–16 cm. tall, several- to many-flowered, glandular-pubescent, especially in the inflorescence; sepals ovate-triangular, 1.5–2 mm. long; petals white, about 3 mm. long; carpels in fruit purplish, about 5 mm. long, the tips divergent.

Circumpolar. Fig. 607.

9. *S. rufidula* (Small) Engl. & Irmscher.   Rusty Saxifrage

Leaves similar to those of *S. nivalis* or *S. reflexa* but bright green and glabrous or essentially so on the upper surface and densely red-tomentose beneath; scapes 5–20 cm. tall, somewhat purplish, pubescent below, inconspicuously so or glabrate above; sepals glabrous, 2–2.25 mm. long; petals white with short claw.

Has been reported from southeastern Alaska—B. C.—Ore.

10. *S. davurica* Willd. ssp. *grandipetala* (Engl. & Irmscher) Hult.

Leaves ascending, the blades flabellate, cuneate at the base, coarsely several-toothed above, glabrous or nearly so, 1–3 cm. long; scapes 6–16 cm. tall, somewhat glandular-pubescent; flowers few–several; sepals 1.5–2 mm. long, purple, reflexed; petals white, up to 5 mm. long; mature carpels erect, 6–8 mm. long.

Eastern Asia—central Alaska. Fig. 608.

11. *S. reflexa* Hook.   Yukon Saxifrage
Micranthes yukonensis Small.

Leaves ovate, coarsely crenate, more or less hirsute or pubescent on both surfaces, 15–50 mm. long; scapes usually more than one, 8–45 cm. tall, glandular-pubescent, many-flowered; sepals 2–2.5 mm. long; filaments dilated in upper portion; fruiting carpels 3-5 mm. long, the tips divergent. Dry situations, Bering Sea—Northwest Territories. Fig. 609.

12. *S. spicata* D. Don.   Spiked Saxifrage
Micranthes galacifolia Small.

Leaves ascending, the blades reniform to oval, 3-9 cm. wide, crenate-dentate with gland-tipped teeth, cordate at base, with petioles 4-18 cm. long; scapes 18-65 cm. tall, glandular-pubescent, the inflorescence spike-like; sepals about 2 mm. long, reflexed; petals yellowish, about 4 mm. long; fruiting carpels 6-10 mm. long.

Bering Str.—Yukon—southwestern Alaska. Fig. 610.

13. *S. lyallii* Engler.   Red-stemmed Saxifrage

Glabrous below, usually glandular in the inflorescence, 8-30 cm. tall; leaves fan-shaped, rounded at the apex, cuneate at the base, regularly serrate on the rounded portion, 1-4 cm. wide; scapes several- to many-flowered; sepals ovate, acute, reflexed, about 2.5 mm. long; petals white, about 4 mm. long; styles in fruit moderately divergent. Seems to hybridize with *S. punctata nelsoniana*.

Alaska Penin.—Alaska Range—Alta.—northwestern Mont.—Wash. Fig. 611.

14. *S. punctata* L.   Brook Saxifrage

Leaves ascending, the blades suborbicular to reniform, 2-6 cm. wide, coarsely several-toothed with crenate or dentate gland-tipped teeth, cordate at the base; scapes 1-5 dm. tall; sepals 1.25-2 mm. long; petals 3-4.5 mm. long, white; fruit purple, 5-8 mm. long. Common and variable, represented in our area by 3 local races.

Ssp. *nelsoniana* (D. Don) Hult. (*S. nelsoniana* D. Don) is characterized by the leaves being pubescent on both surfaces. It is found in the Bering Sea and Arctic regions eastward. Fig. 612.

Ssp. *pacifica* Hult. (*S. aestivalis* auct.) of the Pacific Coast from Unalaska eastward has glabrous leaves sometimes ciliate on the margins and decidedly clavate filaments.

Ssp. *insularis* Hult. has unusually thick glabrous leaves, linear or only slightly clavate filaments, petals usually purplish, the pedicels viscid-pubescent, and occurs in the Alaska Penin., Aleutian and Shumagin Isls.

Entire species is circumboreal.

15. *S. unalaschensis* Sternb.   Unalaska Saxifrage
Micranthes flabellifolia (R. Br.) Small.

Leaves ascending, the blades flabellate, ciliate on the margins, glabrous or somewhat pubescent on the surfaces, the apex with a few teeth that are usually directed forward, the base narrowed and petiole-like; scapes 5-16 cm. tall, sometimes curved, purple, glandular-villous; flowers 1-9; sepals purple, about 2.5 mm. long; petals white or purplish, about 4 mm. long; carpels 2-5, in fruit erect and 7-10 mm. long.

Eastern Asia—Arctic coast—Alaska Penin.—Aleutian Isls. Fig. 613.

16. *S. integrifolia* Hook.   Hooker Saxifrage

Rootstock fibrous-rooted, stoutish; leaves 1–6 cm. long, ovate-elliptic to oblong-elliptic, entire or rarely sinuate-crenate, viscid-hirsutulous, especially on the upper surface, contracted into short, winged petioles below; scapes rather rigid, scabrous, 2–4 dm. tall; inflorescence rather narrow; sepals 1.5 mm. long; petals white about 2.5 mm. long; filaments subulate; fruit depressed, broad.

Reported from Buckland R., Vancouver Isl.—Calif.

17. *S. foliolosa* R. Br.   Foliose Saxifrage
S. comosa (Poir.) Britt.

Leaves crowded on the short caudex, the blades cuneate to oblanceolate with 3–5 teeth at the apex, more or less ciliate, 8–25 mm. long; scapes often more than 1, simple or branched, 6–22 cm. tall; flowers solitary at the end of the scape and often at the end of the branches, the rest of the inflorescence developing bulblets or rosules of small leaves; sepals about 1.5 mm. long; petals white, 4–5 mm. long; carpels in fruit thick, 4–5 mm. long.

Alpine-arctic, circumpolar. Fig. 614.

18. *S. ferruginea* Grah.   Alaska Saxifrage
S. bongardii Presl.
S. brunoniana Wall.

Leaves spatulate or oblanceolate, thick, hirsute on the upper surface and on the margins, sharply toothed above the middle, tapering below into a ciliate petiole, 2–10 cm. long; scapes 1–4 dm. tall, the inflorescence spreading; sepals oblong-ovate, obtuse, 1.5–2 mm. long; petals about 5 mm. long, the 3 upper differing from the lower; filaments dilated at the base. Var. *macounii* Engl. & Irmscher has many of the flowers replaced by bulblets or rosules.

Aleutian Isls.—Alta.—Mont.—Ore. Fig. 615.

19. *S. aleutica* Hult.   Aleutian Saxifrage

A peculiar, densely caespitose plant about 2 cm. tall; leaves densely congested at the end of the branches, fleshy, glabrous, ligulate, entire, 2–5 mm. long; flowers about 7 mm. in diameter; sepals and petals about equal, 2.5 mm. long; filaments filiform; fruit purplish, thick.

Known only from the high peaks of the Aleutians. Fig. 616.

20. *S. serpyllifolia* Pursh.   Thyme-leaved Saxifrage

Tufted; leaves crowded at the base of the stem, linear-spatulate, thickish, obtuse, entire, glabrous, 4–8 mm. long; stems 1-flowered, glandular, 2–6 cm. tall, with 1–3 reduced leaves; sepals ovate, about 2 mm. long; petals bright yellow, 4–7 mm. long; filaments subulate; fruit 5–7 mm. long. Var. *purpurea* Hult. has purplish petals.

Northern Asia—C. Lisburne—Alaska Range—southern Yukon—southeastern Alaska—Aleutians. Fig. 617.

21. *S. hirculus* L.   Yellow Marsh Saxifrage
Leptasea alaskana Small.

Basal leaves numerous, linear-oblong or linear-ovate, glabrate, entire, 1–4 cm. long; stems leafy, more or less pubescent with brown hairs, 8–25 cm. tall, mostly 1-flowered; sepals ciliate, 3–5 mm. long; petals yellow, 8–14 mm. long; carpels in fruit 8–15 mm. long.

Circumpolar with interruptions in distribution. Fig. 618.

22. *S. flagellaris* Willd.   Flagellate Saxifrage

Basal leaves densely crowded, cuneate-spatulate, margined and tipped with spines, 6–16 mm. long, with many filiform runners from their axils; stems 4–15 cm. tall, glandular-pubescent, several-leaved, 1- to 5-flowered; sepals obtuse, glandular, ciliate, 3.5–5 mm. long; petals bright yellow, 7–11 mm. long.

Alpine-arctic, circumpolar. Fig. 619.

23. *S. tolmiei* T. & G.   Tolmie Saxifrage

Stems leafy, trailing, glabrous, 3–10 cm. long; leaves evergreen, obovate, firm, often grooved above, the margins revolute, 4–9 mm. long; scapes 3–9 cm. tall, 1- to 4-flowered, glandular-pubescent; sepals obtuse, 2–2.5 mm. long; petals white, sometimes pinkish, 3–4 mm. long; fruiting carpels 7–10 mm. long.

Wet alpine, central Alaska—Calif. Fig. 620.

24. *S. bronchialis* L. ssp. *funstonii* (Small) Hult.   Spotted Saxifrage
Leptasea funstonii Small.

Tufted; leaves of the caudices crowded, persistent for several years, more or less parchment-like, linear or oblong-lanceolate with spines along the edges and tip, 6–12 mm. long; scapes with a few reduced leaves, 5–15 cm. tall, several- to many-flowered; sepals about 2 mm. long, glabrous or ciliate; petals cream-colored or yellow, spotted, 5–7 mm. long; fruiting carpels 8–10 mm. long. Var. *cherlerioides* (D. Don) Engl. of eastern Asia and the Aleutians is a form with short, very congested leaves, stems 1–4 cm. tall, petals whitish, 3 mm. long, fruit 3–4 mm. long.

Eastern Asia—Yukon, type form Eurasian. Fig. 621.

25. *S. tricuspidata* Retz.   Three-toothed Saxifrage

Tufted; leaves of the caudices densely crowded, persistent, parchment-like, oblong or spatulate, with 3 sharp teeth at the apex and short-ciliate on the margins, 7–20 mm. long; scapes bearing several reduced leaves and several–many flowers; petals white or cream-color, about 6 mm. long; fruit 5–7 mm. long.

Most of Alaska—Ellesmereland—Greenl.—n. shore of L. Superior. Fig. 622.

26. *S. eschscholtzii* Sternb.   Ciliate Saxifrage

Densely matted; leaves crowded, persistent, parchment-like, obovate with ciliate margins, concave above, convex below, about 1 mm. wide and 1.5–2 mm. long; sepals ciliate, about 1 mm. long; petals none; filaments subulate; fruit 2–3 mm. long on peduncles 5–15 mm. long.

Rocky alpine, northeastern Asia—Arctic coast—central Alaska—Alaska Penin. Fig. 623.

27. *S. nudicaulis* D. Don.   Naked-stemmed Saxifrage

Leaf blades 10–25 mm. wide, reniform, cuneate to cordate at the base, 3- to 9-lobed, the lobes triangular to ovate, acute or apiculate; stipules 4–7 mm. long, ciliate; scapes 6–18 cm. tall, few- to several-flowered, the branches subtended by bracts; sepals triangular to lanceolate, 1.5–3 mm. long; petals white, 4–5 mm. long; fruit about 5 mm. long.

Bering Sea region and eastern Asia. Fig. 624.

28. *S. mertensiana* Bong.   Wood Saxifrage

Leaves 3–10 cm. wide, suborbicular with deeply cordate base, glabrate, shallowly lobed, the lobes usually with 3 rounded, gland-tipped teeth; scapes 2–4 dm. tall, glandular-pubescent, especially above, paniculately branched, the many flowers, except the terminal ones, usually replaced by bulblets; sepals 2–3 mm. long, reflexed; petals white, 3–4 mm. long; filaments clavate.

Central Alaska—western Mont.—northern Calif. Fig. 625.

29. *S. oppositifolia* L.   Purple Mountain Saxifrage
Antiphylla oppositifolia (L.) Small.

Tufted, densely leafy, prostrate; leaves 4-ranked, imbricated, keeled, ciliate, obovate to spatulate, 3–5 mm. long; flowers solitary on leafy stalks up to 3 cm. long; sepals ovate, ciliate, 2.5–3 mm. long; petals purplish, rarely whitish, about 8 mm. long; fruiting carpels 8–10 mm. long.

Rocky slopes, circumpolar. Fig. 626.

20. GROSSULARIACEAE (Gooseberry Family)

Shrubs; leaves palmately veined, usually lobed, petioled; flowers racemose or solitary, regular, perfect; sepals and the small petals each 5, rarely 4; stamens 5, alternate with the petals; carpels 2, united into a 1-celled ovary with 2 parietal placentae; styles 2; fruit a berry.

RIBES L.

Characters of the family. (Arabic name for Rheum ribes).

```
1A. Racemes 1- to 3-flowered............................. 1. R. oxycanthoides
2A. Racemes several- to many-flowered.
   1B. Stems with spines or prickles....................... 2. R. lacustre
   2B. Stems unarmed.
      1C. Racemes 12–30 cm. long........................... 3. R. bracteosum
      2C. Racemes less than 10 cm. long.
         1D. Ovary and fruit smooth, fruit red................ 8. R. triste
         2D. Ovary and fruit glandular.
            1E. Lower surface of leaves with resinous glands.
               1F. Fruit glabrous........................... 4. R. hudsonianum
               2F. Fruit puberulent......................... 5. R. howellii
            2E. Leaves not glandular, fruit prickly with stalked glands.
               1F. Fruit black with bloom.................... 6. R. laxiflorum
               2F. Fruit red................................ 7. R. glandulosum
```

1. *R. oxycanthoides* L. Northern Gooseberry
   Grossularia oxycanthoides (L.) Mill.

   Stems usually less than 1 m. tall, usually bristly, with nodal spines hardly 1 cm. long; leaves 2–4 cm. wide, cordate to widely cuneate at the base, more or less pubescent; peduncles and pedicels short, pubescent; sepals white, glabrous, 2.5–4 mm. long; petals two-thirds as long as the sepals; berry reddish-purple when ripe, about 1 cm. in diameter and of good quality.

   Yukon—Newf.—Mich.—B. C. Fig. 627.

2. *R. lacustre* (Pers.) Poir. Swamp Gooseberry
   R. echinatum Lindl.

   Stems 1–2 m. tall, more or less prickly and spiny; leaves pentagonal in outline, 5- to 7-lobed, incised-dentate, 2–7 cm. wide; petioles bristly-ciliate; flowers light green or purplish; berries black, glandular-hispid. This species is intermediate between gooseberries and currants. The fruit is used to limited extent.

   Alaska Penin.—central Alaska—Labr.—Newf.—Penn.—north Calif. Fig. 628.

3. *R. bracteosum* Dougl. Blue Currant

   Stems 1–3 m. tall with thick twigs; leaves cordate-orbicular in outline, 5- to 7-lobed, the lobes acute or acuminate, irregularly serrate with gland-tipped teeth, resinous-dotted beneath, 6–20 cm. long and wide; racemes with foliaceous lower bracts; flowers greenish-white; berries resinous-dotted, black with whitish bloom, 7–10 mm. in diameter, the aroma similar in an intensified degree to that of the black currant formerly grown in gardens.

   South central Alaska—north Calif. Fig. 629.

4. *R. hudsonianum* Rich. Northern Black Currant

   Stems 5–15 dm. tall with light gray twigs; leaves reniform-cordate, broader than long, 3- to 4-lobed, coarsely dentate, resinous-dotted and villous beneath; racemes 3–6 cm. long; bracts setaceous, villous, about equaling the pedicels, deciduous; flowers whitish; berry black, 5–10 mm. in diameter, scarcely edible.

   West Alaska—Hudson Bay—Minn.—B. C. Fig. 630.

5. *R. howellii* Greene. Maple-leaved Currant
   R. acerifolium Howell not C. Koch.

   Resembling the preceding species in general appearance but the leaves thinner and more maple-like; racemes reflexed with upturned, puberulent pedicels; sepals 3–4 cm. long, obtuse; anthers much larger than in *R. hudsonianum*. The plant from Hyder differs from the type in long pedicels and sessile or nearly sessile glands on the ovary. It may be distinct.

   Hyder—Oregon.

6. *R. laxiflorum* Pursh. Trailing Black Currant

Stems more or less decumbent, 5–20 dm. long; leaves nearly orbicular in outline, cordate, rather deeply 5-lobed, glabrous above, puberulent on the veins beneath, 5–10 cm. wide, the lobes acute, doubly serrate; racemes erect or ascending, 6- to 12-flowered, 6–10 cm. long, pubescent and glandular. The berry has a fetide odor but is often used.

Kenai Penin.—central Alaska—northern Calif. Fig. 631.

7. *R. glandulosum* Grauer. Fetid Currant
   R. prostratum L'Her.

Similar to *R. laxiflorum* in habit and leaf characters; odor very fetid; racemes ascending, 7- to 10-flowered, puberulent; pedicels and hypanthium glandular-bristly; berries red, 6–8 mm. in diameter.

Central Alaska—Labr.—Newf.—N. Car.—Wisc.

8. *R. triste* Pall. American Red Currant

Stems 5–15 dm. tall with reddish-brown, shreddy bark on the twigs; leaves reniform-cordate, 3- to 5-lobed, dentate, glabrous above, glabrate or pubescent beneath, 3–10 cm. wide; flowers purplish; racemes 3–6 cm. long; fruit similar in every way to that of the cultivated garden currant.

Northern Asia—Kobuk River—Labr.—Newf.—Mich.—Ore. Fig. 632.

21. ROSACEAE (Rose Family)

Herbs, shrubs or trees; leaves alternate, usually with stipules; flowers regular, usually perfect but sometimes monoecious or dioecious; hypanthium well developed, ranging from flat with ovaries superior to elongated and enclosing the ovaries; sepals and petals each usually 5, the latter sometimes wanting; stamens 1–many, often 20; carpels 1–many, usually distinct; ovules 1–several in each carpel; fruit various.

```
1A. Ovary superior.
  1B. Carpels 1, becoming a drupe......................... 1. Prunus
  2B. Carpels 3–5, becoming dehiscent follicles.
    1C. Carpels more or less united below, shrub............ 2. Physocarpus
    2C. Carpels distinct.
      1D. Flowers dioecious, tall herb..................... 5. Aruncus
      2D. Flowers perfect.
        1E. Leaves simple, shrubs......................... 3. Spiraea
        2E. Leaves twice or thrice 3-cleft................. 4. Luetkea
  3B. Carpels becoming druplets........................... 6. Rubus
  4B. Carpels becoming achenes.
    1C. Carpels enclosed in the hypanthium which becomes
        fleshy in fruit.................................. 7. Rosa
    2C. Carpels not enclosed.
      1D. Achenes borne on a receptacle which becomes
          fleshy in fruit................................ 8. Fragaria
      2D. Achenes borne on a dry receptacle.
        1E. Style articulate with the ovary and deciduous.
          1F. Stamens numerous........................... 9. Potentilla
          2F. Stamens 5.
            1G. Leaves trifoliate........................10. Sibbaldia
            2G. Leaves 2- to 3-ternate...................11. Chamaerhodos
        2E. Styles persistent.
          1F. Flowers borne in a dense spicate or capitate
              inflorescence.............................14. Sanguisorba
```

  2F. Flowers borne singly or in an open inflorescence.
   1G. Leaves simple............................12. *Dryas*
   2G. Leaves pinnate...........................13. *Geum*.
 2A. Ovary inferior.
  1B. Leaves pinnate.......................................15. *Sorbus*
 2B. Leaves simple.
  1C. Ripe carpels bony..................................18. *Crataegus*
  2C. Ripe carpels papery or leathery.
   1D. Cavities of the ovary as many as the pistils........16. *Malus*
   2D. Cavities of the ovary twice as many as the pistils..17. *Amelanchier*

1. PRUNUS

Shrubs or trees; leaves simple, alternate, toothed; flowers perfect, in our species borne in racemes on leafy branches; sepals 5, imbricate; petals 5, imbricate; stamens 15–30, the filaments filiform and distinct; fruit with a fleshy exocarp and smooth bony stone. (Latin name.)

P. melanocarpa (A. Nels.) Shafer.   Rocky Mountain Wild Cherry

Shrub or small tree; leaves glabrous, obovate or oval, usually abruptly acuminate at the apex and rounded at the base, paler beneath; flowers white, 1 cm. or less broad; fruit purple or black, 6–8 mm. in diameter, sweet or slightly astringent.

  Liard Hot Springs—N. Dak.—N. Mex.—Calif.

2. PHYSOCARPUS Maxim.

Shrubs with exfoliating bark; leaves palmately lobed; flowers in terminal corymbs; hypanthium campanulate, 5-lobed, stellate-pubescent; sepals persistent; petals white, spreading; stamens 20–40; follicles opening along both sutures; seed 2–4, obliquely pear-shaped, shining. (Greek, bellows or bladder and fruit.)

P. capitatus (Pursh) Kuntze.   Pacific Ninebark

1–5 m. tall; leaves 3- to 5-lobed, the lobes incised or doubly serrate, sparingly pubescent or glabrate above, sometimes stellate-pubescent beneath, 3–7 cm. long and about as wide; inflorescence rather dense; petals 3–4 mm. long; carpels 8–10 mm. long, ovate, rather long-acuminate.

  Southeast Alaska—Idaho—central Calif.

3. SPIRAEA

Leaves without stipules; flowers small, in racemes, corymbs, or panicles; hypanthium campanulate or turbinate; sepals 5; petals 5; stamens many; carpels usually 5, inserted at the bottom of the hypanthium; ovules 2–several; fruit composed of leathery follicles which open along the ventral suture; seeds linear, tapering to both ends. (Greek, to twist, referring to the follicles of some species.)

Inflorescence conic or spike-like, petals pink................ 1. *S. menziesii*
Inflorescence flat to hemispherical, flowers white............ 2. *S. beauverdiana*

1. *S. menziesii* Hook.   Menzies Spiraea

An erect, branched shrub 10–15 dm. tall with reddish-brown twigs; leaves elliptic to oval, the wider forms being on the more vigorous growth, serrate on the upper half, acute to rounded at either end, glabrous, or pubescent on the veins, 3-8 cm. long; inflorescence very dense, spike-like, 4-15 cm. long, pubescent; sepals ovate, reflexed; petals rose pink, 1.5 mm. long; follicles glabrous.

Southeast Alaska—Idaho—Ore. Fig. 633.

2. *S. beauverdiana* Schneid. Beauverd Spiraea
   S. stevenii (Schneid.) Rydb.

3-12 dm. tall, with reddish twigs; leaves oblong to ovate, glabrate, serrate from near the base, usually rounded at both ends, 2-5 cm. long; inflorescence 2-4 cm. across, puberulent; sepals ovate, acute, reflexed; petals white, about 1.5 mm. long; follicles puberulent.

East Asia—all of Alaska—Mackenzie. Fig. 634.

4. LUETKEA

Decumbent or creeping undershrub with stoloniferous branches; leaves twice or thrice ternately dissected; flowers borne in a raceme; hypanthium hemispheric; sepals and petals each 5; stamens about 20, the filaments subulate and connate at the base; carpels usually 5; ovules several; follicles coriaceous, dehiscent by both sutures; seed linear-lanceolate, acute. (Count F. P. Luetke was commander of a Russian exploring expedition.)

L. pectinata (Pursh) Kuntze. Luetkea

Flowering shoots glabrate below, pubescent above, 5-15 cm. high; leaves crowded at the base of the flowering shoots, alternate above, glabrate, dissected into linear, acute divisions, 1-2 cm. long; racemes 1-5 cm. long; sepals ovate-lanceolate, acute, about 2 mm. long; petals white, 3 mm. or more long; carpels about 4 mm. long.

Alpine meadows, Bering Str.—Canadian Rockies—Ore. Fig. 635.

5. ARUNCUS (L.) Adans.

Perennials with thick rootstocks and twice to thrice ternate-pinnate leaves without stipules; inflorescence a large panicle, the divisions spicate; flowers dioecious; sepals 5, triangular; petals 5, narrow, white; carpels 3-5; ovules several; follicles cartilaginous, dehiscent along the ventral suture, then splitting at the apex, reflexed; seeds few. (Greek, meaning goat's beard.)

A. vulgaris Raf. Goat's Beard
   A. sylvester Kost.
   A. acuminatus (Dougl.) Rydb.

Stem stout, glabrous, 1-2 m. tall; leaves large; leaflets lanceolate, irregularly and doubly serrate, long-acuminate, 3-12 cm. long; panicles terminal and axillary, 1-5 dm. long; flowers small; follicles about 3 mm. long.

Widely distributed in Europe, Asia, and North America. Fig. 636.

6. RUBUS (Tourn.) L.

Perennial herbs, shrubs or trailing vines, often prickly; leaves alternate, simple or pinnate; inflorescence axillary or terminal, the flowers solitary, racemose or panicled, regular, perfect or dioecious; stipules adnate to the petioles; sepals 5, persistent, petals 5, deciduous; stamens many, distinct; carpels few to many, inserted on a convex or elongated receptacle; fruit composed of few to many fleshy druplets. (Latin, ruber, red.)

```
1A. Herbaceous plants.
   1B. Flowers white.
      1C. Leaves simply lobed.............................. 1. R. chamaemorus
      2C. Leaves 3-foliate................................. 2. R. pubescens
      3C. Leaves 5-foliate................................. 3. R. pedatus
   2B. Flowers pink or red.
      1C. Leaves 3-lobed................................... 4. R. stellatus
      2C. Leaves 3-foliate.
         1D. Stem smooth, leaflets small................... 5. R. arcticus
         2D. Stem glandular-hairy, leaflets larger.......... 6. R. alaskensis
2A. Stems woody.
   1B. Stems biennial.
      1C. Stems bristly.................................... 7. R. strigosus
      2C. Stems prickly.................................... 8. R. leucodermis
   2B. Stems perennial.
      1C. Leaves simple.................................... 9. R. parviflorus
      2C. Leaves compound..................................10. R. spectabilis
```

1. R. chamaemorus L. Cloudberry. Baked-apple Berry

Erect from a creeping rootstock, 5–20 cm. tall; leaves 2 or 3, reniform with 3 or 5 rounded lobes, rugose, 3–10 cm. wide; stipules ovate, obtuse; flowers solitary, dioecious; sepals ovate, glandular pubescent; petals white, obovate, 8–12 mm. long; fruit composed of 6–18 rather large druplets the color of a baked apple when ripe and prized by the Indians and Eskimo.

Circumpolar, south to Newf.—N. Hamp.—Vancouver Isl. Fig. 637.

2. R. pubescens Raf. Dwarf Red Blackberry

Stems slender and with trailing shoots 1–10 dm. long; leaves ternate, rarely quinate; leaflets 2–9 cm. long, the lateral obliquely ovate, the terminal rhomboid, sharply and doubly serrate; flowers 1–3; petals small, white or pink; sepals pubescent, reflexed; droplets few, large, red.

Watson Lake—Newf.—New Jersey—Colo.—B. C.

3. R. pedatus Smith. Five-leaved Bramble

A slender trailing vine rooting at the nodes, glabrate; flowering branches very short, 2- to 4-leaved; leaves 3-foliate but the lateral leaflets so deeply cleft as to appear 5-foliate; leaflets thin, obovate or rhombic, irregularly toothed and incised, 1–3 cm. long; stipules ovate, small; flowers usually solitary; sepals foliaceous, ovate-lanceolate; petals white, ovate-oblong, 1 cm. or less long; fruit composed of 1–6 red, oblong druplets.

Woods, climbing over moss or logs, eastern Asia—Yukon—Mont.—Ore. Fig. 638.

4. **R. stellatus** Smith. Nagoon Berry

Plant low, 5–15 cm. tall from a spreading rootstock, simple or branched from the base; leaves reniform in outline, 3-lobed, sometimes divided to near the base, simple or doubly serrate, cordate at the base; stipules obovate, acuminate, strongly veined; flowers solitary; sepals lanceolate, acute, pubescent and often toothed; petals rose-red, clawed, 15–20 mm. long; fruit of high quality, composed of about 15–25 red druplets to which the calyx strongly adheres.

Wet places in coastal districts and occasionally in interior, East Asia—B. C. Fig. 639.

5. **R. arcticus** L. Nagoon Berry. Kneshenaka
    R. acaulis Michx.

Like *R. stellatus* in habit and fruit; less than 1 dm. tall in exposed places, or to 25 cm. tall in sheltered situations; leaves 3-foliate; terminal leaflet ovate to rhombic, unevenly serrate, 2–4 cm. long; lateral leaflets oblique; flowers 1–3; petals dark rose to red; druplets 20–40, red. *R. acaulis* was the name applied to a dwarf form with more rounded leaflets and the hypanthium part of the calyx glabrous or nearly so, the corresponding part in typical *R. articus* being glandular-hairy. Intermediate forms occur.

Circumpolar, south to north Minnesota. Fig. 640.

6. **R. alaskensis** Bailey. Alaska Bramble

Stems 2–5 dm. tall, often woody at the base, 1- to 3-flowered, pubescent; leaves mostly 3-foliate, the petioles pubescent; terminal leaflet broadly ovate, serrate-dentate, thinly pubescent beneath, 4–8 cm. long; lateral leaflets similar but oblique; sepals narrow, 10–15 mm. long, becoming reflexed; petals pink, broadly spatulate, 12–18 mm. long.

Curry—Matanuska—southeastern Alaska. Fig. 641.

7. **R. strigosus** Michx. American Red Raspberry
    R. idaeus L. var. *canadensis* Rich.
    R. idaeus L. ssp. *sachalinensis* (Levl.) Focke.
    R. subarcticus Rydb.

Canes 6–12 dm. tall, brownish red, densely covered with both rough and fine bristles; leaflets 3–5, irregularly and doubly serrate, whitish-pubescent beneath; stipules very narrow and deciduous; petioles and peduncles more or less glandular; sepals triangular-lanceolate, glandular-pubescent; petals white, about 5 mm. long; fruit composed of red druplets, elongate-hemispheric.

Across N. America, south to Conn.—Colo.—B. C., ?eastern Asia. Fig. 642.

8. **R. leucodermis** Dougl. Western Black Raspberry

Stems 1-2 m. tall, glaucous, armed with stout, flat prickles; leaflets 3–5, ovate to lanceolate, doubly serrate, white-tomentose beneath, the veins and petioles prickly; sepals lanceolate, long-acuminate, in fruit spreading or reflexed; petals white, shorter than the sepals; fruit usually dark with white bloom and agreeable flavor.

Southeast Alaska—Mont.—Utah—Calif.

9. R. parviflorus Nutt. Thimbleberry
R. nutkanus Moc.

An unarmed shrub with shreddy bark, 6–16 dm. tall; leaves pentagonal in outline, 3- to 7- but mostly 5-lobed, coarsely and unevenly serrate with gland-tipped teeth, 7–20 cm. wide; sepals broadly ovate, abruptly narrowed into a long, slender appendage; petals white, 16–25 mm. long; fruit convex, red, composed of numerous small druplets.

Southeastern Alaska—S. Dak.—N. Mex. Fig. 643.

10. R. spectabilis Pursh. Salmonberry

Usually more or less prickly, 1–4 m. tall, the bark yellowish-brown and exfoliating; leaflets 3, usually more or less lobed, the lateral ones unsymmetrical, coarsely and unevenly serrate, 2–12 cm. long; stipules linear or subulate, pubescent; flowers solitary; sepals deltoid-lanceolate, pubescent; petals red, 16–22 mm. long; fruit varying from yellow to dark red, 16–25 mm. in diameter, composed of 20–40 druplets.

East Asia—Idaho—Calif. Fig. 644.

7. ROSA (Tourn.) L.

Erect or climbing shrubs; leaves alternate, pinnate; leaflets serrate; stipules adnate; flowers perfect, pink in our species; hypanthium well developed, elongated upward, contracted at the mouth and enclosing the achenes, becoming fleshy in fruit; sepals usually 5; petals normally 5 but may be numerous by transformation of stamens; stamens numerous, inserted on the margin of the hypanthium; carpels numerous, borne on the base and sides of the hypanthium; achenes bony. (The Latin name.)

1A. Fruit 1 cm. or less in diameter.......................... 3. R. woodsii
2A. Fruit more than 1 cm. in diameter.
  1B. Stems with numerous terete pricles.................. 1. R. acicularis
  2B. Stems with few flattened prickles.................... 2. R. nutkana

1. R. acicularis Lindl. Prickly Rose

Bushy, 3–12 dm. tall, usually armed with moderately strong spines interspersed with weaker ones; stipules pubescent and with glandular margins; leaflets 3–9, usually 5, elliptic or oval, regularly serrate, 15–55 mm. long, glabrous above, pale and pubescent beneath; hypanthium glabrous, pyriform, or elliptic to nearly globose, usually with a neck; sepals pubescent, glandular along the margins of the usually more or less foliose tips; petals obcordate, rose pink, 2–3 cm. long; fruit edible. Sometimes hybridizes with the next species.

Has an interrupted circumboreal distribution south to Mass.—Penn.—Colo. Fig. 645.

2. *R. nutkana* Presl. Nootka Rose
R. aleutensis Crepin.

Stems stout, erect, 6–25 dm. tall, usually armed with paired straight or slightly curved prickles; stipules and leaf-rachis glandular, the stipules with glandular-dentate margins; leaflets 5–9, more or less double-serrate, usually rounded at both ends, 15–50 mm. long; sepals 15–30 mm. long, petals typically rose pink, 20–35 mm. long; fruit glabrous, typically globose and neckless.

Coastal districts, Aleutians—Calif. Fig. 646.

3. *R. woodsii* Lindl. Woods Rose

Bushy, 5–15 dm. tall, armed with numerous prickles 4–8 mm. long; stipules narrow below the spreading tips, leaflets 5–9, obovate, somewhat cuneate at the base, slightly petioluled, serrate, glabrous, the under surface glaucous, 1–2 cm. long; flowers solitary or 2 or 3 together; sepals lanceolate, caudate-attenuate, about 15 mm. long, usually glabrous on the back, tomentose on the margin and within; fruit globose or nearly so.

Circle Hot Springs—Alta.—Minn.—Kans.—Utah. Fig. 647.

8. FRAGARIA L.

Acaulescent perennials with thick, scaly rootstocks propagating by runners which root at the joints; bractlets, sepals, and petals each usually 5; flowers usually white; stamens about 20; receptacle hemispheric or conic, bearing the numerous carpels and becoming enlarged and fleshy in fruit; styles filiform but short and attached near the middle of the ovaries. (Latin, signifying fragrance.)

1A. Leaves thick and coriaceous........................... 1. *F. chiloensis*
2A. Leaves thinner.
  1B. Pubescence of stems and petioles spreading or slightly
      reflexed .... 2. *F. bracteata*
  2B. Pubescence ascending or appressed.................. 3. *F. glauca*

1. *F. chiloensis* (L.) Duch. Beach Strawberry

Rather stout; petioles, peduncles and inflorescence silky-pubescent with spreading or reflexed hairs; leaflets thick, cuneate-obovate or the lateral rhombic, crenate-dentate, rugose above, silky-pubescent beneath, 2–4 cm. long; peduncles shorter than the leaves; sepals acuminate; flowers 2–3 cm. broad; fruit ovoid, up to 25 mm. long, soft and sweet; achenes nearly superfical.

Near the coast, Aleutians—Calif.—Peru—Patagonia, and in Hawaii. Fig. 648.

2. *F. bracteata* Heller. Bracted Strawberry

Rootstock short; leaves thin, silky when young, nearly glabrous in age; leaflets broadly ovate, coarsely serrate, 2–4 cm. long; scapes slender, equaling or exceeding the leaves, usually with a unifoliate bract; flowers 15–20 mm. broad; sepals triangular-lanceolate, longer than the lanceolate bractlets, very acute; fruit ovoid, the achenes nearly superfical.

Hyder—Mont.—N. Mex.—Calif. Fig. 649.

3. *F. glauca* (Wats.) Rydb. Yukon Strawberry
F. yukonensis Rydb.
F. platypetala of reports from Alaska.

Rather slender; petioles and peduncles appressed-villous; leafllets rather thin, obovate, cuneate at the base, sharply and deeply toothed, glabrous above, appressed silky beneath, 15–55 mm. long; scapes leafy-bracted, usually shorter than the leaves; flowers less than 15 mm. wide; fruit subglobose, about 1 cm. in diameter; achenes in shallow pits.

Central Alaska—Gt. Slave L.—Black Hills—N. Mex. Fig. 650.

9. POTENTILLA L.

Herbs or rarely shrubs with alternate, compound leaves; flowers regular, perfect; hypanthium concave to hemispheric; bractlets, sepals and petals each 5; stamens usually many; receptacle hemispheric or conic, bearing many carpels; styles terminal or lateral; fruit composed of many achenes on a dry receptacle. (Latin, powerful, from medicinal properties of some species.)

1A. Petals purple, short. (*Comarum* L.) .... 1. *P. palustris*
2A. Petals white or cream color. (*Drymocallis* Fourr.) .... 2. *P. arguta*
3A. Petals yellow.
  1B. Plant shrubby. (*Dasiphora* Raf.) .... 3. *P. fruticosa*
  2B. Plant herbaceous.
    1C. Plant stoloniferous, flower solitary. (*Argentina* Lam.)
      1D. Bractlets toothed or divided, achenes grooved ...... 4. *P. anserina*
      2D. Bractlets entire, achenes not grooved .... 5. *P. pacifica*
    2C. Plants lacking runners, flowers in cymes.
      1D. Leaves odd-pinnate.
        1E. Leaves with 3–7 pairs of leaflets.
          1F. Leaves silky-tomentose on both sides .... 6. *P. hippiana*
          2F. Leaves green or grayish on upper side .... 7. *P. pennsylvanica*
        2E. Lower leaves with 2 or 3 pairs of leaflets.
          1F. Style filiform.
            1G. Leaflets pinnatifid ....11. *P. multifida*
            2G. Leaflets toothed ....12. *P. rubricaulis*
          2F. Style enlarged and glandular at the base.
            1G. Leaves silky-pubescent on both sides ....10. *P. pulchella*
            2G. Leaves tomentose beneath, green above.
              1H. Leaflets pinnatifid almost to the midrib.... 9. *P. virgulata*
              2H. Leaflets pinnatifid ½–¾ way to midrib.... 8. *P. pectinata*
      2D. Leaves palmately 5- to 7-foliate.
        1E. Leaflets toothed to the base ....14. *P. gracilis*
        2E. Leaflets toothed on upper half only ....13. *P. diversifolia*
      3D. Leaves trifoliate.
        1E. Leaflets cleft to the middle or lower (see also *P. vahliana*).
          1F. Petals 2–4 mm. long ....16. *P. elegans*
          2F. Petals 5–8 mm. long ....15. *P. biflora*
        2E. Leaflets toothed.
          1F. Leaves hirsute on lower surface.
            1G. Plant erect ....18. *P. monspeliensis*
            2G. Plant spreading ....17. *P. emarginata*
          2F. Leaves tomentose or densely sericeous on lower surface.
            1G. Stems 1 dm. or less tall, 1- to 3-flowered.
              1H. Petals obcordate ....20. *P. uniflora*
              2H. Petals obreniform ....21. *P. vahliana*
            2G. Stems normally 1–2 dm. tall, several-flowered.
              1H. Flowers 2–3 cm. in diameter ....19. *P. villosa*
              2H. Flowers 15 mm. or less in diameter.
                1J. Leaves deeply dissected ....22. *P. hookeriana*

2J. Leaves coarsely dentate.................23. *P. nivea*
  3G. Stems more than 25 cm. tall.................24. *P. chamissonis*

1. *P. palustris* (L.) Scop.   Purple or Marsh Cinquefoil
Comarum palustre L.

Aquatic or marsh perennial with creeping rootstocks; stems ascending, more or less hirsute and glandular-pubescent above; leaves pinnate, leaflets 3–7, usually 5, green above, pale beneath, oblong or oval, sharply serrate, 2–6 cm. long; bractlets small and narrow; sepals purple, ovate, acuminate, 8–15 mm. long; petals much shorter than the sepals; style lateral; achene smooth with purplish apex.

Circumboreal, south to Penn.—Wyo.—Calif. Fig. 651.

2. *P. arguta* Pursh.   Tall or Glandular Cinquefoil
Drymocallis arguta (Pursh) Rydb.

Rootstock stout and woody; stems stout, erect, 3–10 dm. tall, striate, hirsute, glandular or viscid; basal leaves 7- to 11-foliate; leaflets ovate, oval or rhomboid, the terminal one cuneate, the lateral ones oblique, all sharply incised-dentate; stem leaves reduced; flowers in a dense cyme, 12–18 mm. in diameter; hypanthium, bractlets and calyx glandular viscid; petals whitish, drying yellowish, a little longer than the sepals.

Yukon—N. B.—Va.—Colo. Fig. 652.

3. *P. fruticosa* L.   Shrubby Cinquefoil. Yellow Rose
Dasiphora fruticosa (L.) Rydb.

A much-branched shrub with shreddy bark, 2–12 dm. tall; leaves pinnate, silky pubescent, especially beneath; leaflets usually 5, oblong or linear-oblong, entire and usually with more or less revolute margins, 10–25 mm. long; petals 10–15 mm. long, much longer than the sepals; achenes dark, receptacle with long brown hairs.

Circumpolar, south to N. Jer.—Minn.—N. Mex.—Calif. Fig. 653.

4. *P. anserina* L.   Common Silverweed
Argentina anserina (L.) Rydb.

Leaves 1–2 dm. long; leaflets 9–31 with smaller ones interspersed, 1–4 cm. long, oblong or oblong-lanceolate, white-silky beneath, sparingly silky to green and glabrate above; peduncles 3–15 cm. long; petals 7–15 mm. long; achenes corky, grooved on the upper end. Var. *sericea* Hayne (*Argentina argentea* Rydb.) has the upper surface of the leaves silky-tomentose.

Interrupted circumboreal, south to N. Jer.—N. Mex. Fig. 654.

5. *P. pacifica* Howell.   Pacific Silverweed
P. yukonensis Hult.
P. egedii Wormskj. var. *groenlandica* (Tratt.) Polunin.
Argentina occidentalis Rydb.
A. subarctica Rydb.

Resembling the preceding in appearance; leaves up to 4 dm. long including petiole; leaflets up to 6 cm. long, glabrous or nearly so above, silky-tomentose beneath; peduncles up to 3 dm. long; petals up to 15 mm. long. The vigorous form on the Pacific Coast gives way gradually to the diminutive form of the Arctic Coast which often has leaves and peduncles only a few centimeters high.

Mostly along beaches, circumpolar, south to Calif. Fig. 655.

6. *P. hippiana* Lehm.   Wooly Cinquefoil

Stems erect, 3–6 dm. tall, silky canescent; lower leaves 5-to 11-foliate; leaflets oblanceolate or oblong, obtuse, narrowed or cuneate at the base, 15–50 mm. long, deeply toothed; sepals ovate-lanceolate, 5–7 mm. long; bractlets nearly equaling the sepals but narrower; petals 6–8 mm. long.

Central Alaska—Minn.—Nebr.—Ariz.

7. *P. pennsylvanica* L.   Pennsylvania Cinquefoil

Stems erect or ascending, 4–8 dm. tall in the typical form, more or less tomentose; leaves pinnately 5- to 15-foliate; leaflets oblong or oblanceolate, cleft one-half way to the midrib into oblong divisions, grayish tomentose and veiny beneath, glabrous or nearly so above; bractlets about equaling the sepals; petals longer than the sepals; achenes smooth or more often somewhat rugulose. Var. *strigosa* Pursh is generally lower, 3–5 dm. tall, leaflets deeply divided into narrow lobes with revolute margins. Var. *glabrata* Wats. has stem and leaves nearly glabrous.

Asia, east central Alaska—Hudson Bay—Kans.—N. Mex. Fig. 656.

8. *P. pectinata* Fisch.   Coast Cinquefoil

Stems usually clustered from a woody rootstock, finely pubescent, 2–5 dm. tall; stipules large, foliaceous and lobed; leaves mostly 5-foliate; leaflets obovate or oblong, cut into narrow lobes with revolute margins, 2–5 cm. long; flowers about 15 mm. in diameter; bractlets lanceolate with narrowed bases; sepals lanceolate with broad bases, a little longer than the bractlets; achenes smooth or minutely rugulose.

Skagway and in eastern North America. Fig. 657.

9. *P. virgulata* A. Nels.

Caudex short and with a taproot; stems 2–5 dm. tall; leaflets ovate, 1–4 cm. long, sparingly hairy above, white pubescent beneath, dissected into narrowly linear divisions with revolute margins; bractlets linear, about equaling the lanceolate sepals; petals somewhat exceeding the sepals; achenes smooth.

Seward Penin.—Wyo.—Utah. Fig. 658.

10. *P. pulchella* R. Br.

Densely caespitose and silky-hirsute with white or yellowish hairs; stems spreading, 1- to few-flowered, less than 1 dm. long; leaves usually 5-foliate; leaflets obovate-cuneate, deeply dissected into linear segments; bractlets oblong, nearly as long as the ovate sepals; petals 5–6 mm. long, a little exceeding the sepals; styles short.

Wrangel Isl.—Ellsmereland—Spitzbergen — Nova Zemlya — Labr. — Seward Penin.—?Kiska Isl.

11. P. multifida L.   Cut-leaved Cinquefoil

Stems, several to many, arising from a woody caudex, ascending or spreading, somewhat appressed-strigose, 1-3 dm. long; leaflets pectinately divided to very near the midrib into narrow, linear divisions with more or less revolute margins, smooth above, tomentose beneath; bractlets slightly shorter and petals slightly longer than the 3-4 mm. long sepals; style short; achenes smooth or somewhat rugose.

Circumpolar, south to Great Slave L.—southern Alaska. Fig. 659.

12. P. rubricaulis Lehm.   Red-stemmed Cinquefoil

Stems several, ascending or prostrate, often tinged with red, 1-2 dm. long, pubescent with spreading hairs; leaflets glabrate above, white tomentose beneath, 1-3 cm. long, obovate or oblanceolate, pinnately cleft into lanceolate, acute teeth; stem leaves usually ternate; cymes 5- to 9-flowered; petals obcordate, a little longer than the sepals.

Reported from Herschel Isl., Mackenzie—Ellsmereland—Great Bear Lake.

13. P. diversifolia Lehm.   Diverse-leaved Cinquefoil
P. glaucophylla of reports from Alaska.

Stems 1-few from a woody caudex, 2-5 dm. tall; leaflets pubescent when young, often glabrate in age, obovate or oblanceolate, toothed or lobed on the upper half, 2-5 cm. long; stipules of basal leaves lanceolate and scarious, of upper leaves wider and foliaceous; bractlets shorter than the sepals; petals obcordate, 5-9 mm. long; styles long, filiform.

Alaska Range—S. Dak.—Colo.—Calif. Fig. 660.

14. P. gracilis Dougl.
P. alaskana Rydb.
P. blaschkeana and P. nuttallii of reports from Alaska.

Stems pubescent, branched above, 4-9 dm. tall; basal leaves long-petioled; leaflets obovate, cut one-half way to the midrib into narrow lobes, pubescent, sometimes silky, especially beneath, 3-12 cm. long; inflorescence silky; sepals 8-10 mm. long, longer than the bractlets; petals about equaling the sepals; achenes smooth; base of style dilated.

Kodiak Isl.—Wiseman—Alta.—Mont.—Calif. Fig. 661.

15. P. biflora Willd.   Two-flowered Cinquefoil

Almost acaulescent, caespitose, silky pubescent alpine plant; stipules linear-lanceolate; terminal leaflet split nearly to the base into 3 linear divisions, the lateral into 2 such divisions, all pubescent beneath, glabrate above in age, the margins revolute; scapes 3-10 cm. tall, 1- or 2-flowered; bractlets and sepals about equal; achenes nearly 2 mm. long; receptacle with long silky pubescence.

Eastern Asia—Mackenzie River—Alaska Range. Fig. 662.

16. *P. elegans* C. & S. Pretty Cinquefoil

Densely caespitose or pulvinate; stems 15–30 mm. tall, 1-flowered; leaves short petioled, the leaflets 3–6 mm. long, sparsely villous-pilose; petals slightly exceeding the sepals and bracts. A very small and delicate species.

Eastern half of Asia, extending to Mackenzie.

17. *P. emarginata* Pursh. Arctic Cinquefoil
P. nana Willd.

Caespitose; 2–15 cm. tall; leaflets sessile, softly hirsute, 3- to 9-toothed, 5–15 mm. long; stems 1- or 2-flowered; bractlets 4–5 mm. long, about equaling the ovate, acute sepals; petals broadly obcordate, 5–9 mm. long; style filiform, short; achenes glabrous.

Circumpolar, south to Labr.—southern Alaska—Aleutians. Fig. 663.

18. *P. monspeliensis* L. Rough Cinquefoil
P. norvegica ssp. *monspeliensis* (L.) Achers. & Graebn.

Stems erect from an annual or biennial root, branched, hirsute, 2–8 dm. tall; stipules foliaceous, entire or dentate; leaves trifoliate, rarely 5-foliate on young, vigorous growth; leaflets variable, usually obovate, deeply serrate, pubescent with spreading hairs, 2–6 cm. long; flowers in rather dense, leafy-bracted cymes; bractlets and sepals lanceolate, acute; petals nearly as long as the sepals; achenes rugulose.

Widespread in our area, Alaska—Labr.—Mexico—Calif. Fig. 664.

19. *P. villosa* Pall. Villous Cinquefoil

Silky-villous throughout; stems 15–30 cm. tall, 1- to several-flowered; leaflets with prominent veins, greenish above, silvery beneath, obovate, deeply crenate-dentate with rounded teeth, 1–5 cm. long; bractlets acute; sepals acute, broader and slightly longer than the bractlets, 6–8 mm. long; petals 8–12 mm. long; achenes nearly smooth but generally with a few lines.

Eastern Asia—Seward Penin.—Alaska Range—Aleutians. Fig. 665.

20. *P. uniflora* Ledeb. One-flowered Cinquefoil

Silky-pubescent, stems 3–12 cm. tall, usually 1-flowered but sometimes 2-flowered; leaflets silky or glabrate above, white-tomentose beneath, deeply cut from the apex, the terminal one cuneate-obovate, the lateral ones rhombic, 1–2 cm. long; bractlets and sepals silky, the bractlets obtuse, the sepals acute, 4–5 mm. long; petals obcordate, 6–8 mm. long.

Lena River, Siberia—Alta.—Mont.—Colo.—Ore.—Kamchatka. Fig. 666.

21. *P. vahliana* Lehm. Vahl Cinquefoil

Caudex woody, covered with old remains of stipules and petioles; whole plant covered with yellowish villous hairs; leaves crowded, short petioled; leaflets usually 1 cm. or less long, cuneate, coarsely and deeply dentate at the apex; bractlets broadly ovate or elliptic, often obtuse; petals usually broader than long and overlapping.

Wrangel Isl.—Ellsmereland—Labr.—St. Matthew Isl. Fig. 667.

22. *P. hookeriana* Lehm. Hooker Cinquefoil

Caespitose; stems 1-2 dm. tall, tomentose; basal leaves on petioles 1-3 cm. long; leaflets 1-2 cm. long, deeply cleft into oblong lobes, silky villous above, densely tomentose beneath; bractlets almost as long as the sepals which are about 4 mm. long; petals obcordate, slightly exceeding the sepals.

Urals—Victoria Land—Mont. Fig. 668.

23. *P. nivea* L. Snow Cinquefoil

Caespitose, the caudex covered with the brown stipules and old leaves; stems several, 10-25 cm. tall, more or less tomentose, few-leaved; basal leaves on petioles 2-5 cm. long; leaflets oblong-cuneate or obovate, 15-30 mm. long, glabrate or slightly villous above, densely white-tomentose beneath, coarsely and deeply crenate; sepals ovate-lanceolate, longer than the bractlets and shorter than the petals which are narrowly obcordate.

Circumboreal, Yukon, south to Colo.—Nevada. Fig. 669.

24. *P. chamissonis* Hult. Chamisso Cinquefoil

Stems several, 2-5 dm. tall; lower leaves long-petioled, 3-foliate or a few 5-foliate; leaflets obovate, the lateral sessile, the terminal long petiolulate, deeply serrate-dentate; inflorescence many-flowered; bracts linear to lanceolate, shorter than the narrowly triangular sepals; achenes about 1 mm. long, the style being of about same length and papillose at the base.

Southern Yukon — Quebec — Greenland — Spitzbergen — northern Scandinavia.

10. SIBBALDIA L.

Low, tufted perennials with woody caudices; leaves ternate; flowers in cymes on scape-like, nearly leafless stems; bractlets, sepals, and petals each 5; petals obovate, yellow, shorter than the sepals; stamens 5, inserted alternate with the petals on the wooly edge of the hypanthium; carpels 5-20; styles lateral; achenes glabrous. (Robert Sibbald was a Scotch naturalist.)

S. procumbens L. Sibbaldia

Stems pubescent, less than 1 dm. tall; leaflets more or less appressed-pubescent, obovate, cuneate at the base, 2- to 5- but usually 3-lobed at the apex, 1-3 cm. long; sepals slightly longer than the bractlets, acute or acuminate.

Alpine-arctic, circumboreal, south to Newf.—N. Hamp.—Colo.—Calif. Fig. 670.

11. CHAMAERHODOS Bunge.

Perennial or biennial herbs; leaves ternately divided; flowers small, perfect, borne in cymes; bractlets wanting; hypanthium cup-shaped, small; sepals and petals each 5, stamens 5, opposite the petals; pistils 5–20; style filiform, basal. (Greek, a low rose.)

C. *nuttallii* (T. & G.) Pickering. American Chamaerhodos
C. *erecta* (L.) Bunge ssp. *nuttallii* (T. & G.) Hult.

Usually much branched, hirsute and glandular; basal leaves 2- to 4-ternately divided into linear or oblong segments; stem leaves diminishing in size and complexity upward; flowers numerous; hypanthium 2–3 mm. in diameter; sepals triangular-lanceolate, about equaling the white petals.

Yukon—L. Athabasca—Manitoba—Minn.—Colo. Fig. 671.

12. DRYAS L.

Low tufted or matted subshrubs; leaves alternate, petioled, simple, more or less rugose, white-tomentose beneath; flowers solitary on naked peduncles; bractlets wanting; sepals 7–10, persistent; petals 7–10, longer than the sepals, often persistent; stamens numerous; carpels numerous; style terminal, elongating and becoming plumose in fruit. (Latin name of a Greek wood nymph.)

```
1A. Sepals ovate or ovate-lanceolate, petals yellow
     and   ascending......................................... 1. D. drummondii
2A. Sepals linear or linear-lanceolate, petals whitish, spreading.
  1B. Leaf-blades crenate, strongly rugose.................. 2. D. octopetala
  2B. Leaf-blades entire or with a few teeth, not
      conspicuously   rugose............................. 3. D. integrifolia
```

1. *D. drummondii* Rich. Drummond Mountain Avens

Often forming mats several decimeters in diameter; leaves elliptic, narrowed at the base, the margins slightly revolute, 1–3 cm. long; peduncles 5–20 cm. tall, tomentose; sepals black glandular-pubescent, about 5 mm. long; petals obovate, about twice as long as the sepals; achenes with plumes up to 4 cm. long.

Central and southern Alaska—Great Bear L.—Mont.—Ore., and in Ontario and Quebec. Fig. 672.

2. *D. octopetala* L. Eight-petaled Mountain Avens

Densely tufted; leaves elliptic, glabrous and rugose above, the margins revolute, rounded at the apex, rounded or subcordate at the base, 1–3 cm. long; peduncles 3–15 cm. long, tomentose, often black hairy on upper part; sepals black glandular-pubescent, about 7 mm. long; petals about 1 cm. long; achenes with plumes up to 3 cm. long. Variable and consists of several races or varieties. Hybridizes with *D. integrifolia*.

Circumpolar, south to Colo. Fig. 673.

3. *D. integrifolia* Vahl. Entire-leaved Mountain Avens

Similar to *D. octopetala*; leaves only slightly rugose, ovate or ovate-lanceolate, the revolute margins entire or with a few teeth near the base, the apex sometimes acute; sepals acute. Var. *sylvatica* Hult. is a shade form with narrower leaves up to 45 mm. long; peduncles in fruit up to 20 cm. long.

Eastern Asia—Ellesmereland—Greenl.—Newf.—N. Hamp.—B. C. Fig. 674.

13. GEUM L.

Perennials; leaves pinnate, in some species the terminal leaflet much the largest; flowers yellow or whitish; bractlets, sepals and petals each 5; stamens many, filaments capillary; carpels many on a conical or clavate receptacle; style persistent. (The ancient Latin name.)

1A. Style conspicuously bent and geniculate above.......... 1. *G. macrophyllum*
2A. Style not conspicuously bent or geniculate (*Sieversia* Willd.).
  1B. Style not much elongated in fruit..................... 2. *G. rossii*
  2B. Style elongating in fruit and plumose below.
    1C. Basal leaves with terminal leaflet much the largest... 3. *G. calthifolium*
    2C. Basal leaves pinnate with leaflets of nearly same size.
      1D. Leaflets 5-7.. 4. *G. pentapetalum*
      2D. Leaflets 11-17..................................... 5. *G. glaciale*

1. *G. macrophyllum* Willd.   Large-leaved Avens

Stem more or less hirsute, 4–9 dm. tall; basal leaves interruptedly pinnate with a large terminal cordate, doubly crenate-dentate leaflet 5–10 cm. broad; stem leaves 3-foliate or deeply 3-lobed; all leaflets more or less hirsute on both sides; petals ovate, longer than the reflexed calyx lobes; receptacle and ovary pubescent; style curved and jointed, the lower portion of upper joint pubescent; achenes hooked. Ssp. *perincisum* (Rydb.) Hult. has narrower, more deeply incised, acute leaflets with longer, more acute teeth.

The species in the coast regions and the subspecies in interior Alaska, eastern Asia—Newf.—N. Hamp.—Colo.—Ariz.—Calif. Fig. 675.

2. *G. rossii* (R. Br.) Ser.   Ross Avens
    Sieversia rossii R. Br.

Stems arising from a large, upright, woody caudex, 7–25 cm. tall; basal leaves interruptedly pinnate, 5–10 cm. long including petiole; larger leaflets 9–15, variously incised and toothed, pubescent on the margins, 7–15 mm. long; stems with about 3 reduced leaves, 1- or 2-flowered; sepals and bractlets lanceolate, pubescent; petals bright yellow, about 1 cm. long and broad.

Alpine-arctic, eastern Asia—Melville Isl.—Yukon—Aleutians. Fig. 676.

3. *G. calthifolium* Menz.   Caltha-leaved Avens
    Sieversia calthifolia (Menz.) D. Don

Hirsute; rootstock thick, nearly horizontal; stems 1–3 dm. tall, scape-like, with a few reduced leaves; basal leaves of one large, cordate-reniform, doubly crenate, often slightly lobed leaflet, 3–10 cm. wide, and a few much reduced lateral ones; flowers 1–few; sepals lanceolate, acute, hirsute, 8–10 mm. long; petals broad, usually emarginate, 8–14 mm. long; achenes hirsute, the developed plumose style 10–15 mm. long.

Coast regions, eastern Asia—Aleutians—Vancouver Isl. Fig. 677.

G. calthifolium × *G. rossii* (*Sieversia macrantha* Kearney) occurs where the ranges of the two species overlap. Stems 1–4 dm. tall, more or less pubescent, branched above; basal leaves up to 14 cm. long; leaflets 7–13, the upper one deeply lobed, the lowermost reduced, all irregularly serrate; flowers and fruit intermediate between the parents.

4. *G. pentapetalum* (L.) Makino.   Low Avens
Sieversia pentapetala (L.) Greene

Base more or less suffruticose; leaves glabrous, crowded at the end of the branches; leaflets 5–7, cuneate or ovate-lanceolate, toothed toward the apex, 5–15 mm. long; peduncles 3–10 cm. long; bractlets shorter than the sepals; sepals ovate-lanceolate, acuminate, 6–8 mm. long; petals about 1 cm. long, very light yellow.

Japan—eastern Siberia—Aleutians. Fig. 678.

5. *G. glaciale* Adams.   Glacier Avens
Sieversia glacialis (Adams) Spreng.

Rootstocks short, thick, dark purplish-brown; basal leaves sparsely pilose above, densely so beneath with soft yellowish hairs; leaflets many, mostly 8–12 mm. long, often toothed, tipped with long hairs, the terminal one larger and lobed; stem leaves few and small; stems usually 1-flowered, 1–2 dm. tall, bractlets lanceolate, shorter than the sepals; sepals acute, 7–8 mm. long; petals rather light yellow, longer than the sepals.

Bering Sea and Arctic coasts, Lena R.—Mackenzie R. Fig. 679.

14. SANGUISORBA L.

Perennials with thick rootstocks; leaves odd-pinnate; flowers small, borne in dense spikes on long, naked peduncles; stipules adnate; leaflets toothed; hypanthium urn-shaped, angled, constricted at the mouth; sepals 4, petaloid; petals none; stamens 4–12 or more; carpels 1–3; style filiform, terminal; achenes usually 1, enclosed in the hypanthium. (Latin, blood and absorb.)

1A. Stamens scarcely or not at all exceeding the sepals, the
    filaments filiform..................................... 1. *S. officinalis*
2A. Stamens longer than the sepals, filaments flattened.
    1B. Flowers purplish................................... 2. *S. menziesii*
    2B. Flowers greenish or whitish........................ 3. *S. sitchensis*

1. *S. officinalis* L.   Official Great Burnet
S. microcephala Presl of some reports.

Glabrous, rather slender, 3–12 dm. tall; leaflets 7–13, oval or ovate, regularly serrate with gland-tipped teeth, on petiolules less than 1 cm. long, 1–6 cm. long; flowers dark purple in spikes 1–3 cm. long and about

1 cm. thick; sepals ovate, often minutely pubescent on the back; hypanthium and fruit 4-winged.

Bogs and wet soil, Bering Str.—Yukon, Eurasia. Fig. 680.

2. *S. menziesii* Rydb.   Menzies Great Burnet

Stems slender, 3–10 dm. tall; leaflets 9–15, rounded oval to ovate, 2–6 cm. long, coarsely serrate with broadly ovate teeth; petiolules 6–25 mm. long; spikes 1–3 cm. long; sepals dark purple, oval, about 2.5 mm. long; filaments 5–7 mm. long.

Southern Alaska—Wash. Fig. 681.

3. *S. sitchensis* C. A. Mey.   Sitka Great Burnet
S. latifolia (Hook.) Cov.

Leafy, 4–12 dm. tall; leaflets 7–21, ovate or elliptic, serrate with sharp-pointed teeth, cordate, 1–7 cm. long; spike dense, 2–10 cm. long, 1 cm. or more thick; flowers greenish-white, sometimes tinged with purple; sepals oval; stamens 4, long-exerted.

Wet soil, Arctic Circle—Idaho—Ore.—eastern Asia. Fig. 682.

15. SORBUS (Tourn.) L.

Trees or shrubs; ours with alternate, pinnate leaves; stipules deciduous, flowers small, perfect, regular, white, borne in terminal compound cymes; sepals 5, deciduous; styles usually 3, distinct; ovules 2 in each cell of the ovary; fruit a red berry-like pome. (The ancient Latin name for the pear or service-tree.)

1A. Tree, up to 15 m. tall..................................... 4. *S. aucuparia*
2A. Shrubs, 4 m. or less tall.
   1B. Leaflets usually 7 or 9............................. 1. *S. sambucifolia*
   2B. Leaflets usually 9 or 11............................ 2. *S. sitchensis*
   3B. Leaflets 11–15.................................... 3. *S. scopulina*

1. *S. sambucifolia* (C. & S.) Roem.   Elder-leaved Mountain Ash

1–2 m. tall; leaflets 7–11, 2–7 cm. long, lanceolate to ovate-lanceolate, acuminate, usually broadest at the asymmetrical base, the margins sharply serrate almost to the base; inflorescence round-topped, 8- to 15-flowered; flowers 10–15 mm. in diameter; sepals triangular, somewhat ciliolate, stamens about as long as the petals; styles 5; fruits ellipsoid, glaucesent, 10–15 mm. in diameter.

An east Asian species occurring in the western Aleutians. Fig. 683.

2. *S. sitchensis* Roem.   Sitka Mountain Ash

Usually about 1 m. tall in alpine situations but up to 4 m. at lower elevations; leaflets oval or oblong, 3–7 cm. long, the apex rounded or slightly acutish, the margins serrate on the upper one-third to two-thirds; inflorescence round-topped, 15- to many-flowered; flowers 6–9 mm. broad, fragrant; sepals ciliolate; top of ovary pubescent; fruit subglobose or ellipsoid, red, becoming orange and finally purplish, 8–10 mm. in diameter.

Pacific coast of Alaska—Mont.—B. C. Fig. 684.

3. **S. scopulina** Greene. Western Mountain Ash
   S. alaskana G. N. Jones not Hollick.
   S. andersonii G. N. Jones

   1–4 m. tall; leaflets elliptic or elliptic-lanceolate, acute or acuminate, serrate from near the base, 3–8 cm. long; inflorescence many-flowered; fruit bright red, subglobose, 8–10 mm. in diameter.
   Bering Sea—L. Athabasca—Black Hills—N. Mex.—Calif. Fig. 685.

4. **S. aucuparia** L. European Mountain Ash. Rowan Tree

   Leaflets 9–15, oblong-lanceolate, acute, 3–5 cm. long, upper two-thirds serrate, entire toward the base; inflorescence usually 75–100-flowered; fruit scarlet, subglobose, 9–11 mm. in diameter.
   A native of Europe but spreading rapidly from cultivation.

16. MALUS Juss.

Trees or shrubs; leaves toothed or lobed; flowers perfect, regular, showy, white or pink; flowers in small cymes; sepals 5; petals 5, rounded and clawed; styles 2–5, united at the base; ovary 2- to 5-celled, with 2 ovules in each cell; carpels papery or leathery, enclosed in the enlarged hypanthium, forming a pome usually depressed at the base. (Greek, apple.)

M. fusca (Raf.) Schneider. Western Crab-apple
   M. diversifolia (Bong.) Roem.
   Pyrus diversifolia Bong.
   Pyrus rivularis Dougl.

   A shrub or small tree, 2–5 m. tall; young growth pubescent; leaves ovate, variable, serrate, sometimes more or less lobed, glabrous above, pubescent beneath, acute, 3–8 cm. long; petals white, about 1 cm. long; calyx pubescent, not persisting in fruit; fruit usually oblong, sometimes subglobose, about 1 cm. long, acid but not astringent.
   Near the coast, southern Alaska—Calif. Fig. 686.

17. AMELANCHIER Medic.

Shrubs or trees; leaves simple; flowers racemose, white; sepals 5, reflexed, persistent; stamens many, inserted on the throat of the calyx; styles 3–5; ovary 3- to 5-celled becoming twice as many celled by intrusion of false partitions from the back; ovules solitary in each cell; fruit berry-like. (The Savoy name of the Medlar.)

Leaves about as broad as long.............................. 1. *A. alnifolia*
Leaves distinctly longer than broad........................ 2. *A. florida*

1. **A. alnifolia** Nutt. Northwestern Service-berry

   A low shrub, 1–2 m. tall; leaves thick and firm, nearly orbicular or round-oval, 2–4 cm. long, glabrous above, tomentose beneath when young; sepals densely wooly; petals oblanceolate-oblong, about 1 cm. long; fruit about 8 mm. in diameter.
   Central Alaska—Sask.—Nebr.—Colo. Fig. 687.

2. **A. florida** Lindl. Pacific Service-berry

A shrub or tree, 2-5 m. tall; leaves oblong, usually entire near the rounded base, serrulate toward the rounded apex, 2-5 cm. long; racemes 4-8 cm. long; sepals lanceolate, acute, glabrous or slightly pubescent; petals oblanceolate, 12-15 mm. long; fruit purple, juicy, 8-10 mm. in diameter.

Alaska Penin.—Oregon. Fig. 688.

18. CRATAEGUS L.

Shrubs or small trees, usually armed with spines; leaves simple, alternate, toothed, often lobed; flowers in corymbs, usually white; sepals 5; petals 5; stamens 5-25; carpels 1-5, separate; fruit a drupe-like pome containing 1-5 bony nutlets. (Greek, meaning strong, from the toughness of the wood.)

C. douglasii Lindl. Black Hawthorn

Spines 15-25 mm. long; leaves variable, doubly serrate above the cuneate base, often slightly lobed, 2-8 cm. long, glabrous beneath, pubescent above, at least on the midrib and veins; corymbs usually many-flowered; petals orbicular, 4-5 mm. long; fruit black.

Hyder—Mich.—N. Mex.—California. Fig. 689.

22. FABACEAE (Pea Family)

Herbs or woody plants; leaves mostly compounds, alternate and with stipules; flowers perfect, irregular and zygomorphic; calyx of 4 or 5 more or less united sepals; petals 5, the upper, called the standard or banner, enlarged and enclosing the others in the bud, the two lowermost united to form the keel and enclose the pistil and stamens, the two lateral form the wings; stamens usually 10, and diadelphious, 9 being united by their filaments, the other being free; ovary superior, 1- or sometimes 2-celled by intrusion of the sutures; ovules 1-many; fruit a legume, or a loment by constriction between the seeds. Members of this family are usually known as legumes.

```
1A. Leaflets 3.
   1B. Flowers in dense heads............................. 1. Trifolium
   2B. Flowers not in dense heads.
      1C. Pods rugose, ovoid................................ 2. Melilotus
      2C. Pods coiled or curved............................. 3. Medicago
2A. Leaflets more than 3.
   1B. Leaves palmately compound.
      1C. Leaflets serrulate................................. 1. Trifolium
      2C. Leaflets entire.................................... 4. Lupinus
   2B. Leaves pinnately compound.
      1C. Leaves usually with tendrils.
         1D. Styles filiform with a tuft or ring of hairs at
             the apex...................................... 8. Vicia
         2D. Styles flattened upward, hairy down inner side.... 9. Lathyrus
      2C. Leaves without tendrils.
         1D. Fruit a loment................................. 7. Hedysarum
         2D. Fruit on ordinary pod.
            1E. Keel of the corolla acute or subulate at the apex. 6. Oxytropis
            2E. Keel of the corolla obtuse at apex............... 5. Astragalus
```

TRIFOLIUM (Tourn.) L.

Herbs; leaves denticulate; flowers white, pink, purple, red, or yellow, in dense heads or spikes; calyx pedicelled, with 5 subulate teeth; corolla persistent, the wings narrow and longer than the keel; pod flattened or terete, included in the persistent corolla, 1- to 6-seeded. (Latin three, and leaf.) With the exception of No. 8 all the species are introduced and only Nos. 4, 5, and 6 are common.

```
1A. Leaves mostly 5-foliate.............................. 1. T. lupinaster
2A. Leaves trifoliate.
  1B. Heads involucrate.
    1C. Involucre cup-shaped............................ 7. T. microcephalum
    2C. Involucre rotate.
      1D. Perennial, corolla 12 mm. long................. 8. T. fimbriatum
      2D. Annual, corolla 6-8 mm. long................... 9. T. variegatum
  2B. Heads without an involucre.
    1C. Annuals, flowers yellow.
      1D. Heads 10- to 20-flowered...................... 3. T. dubium
      2D. Heads 20- to 40-flowered...................... 2. T. procumbens
    2C. Biennials or perennials.
      1D. Peduncles terminal or subterminal.............. 4. T. pratense
      2D. Peduncles axillary.
        1E. Stems prostrate, rooting at the nodes........ 5. T. repens
        2E. Stems ascending.............................. 6. T. hybridum
```

1. T. lupinaster L.   Lupine Clover

Perennial; stems erect or ascending, appressed pubescent, 3–5 dm. tall; leaflets linear-elliptic, acute, finely setose-serrulate, 2–4 cm. long; heads about 3 cm. thick; calyx pubescent, the tube about 3 mm. long, the teeth 5–8 mm. long; corolla pink, about 15 mm. long.

Escaped near Fairbanks and along the Yukon. Native of Eurasia. Fig. 690.

2. T. procumbens L.   Low Hop Clover

Stems decumbent, 15–50 cm. long; leaflets obovate, cuneate at the base, rounded, truncate or emarginate at the apex, denticulate toward the apex, 10–15 mm. long, the terminal one stalked; flowers yellow, reflexed and brown in age, the standard broad and striate.

Native of Europe.

3. T. dubium L.   Shamrock

Similar to T. procumbens but the leaflets are more distinctly cuneate, the standard narrower and only faintly striate and the whole plant more slender. This is claimed to be the true Shamrock.

Native of Europe.

4. T. pratense L.   Red Clover

Stems more or less pubescent, branching, ascending, 2–7 dm. tall; stipules strongly veined, subulate-tipped; leaflets rounded or retuse at apex, minutely denticulate, 2–5 cm. long, often with dark spot near middle; heads ovoid, usually sessile; flowers rose-red, about 12 mm. long; calyx long-hairy.

Extensively naturalized, native of Eurasia.

5. *T. repens* L.   White or Dutch Clover

Stems creeping, glabrous; leaves long-petioled; stipules small, membranous, acute; leaflets broadly obovate, more or less emarginate at the apex, 8–25 mm. long; heads long-peduncled; flowers pedicelled, 8–12 mm. long, reflexed in fruit.

Extensively naturalized, native of Europe. Fig. 691.

6. *T. hybridum* L.   Alsike Clover

Perennial; stems erect or ascending, 2–7 dm. tall; heads long-peduncled; flowers pink to nearly white, pedicelled and reflexed in fruit; calyx teeth subulate. This is a natural species and not a hybrid as the name would indicate.

Extensively naturalized, native of Europe. Fig. 692.

7. *T. microcephalum* Pursh.   Small-headed Clover

Annual, stem sparingly villous, branched from the base, 2–4 dm. long; leaflets 5–15 mm. long, obcordate or cuneate-obovate, emarginate, serrate; involucral lobes 7–10, with scarious web-like margins; heads 5–10 mm. long; calyx pubescent, corolla rose to white, 6 mm. long.

Manley Hot Springs, B. C.—Mont.—Lower Calif.

8. *T. fimbriatum* Lindl.   Coast or Cow Clover

With slender, creeping rootstocks; stems decumbent, branching from the base, 1–4 dm. long; leaflets obovate to oblanceolate, finely setose-serrulate, 10–25 mm. long; involucre about 15 mm. broad deeply and lacinately lobed; heads 2–3 cm. broad; corolla about 12 mm. long, white or light purple, the wings reddish-purple.

Loring, B. C.—Calif.

9. *T. variegatum* Nutt.   White-tipped Clover

Stems glabrous, decumbent or ascending, 2–10 dm. long; leaflets variable, the lower small, cuneate, obcordate, the upper obovate or oblong-lanceolate, 5–15 mm. long, setose-serrulate; heads 6–12 mm. broad; involucre lobed and deeply lacinate-toothed; corolla purple, white-tipped, 6–8 mm. long.

St. Michael, B. C.—Calif.

2. MELILOTUS (Tourn.) Hill

Our species sweet-scented herbs; flowers borne in spike-like racemes; calyx teeth nearly equal; pod ovoid, short and thick, indehiscent or nearly so. (Greek, honey and lotus.) Both species have become established at Fairbanks and Palmer in Alaska and at Mayo in Yukon. They are native to Eurasia.

Flowers white... 1. *M. alba*
Flowers yellow.. 2. *M. officinalis*

1. *M. alba* Desv.   White Sweet Clover

Stems erect, branched, 3–8 dm. tall; leaflets narrowly oblong- obovate, denticulate, 15–25 mm. long, narrowed at the base; flowers numerous, 4–6 mm. long; pod about 3 mm. long. Fig. 693a.

2. *M. officinalis* (L.) Lam.   Yellow Sweet Clover

Similar to the preceding; leaflets somewhat broader and more sharply denticulate. Fig. 693b.

3. MEDICAGO (Tourn.) L.

Herbs with yellow or purple flowers in axillary heads or racemes; leaflets toothed; calyx with slender nearly equal lobes; pods curved or spirally coiled, in some species spiny. (Greek, from Medea.) Our species are escapes from cultivation or introduced weeds and are not common.

```
1A. Flowers purple........................................... 1. M. sativa
2A. Flowers yellow.
  1B. Pods simply twisted................................. 2. M. falcata
  2B. Pods reniform...................................... 3. M. lupulina
  3B. Pods spirally coiled................................ 4. M. hispida
```

1. *M. sativa* L.   Alfalfa

Perennial; much branched, partly decumbent or ascending; leaflets oblanceolate, truncate or retuse and toothed at the apex, 1–3 cm. long; corolla 8–10 mm. long; pod pubescent, spirally twisted into 2 or 3 coils.

Native of Europe. Fig. 694a.

2. *M. falcata* L.   Yellow-flowered Alfalfa

Branched, decumbent or ascending perennial, 3–5 dm. tall; leaflets obovate-cuneate, toothed at the rounded, mucronate apex, 7–20 mm. long; flowers 7–10 mm. long; pod nearly straight but twisted, reticulated and finely pubescent, about 12 mm. long.

Near Fairbanks, native of Europe. Fig. 694b.

3. *M. lupulina* L.   Nonsuch. Hop Clover

Annual, branched from the base, the branches decumbent or spreading, more or less pubescent throughout; leaflets cuneate, rounded, toothed, notched, mucronulate at the apex, the nerves ending in teeth; flowers in small head-like racemes, about 3 mm. long; pod pubescent, reticulated.

Occasionally adventive in our area, native of Eurasia. Fig. 694c.

4. *M. hispida* Gaertn.   Burr Clover

Annual; stems glabrous or with a few appressed hairs, branched from the base, spreading or ascending, 2–8 dm. long; leaflets obovate or obcordate, 8–20 × 5–15 mm., crenulate; pods coiled, reticulate, armed on the edges with hooked prickles.

Native of Eurasia.

4. LUPINUS (Tourn.) L.

Our species all perennials; flowers showy, in terminal recemes; leaves 5- to 15-foliate; calyx 2-lipped, the upper lip of 2 partly and the lower of 3 partly or wholly united sepals; corolla in ours blue, rarely white, often tinted with other colors; standard broad with reflexed margins; wings curved; keel sickle-shaped; stamens monodelphous; anthers alternately oblong and roundish; pod a flat 2-valved legume. (Latin, Lupus, a wolf.)

1A. Leaves green, thinly pubescent or glabrous above.
  1B. Leaflets 9–15... 3. *L. polyphyllus*
  2B. Leaflets 5–10.
    1C. Leaflets with acute tips........................... 1. *L. arcticus*
    2C. Leaflets with rounded tips........................ 2. *L. nootkatensis*
2A. Leaflets canescent on both sides.
  1B. Flowers subsessile.................................... 4. *L. lepidus*
  2B. Flowers with pedicels 4–7 mm. long............. 5. *L. sericeus*

1. *L. arcticus* Wats.   Arctic Lupine

Stems in clumps, 2–5 dm. tall; leaflets narrowly oblanceolate or linear obovate, appressed pubescent beneath, acute, often mucronate, 2–8 cm. long; stipules subulate; flowers often shaded pink or white, 15–18 mm. long; wings and standard nearly equal; calyx villous, the upper lip gibbous, 5–6 mm. long; lower lip 7–8 mm. long; pods with brown pubescence; seed brown, mottled. A variable group from which forms have been described as species.

Bering Sea—Arctic Archipelago—B. C. Fig. 695.

2. *L. nootkatensis* Donn.   Nootka Lupine

Stems clustered, branched, varying from glabrous to densely villous, 2–10 dm. tall; leaflets 6–8, obovate or oblanceolate, 2–6 cm. long; racemes rather dense, up to 25 cm. long; flowers often shaded pink or white, rarely pure white, 13–18 mm. long; upper lip of calyx 8 mm. long, the lower lip 10 mm. long; wings and standard subequal; pod 3–4 cm. long. *L. kiskensis* C. P. Smith appears to be a very depauperate form of this species.

Mostly along the coast but extending to the Alaska Range, Attu Island to Vancouver Island. Fig. 696.

3. *L. polyphyllus* Lindl.   Large-leaved Lupine

Stems stout, 6–15 dm. tall; leaflets narrowly oblanceolate, appressed-pubescent beneath, acute, 6–12 cm. long; racemes up to 50 cm. long; calyx gibbous on upper side, upper lip 4–5 mm. long, lower lip 5–6 mm. long; corolla blue, purple or reddish-purple; wings longer than the standard, about 15 mm. long; pods densely pubescent with long brown hairs.

Seward to Mt. McKinley Park, Vancouver Island—Mont.—Calif. Fig. 697.

4. *L. lepidus* Dougl.   Prairie Lupine

Stems somewhat decumbent at base, densely silky, 15–40 cm. tall; leaves long-petioled; leaflets 5–9, oblanceolate, 12–25 mm. long, usually folded; racemes 8–16 cm. long; flowers 10–13 mm. long; pods silky, 1–2 cm. long.

South Yukon—Hyder—Idaho—Calif. Fig. 698.

5. *L. sericeus* Pursh.   Silky Lupine

Appressed silky throughout; stems erect, 3–6 dm. tall; leaflets 5–9, oblanceolate, acute, 3–8 cm. long; racemes up to 15 cm. long; flowers about 1 cm. long; pod 2–3 cm. long, yellow.

Whitehorse, B. C.—Mont.—S. Dak.—Ore. Fig. 699.

5. ASTRAGALUS (Tourn.) L.

Herbs, ours all perennial and with evident stems; leaves usually odd-pinnate; flowers violet-purple, white or yellowish, borne in spikes or racemes; calyx tubular with nearly equal teeth; petals clawed, the standard erect, the keel blunt, about equaling the wings. (The Greek name of some legume.) In addition to the species here described, forms have been collected that may represent undescribed species.

```
1A. Pod sickle-shaped.................................... 1. A. nutzotinensis
2A. Pod straight or nearly so.
  1B. Pod wholly 1-celled.
    1C. Pod compressed laterally.
      1D. Pod glabrous.................................... 2. A. tenellus
      2D. Pod with black hairs............................ 3. A. amblyodon
    2C. Pod slightly or not at all compressed.
      1D. Pod stipitate.
        1E. Pod glabrous.................................. 4. A. americanus
        2E. Pod black-hairy............................... 5. A. umbellatus
      2D. Pod sessile.
        1E. Pod more than 15 mm. long..................... 6. A. polaris
        2E. Pod less than 1 cm. long...................... 7. A. yukonis
  2B. Pods with the lower suture inflexed.
    1C. Septum incomplete.
      1D. Pod not sulcate on lower suture, both sutures prominent.
        1E. Pod compressed, nearly glabrous............... 8. A. aboriginorum
        2E. Pod more turgid, black-hairy.
          1F. Pod subsessile.............................. 9. A. eucosmus
          2F. Pod stipitate.
            1G. Corolla 8–12 mm. long....................10. A. macounii
            2G. Corolla 12–15 mm. long...................11. A. harringtonii
      2D. Pod sulcate on lower suture.
        1E. Pod erect or ascending......................12. A. williamsii
        2E. Pod drooping, stipitate......................13. A. alpinus
    2C. Septum complete or nearly so.
      1D. Partition of pod complete.....................15. A. hypoglottis
      2D. Partition of pod not completely joined with upper
          suture ......................................14. A. vicifolius
```

1. *A. nutzotinensis* Rousseau.   Sickle-pod Milk Vetch
   A. falciferous Hult.
   Gynophoraria falcata Rydb.

Stems weak and trailing, 3–20 cm. long; leaflets 9–19, obovate, elliptic or ovate, glabrate above, hirsute beneath; peduncles 1- to 4-flowered; flowers tinted lilac, 12–15 mm. long; calyx minutely black-hairy, the subulate teeth nearly as long as the tube; pod minutely black-hairy, 3–5 cm. long.

Chickaloon—Mt. McKinley Park—Yukon. Fig. 700.

2. **A tenellus** Pursh. Loose-flowered Milk Vetch
Homalobus tenellus (Pursh) Britt.

Caespitose; stems 3–5 dm. tall, sparingly strigose; leaflets 11–21, linear or oblong, obtuse at the apex, 1–2 cm. long, 1–3.5 mm. wide, glabrous on both sides or with a few hairs beneath; racemes several- to many-flowered; calyx-tube about 2 mm. long, the teeth slightly shorter; corolla ocroleuceus, 6–10 mm. long; pod stipitate, 8–10 × 3 mm., reticulate.

Yukon—lower Mackenzie—Manitoba—Colo.—Nev. Fig. 701.

3. **A. amblyodon** Kearney.
Homalobus amblyodon (Kearney) Rydb.

Stems caespitose, decumbent or prostrate, 1 dm. or less long; leaflets 5–13, oval or obovate, retuse at the apex, glabrous above, sparingly strigose beneath, 3–5 mm. long; peduncles few-flowered; calyx black-hairy, the tube about 3 mm. long, the teeth scarcely 1 mm. long; corolla, 10–12 mm. long.

Alaska Penin. and Mt. McKinley Park.

4. **A. americanus** (Hook.) M. E. Jones. Arctic Milk Vetch
Phaca americana (Hook.) Rydb.

Erect, 3–10 dm. tall, glabrous below, slightly pubescent above; leaflets oval, elliptic or oblong, obtuse, glabrous above, somewhat pubescent beneath, 2–5 cm. long; calyx about 4 mm. long, nearly glabrous, the margin ciliate, the teeth short; pod glabrous, its stipe about 5 mm. long, the body about 2 cm. long.

Central Alaska—Great Slave Lake—Que.—Wyo.—B. C. Fig. 702.

5. **A. umbellatus** Bunge. Hairy Arctic Milk Vetch
A. littoralis (Hook.) Cov. & Kearney.
Phaca littoralis (Hook.) Rydb.

Stems more or less pubescent, 5–25 cm. tall; leaflets 7 or 9, oblong to ovate, glabrous above, pubescent beneath, 12–25 mm. long; peduncles 5- to 15-flowered; flowers yellowish, about 15 mm. long; calyx 5–8 mm. long, the teeth triangular, short, pubescent; pod short-stipitate, 15–20 mm. long, covered with short, black pubescence.

Alaska and Yukon except the southeastern coast, Eurasia. Fig. 703.

6. **A. polaris** (Seem.) Benth. Polar Milk Vetch
Phaca polaris (Seem.) Rydb.

Stems slender, decumbent or creeping, 1–8 cm. long; leaflets 11–15, ovate or obovate, 3–10 × 3–5 mm., notched at the apex; racemes 1- to 5-flowered; calyx black-hairy, the teeth triangular; corolla purple, about 15 mm. long; pod minutely strigulose, inflated, membranous, 20–30 × 10–15 mm.

Cape Vancouver—Point Hope—Wiseman. Fig. 704.

7. **A. yukonis** M. E. Jones. Yukon Milk Vetch

Stems very slender, decumbent or ascending, 1–3 dm. long; leaflets 7–15, 4–12 × 1.5–3.5 mm., glabrous above, strigose beneath; flowers 7–10 mm. long, the tips light purple, calyx black-hairy, the tube about 2.5 mm. long, the subulate teeth 1.5 mm. long; pod black-hairy, 5–7 mm. long.

Central Alaska—Yukon—western Mackenzie. Fig. 705.

8. **A. aboriginorum** Rich. Indian Milk Vetch
  A. linearis (Rydb.) Pors.
  Atelophragma aboriginum (Rich.) Rydb.

Caespitose, stems erect or decumbent at the base, 15–40 cm. tall; leaflets 9–13, oblong, lance-oblong or linear, more or less villous beneath, villous to glabrate above, 8–20 mm. long; peduncles longer than the leaves; racemes short in anthesis, elongated and lax in fruit; calyx black-hairy, the teeth subulate, nearly equaling the tube; corolla 8–10 mm. long, white, tinged with violet; pods long-stipitate, glabrous when mature, the body 15–25 mm. long, acute at both ends. A variety with black-strigulose pods is var. *muriei* Hult.

Seward Penin.—the Arctic Archipelago—Black Hills—Colo.—Nev. and the Gaspe Penin., Que. Fig. 706.

9. **A. eucosmus** Robin. Pretty Milk Vetch
  Atelophragma elegans (Hook.) Rydb.

Stems glabrous or nearly so, somewhat branched, 25–55 cm. tall; leaflets usually 13 or 15, oblong or linear-oblong, 10–25 mm. long, glabrous above, strigose beneath; corolla about 8 mm. long, purple; pod usually black-hairy but sometimes white-hairy.

Seward Penin.—Great Bear Lake—Labr.—Newf.—Colo.—B. C. Fig. 707.

10. **A. macounii** Rydb. Macoun Milk Vetch
  Atelophragma collieri Rydb. in part.

Stems 3–6 dm. tall, branched, somewhat angled, glabrous or nearly so; leaflets 11–19, elliptic or ovate, 15–30 × 5–10 mm., glabrous and dark green above, paler and sparingly pilose beneath; calyx black-hairy, the teeth subulate and much shorter than the tube; corolla about 12 mm. long, nearly white; pods about 2 cm. long with a stipe about 5 mm. long, acute at each end.

Kobuk River—Great Bear Lake—Idaho—Colo. Fig. 708.

11. **A. harringtonii** Cov. & Standl. Harrington Milk Vetch

Stems branching, 2–5 dm. tall; leaflets 9–15, oblong, elliptic, or ovate, obtuse at the apex, glabrous or sparingly pubescent above, decidedly pubescent beneath, 1–4 cm. long; calyx pubescent with black hairs, the subulate teeth about the same length as the tube; pods about 15 mm. long, covered with mostly black pubescence.

Coastal region of Alaska north to Deering. Fig. 709.

12. *A. williamsii* Rydb. Williams Milk Vetch
Atelophragma williamsii Rydb.

Stems ascending or erect, 3–6 dm. tall, more or less 4-angled; leaflets 9–13, oval to linear, 15–35 × 4–12 mm., obtuse or retuse at the apex, glabrous or nearly so; racemes compact but elongating in fruit; calyx black-hairy, the tube about 3 mm. long, the teeth short; petals ochroleucous, the keel purplish-tipped; pods erect, subsessile, 10–14 mm. long, in age glabrous and reticulate, deeply sulcate on lower suture.

Central Alaska—Yukon. Fig. 710.

13. *A. alpinus* L. Alpine Milk Vetch

Ascending or decumbent, branched, 1–4 dm. tall; leaflets 11–29, ovate, elliptic or obovate, glabrate above, pilose beneath, 6–15 mm. long; flowers violet, about 12 mm. long; calyx black-hairy, the teeth triangular-subulate, nearly as long as the tube; pod stipitate, densely black-hairy. A variable and widely distributed species occurring in several forms or races.

Circumpolar, south to Vt.—Colo.—Idaho. Fig. 711.

14. *A. vicifolius* Hult. Vetch-leaved Milk Vetch

Caespitose; stems 15–50 cm. tall, angled, strigose; leaflets 13–17, linear-oblong, 10–30 × 2–6 mm., strigose pubescent beneath, usually glabrous above; racemes dense, many-flowered; calyx pubescent with black hairs usually with some white ones intermixed, the tube 4–5 mm. long, the teeth short; corolla about 15 mm. long, ochroleucous with purple tip; pod white-hairy, about 6 mm. long.

Central Alaska—Yukon. Fig. 712.

15. *A. hypoglottis* L. Purple Milk Vetch
A. agrestis Dougl.
A. tarletonis Rydb.

Caespitose; stems branched, angled, decumbent or ascending, 1–3 dm. tall; leaflets 15–25, oblong to elliptic, emarginate, 6–15 mm. long, rather densely pubescent beneath, less so above; flowers purple, in dense heads; calyx black-pubescent; pods short, sessile, densely pilose.

Yukon—Great Slave Lake—Hudson Bay—Minn.—Calif. Eurasia. Fig. 713.

6. OXYTROPIS DC.

Tufted perennial, nearly acaulescent herbs resembling *Astragalus;* leaves odd-pinnate; flowers racemose or spicate, sometimes reduced to 1; calyx teeth nearly equal; petals clawed, the keel erect, its apex mucronate, acuminate, or appendaged; pod 2-valved, 1-celled or more often 2-celled by the intrusion of the ventral suture. (Greek, sharp keel.) A very critical and confusing group, forms occurring that can scarcely be assigned to any of the following species.

1A. Leaves unifoliate or trifoliate.......................... 3. *O. mertensiana*
2A. Leaves pinnate.
  1B. Calyx lobes glandular.
    1C. Stipules long-ciliate on the margins................ 4. *O. leucantha*
    2C. Stipules also pubescent on the back.
      1D. Calyx mostly white-hairy, pods abruptly pointed.. 5. *O. viscida*
      2D. Calyx mostly black-hairy, pods more long-
        acuminate .... 6. *O. viscidula*
  2B. Calyx lobes not glandular.
    1C. Peduncles rarely more than 2-flowered.
      1D. Old stipules stiff, castaneous...................... 7. *O. kokrinensis*
      2D. Old stipules membranous and thin, light- or
        grayish-brown .... 8. *O. nigrescens*
    2C. Peduncles 2- to 5-flowered......................... 9. *O. scammaniana*
    3C. Peduncles mostly more than 5-flowered.
      1D. Flowers blue or purplish.
        1E. Pods reflexed.
          1F. Racemes strongly elongated in fruit........... 1. *O. deflexa*
          2F. Racemes short and head-like................. 2. *O. foliolosa*
        2E. Pods ascending.
          1F. Scapes 5- to 8-flowered......................10. *O. roaldi*
          2F. Scapes up to 15-flowered....................11. *O. ?erecta*
      2D. Flowers yellowish.
        1E. Old stipules dark castaneous brown.............13. *O. maydelliana*
        2E. Old stipules yellowish or light brown...........13. *O. gracilis*
3A. Leaves with verticillate leaflets (see also 11. *O. ?erecta*).
  1B. Flowers blue..14. *O. splendens*
  2B. Flowers yellowish...................................15. *O. varians*

1. *O. deflexa* (Pall.) A.DC.   Deflexed-podded Oxytrope
   O. retrorsa Fern.

Silky-pubescent, with short stems; peduncles 15–40 cm. tall; leaflets 23–45, crowded, lanceolate or ovate, silky, rounded at the base, acute at the apex, 5–15 mm. long; fruiting racemes 4–12 cm. long; flowers dingy white with bluish apex, 6–9 mm. long; calyx teeth subulate, about as long as the tube; pod pubescent with soft white or brown hairs, nearly 15 mm. long, strongly deflexed and the ventral suture deeply intruded.

Circumpolar, south to the Black Hills, N. Mex., Idaho, and B. C. Fig. 714.

2. *O. foliolosa* Hook.   Foliose Oxytrope

Resembling *O. deflexa* but aculescent; leaflets 15–29, appressed pilose, ovate, 2–10 mm. long; spike compact, 1–3 cm. long, 2- to 10-flowered; calyx campanulate, black-pilose, the lance-subulate lobes about equaling the tube; corolla deep violet, 8–10 mm. long; pod stipitate within the calyx, 10–15 mm. long, black-hirsute.

Yukon—Hudson Str.—Newf.—Colo.

3. *O. mertensiana* Turcz.   Mertens Oxytrope

Less than 1 dm. tall; leaves usually reduced to 1 leaflet but sometimes trifoliate; leaflets linear-elliptic, ciliate on the margins, acute at both ends, 15–30 mm. long; peduncles 1- to 3-flowered, more or less villous; flowers purple, about 14 mm. long; calyx black-wooly, about 15 mm. long, short-beaked.

Central and northwest Alaska, northeast Asia. Fig. 715.

4. *O. leucantha* (Pall.) Bunge.
   O. borealis Hook.
   Caespitose, hirsute and glandular; leaflets 17–25, the margins revolute, upper surface glabrous, lower surface ciliate, 5–10 mm. long; peduncles 5–30 cm. tall; heads short and compact, usually 6- to 12-flowered; calyx densely black and white long-hairy, the tube about 6 mm. long, the teeth 3–4 mm. long; corolla violet blue, about 15 mm. long; pod black-hairy.
   Eastern Asia and Bering Sea region—western Yukon. Fig. 716.

5. *O. viscida* Nutt.   Vicid Oxytrope
   Aragallus viscidus (Nutt.) Greene.
   Caespitose; leaflets up to 57 in number, usually acute, 3–18 mm. long; scapes 8–30 cm. tall, erect, hirsute; spikes 3–8 cm. long; calyx villous, the tube about 5 mm. long, the teeth 3 mm. long; corolla violet or whitish, about 12 mm. long; pod minutely pubescent, 12–15 mm. long.
   Central Alaska—Mont.—Colo.—Nev. Fig. 717.

6. *O. viscidula* (Rydb.) Tidestrom.
   Aragallus viscidulus Rydb.
   Caespitose; leaflets 17–31, 5–13 mm. long, sparingly villous and glandular; scapes 6–22 mm. tall; spikes 2–5 cm. long; calyx soft-hairy, the tube 4–5 mm. long, nearly equaled by the teeth; corolla purple with yellowish base, about 12 mm. long; pod 10–15 mm. long, finely black-hairy.
   Katmai—Wiseman—Great Slave Lake—Colo.—Utah. Fig. 718.

7. *O. kokrinensis* Pors.   Kokrines Mountains Oxytrope
   Caudices long, densely covered by long-persisting, ferrugineous stipules with attached petioles; free part of stipules silky villous, in age merely ciliate to almost glabrous, long-triangular, acute; leaves long-petioled, 3–5 cm. long with 3 or 4 pairs of revolute leaflets, long, silky-villous; scapes barely exceeding the leaves, usually 2-flowered; calyx purplish-brown, villous, the teeth subulate, one half as long as the tube; corolla purple, 10–15 mm. long; pod stipitate within the calyx, 20–25 × 6–8 mm., with short grayish-black, appressed pubescence.
   Kokrines Mountains.

8. *O. nigrescens* (Pall.) Fisch.   Blackish Oxytrope
   Densely caespitose with branching caudex a few centimeters long; leaves and peduncles silky-canescent with white hairs, leaflets 7–13, ovate to lanceolate, 3–8 mm. long; stipules scarious, the lobes lanceolate and ciliate margined; scapes 1- or 2-flowered; calyx densely black-hairy, the lobes narrow; corolla blue or purplish, 15–25 mm. long; pods 2–3 cm. long, inflated. This species is represented in our area by 2 subspecies. Ssp. *bryophila* (Greene) Hult. (*Aragallus bryophilus* Greene) has the free part of the stipules long attenuate. Ssp. *pygmaea* (Pall.) Hult. (*O. pygmaea* (Pall.) Fern.) is more pulvinate in habit and has shorter, blunt stipules.

The type form Asiatic, ssp. *bryophila* in central and western Alaska, ssp. *pygmaea* in northern and central Alaska to Hudson Bay. Fig. 719.

9. *O. scammaniana* Hult.   Scamman Oxytrope
O. arctica Am. auct.

Densely tufted; leaves 15–40 mm. long; leaflets 7–19, 4–8 × 1–3 mm., hirsute, at least beneath, often ending in a tuft of hairs; scapes 2–7 cm. tall, 2- to 5-flowered; Calyx densely black-hairy, the tube 3–4 mm. long, the teeth 2–2.5 mm. long; corolla violet, 11–15 mm. long; pod black-hairy and with incurved tip, about 15 mm. long.

Central and eastern Alaska. Fig. 720.

10. *O. roaldi* Ostenf.   Roald Oxytrope

Loosely caespitose; free part of stipules deltoid, hyaline, pubescent on back with long, appressed, white or yellowish appressed hairs; leaves up to 9 cm. long; leaflets 11–19, 3–10 mm. long, serico-villous; scapes 1 dm. or less tall, 3- to 8-flowered; calyx black-hairy; corolla purplish, about 2 cm. long.

Rare, arctic Asia, Alaska and Yukon—Victoria Island and Coronation Gulf.

11. *O. ?erecta* Kom.

Stipules hairy, the free part long-attenuate, hyaline, reticulate; leaves up to 13 cm. long, pinnate or the leaflets verticillate; leaflets acute, 12–20 mm. long, about 3 mm. wide, sparingly silky on both sides; bracts linear-setaceous, up to 13 mm. long, white-hairy; calyx with long white and short black hairs, the tube 6–8 mm. long, the teeth about 5 mm. long; corolla violet, about 2 cm. long; ovary pubescent, slightly stipitate.

North slope of Brooks Range. *O. erecta* is a Kamchatka species and this may prove to be an unnamed form.

12. *O. maydelliana* Trautv.   Maydell Oxytrope

More or less villous-pubescent throughout with rather long, white hairs except the calyx on which the hairs are black or mixed; leaflets 11–19, ovate-lanceolate, 4–16 mm. long; scapes several- to many-flowered; inflorescence a bracted spike, often head-like; flowers yellowish, 12–18 mm. long; pod generally with both black and white hairs, 12–15 mm. long, tipped with the beak-like persistent style.

Eastern Asia—arctic Alaska—Baffin Land—Labr.—Alaska Penin. Fig. 721.

13. *O. gracilis* (A. Nels.) K. Schum.   Northern Yellow Oxytrope
O. campestris Am. auct.

Leaflets 21–31, more or less appressed-silky, usually acute, 10–25 mm. long; scapes 2–4 dm. tall; spikes 4–12 cm. long; calyx silky, the tube 6–8 mm. long, the linear-subulate teeth 2.5–4 mm. long; corolla yellowish, about 15 mm. long; pod black-hirsute, often with some white hairs, about 2 cm. long.

Central and south Alaska—Alta.—Man.—S. Dak.—Idaho. Fig. 722.

14. *O. splendens* Dougl.   Showy Oxytrope
O. richardsonii Hook.

Silky-villous throughout; leaflets 20–50, in whorls of 2–4, oblong-lanceolate, 8–20 mm. long; scapes 1–3 dm. tall; calyx densely villous, the teeth narrow, less than half the length of the tube; corolla violet blue, 10–15 mm. long; pod long-villous, long-pointed, about 15 mm. long.

East Alaska—Yukon—Sask.—Minn.—Mont.—B. C. Fig. 723.

15. *O. varians* (Rydb.) Hult.   Variable Oxytrope
Aragallus varians Rydb.

Leaflets 25–50, many of them in verticels, densely silky-villous when young, 1–2 cm. long; scapes 15–40 cm. tall; spikes 4–12 cm. long; calyx silky-villous, the tube about 5 mm. long, the teeth 2–3 mm. long; corolla about 12 mm. long, yellowish; pod villous and with short black hairs, 12–15 mm. long.

Eastern and south central Alaska to Yukon—Victoria Land—Great Bear Lake. Fig. 724.

7. HEDYSARUM (Tourn.) L.

Perennials; flowers showy, in axillary racemes; calyx bracteolate and with 5 nearly equal teeth; standard obovate or obcordate, clawed; keel longer than the wings, obliquely truncate; pod flat, divided transversely into rounded or rhombic, 1 seeded internodes forming what is known as a loment.

Calyx teeth ovate, acute, shorter than the tube........ 1. *H. alpinum americanum*
Calyx teeth subulate, longer than the tube............ 2. *H. mackenzii*

1. *H. alpinum* L. ssp. *americanum* (Michx.) Fedtsch.
                                             American Hedysarum
H. boreale Nutt.
H. auriculatum Eastw.

Stems erect or ascending, glabrous or nearly so, often strigose above, 2–7 dm. tall; leaflets 9–21, variable, often mucronulate, rounded at the base, 15–30 mm. long, sparingly hairy beneath; racemes long; flowers violet purple to white, numerous, deflexed, 12–18 mm. long; pod of 3–5 strongly reticulated, mostly oval joints 6–10 mm. long. The var. *grandiflorum* Rollins (*H. truncatum* Eastw.) has flowers longer than 16 mm.

The whole species circumpolar, ssp. *americanum* south to B. C., Wyo., S. Dak., N. Hamp. and Maine. Fig. 725.

2. *H. mackenzii* Rich.   Wild Sweet Pea

Minutely pubescent or strigose; stems 3–6 dm. tall; leaflets 9–17, elliptic, 1–3 cm. long, glabrate above, grayish strigose beneath; flowers fragrant, violet purple, 17–20 mm. long; pods normally about 6-jointed, minutely strigose and cross-reticulated, the internodes nearly orbicular, 5–7 mm. long.

North Asia—Banks and Victoria Lands—Man.—Alta—Ore. Also in Que. and Newf. Fig. 726.

8. VICIA L.

Climbing or trailing herbs; leaves pinnate, bearing tendrils at the tip; flowers axillary, solitary or more often borne in racemes; calyx somewhat oblique, 5-toothed, the upper 2 shorter; standard obovate or oblong, clawed; wings adherent to the keel; style slender, with a tuft or ring of hairs at the summit; pod flat, 2-valved, dehiscent, few- to several-seeded. (The classical Latin name.)

```
1A. Flowers solitary or in pairs........................... 5. V. angustifolia
2A. Flowers in racemes.
   1B. Racemes short, 2- to 8-flowered..................... 1. V. americana
   2B. Racemes elongated, many-flowered.
      1C. Plant villous with spreading hairs................ 3. V. villosa
      2C. Glabrous or with appressed pubescence.
         1D. Flowers 9-12 mm. long........................ 4. V. cracca
         2D. Flowers more than 12 mm. long................ 2. V. gigantea
```

1. V. americana Muhl.   American Vetch

Perennial; stems 6–10 dm. long, sparsely pubescent; leaflets 8–16, nearly elliptic, cuspidate and often with a few serrations, 15–45 mm. long; calyx teeth lanceolate; corolla purple, 15–18 mm. long; pod glabrous, 3–4 cm. long.

Central Alaska—Great Slave Lake—Dela.—Va.—Mo.—Texas—Ariz.—Calif. Fig. 727.

2. V. gigantea Hook.   Sitka Vetch
V. sitchensis Bong.

A vigorous perennial; stems slightly pubescent below, more so toward the summit; leaves 15–30 cm. long including the tendrils; leaflets 14–32, ovate-oblong, obtuse or rounded and mucronulate at the apex, 18–60 mm. long, joined to the rachis by short stalk; peduncles 5- to 16-flowered; flowers purple or ochroleucous, 12–16 mm. long; pod stipitate, glaucous, blackish, about 45×15 mm.

Along the coast, Cook Inlet—central Calif. Fig. 728.

3. V. villosa Roth.   Hairy Vetch

Annual or biennial; stems villous with spreading hairs, up to 15 dm. long; leaflets linear to oblong-linear, 15–30 mm. long; flowers 15–20 mm. long; pod up to 3 mm. long.

Escaped from cultivation at Palmer. Native of Eurasia.

4. V. cracca L.   Cow Vetch

Perennial; stems slender, usually finely pubescent, up to 1 m. long; leaflets 8–24, linear to lance-oblong, 12–20 mm. long; racemes elongated and densely-flowered; flowers violet; pods glabrous, 18–24 mm. long.

Escaped at Fairbanks and Palmer. Native of Eurasia.

5. *V. angustifolia* (L.) Reich. — Narrow-leaved Vetch

Stems slender, glabrous or puberulent, 3–6 dm. long; leaflets 4–16, 8–35 × 2–4 mm.; flowers 2 or more usually 1 in the upper axils, purple; pod linear, glabrous, 25–50 × 5–7 mm.

Introduced at Sitka. Native of Europe.

9. LATHYRUS (Tourn.) L.

Ours perennial herbaceous vines with horizontal rootstocks; leaves pinnate and tendril-bearing; flowers in racemes; calyx oblique or gibbous at the base, the teeth nearly equal or the upper shorter; standard obovate, emarginate and clawed; wings oblique, adherent to the shorter keel; stamen diadelphious above, monodelphious below; style curved, hairy along the inner side; pod linear, flattened, continuous between the seeds. (Ancient Greek name of some legume.)

1A. Stipules nearly as large as the leaflets................... 1. *L. maritimus*
2A. Stipules much smaller than the leaflets.
  1B. Stems winged, leaflets narrow....................... 2. *L. palustris pilosus*
  2B. Stems simply keeled, leaflets wider..................... 3. *L. venosus*

1. *L. maritimus* (L.) Bigel. — Beach Pea
L. japonicus Willd.

Stems glabrous or in northern forms somewhat pubescent, 2–6 dm. long; stipules ovate, sagittate-hastate, acute, 2–4 cm. long; leaflets 3–6 pairs, oblong-elliptical, obtuse and mucronate at apex; tendrils often branched; flowers purple, showy, 18–25 mm. long; pod linear-oblong, 4–5 cm. in length.

Near the sea, circumpolar and widely distributed. Fig. 729.

2. *L. palustris* L. ssp. *pilosus* (Cham.) Hult. — Wild Pea

Stems slender, somewhat pubescent, angled and winged, 4–10 dm. long; leaflets 2–4 pairs, 25–50 cm. long, 3–10 mm. wide, mucronate at the apex; tendrils mostly branched; peduncles 2- to 6-flowered; corolla purple, 14–20 mm. long; pods linear, slightly pubescent, 40–60 × 6 mm.

The whole species circumboreal. the ssp. south to N. Car. and Okla. Fig. 730.

3. *L. venosus* Muhl. — Veiny Pea

Stems slender, sparingly pubescent, 5–10 dm. long; leaflets 6–14, elliptic, 2–5 cm. long; stipule small, entire; tendrils well developed; peduncles about as long as the leaves, 5- to 10-flowered; calyx pubescent, the teeth shorter than the tube; corolla 15–18 mm. long, the standard purplish, the other petals whitish; pods glabrous, linear, 4–5 cm. long.

Hyder—Sask.—Ont.—Penn.—Ga.—La.—Kans.—Mont.

PLATE XXIV

Scale marked in millimeters
FIG.
586. *Parnassia fimbriata* Konig. Petal and leaf.
584. *Sedum roseum* (L.) Scop. Fruit and leaves.
585. *Sedum stenopetalum* Pursh. Fruit and cluster of leaves.
586. *Parnassia fimbriata* Kong. Petal and 1 eaf.
587. *Parnassia palustris* L. Petal, staminodium, and leaf.
588. *Parnassia kotzebuei* C. & S. Flower and leaf.
589. *Chrysosplenium tetrandrum* Th. Fries. Fruit, bract, and leaf.
590. *Chrysosplenium wrightii* Franch. & Sav. Fruit, leaf, and base of petiole.
591. *Mitella pentandra* Hook. Flower and leaf.
592. *Mitella nuda* L. Flower and leaf.
593. *Tellima grandiflora* (Pursh) Dougl. Flower and leaf.
594. *Tolmiea menziesii* (Pursh) Torr. & Gray. Flower, fruit, and leaf.
595. *Heuchera glabra* Willd. Flower and leaf.
596. *Boykinia richardsonii* (Hook.) Gray. Flower and leaf.
597. *Tiarella trifoliata* L. Flower and leaf.
598. *Tiarella unifoliata* Hook. Fruit and leaf.
599. *Leptarrhena pyrolifolia* (D. Don) Sér. Fruit and leaf.
600. *Saxifraga bracteata* D. Don. Fruit and lower leaf.
601. *Saxifraga rivularis* L. Fruit and lower leaf.
602. *Saxifraga cernua* L. Stem bulblets and lower leaf.
603. *Saxifraga radiata* Small. Fruit and lower leaf.
604. *Saxifraga adscendens oregonensis* (Raf.) Bacigalupi. Flower and leaves.
605. *Saxifraga caespitosa sileniflora* (Sternb.) Hult. Fruit and leaf.
606. *Saxifraga hieracifolia* Wallst. & Kit. Fruit and leaf.
607. *Saxifraga nivalis* L. Fruit and leaf.
608. *Saxifraga davurica grandipetala* (Engl. & Irmscher) Hult. Flower and leaf.

FLORA OF ALASKA

PLATE XXIV

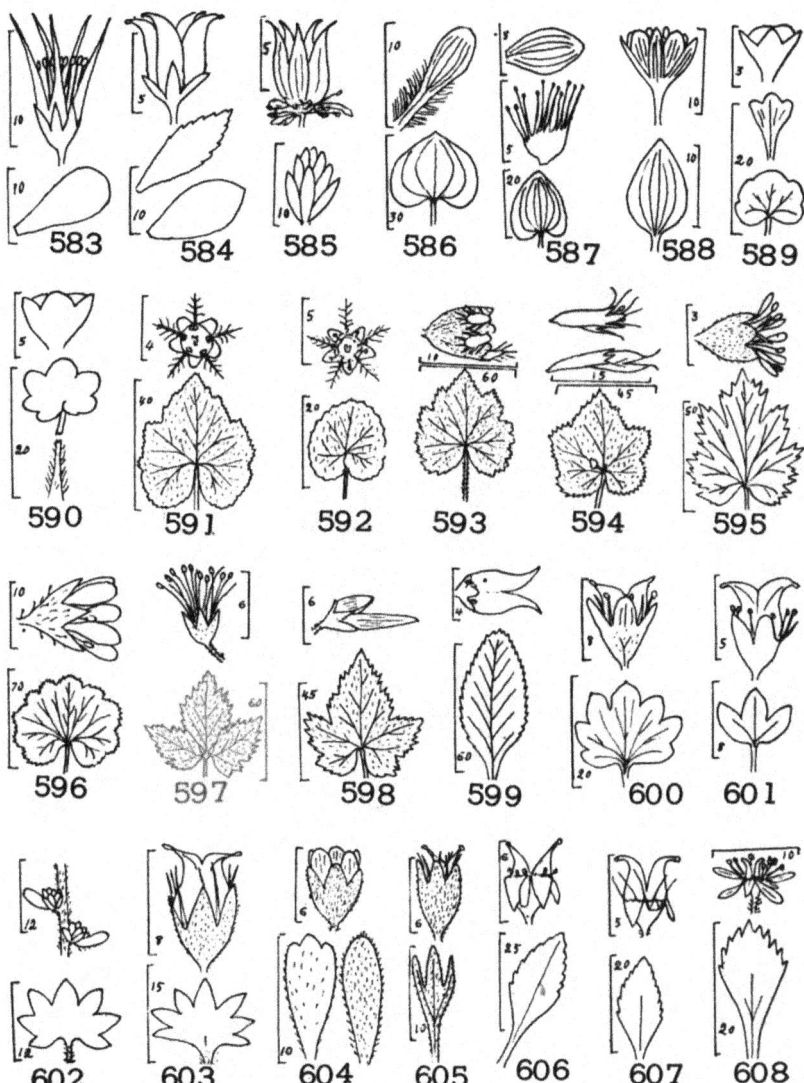

PLATE XXV

Scale marked in millimeters
FIG.
609. *Saxifraga reflexa* Hook. Fruit and leaf.
610. *Saxifraga spicata* D. Don. Fruit and leaf.
611. *Saxifraga lyallii* Engler. Fruit and leaf.
612. *Saxifraga punctata nelsoniana* (D. Don) Hult. Fruit and leaf.
613. *Saxifraga unalaschkensis* Sternb. Fruit and leaves.
614. *Saxifraga foliolosa* R. Br. Fruit, bulblet, and leaf.
615. *Saxifraga ferruginea* Grah. Fruit and leaf
616. *Saxifraga aleutica* Hult. Fruit and leaf.
617. *Saxifraga serpyllifolia* Pursh. Flowering plant.
618. *Saxifraga hirculus* L. Fruit and leaf.
619. *Saxifraga flagellaris* Willd. Flower and leaf.
620. *Saxifraga tolmiei* Torr. & Gray. Fruit and leaf.
621. *Saxifraga bronchialis funstonii* (Small) Hult. Fruit and leaf.
622. *Saxifraga tricuspidata* Retz. Fruit and leaf.
623. *Saxifraga eschscholtzii* Sternb. Flowering branch.
624. *Saxifraga nudicaulis* D. Don. Fruit and leaf.
625. *Saxifraga mertensiana* Bong. Flower and leaf.
626. *Saxifraga oppositifolia* L. Fruit and portion of stem.
627. *Ribes oxycanthoides* L. Leaf and fruit.
628. *Ribes lacustre* (Pers.) Poir. Fruit and leaf.
629. *Ribes bracteosum* Dougl. Fruit and leaf.
630. *Ribes hudsonianum* Rich. Fruit and leaf.
631. *Ribes laxiflorum* Pursh. Fruit and leaf.
632. *Ribes triste* Pall. Fruit and leaf.

FLORA OF ALASKA

PLATE XXV

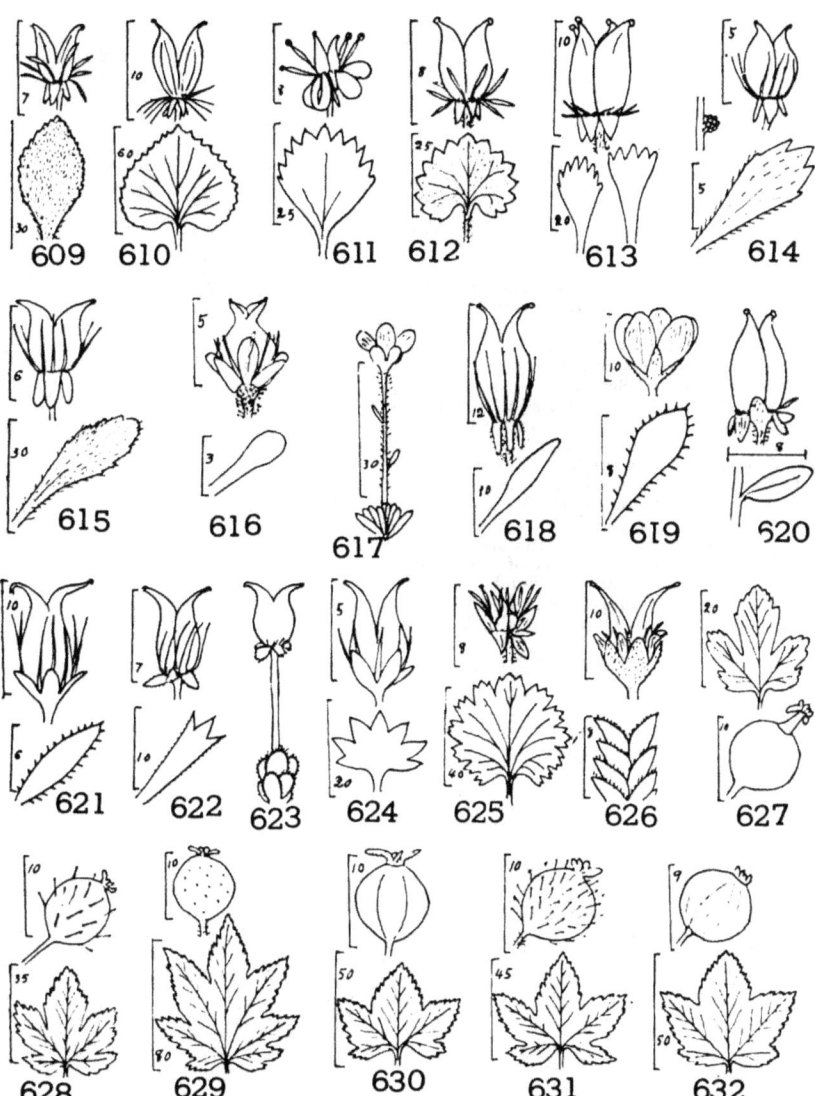

PLATE XXVI

Scale marked in millimeters
FIG.
633. *Spiraea menziesii* Hook. Leaf, flower, and petal
634. *Spiraea beauverdiana* Schneid. Leaf, fruit, and petal.
635. *Luetkea pectinata* (Pursh) Kunze. Leaf and flower.
636. *Aruncus vulgaris* Raf. Seed, fruit, and part of leaf.
637. *Rubus chamaemorus* L. Fruit, druplet, and leaf.
638. *Rubus pedatus* Smith. Fruit and leaf.
639. *Rubus stellatus* Smith. Sepal, stamen, petal, and leaf.
640. *Rubus arcticus* L. Fruit and leaf.
641. *Rubus alaskensis* Bailey. Flower and leaflet.
642. *Rubus strigosus* Michx. Flower and fruit.
643. *Rubus parviflorus* Nutt. Fruit and leaf.
644. *Rubus spectabilis* Pursh. Part of leaf and fruit.
645. *Rosa acicularis* Lindl. Part of leaf and fruit.
646. *Rosa nutkana* Presl. Fruits and part of leaf.
647. *Rosa woodsii* Lindl. Part of leaf and fruit.
648. *Fragaria chiloensis* (L.) Duch. Part of leaf and fruit.
649. *Fragaria bracteata* Heller. Leaflet, calyx with bracts, and section of petiole.
650. *Fragaria glauca* (Wats.) Rydb. Leaflet, fruit, and section of petiole.

FLORA OF ALASKA

PLATE XXVI

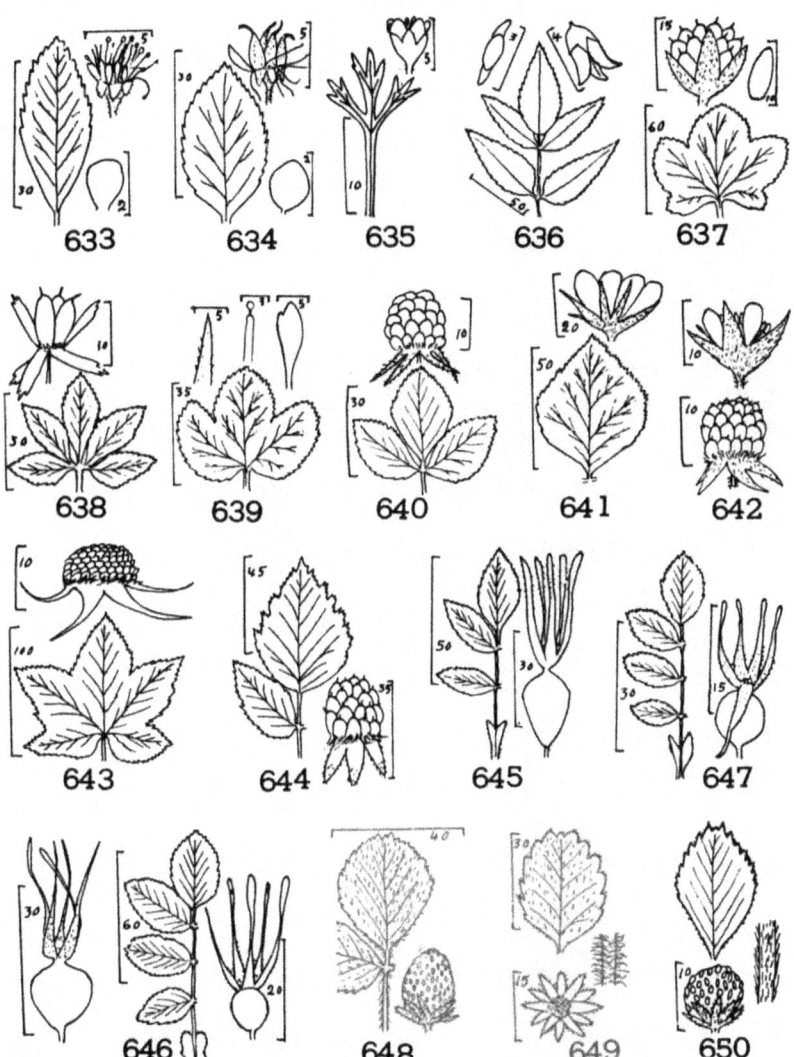

PLATE XXVII

Scale marked in millimeters
FIG.
651. *Potentilla palustris* (L.) Scop. Part of leaf, petal, and achene.
652. *Potentilla arguta* Pursh. Part of leaf and achene.
653. *Potentilla fruticosa* L. Leaf and achene.
654. *Potentilla anserina* L. Part of leaf and achene.
655. *Potentilla pacifica* Howell. Achene.
656. *Potentilla pennsylvanica* L. Leaflet, calyx with bracts, and achene.
657. *Potentilla pectinata* Fisch. Leaf, calyx with bracts, and achene.
658. *Potentilla virgulata* A. Nels. Part of leaf, calyx with bracts, and achene.
659. *Potentilla multifida* L. Part of leaf, calyx with bracts, and achene.
660. *Potentilla diversifolia* Lehm. Leaf, calyx with bracts, and achene.
661. *Potentilla gracilis* Dougl. Leaflet, achene, and calyx with bracts.
662. *Potentilla biflora* Willd. Leaf, back of flower, and achene.
663. *Potentilla emarginata* Pursh. Leaf, calyx with bracts, and achene.
664. *Potentilla monspeliensis* L. Part of leaf, stipule, calyx with bracts, and achene.
665. *Potentilla villosa* Pall. Leaf, calyx with bracts, and achene.
666. *Potentilla uniflora* Ledeb. Leaf, calyx with bracts, and petal.
667. *Potentilla vahliana* Lehm. Leaf, calyx with bracts, petal, and achene.
668. *Potentilla hookeriana* Lehm. Leaf, calyx with bracts, and achene.
669. *Potentilla nivea* L. Part of leaf, calyx with bracts, stipule and achene.
670. *Sibbaldia procumbens*. Part of leaf, flower, and achene.

FLORA OF ALASKA

PLATE XXVII

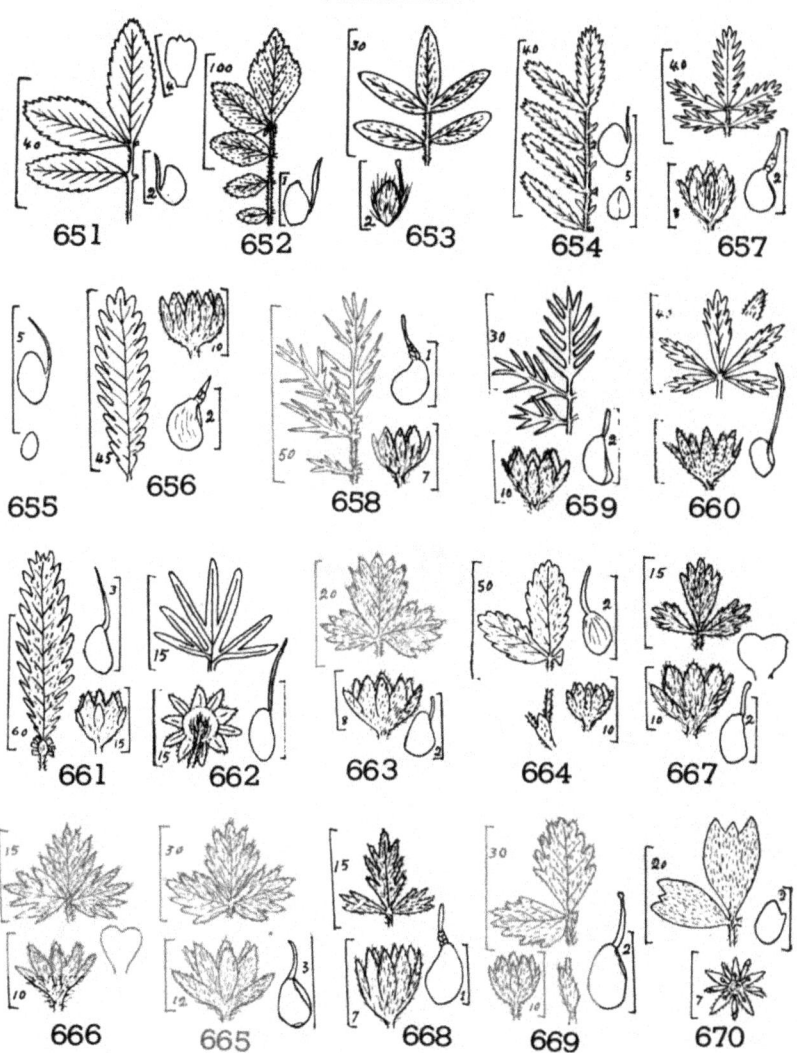

PLATE XXVIII

Scale marked in millimeters

FIG.
671. *Chamaerhodos nuttallii* (T. & G.) Pickering. Leaf, achene, and flower.
672. *Dryas drummondii* Rich. Leaf and achene.
673. *Dryas octopetala* L. Leaf and achene.
674. *Dryas integrifolia* Vahl. Leaves and achene.
675. *Geum macrophyllum* Willd. Leaf and achene.
676. *Geum rossii* (R. Br.) Sér. Leaflet and achene.
677. *Geum calthifolium* Menz. Leaf and achene.
678. *Geum pentapetalum* (L.) Makino. Leaf with stipules and achene.
679. *Geum glaciale* Adams. Leaf and achene.
680. *Sanguisorba officinalis* L. Part of leaf and flower.
681. *Sanguisorba menziesii* Rydb. Part of leaf and flower.
682. *Sanguisorba sitchensis* C.A.Mey. Part of leaf and flower.
683. *Sorbus sambucifolia* (C. & S.) Roem. Leaflet and flower.
684. *Sorbus sitchensis* Roem. Leaves.
685. *Sorbus scopulina* Greene. Leaflet and fruit.
686. *Malus fusca* (Raf.) Schneider. Leaves and fruit.
687. *Amelanchier alnifolia* Nutt. Leaf and flower.
688. *Amelanchier florida* Lindl. Leaf and fruit.
689. *Crataegus douglasii* Lindl. Leaves and fruit.
690. *Trifolium lupinaster* L. Leaf and calyx.

FLORA OF ALASKA 335

PLATE XXVIII

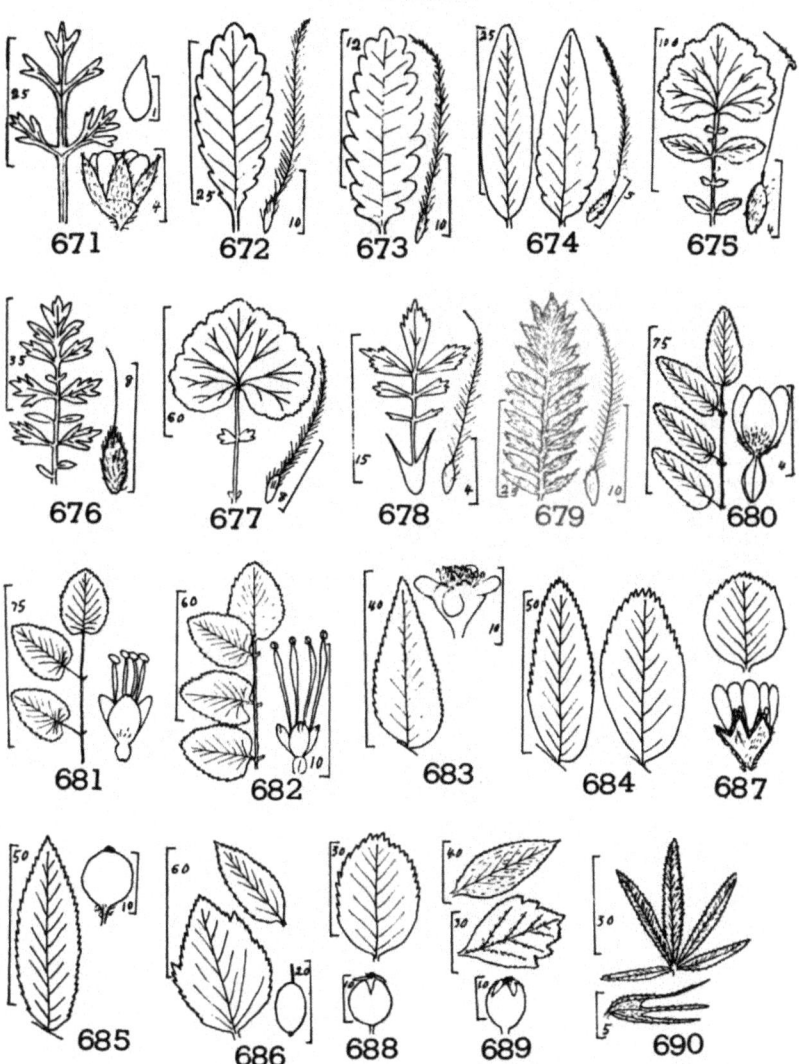

PLATE XXIX

Scale marked in millimeters

FIG.
691. *Trifolium repens* L. Leaf and flower.
692. *Trifolium hybridum* L. Leaf and flower.
693. Pods of *Melilotus*. (a) *M. alba* Desv. (b) *M. officinalis* Lam.
694. Pods of *Medicago*. (a) *M. sativa* L. (b) *M. falcata* L. (c) *M. lupulina* L. (d) *M. hispida* Gaertn.
695. *Lupinus arcticus* Wats. Flower, keel, part of leaf, and pod.
696. *Lupinus nootkatensis* Donn. Part of leaf, keel, and pod.
697. *Lupinus polyphyllus* Lindl. Leaflet, pod, and keel.
698. *Lupinus lepidus* Dougl. Leaf and keel.
699. *Lupinus sericeus* Pursh. Leaflet, flower, and keel.
700. *Astragulus nutzotenensis* Rousseau. Pod and leaf.
701. *Astragalus tenellus* Pursh. Part of leaf and pod.
702. *Astragalus americanus* (Hook.) Jones. Part of leaf and pod.
703. *Astragalus umbellatus* Bunge. Part of leaf and pod.
704. *Astragalus polaris* (Seem.) Benth. Part of leaf and pod.
705. *Astragalus yukonis* Jones. Leaflets and pod.
706. *Astragalus aboriginorum* Rich. Part of leaf and pod.
707. *Astragalus eucosmus* Robins. Part of leaf and pod.
708. *Astragalus macounii* Rydb. Part of leaf and pod.
709. *Astragalus harringtonii* Cov. & Standl. Part of leaf and pod.
710. *Astragalus williamsii* Rydb. Part of leaf and pod.

PLATE XXIX

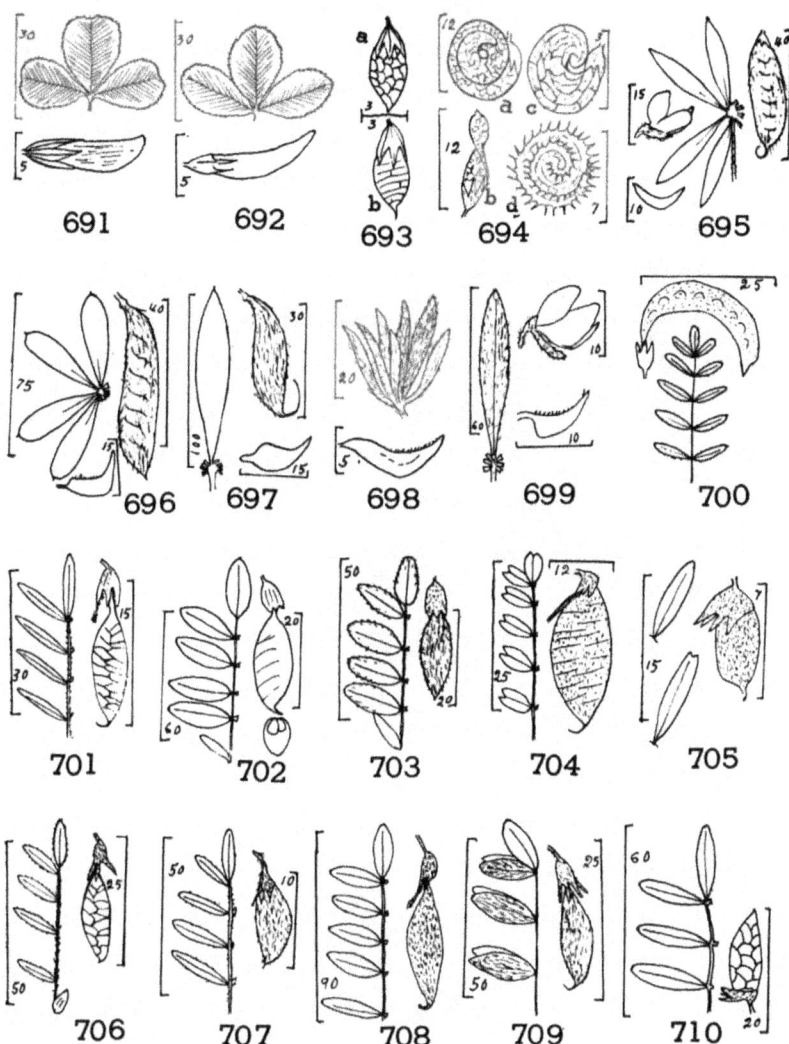

PLATE XXX

Scale marked in millimeters
FIG.
711. *Astragalus alpinus* L. Part of leaf and pod.
712. *Astragalus vicifolius* Hult. Part of leaf and pod.
713. *Astragalus hypoglottis* L. Part of leaf and pod.
714. *Oxytropis deflexa* (Pall) DC. Part of leaf, pod, and stipule.
715. *Oxytropis mertensiana* Turcz. Leaf, pod, and stipule.
716. *Oxytropis leucantha* (Pall.) Bunge. Part of leaf, pod, and stipules. These are parts illustrated in the other species of *Oxytropis*.
717. *Oxytropus vicida* Nutt.
718. *Oxytropis vicidula* (Rydb.) Tidestrom.
719. *Oxytropis nigrescens bryophila* (Greene) Hult.
720. *Oxytropis scammaniana* Hult.
721. *Oxytropus maydeliana* Trautv.
722. *Oxytropis gracilis* (A.Nels.) K. Schum.
723. *Oxytropis splendens* Dougl.
724. *Oxytropis varians* (Rydb.) Hult.
725. *Hedysarum alpinum americanum*. (Michx.) Fedtsch. Part of leaf and fruit.
726. *Hedysarum mackenzii* Rich. Part of leaf and fruit.
727. *Vicia americana* Muhl. Flower, leaflet, and stipule.
728. *Vicia gigantea* Hook. Pod, leaflet, and stipule.
729. *Lathyrus maritimus* (L.) Bigel. Leaflet, stipule, and pod.
730. *Lathyrus palustris pilosus* (Cham.) Hult.

FLORA OF ALASKA 339

PLATE XXX

23. GERANIACEAE (Geranium Family)

Herbs with stipulate leaves; flowers perfect, regular, axillary, solitary or clustered; sepals and petals usually 5 each; stamens distinct; anthers 2-celled, versatile; ovary of 5 carpels separating elastically at maturity with long styles attached to a central axis.

Carpels rounded, anthers 10.......................... 1. *Geranium*
Carpels spindle-shaped, anthers 5......................2. *Erodium*

1. GERANIUM (Tourn.) L.

Leaves palmately lobed, cleft or divided; sepals and petals imbricated; ovary 5-lobed, 5-celled, beaked; style compound; ovules 2 in each cell but the carpels 1-seeded. (Greek, a crane, from the beaked fruit.)

1A. Perennials, petals 1 cm. or more long.
  1B. Flowers bluish or rose-purple................. 1. *G. erianthum*
  2B. Flowers white............................... 2. *G. sanguineum*
2A. Annuals or biennials.
  1B. Leaves divided to the base..................... 5. *G. robertianum*
  2B. Leaves not divided entirely to the base.
    1C. Beak short-pointed, inflorescence compact..... 4. *G. carolinianum*
    2C. Beak long-pointed, inflorescence loose......... 3. *G. bicknellii*

1. *G. erianthum* DC.   Northern Geranium.

Stems appressed-pubescent, 2–8 dm. tall; leaves cordate to reniform in outline, 5- to 7-parted, the divisions lobed and toothed, pubescent on both sides or at least beneath, 5–15 cm. broad; sepals oblong, silky-pilose, ending rather abruptly in an awn 2–3 mm. long; petals 15–20 mm. long, pubescent at the base; style column 20–30 mm. long, finely villous.

Wet soil, especially alpine meadows, northeast Asia—south half of Alaska—B.C. Fig. 731.

2. *G. sanguineum* L.

Stems spreading, pubescent with widely spreading hairs, 2–4 dm. tall; leaves pentagonal in outline, 3–6 cm. broad, 3- to 7-parted, the divisions incised, ciliate-pubescent on the margins and the veins beneath; peduncles usually 1-flowered; sepals elliptic-lanceolate, ciliate on the margins and ribs, about 1 cm. long with an awn about 3 mm. long; petals white, 12–20 mm. long; style column in fruit 2 cm. or more long.

The plant here described differs from the typical form of Europe in the white flower, smaller leaves and more spreading habit. It may be a horticultural variety introduced many years ago at Sawmill Creek near Sitka where it was found by G. Turner and removed to the garden of the Alaska Pioneers Home at Sitka. Fig. 732.

3. *G. bicknellii* Britt. Bicknell Crane's-bill.

Stem erect, 4–12 dm. tall and with ascending branches; leaves pentagonal or the lower orbicular in outline, divided nearly to the base, the divisions again cleft or incised, 2–6 cm. broad; peduncles usually 2-flowered; sepals lanceolate, ending in an awn 1–2 mm. long; petioles, peduncles, pedicels and sepals glandular-pubescent; petals about same length as the sepals, rose-purple.

Central Alaska—Lake Athabasca—Newf.—N. Y.—Utah—Wash. Fig. 733.

4. *G. carolinianum* L. Carolina Crane's-bill

Somewhat resembling *G. bicknellii* but of lower, more spreading growth, 15–40 cm. tall; branches wide-spreading, especially from the base; leaves somewhat less divided than in the preceding.

An occasional introduced weed, B.C.—Ont.—Bermuda—Jamaica—Mex.—Calif.

5. *G. robertianum* L. Herb Robert.

Stem weak, extensively branching, 15–45 cm. tall; leaves thin, the divisions lobed or toothed; peduncles 2-flowered, the pedicels divaricate; sepals awn-pointed; petals red-purple, about 1 cm. long.

Escaped at Juneau, Manitoba—N. S.—N. J.—Mo.

2. ERODIUM L'Her.

Stems generally with jointed nodes; flowers in axillary umbels, nearly regular; sepals usually awn-tipped; stamens 5, alternating with 5 staminodia; style column very elongate. (Greek, a heron, from the resemblance of the fruit to its beak.)

E. cicutarium (L.) L'Her. Alfilaria.

Annual, branched, 15–40 cm. tall; leaves pinnate, the segments pinnatifid or incised; peduncles and pedicels more or less hirsute, the umbels

2- to 12-rayed; sepals 6–7 mm. long, sharp-tipped; petals slightly longer than the sepals, pink.

An occasional weed, adventive from Europe.

24. LINACEAE (Flax Family)

Herbs with alternate leaves; flowers perfect, regular, borne in racemes or panicles; stamens monodelphous; fruit a capsule opening by twice as many valves as there are carpels.

LINUM L.

Leaves narrow, sessile, entire; sepals 5, persistent; petals in ours blue or rarely white, fugaceous; stamens alternate with the petals, their filaments united at the base, each sinus with a short staminodium; styles 5; seed flat. (Classical Latin name.)

L. perenne L. ssp. *lewisii* (Pursh) Hult.   Lewis Wild Flax.
L. lewisii Pursh.

Stems 2–7 dm. tall, often several arising from a perennial woody root; leaves linear, ascending, sharply acute at apex, 1–3 cm. long; petals 15–20 mm. long.

Central Alaska—Victoria Land—Texas—northern Mex. Fig. 734.

25. BALSAMINACEAE (Jewel-weed Family)

Succulent herbs with swollen nodes; leaves simple; flowers irregular, perfect; sepals usually 3, the posterior one petaloid and strongly saccate or spurred; petals 5 or by union only 3; stamens 5, the anthers more or less united around the stigma; pod a 5-celled capsule, elastically dehiscent.

IMPATIENS L.

Lateral petals united with the posterior ones, hence 2-lobed; each cell of the capsule with few to several seeds, bursting violently when touched, which property is responsible for the name.

I. noli-tangere L.   Western Touch-me-not.
I. occidentalis Rydb.

Annual, stems light green, 6–12 dm. tall; leaves oval, thin, crenate-dentate, acuminate; lateral sepals acuminate, nerved; posterior sepal conic trumpet-shaped with curved spur, about 2 cm. long, pale yellow.

Bering Sea—central Alaska—Lake Athabasca—Wash. Also Eurasia. Fig. 735.

26. CALLITRICHACEAE (Water Starwort Family)

Small aquatic plants with capillary stems; leaves opposite, entire; flowers minute, axillary, perfect or monoecious; calyx none, but the flower usually subtended by 2 bracts; corolla none; stamen 1; anther 2-celled; styles 2, filiform; ovary 4-celled; fruit leathery, 4-lobed, 4-seeded.

CALLITRICHE L.

The only genus. (Greek, beautiful hair.)

1A. Fruit winged, leaves linear, all alike and submerged..1. *C. autumnalis*
2A. Fruit not winged, floating leaves spatulate.
  1B. Styles shorter than the fruit......................2. *C. verna*
  2B. Styles longer than the fruit........................3. *C. bolanderi*

1. *C. autumnalis* L.   Northern Water-Starwort.

Leaves clasping at the base, retuse at the apex, 10–16 mm. long; fruit 1.5–2 mm. long, nearly as wide, its lobes separated by a deep groove, broadly winged on the margins.

Circumpolar, south to N. Y.—Colo.—Ore.

2. *C. verna* L.   Vernal Water-Starwort.
    C. palustris L.

Stems usually floating, 2–30 cm. long; submerged leaves linear, 1-nerved, obtuse or emarginate at the apex, 8–25 mm. long; floating leaves ovate, usually crowded, petioled, 5–10 mm. long; fruit obovoid, about 1 mm. long and broad, slightly notched, grooved, keeled or slightly winged above.

Circumpolar, south to Florida and Calif. Fig. 736.

3. *C. bolanderi* Hegelm.   Bolander Water-Starwort.

Plants entirely aquatic; submerged leaves linear, notched at the apex, 12–50 mm. long; emersed leaves forming a rosette floating on the surface, roundish-ovate to spatulate, up to 1 cm. long; fruit obcordate, the semipersistent styles 1.25–2 times as long as the fruit. Resembles *C. verna*.

Alaska along the coast to Calif.

27. EMPETRACEAE (Crowberry Family)

Low evergreen heath-like shrubs; leaves small, narrow, channelled, nearly sessile, jointed at the base; flowers small, monoecious, dioecious or polygamous; sepals 3; petals 2 or 3 or none; stamens usually 3; anthers 2-celled; pistillate flowers with a 2- to several-celled ovary; fruit a berry-like drupe.

EMPETRUM L.

Depressed, branched, spreading shrub with densely leafy branches; flowers polygamous, axillary; sepals 3, petaloid; petals 3; stamens 3; ovary 6- to 9-celled; stigma with 6–9 toothed lobes; fruit with 6–9 nutlets. (Greek, upon a rock.)

E. nigrum L.   Crowberry.

Leaves linear-oblong, 4–7 mm. long, the groove on the lower surface caused by the revolute margins; flowers small, purplish; sepals and petals spreading; fruit globose, 4–6 mm. in diameter.

Very common throughout our area, circumboreal, south to the Great Lakes. Fig. 737.

28. ACERACEAE (Maple Family)

Trees or shrubs; leaves opposite, simple or compound; flowers perfect, polygamous, monoecious, or dioecious; sepals 4 or 5, rarely none, often colored; petals of same number, inserted on the margin of the indistinct disc, or none; ovary 2-celled with 2 ovules in each cell; fruit composed of 2 winged carpels united below.

ACER (Tourn.) L.

Leaves petioled, in ours more or less palmately cleft, lobed or parted; flowers polygamous or rarely perfect, in axillary or terminal racemes or corymbs; fruit or 2 samaras with reticulate wings. Most species in spring furnish a sweet watery sap. (Classical name.)

Mature carpels glabrous.................................1. *A. glabrum* var. *douglasii*
Mature carpels hairy...................................2. *A. macrophyllum*

1. *A. glabrum* Torr. var. *douglasii* (Hook.) Dipp.   Douglas Maple.
   A. douglasii Hook.

Shrub or small tree up to 10 m. tall; twigs purplish or red; leaves long-petioled, 3–10 cm. long and broad, 3- to 5-lobed, sharply serrate, pale underneath, the lobes sharp-pointed; flowers yellowish-green, appearing with the leaves; samaras 25–35 mm. long the wings ascending.

Southeast Alaska—Alta.—Wyo.—Ore. Fig. 738.

2. *A. macrophyllum* Pursh.   Broad-leaved Maple.

A tree 20–30 m. tall; leaves 1–3 dm. broad, cordate, deeply 5-lobed, the sinuses rounded; flowers appearing just before the leaves; sepals and petals about equal, greenish-yellow; samaras 35–45 mm. long.

Along the cost, southeast Alaska—Calif.

29. VIOLACEAE (Violet Family)

Low, usually perennial herbs (some tropical species shrubs); leaves simple, alternate or basal and with stipules; flowers perfect, irregular; sepals 5; petals 5, the lower one spurred and saccate at the base; stamens 5, the anthers united or connivent; ovary 1-celled with 3 parietal placentae; capsule loculicidal.

VIOLA (Tourn.) L.

Flowers solitary, scapose or axillary; the showy flowers produced early in the season and occasionally late in the fall; most species produce small, inconspicuous cleistogamous flowers through the summer which are far more productive of seed; sepals persistent, auricled at the base; two lower stamens with nectariferous projections extending into the spur. The margins of the leaves in all our species are more or less crenate-dentate. (The Latin name.)

1A. Plant acaulescent or nearly so.
  1B. Plant with stolons..............................7. V. *epipsila repens*
  2B. Plant not stoloniferous.
    1C. Flowers white...............................6. V. *renifolia*
    2C. Flowers violet or lilac.
      1D. Petals beardless............................8. V. *selkirkii*
      2D. Lateral petals only bearded.................9. V. *langsdorfii*
      3D. All the petals bearded......................5. V. *nephrophylla*
2A. Plants with evident stems.
  1B. Petals white on inner surface..................3. V. *rugulosa*
  2B. Petals violet.
    1C. Leaves 10–25 mm. wide, spur long............10. V. *adunca*
    2C. Leaves 2–5 cm. wide, spur short...............9. V. *langsdorfii*
  3B. Petals yellow.
    1C. Petals beardless............................4. V. *biflora*
    2C. Petals bearded.
      1D. Stems 3–10 cm. long........................2. V. *orbiculata*
      2D. Stems 10–30 cm. long.......................1. V. *glabella*

1. *V. glabella* Nutt.   Stream Violet.

Rootstock horizontal or nearly so; base leaves 1–3; stems 1–4, 2- to 4-leaved at the top; leaves reniform to broadly ovate, cordate at the base, short-pointed, nearly glabrous, 2–7 cm. wide; flowers axillary; corolla yellow with dark markings in the throat; spur saccate, short; seed dark, about 2 mm. long.

Aleutians—Mont.—Calif. Fig. 739.

2. *V. orbiculata* Geyer.   Western Round-leaved Violet.

Rootstock stout and rough; basal leaves nearly round in outline, often pointed, with scattered hairs on upper surface; stems bearing somewhat reduced leaves and both petaliferous and cleistogamous flowers; stipules brown, scarious; seeds brown, 2 mm. long.

Near Ketchikan, B. C.—Mont.—Ore.

3. *V. rugulosa* Greene.   Tall-stemmed White Violet.

Stems 2–6 dm. tall; basal leaves on long petioles, up to 1 dm. broad, broadly ovate to reniform, abruptly short-pointed, hairy beneath and sometimes on the veins above; stem leaves narrower and smaller; flowers axillary, the petals white with yellow base, often drying purplish; seeds brown, 2 mm. long. Produces underground runners.

Hot Springs, Liard River—Minn.—Iowa—Colo.—Wash.

4. *V. biflora* L.   Two-flowered Violet.

Stems several from a rather short, fleshy rootstock, 2- to 3-leaved; leaves orbicular to reniform with hairy margins, 2–4 cm. wide; sepals narrow, ciliate; flowers small.

Bering Sea across Alaska and in Colo. and Eurasia. Fig. 740.

5. *V. nephrophylla* Greene.   Northern Bog Violet.

Glabrous or nearly so; leaves ovate to reniform; petals large, bluish violet with white, bearded bases; cleistogamous flowers on erect peduncles; seeds olive brown, 2 mm. long.

Hot Springs, Liard River—Newf.—Wis.—Colo.—Calif.

6. *V. renifolia* Gray.   Kidney-leaved White Violet.

Often pubescent throughout but may be glabrate, especially on upper surface of the leaves; leaves reniform, sometimes ending in a point; petals white, all beardless, the 3 lower veined with purple; cleistogamous flowers on horizontal peduncles until the capsule ripens. Our form is the var. *brainerdii* (Greene) Fern. which has the upper surface of the leaves glabrous.

Tyonek and Watson Lake—Labr.—Newf.—Penn.—Mich.—Colo. Fig. 741.

7. *V. epipsila* Ledeb. ssp. *repens* (Turcz.) W. Bckr.

Northern Marsh Violet.

V. achyrophora Greene.
V. palustris Auct.

Spring flowers 1 or 2, arising with the 1 or 2 leaves from the end of a creeping rhizome; leaves ovate to reniform; stipules glabrous, purplish, glandular-tipped; petals violet to lavender or white, with or without beards; seed less than 2 mm. long.

The species is circumpolar. Fig. 742.

8. *V. selkirkii* Pursh.   Great-spurred Violet.

5–10 cm. tall; leaves broadly ovate-cordate, the upper surface sparsely hairy, the basal sinus narrow; petals pale violet, the spur 3–5 mm. long with enlarged end; cleistogamous flowers on ascending peduncles.

Circumboreal but of very scattered distribution. Fig. 743.

9. *V. langsdorfii* Fisch.   Alaska Violet.

Glabrous; nearly stemless early in the season but with stems a few centimeters to one decimeter or more long later; leaves variable, reniform to oval, 2–5 cm. broad; stipules sharp-pointed, sometimes toothed; style small, beaked; corolla dark violet blue with short, light spur; petals 15–22 mm. long, the lateral ones bearded; capsules 10–15 mm. long. Our most showy violet.

Northeast Asia—central Alaska—Yukon—Calif. Fig. 744.

10. *V. adunca* Smith.   Hook-spurred Violet.

Stems from a woody rootstock, short at first flowering; leaves subcordate to ovate, obtuse, more or less finely puberulent, 1–3 cm. wide; stipules narrow and pointed with setulose teeth near the base; petals light to deep violet, 12–18 mm. long; spur 5–7 mm. long, hooked or straight; capsules brown-spotted.

Widely scattered, Kodiak Isl.—Great Bear Lake—Labr.—N. B.—N. Mex.—Calif. Fig. 745.

30. ELAEAGNACEAE (Oleaster Family)

Shrubs or trees with silvery, stellate or scaly pubescence; leaves entire; flowers perfect, polygamous or dioecious, borne in axillary clusters; hypanthium enclosing the ovary and becoming berry-like in fruit; sepals 4, deciduous; corolla none; stamens 4 or 8; ovary 1-celled, 1-ovuled; fruit a drupe.

Stamens 4, leaves alternate..........................1. *Elaeagnus*
Stamens 8, leaves opposite.......................... 2. *Shepherdia*

1. ELAEAGNUS (Tourn.) L.

Shrubs or trees; flowers in axillary clusters of 1-4, perfect or polygamous; perianth constricted over the top of the ovary, the upper part campanulate, four-lobed; stamens borne on the upper part of the perianth. (Greek, sacred olive.)

E. commutata Bernh. *Silverberry.*
E. argentea Pursh.

A shrub 1-4 m. tall; leaves elliptic to ovate, 2-10 cm. long, silvery-scurfy on both sides; flowers fragrant, silvery on the outside, yellowish inside; fruit ellipsoid, silvery, 8-12 mm. long, dry and mealy, edible.
Central Alaska—Que.—Minn.—S. Dak.—Utah. Fig. 746.

2. SHEPHERDIA Nutt.

Shrubs with silvery or reddish-brown scaly or stellate pubescence; flowers small, borne in clusters at the nodes of the preceding season's twigs; perianth an 8-lobed disc, in the staminate flower the stamens alternate with the lobes of the disc. (John Shepherd was an English botanist.)

S. canadensis (L.) Nutt. Soapberry, Soopolallie.
Lepargyraea canadensis (L.) Greene.

Branching shrub, 1-2 m. tall; young twigs and buds with reddish-brown scales; leaves ovate, 15-60 mm. long, sparingly stellate above, densely stellate-pubescent below mingled with brownish scales; flowers yellow; fruit red, 5-6 mm. long. Native Indians mix the berries with sugar and water and beat it into a froth much relished by them.
Noatak—Mackenzie delta—Newf.—N. Y.—Utah—Ore. Fig. 747.

31. ONAGRACEAE (Evening-Primrose Family)

Herbs (some exotic species shrubs); leaves simple; flowers perfect, axillary or borne in terminal racemes; hypanthium sometimes elongate, enclosing the ovary; sepals and petals usually 4; stamens as many or twice as many as the petals; ovary usually 4-celled; fruit a capsule or nut-like.

Flowers 2-merous, fruit bristly..........................1. *Circaea*
Petals 4, stamens 8, fruit a many-seeded capsule........2. *Epilobium*

1. CIRCAEA L.

Low slender perennials with succulent stems; leaves opposite, dentate, petioled; flowers small, white, borne in terminal and axillary racemes; sepals 2; petals 2, notched; fruit indehiscent, 1- to 2-celled; 1- to 2-seeded. (Circe of mythology was an enchantress.)

C. *alpina* L.   Enchanter's Nightshade.

Stem 5–25 cm. tall; leaves cordate, sharply dentate, 2–5 cm. long; pedicels 3–4 mm. long, reflexed in fruit; fruit narrowly obovoid, about 2 mm. long, covered with soft hooked hairs.

Wet woods, circumpolar, central Alaska south to Ga., Iowa, and Calif. Fig. 748.

2. EPILOBIUM (Gesn.) L.

Annuals or perennials; leaves sessile or short-petioled, entire or toothed; flowers perfect, solitary, axillary or borne in spike-like racemes; sepals 4; petals 4, purple, pink or white, in one species yellow, often notched; stigma club-shaped or 4-lobed; seeds numerous, each with a silky coma. Our species are all perennial by creeping horizontal stems, in some species below the surface (sobols) and in some on the surface, the latter sometimes shortened to form rosettes of thick, fleshy leaves at or near the base of the stem. The seeds of many species are finely papillose, but it takes a strong magnification to determine this character. Hybrids seem to occur. The small-flowered section constitutes a very confusing group. (Greek, upon a pod.)

1A. Leaves all alternate, flowers large and showy.
  1B. Plant erect, high-growing...................... 1. *E. angustifolium*
  2B. Plant decumbent, lower-growing............... 2. *E. latifolium*
2A. At least some of the leaves opposite, flowers smaller.
  1B. Petals yellow................................... 3. *E. luteum*
  2B. Petals pink, purple or white.
    1C. Stigma 4-lobed.............................. 4. *E. treleaseanum*
    2C. Stigma entire.
      1D. Petals 3–5 mm. long.
        1E. Leaves narrowly linear.................... 5. *E. davuricum*
        2E. Leaves wider.
          1F. Plants less than 25 cm. tall.
            1G. Stems simple, curved.................. 6. *E. anagallidifolium*
            2G. Stems usually branched................ 7. *E. leptocarpum*
          2F. Plants usually more than 3 dm. tall.
            1G. Leaves narrowly lanceolate or linear-
                lanceolate, entire.................. 8. *E. palustre*
            2G. Leaves wider, thick, prominently
                toothed........................... 9. *E. adenocaulon*
      2D. Petals 5–10 mm. long.
        1E. Tall plants, up to 8 dm. tall................ 10. *E. glandulosum*
        2E. Plants usually 2–4 dm. tall.

1F. Plants stoloniferous.
  1G. Leaves thick, middle stem leaves sessile,
      sharply toothed...................... 11. *E. beringianum*
  2G. Leaves thin, the middle ones petioled,
      sparsely toothed...................... 12. *E. lactiflorum*
 2F. Plants soboliferus.
  1G. Stems simple......................... 13. *E. hornemannii*
  2G. Stems often branched................ 14. *E. sertulatum*

1. *E. angustifolium* L.   Fireweed.
   Chamaenerion angustifolium (L.) Scop.
   C. spicatum (Lam.) S. F. Gray.

   Stems usually simple, 5–25 dm. tall, glabrous below, puberulent above; leaves lanceolate or linear-lanceolate, paler beneath, 5–15 cm. long; flowers in terminal spike-like racemes; petals 10–18 mm. long, rose-purple or occasionally white or pink; style longer than the stamens, deeply cleft; capsule 5–8 cm. long.

   Very common, circumpolar, south to N. Car., Texas, Ariz., and Calif. Fig. 749.

2. *E. latifolium* L.   Dwarf Fireweed. Riverweed.
   Chamaenerion latifolium (L.) Sweet.

   Branched from the base, glabrate below, often canescent above, 1–5 dm. tall; leaves ovate-lanceolate, thick, pale, 2–7 cm. long, entire or with a few small teeth; inflorescence short; petals 15–25 mm. long, rose, pale purple or white; style shorter than the petals; capsule 5–8 cm. long. Favorite habitat is a sandy or gravelly deposit along streams.

   Interrupted circumpolar in distribution, south to Gaspé Peninsula, Penn., S. Dak., Colo. and Ore. Fig. 750.

3. *E. luteum* Pursh.   Yellow Willow-herb.

   Decumbent or ascending, 2–8 dm. tall, the stems terete, glabrous below, pubescent on decurrent lines above; leaves sessile, ovate-lanceolate, glandular-toothed, 3–8 cm. long; inflorescence glandular-pubescent; sepals 1 cm. or more long; petals 12–18 mm. long, style exerted, stigma 4-lobed; capsule 4–6 cm. long.

   Wet places, Aleutians—Alta.—Wash. Fig. 751.

4. *E. treleaseanum* Lévl.   Trelease Willow-herb.

   Stems about 2 dm. tall; leaves wide ovate, abruptly contracted at the base; capsule glabrous; petals pink. Resembles *E. luteum* except for flower color.

   Rare, eastern Aleutians, Shumigan Isls. and Selkirk Mts., B. C.

5. *E. davuricum* Fisch.   Davurian Willow-herb.

   Stems simple, slender, 1–4 dm. tall; leaves 8–25 mm. long, 1–3 mm. wide; flowers few; petals pale, 2–3 mm. long; capsules erect, 3–5 cm. long, nearly glabrous when mature; seeds papillose.

Wet places, Seward Penin.—Baffin Land—Newf.—Hudson Bay, and in Eurasia. Fig. 752.

6. **E. anagallidifolium** Lam. Pimpernel Willow-herb.

Stem strongly curved when young, more erect at maturity, often tufted, 5–15 cm. tall, pubescent in decurrent lines; leaves 6–25 mm. long, oval, obtuse, narrowed into a short petiole, often with a few short teeth; flowers 1–5, grouped at the top, nodding; petals lilac or rose, 4–5 mm. long; capsules erect, 2–4 cm. long. Var. *pseudo-scaposum* (Hausskn.) Hult. is a form with long-peduncled capsules.

Circumpolar, south to Maine, Colo., and Calif. Fig. 753.

7. **E. leptocarpum** Hausskn. Thin-capsuled Willow- herb.

Stems 5–25 cm. tall, usually much branched, but in var. *macounii* Trel. nearly simple; leaves lanceolate, toothed, up to 25 mm. long; flowers one to several on each branch, usually many for the entire plant; petals 3–5 mm. long; capsules 2–4 cm. long on long peduncles; seeds papillose; coma dingy.

Aleutians—Ore. and in Newf. Fig. 754.

8. **E. palustre** L. Swamp Willow-herb.

Stems 2–6 dm. tall, simple or branched, canescent above with incurved hairs; leaves opposite below, often alternate above, narrow, 3–6 cm. long, usually shorter than the internodes; petals 3–5 mm. long, white or pink; capsule 4–8 cm. long, canescent; seed about 1 mm. long.

Circumpolar, south to Dela., Colo., and Wash. Fig. 755.

9. **E. adenocaulon** Hausskn. Northern Willow-herb.

A stout, usually branched, weedy plant 3–9 dm. tall, pubescent above, glandular in the inflorescence; leaves mostly lanceolate, the middle ones short-petioled, glandular-serrate, 3–8 cm. long; petals 3–4 mm. long; capsules slender, 3–8 cm. long; seeds papillose.

Central Alaska—Gt. Slave Lake—Newf.—Penn.—Mo.—N. Mex.—Ore. Fig. 756.

10. **E. glandulosum** Lehm. Glandular Willow-herb.

Stems stout, 3–9 dm. tall, somewhat angled, glabrate below, crisp-hairy and glandular above; leaves ovate or ovate-lanceolate, dentate, acute, sessile, 4–10 cm. long; petals purplish, 5–8 mm. long; capsule pubescent, 5–8 cm. long; seeds papillose.

North Asia—central Alaska—Labr.—Newf.—Gaspé Penin.—Wyo.—Ore. Fig. 757.

11. **E. behringianum** Hausskn. Bering Willow-herb

Stems usually simple, glabrous except along the decurrent lines, often nodding at the apex, 2–4 dm. tall; leaves ovate, the lower ones petioled, up to 45 mm. long, more or less toothed; petals rose-purple, pink or white, 6-9 mm. long; capsule nearly glabrous, up to 5 cm. long; seed smooth.

Along the coast, east Asia—Aleutians—St. Matthew Isl.—southeast Alaska and in Labr. and Newf. Fig. 758.

12. *E. lactiflorum* Hausskn.   Thin-leaved Willow-herb.

Stems simple, rather slender, 15-35 cm. tall; leaves distant, thin; ascending, entire or denticulate with small distant teeth, 2-4 cm. long, the middle ones petioled; petals about 5 mm. long, usually pink in our form; capsules 3-5 cm. long; seed smooth.

East Alaska—Greenl.—N. Hamp.—Colo.—Calif. and in Europe.

13. *E. hornemannii* Rchb.   Hornemann Willow-herb.
E. bongardii Hausskn.

Variable; stems usually unbranched, 1-3 dm. tall, pubescent along the decurrent lines; leaves ovate, obtuse, denticulate, or nearly entire, the middle ones petioled, 1-5 cm. long; flowers few; petals pink or rose-purple, 4-7 mm. long; pods erect, glabrous or nearly so, 4-6 cm. long; seeds papillose to nearly smooth.

Aleutians—Nome—Yukon—Minn.—Colo.—Calif. and Labr.—Newf.—N. Hamp. Also in Eurasia. Fig. 759.

14. *E. sertulatum* Hausskn.

Stems 12-30 cm. tall, often branched, pubescent along the decurrent lines; leaves ovate, petioled, often crowded at the top of the stem, more scattered below, rather thick and subcoriaceus, denticulate, 15-40 mm. long; petals about 5 mm. long; pod 3-5 cm. long, nearly glabrous to sparsely pubescent; seeds smooth.

Kamchatka—Nome—southeast Alaska—Aleutians.

32. HALORAGIDACEAE (Water-Milfoil Family)

Aquatic or marsh plants, mostly perennials with whorled leaves; flowers perfect or monoecious, borne in the axils of the leaves, in some cases appearing spicate; sepals 2-4; petals 2-4 and small, or wanting; stamens 1-8; ovary inferior, 1- to 4-celled; angled or winged; fruit a nutlet or drupe.

Leaves entire, stamens and styles each 1............. 1. *Hippuris*
Leaves dissected, flowers 4-merous................... 2. *Myriophyllum*

1. HIPPURIS L.

Calyx adherent to the ovary and with a minute, entire limb; petals none; style filiform, lying in a groove of the anther; fruit 1-celled, 1-seeded. (Greek, horse and tail.)

1A. Leaves linear or lanceolate.
1B. Leaves in whorls of 5-6, small, delicate alpine... 1. *H. montana*

2B. Leaves in whorls of 5–12, aquatic................2. *H. vulgaris*
2A. Leaves obovate or oblanceolate...................3. *H. tetraphylla*

1. *H. montana* Ledeb. Mountain Mare's-tail.

Stems weak, 4–8 cm. tall; leaves linear, acute or mucronate, 3–8 mm. long, 1 mm. or less broad; flowers sometimes monoecious; fruit 1 mm. or less long, minutely granulate.

Wet alpine meadows, Aleutians—Wash. Fig. 760.

2. *H. vulgaris* L. Common Mare's-tail.

Stems usually partly immersed, 2–6 dm. long; leaves linear, acute, 1–2 cm. long on emersed stems, often much longer on the immersed parts of the stems; stamen with a short filament and a large anther opening by side slits; fruit ovoid, about 2 mm. long, minutely granulate.

Circumpolar, south to N. S., N. Y., N. Mex., Calif. Also in Patagonia and Tierra del Fuego. Fig. 761.

3. *H. tetraphylla* L.f Four-leaved Mare's-tail.

Stems 1–4 dm. long; leaves obovate or oblanceolate, entire, in whorls of 4–6, 8–16 mm. long; fruit rugose, about 2 mm. long.

Circumpolar, south to Gaspé Penin., Hudson Bay, B. C. Fig. 762.

2. MYRIOPHYLLUM (Vaill.) L.

Stems slender, usually floating; immersed leaves finely dissected into filiform divisions, the emersed ones entire or pectinately lobed; flowers axillary or in terminal spikes, the upper staminate, the lower pistillate, the intermediate perfect; stamens 4–8; ovary 2- to 4-celled with 1 ovule in each cavity. (Greek, myriad-leaved.)

Floral bracts verticillate.............................1. *M. spicatum*
Floral bracts alternate...............................2. *M. alterniflorum*

1. *M. spicatum* L. Spiked Water-Milfoil.

Stems 2–10 dm. long; leaves verticillate in 4's or 5's, pinnatifid into fine capillary divisions; floral leaves ovate, toothed or more often pectinately pinnatifid, 1–2 times the length of the flowers; flowers in an interrupted spike; petals 4, deciduous; fruit about 3 mm. broad, slightly broader than long; carpels rounded on the back. Most of our material belongs to the form described as *M. exalbescens* Fern. (*M. spicatum* ssp. *exalbescens* [Fern.] Hult.) in which the stems have a tendency to whiten on drying.

Circumpolar, south to Mass., Ga., Colo., and Calif. Fig. 763.

2. *M. alterniflorum* DC. Loose-flowered Water-Milfoil.

Submerged leaves in whorls of 3–5, usually less than 1 cm. long; floral leaves ovate or linear, entire or minutely toothed, smaller than the flowers.

Reported from the Buckland River and Mackenzie District. Otherwise known from Greenland to Mass. and Minn.

33. ARALIACEAE (Ginseng Family)

Aromatic herbs, shrubs or trees; leaves alternate or whorled, simple or compound; flowers regular, perfect or polygamous, inconspicuous; sepals 5; petals and stamens usually 5 each; ovary 2- to 5-celled; ovules solitary in each cavity; fruit a berry or drupe.

Herb, leaves compound.............................. 1. *Aralia*
Prickly shrub, leaves simple........................... 2. *Oplopanax*

1. ARALIA (Tourn.) L.

Perennial herbs, shrubs or trees; leaves pinnately or ternately decompound; flowers in a compound umbel in our species; calyx truncate or 5-toothed; styles 5; fruit a small berry enclosing up to 5 seeds.

A. nudicaulis L.   Wild Sarsaparilla.

Nearly acaulescent with a long rootstock; leaf 1, ternate, divisions each bearing 3–5 leaflets; leaflets acuminate, finely serrate, 5–13 cm. long; scapes shorter than the leaves, bearing a compound umbel with 3–5 primary rays; flowers greenish; fruit globose, black, 5-lobed when dry.

Woods, Hot Springs of the Liard River—Mack.—Newf.—Ga.—Colo.—Idaho—B. C.

2. OPLOPANAX Miq.

Very prickly shrubs; leaves large, palmately lobed; flowers in panicled umbels; calyx teeth nearly obsolete; petals 5, greenish; stamens 5, the filaments filiform, the anthers oblong; ovary bicarpellary; fruit flattened.

O. horridus (Sm.) Miq.   Devil's Club.
Echinopanax horridum (Sm.) Dec. & Planch.
Fatsia horrida (Sm.) B. & H.

A straggling shrub 1–5 m. tall, armed with numerous prickles; leaves orbicular in outline with prickles on the petiole and veins, 2–7 dm. wide, palmately 3- to 7-lobed, further incised, sharply and unevenly serrate, cordate at the base; inflorescence terminal, 1–3 dm. long; fruit scarlet, 5–7 mm. long.

Japan and Korea—south central Alaska—Mich.—Mont.—Ore. Fig. 764.

34. AMMIACEAE (Carrot Family)

Herbs, usually with hollow stems; leaves compound or decompound, rarely simple, the petioles dilated at the base and sheathing the stem; flowers small, perfect or polygamous, in simple or compound umbels, the umbels usually subtended by bracts forming involucres for the primary umbels and involucels for the secondary umbels; calyx adhering to the ovary, its limb 5-toothed or obsolete; petals 5; stamens 5, inserted on the epigenous disc; anthers versatile; pistils of 2 united carpels, each

1-ovuled, the 2 distinct styles borne on more or less thickened bases (stylopodia); fruit of 2 distinct carpels separating at maturity; the inner faces form the commissure; each carpel usually with 5 primary ribs and often secondary ribs between them, the space between them called the intervals. The pericarp usually has oil tubes in the intervals and on the commissural side. Some of the ribs are often winged. This family is often known as the UMBELLIFERAE. Determinations are easiest with mature fruit.

```
1A. Fruit bristly.
   1B. Fruit globose or ovoid.......,..................... 1. Sanicula
   2B. Fruit linear....................................... 2. Osmorrhiza
2A. Fruit smooth or slightly pubescent.
   1B. Leaves reduced to hollow, septate petioles....... 3. Lilaeopsis
   2B. Leaves normal.
      1C. Leaves simple, linear-lanceolate................ 4. Bupleurum
      2C. Leaves compound.
         1D. Fruit flattened dorsally (parallel to the commissure).
            1E. Flowers yellow.......................... 5. Pastinaca
            2E. Flowers white.
               1F. Leaf segments small.
                  1G. Fruit 4-6 mm. long.................... 6. Conioselinum
                  2G. Fruit 2.5-3 mm. long.................. 7. Cnidium
               2F. Leaf segments large.
                  1G. Plant pubescent...................... 8. Heracleum
                  2G. Plant glabrous or nearly so............ 9. Angelica
         2D. Fruit terete or only slightly compressed.
            1E. Stems erect or ascending.
               1F. Ribs of fruit thick and corky............ 9. Angelica
               2F. Ribs of fruit thin and acute............. 10. Ligusticum
            2E. Stems prostrate or spreading.
               1F. Plant tomentose........................11. Glehnia
               2F. Plant glabrous or nearly so.............. 12. Oenanthe
         3D. Fruit flattened laterally.
            1E. Leaves decompound........................13. Cicuta
            2E. Leaves simply pinnate....................14. Sium
```

1. SANICULA (Tourn.) L.

Glabrous perennials; leaves alternate, palmately 3- to 7-lobed; flowers yellowish, in irregularly compound few-flowered umbels; calyx teeth foliaceous, lanceolate; fruit globose or ovoid, without ribs but covered with hooked bristles. (Latin, to heal.)

S. *marylandica* L.   Black Snakeroot.

Stems 3–12 dm. tall; basal leaves large, long petioled, 3- to 5-divided to the base and the lateral divisions 2-cleft, all divisions irregularly serrate or dentate, often incised; pistillate flowers sessile, the staminate pedicelled; fruit 6–7 mm. long.

Hot Springs, Liard River—Newf.—Ga.—Colo.—Wash.

2. OSMORRHIZA Raf.

Perennials from aromatic, clustered, fleshy roots; leaves ternately decompound; leaflets ovate or lanceolate, toothed or incised; involucres and involucels small or obsolete; umbels few-rayed, long-peduncled;

calyx teeth obsolete; stylopodium conic; fruit narrow, attenuate at the base, bristly on the ribs; oil tubes obsolete. (Greek, a scent and root.)

1A. Fruit clavate...1. *O. obtusa*
2A. Fruit beaked.
  1B. Beak short, flowers purplish....................2. *O. purpurea*
  2B. Beak about 2 mm. long, flowers white or greenish.3. *O. chilense*

1. *O. obtusa* (Coult. & Rose) Fern.   Blunt-fruited Sweet-Cicely.
   Washingtonia obtusa Coult. & Rose.

Stems 2-7 dm. tall; leaves biternate or ternate-pinnate; leaflets 15-50 by 10-30 mm., rays of the umbels 2-5, divergent or reflexed, 2-5 cm. long; pedicels 2-5, divergent, 10-35 mm. long; fruit 12-17 mm. long, obtuse or abruptly acute at the apex, densely hispid at the base.
   Central Pacific coast of Alaska—Labr.—Newf.—Colo.—Ariz.—Calif. Fig. 765.

2. *O. purpurea* (Coult. & Rose) Suksd.   Sitka Sweet-Cicely.
   Washingtonia purpurea Coult. & Rose.

Stems rather slender, 2-7 dm. tall; leaves 1- to 3-ternate; leaflets lanceolate to ovate, 15-70 by 5-40 mm., acute or acuminate, serrate to incised or lobed, usually hispidulous on the veins and margins; rays of the umbel 2-6, 20-75 mm. long; pedicels 5-25 mm. long; flowers purple or greenish-purple; styles 0.5-1 mm. long; fruit 10-13 mm. long, hispid at base only.
   Kodiak along the coast to Oregon. Fig. 766.

3. *O. chilense* Hook. & Arn.   Chile Sweet-Cicely.
   Washingtonia divaricata Coult. & Rose.

Stems 3-10 dm. tall; foliage somewhat pubescent or nearly glabrous; leaflets thin, 2-8 cm. long; umbels 3- to 7-rayed; fruit strongly beaked at the top, 12-20 mm. long, densely hispid at the base.
   Aleutians—Que.—N. Hamp.—Colo.—Ariz.—Calif. and in temperate South America. Fig. 767.

3. LILAEOPSIS Greene.

Small, creeping, glabrous perennial; flowers white, in simple umbels on scapes; fruit globose, somewhat flattened laterally; lateral ribs corky-thickened. (Greek, resembling the genus *Lilaea*.)

L. occidentalis Coult. & Rose.

   Stems rooting at the nodes; leaves 2-4 cm. long, linear, terete; peduncles shorter than the leaves; umbels 5- to 12-rayed; fruit ovoid, 2 mm. long.
   Reported from southern Alaska but the report needs confirmation. Vancouver Isl.—central Calif.

4. BUPLEURUM L.

Leaves simple, entire, clasping or perfoliate; involucre present in our species; involucels of 5 or more conspicuous ovate bractlets; calyx teeth obsolete; stylopodium flat, prominent; style short; fruit oblong, flattened laterally, with slender equal ribs. (Greek, ox-ribbed, from the veining of the leaves, not evident in our species.)

B. *americanum* Coult. & Rose.   American Thorough-wort.

Perennial with a woody caudex; stems 1-3 dm. tall; basal leaves linear-lanceolate, 4-15 cm. long with parallel veins; stem leaves lanceolate and clasping; involucre and involucels prominent; flowers yellow or purplish; fruit oblong, 3-4 × 2-2.5 mm.

Bering Sea—Yukon and in south Alta., Mont., Idaho, Wyo. Fig. 768.

5. PASTINACA L.

Tall glabrous biennial; leaves pinnate, the leaflets broad; flowers yellow, in large compound umbels; fruit oval, much flattened dorsally, the lateral ribs broadly winged. (Latin, pastus, food.)

P. *sativa* L.   Common Garden Parsnip.

Root fleshy, fusiform; stems stout, 5-15 dm. tall, grooved; leaflets ovate or oval, sessile, dentate and usually lobed, 2-10 cm. long; fruit 5-7 × 4-6 mm.

Naturalized at Manly Hot Springs. Native of Europe but widely introduced as a weed.

6. CONIOSELINUM Fisch.

Tall, stout, glabrous perennials with thick roots; leaves ternate, then pinnately decompound, the leaflets lobed or toothed; flowers white, in compound umbels; involucres small or wanting; involucels composed of narrow, linear bractlets; calyx-teeth obsolete; stylopodium slightly conic; fruit flattened dorsally or nearly terete; ribs prominent. (*Conium* and *Selinum* are related genera.)

Lateral wings of the fruit much longer than the dorsal..1. *C. benthami*
All the ribs with wings nearly equal...................2. *C. cnidifolium*

1. C. *benthami* (Wats.) Fern.   Western Hemlock-Parsley.
   C. *gmelini* Coult. & Rose, not Steud.

Stems from tapering roots, 5-12 dm. tall; glaucous below but strigose in the inflorescence; ultimate leaf segments variable but broader than in *C. cnidifolium;* bractlets linear-subulate, longer than the pedicels; fruit 5-6 mm. long. A low form in the Bering Sea region is only 1-3 dm. tall. May not be specifically distinct from *C. chinense* (L.) B.S.P.

Along the coast, east Asia—Point Hope and the Aleutians to Ore. Fig. 769.

2. C. cnidifolium (Turcz.) Pors.   Dawson Hemlock-Parsley.
   C. dawsonii Coult. & Rose.

Stems 4–10 dm. tall; ultimate leaf segments small and narrow with acute tips; bracts with foliose divided tips, deciduous; bractlets longer than the pedicels, ending in a long attenuation; fruit 4–5 mm. long.

Siberia across Alaska to Mackenzie. Fig. 770.

7. CNIDIUM Cusson.

Stems from slender taproots, slender, erect and branching; leaves pinnately dissected; petioles sheathing; inflorescence of loose compound umbels; involucre usually wanting; involucels of several slender bractlets; rays numerous; flowers white; petals obovate with inflexed tips; fruit ovoid, slightly flattened dorsally; ribs prominently corky-winged.

C. ajanense (Reg. & Tiling) Drude.

Leaves resembling Conioselinum but less complex and ovate in outline; lateral wings of the carpels markedly longer than the dorsal; involucels usually shorter than the pedicels.

Central Yukon River district and eastern Asia.

8. HERACLEUM L.

Tall, stout, leafy-stemmed perennial; leaves large, ternately compound; leaflets large and broad; flowers white, borne in large compound umbels; calyx teeth small or obsolete; stylopodium conic; fruit flattened dorsally, obovate, the lateral ribs with broad wings. (Named for Hercules of mythology.)

H. lanatum Michx.   Cow Parsnip.

Very stout, 10–25 dm. tall, tomentose-pubescent; leaflets 1–3 dm. broad, stalked, palmately cleft and incised; base of petiole much dilated and wooly; bractlets subulate; fruit ovate or obcordate with conspicuous oil tubes, 9–12 mm. long.

East Asia—Alaska—Lake Athabaska—Labr. — Ga. — Ariz. — Calif. Fig. 771.

9. ANGELICA L.

Ours stout, fistulose perennials from stout taproots; leaves ternately-pinnately compound, the leaflets broad, sometimes lobed; inflorescence of large, compound umbels; flowers white, pinkish or greenish; calyx teeth minute or obsolete; fruit somewhat flattened dorsally. (Named for supposed medicinal virtues.)

Oil tubes numerous, seed free in the pericarp at maturity. 1. A. lucida
Oil tubes few, seed adhering to the pericarp.............. 2. A. genuflexa

1. A. lucida L.   Sea Coast Angelica.
   Coelopleurum gmelini (DC.) Ledeb.

Stems leafy, 5–12 dm. tall; leaves mostly tri-ternate, the petioles with much-inflated bases; leaflets rather thick, mostly ovate, coarsely and unevenly serrate, 3–8 cm. long; rays up to 50 or more; fruit ellipsoid, 7–9 mm. long; pedicels 8–16 mm. long.

East Asia across Alaska and Yukon to Labr.—N. Y.—Calif. Fig. 772.

2. *A. genuflexa* Nutt. Bent-leaved Angelica.

Stems 4–18 dm. tall, glabrous below the inflorescence; leaves ternate or biternate, the divisions pinnate, the primary divisions usually strongly deflexed; leaflets ovate or lanceolate, acuminate, irregularly but sharply serrate, 3–8 cm. long; fruit oblong, 4–5 mm. long.

Mostly along the coast, east Asia to Calif. Fig. 773.

10. LIGUSTICUM L.

Glabrous perennials; bracts often deciduous; bractlets narrow; stylopodium conical; fruit oblong or ellipsoid, only slightly flattened laterally; ribs all prominent and nearly equal. (Named for Liguria in Italy.)

Low alpine plant, leaves pinnate.....................1. *L. mutellinoides*
Tall marine plant, leaves biternate....................2. *L. hultenii*

1. *L. mutellinoides* (Crantz) Willar ssp. *alpinum* (Ledeb.) Thellung.
   L. macounii Coult. & Rose.
   Podistera macounii Mathias & Constance.

Leaves 4–10 cm. long including the petioles; leaflets 3–7, broadly ovate, 2- to 3-lobed, again cleft or toothed, 3–15 mm. long; flowers yellowish-green, in few-rayed umbels; bracts and bractlets rather narrow and only occasionally toothed; pedicels very short; fruit ovoid, 3–4 mm. long.

Eurasia, in Alaska from Bering Sea to Eagle. Fig. 774.

2. *L. hultenii* Fern. Hultén Sea Lovage.
   L. scoticum of reports.

Stems more or less branched, 2–7 dm. tall; leaves mostly biternate, thick; leaflets broadly ovate, 15–60 mm. long, coarsely serrate; inflorescence glabrous; flowers white or pinkish; rays 2–5 cm. long; pedicels 5–10 mm. long; fruit oblong, 6–10 mm. long. Closely related to *L. scoticum* of the Atlantic coasts and may be only a geographic race of that species.

East Asia and the coasts of Alaska south to Vancouver Island. Fig. 775.

11. GLEHNIA Schmidt.

Low, spreading or prostrate, subcaulescent, pubescent perennials; leaves coriaceous, once or twice ternate or ternate-pinnate; leaflets oblong-ovate or cuneate with crenate-dentate margins; flowers white, the calyx teeth inconspicuous; fruit globose to ovoid-oblong, the ribs all corky-winged, the wings broadest at the base.

G. *littoralis* Schmidt ssp. *leiocarpa* (Mathias) Hult.
G. *leiocarpa* Mathias.

Leaflets 5–50 × 4–30 mm., hirtellous on the rachis and nerves above, tomentose beneath; inflorescence densely villous, usually shorter than the leaves; fruit 4–12 mm. long, nearly glabrous.

Port Hobron to Calif., the species in east Asia.

12. OENANTHE L.

Glabrous aquatic or marsh plants; leaves bipinnate or ternate-pinnate; flowers white, borne in compound umbels; calyx lobes evident; stylopodium conical or hemispherical; petals lobed or with an inflexed point; fruit ellipsoidal, terete, or slightly flattened laterally; oil tube solitary in the intervals, 2 on the commissural side. (Greek, wine and flower.)

O. *sarmentosa* Presl.   Water Parsley.

Stems weak and reclining, 5–10 dm. long; leaflets ovate or lanceolate in outline, 1–5 cm. long, deeply toothed; rays 4-angled; bracts few; bractlets many, narrow; fruit short-pedicelled, 2.5–3.5 mm. long.

Along the coast, south Alaska—central Calif. Fig. 776.

13. CICUTA L.

Tall, stout, glabrous or glaucous, poisonous perennials with short, more or less chambered rootstocks; leaves pinnate or pinnately compound; leaflets serrate; flowers white, borne in large, compound umbels; bracts few or none; bractlets several, slender; calyx teeth prominent, acute; stylopodium low; fruit flattened laterally; ribs corky, the lateral strongest; oil-tubes solitary in the intervals, 2 on the commissural side. The poison is largely concentrated in the rootstocks and seed. (The ancient Latin name.)

1A. Fruit oblong, longer than wide.................... 1. *C. maculata*
2A. Fruit orbicular, leaflets ovate to lanceolate........ 2. *C. douglasii*
3A. Fruit shorter than wide, leaflets linear or linear-
    lanceolate.. 3. *C. mackenziana*

1. *C. maculata* L.   Spotted Water Hemlock.

Stems 10–25 dm. tall; leaves bipinnate; leaflets sharply serrate, 3–8 cm. × 5–20 mm.; fruit about 3.25 × 2.75 mm., not constricted at the commissure; pedicels 5–15 mm. long.

Central Alaska—Que.—N. Car.—Texas. Fig. 777.

2. *C. douglasii* (DC.) Coult. & Rose.   Western Water Hemlock.

Stems 8–20 dm. tall; leaves usually bipinnate; leaflets ovate or lanceolate, deeply serrate to incised, 3–10 cm. long, veins prominent beneath; fruit 2–2.5 mm. long and wide, constricted at the commissure.

South Alaska—Alta.—Mont.—N. Mex.—Calif. Fig. 778.

3. **C. mackenzieana** Raup   Mackenzie Water Hemlock

Stems 5–15 dm. tall; leaves once to thrice pinnate; leaflets linear-lanceolate, 3–10 cm. × 2–10 mm., remotely serrate with forward-pointing teeth; fruit about 2 mm. long, 2–2.5 mm. wide, constricted at the commissure.

Bering Sea—Mackenzie. Fig. 779.

14. SIUM (Tourn.) L.

Perennial marsh plants; leaves pinnate; leaflets serrate or pinnatifid; flowers in large compound umbels; involucre and involucels of numerous narrow bracts and bractlets; fruit oval in outline, glabrous, with prominent and corky ribs. (Greek name of a marsh plant.)

S. suave Walt.   Hemlock Water Parsnip.
S. cicutaefolium Schrank.

Stems stout, 6–15 dm. tall; leaflets linear or lanceolate, sharply serrate, 3–10 cm. long, or if growing in water more or less dissected: fruit ovate, about 3 mm. long.

Collected at Galena, east Asia—Newf.—Va.—central Calif.—B. C. Fig. 780.

35. CORNACEAE (Dogwood Family)

Herbs, shrubs or trees; leaves simple, alternate, opposite or whorled, usually entire; flowers perfect or unisexual, usually borne in cymes or heads; calyx adherent to the ovary, the flowers 4- or 5-merous; fruit a drupe, the stone 1- or 2-celled, 1- or 2-seeded.

CORNUS (Tourn.) L.

Flowers perfect, small, white or purplish; calyx small, 4-toothed; fruit a white or red drupe. (Greek, horn, from the toughness of the wood of some species.)

1A. Shrub, flowers in cymes........................... 3. *C. stolonifera*
2A. Perennial herbs, flowers in heads subtended by white petaloid involucral bracts.
  1B. Leaves whorled at the summit of the stem....... 1. *C. canadensis*
  2B. Leaves opposite................................2. *C. suecica*

1. **C. canadensis** L.   Bunchberry.

Stems 1–3 dm. tall from creeping rootstocks, with a whorl of 6 leaves at the summit and occasionally 1 or 2 pairs of smaller leaves or bracts below; leaves ovate or oval, acute at both ends, 3–7 cm. long, the two opposite ones being larger and broader than the intermediate ones; floral bracts usually 4, white or sometimes blotched with red; petals white or purplish, one of them bristly-tipped; fruit a bunch of orange-red drupes; stone globose.

Very common, east Asia and all the northern part of North America south to Va. and Calif. Fig. 781. Where the range of this species overlaps the range of the next, hybrids are found. These are intermediate between the two species and were described as *C. unalaskensis* Ledeb.

2. *C. suecica* L. Lapland Cornel.

Leaves usually 3 pairs below the inflorescence, 2- to 6-leaved branches later arising on either side of the peduncle; leaves smaller than in *C. canadensis*, 5- to 7-veined; floral bracts usually 4, ovate; petals dark purple; drupes globose or ovoid, rose-red; stone slightly flattened and channeled on both sides.

Distribution interrupted circumpolar south to Que. and Calif. Fig. 782.

3. *C. stolonifera* Michx. Red Osier Dogwood.
C. stolonifera var. *baileyi* (Coult. & Evans) Drescher.
Svida instolonea A. Nels.

A branching shrub 1-3 m. tall; young branches and inflorescence appressed-pubescent; leaves only slightly paler beneath, thin, oval, ovate or elliptic, entire, acute or acuminate, strigose on both sides; corolla white; petals about 3 mm. long; fruit white; stone flattened, about 5 mm. long.

Central Alaska—Labr.—Newf.—Va.—Mexico—Calif. Fig. 783.

36. PYROLACEAE (Wintergreen Family)

Rather low, evergreen perennials; leaves thick and leathery, usually clustered at the base of the stems; flowers perfect, often slightly irregular; sepals 4 or 5, persistent; corolla of 4 or 5 wax-like petals; stamens twice as many as the petals; ovary superior, 4- to 5-celled; styles united; stigmas 5-lobed; capsule loculicidal with many minute seeds.

1A. Stems leafy, style very short..................... 1. *Chimaphila*
2A. Stems scapose, styles evident.
  1B. Flowers solitary................................ 2. *Moneses*
  2B. Flowers borne in racemes...................... 3. *Pyrola*

1. CHIMAPHILA Pursh.

Stems decumbent with ascending leafy branches; leaves opposite or whorled, thick and shining; flowers borne in terminal corymbs; petals 5, orbicular, concave; capsule erect, globose, 5-celled. (Greek, winter-loving, from the evergreen leaves.)

C. umbellata (L.) Bart. ssp. *occidentalis* (Rydb.) Hult.
Pipsissewa, Prince's Pine.

Stems 1-2 dm. tall; leaves whorled, oblanceolate, cuneate at the base, rounded or acute at the apex, sharply serrate, 3-7 cm. long; flowers 3-7; petals reddish; capsule depressed-globose, 5-6 mm. in diameter.

Southeast Alaska—N. S.—Ga.—Mex.—Calif. Fig. 784.

2. MONESES Salisb.

Low glabrous perennial; rootstock slender; leaves coriaceous, serrate, crowded at the end of the stem; flower nodding at the end of a long peduncle; petals white or tinted rose; ovary globose; stigma peltate, usually 5-lobed; capsule subglobose. (Greek, single-delight.)

M. *uniflora* (L.) Gray.   Single Delight. Wax Flower.
Pyrola uniflora L.

Peduncles 4–12 cm. long; leaf blades 5–15 mm. long, crenate, rounded at the tip, rounded or cuneate at the base; sepals ovate, ciliolate, about 3 mm. long; petals ovate, 6–9 mm. long; capsule 6–8 mm. in diameter. A robust form with leaves 8–25 mm. in diameter has been described as *M. reticulata* Nutt. It may be regarded as a variety.

Woods, circumboreal south to Penn.—Colo.—Calif. Fig. 785.

3. PYROLA (Tourn.) L.

Glabrous perennials with stoloniferous rootstocks; leaves thick, mostly basal; flowers in racemes, nodding; sepals 5, petals 5, concave, deciduous, spreading or connivent; anthers erect in bud, emarginate or 2-beaked at the base and generally reversed at flowering time; ovary 5-celled; stigma 5-lobed; fruit a 5-lobed capsule opening from the base. (Latin, diminutive of Pyrus, the pear, in reference to the leaves.)

```
1A. Style curving, stigma narrower than the style.
    1B. Sepals little if at all longer than broad.........1. P. chlorantha
    2B. Sepals much longer than broad.
        1C. Petals white or greenish....................2. P. grandiflora
        2C. Petals pink to purple.......................3. P. asarifolia
2A. Style straight; stigma capitate, broader.
    1B. Style included..................................4. P. minor
    2B. Style exerted..................................5. P. secunda
```

1. *P. chlorantha* Swartz.   Greenish-flowered Wintergreen.

Scapes 1–3 dm. tall, 3- to 10-flowered; leaves obscurely crenulate or entire, orbicular or broadly oval, rounded at both ends or sometimes mucronate at the apex, 1–3 cm. long; sepals triangular-ovate, about 1.5 mm. long; petals greenish-white, about 6 mm. long; capsule about 7 mm. in diameter.

Woods, circumpolar, south to D. C., Ariz., and Calif. Fig. 786.

2. *P. grandiflora* Radius.   Large-flowered Wintergreen.
P. borealis Rydb.
P. gormanii Rydb.
P. occidentalis Rydb.

Scapes 8–20 cm. tall; leaves orbicular or oval with light-colored veins, 15–50 mm. long; flowers 5–10, 15–22 mm. wide; sepals usually pinkish, about 3 mm. long; petals whitish 6–9 mm. long; fruit about 8 mm. in diameter.

Circumpolar, high arctic to Kenai Penin. and Quebec. Fig. 787.

3. *P. asarifolia* Michx.   Liver-leaf Wintergreen.
P. uliginosa Torr.

Scapes 15–40 cm. tall; leaves oval, orbicular or reniform, sometimes broader than long, crenulate, shining, 2–7 cm. long; sepals acute or acuminate, about 3 mm. long; petals pink to purplish, oval, about 7 mm. long; capsules about 6 mm. in diameter. The typical form has leaves somewhat cordate at the base, the more common var. *incarnata* (DC.) Fern. has leaves with rounded or slightly cuneate bases.

Woods, Siberia and Japan—Yukon—N. S.—Mass.—Mich.—S. Dak.—N. Mex.—Calif. Fig. 788.

4. *P. minor* L.   Lesser Wintergreen.
Erxlebania minor (L.) Rydb.

Scapes 5–20 cm. tall; leaves orbicular or oval, slightly crenulate, obtuse or mucronate at the apex, rounded at the base, 15–40 × 10–30 mm.; sepals triangular-ovate, about 1.5 mm. long; petals white or more usually pink, orbicular, 4–5 mm. long; style short, included; capsule about 5 mm. in diameter.

Woods, circumpolar, south to Aleutians, Conn., Colo., and Calif. Fig. 789.

5. *P. secunda* L.   One-sided Wintergreen.
Ramischia secunda (L.) Gercke.

Scapes usually several from a much-branched rootstock, 8–20 cm. tall; leaves elliptical, 15–35 mm. long, crenulate, acute or mucronate at the tip, rounded or narrowed at the base; flowers greenish-white in a one-sided raceme; pedicels short; calyx lobes triangular, obtuse, very short; petals oval, about 4 mm. long; capsule subglobose, about 4 mm. long. Var. *obtusata* Turcz. is a smaller growing form of central and northern Alaska.

Woods, circumpolar, south to N. J., N. Mex., and Calif. Fig. 790.

37. MONOTROPACEAE (Indian Pipe Family)

Saprophytes growing in humus or root parasites without chlorophyll; leaves reduced to scales; flowers perfect, usually drooping; calyx 2- to 6-parted; sepals erect, deciduous; petals distinct or partly united; stamens 6–12, anthers 2-celled or confluently 1-celled; ovary superior. 4- to 6-lobed, 1- to 6-celled; fruit a 1- to 6-celled loculicidal capsule with numerous seeds.

Flower solitary; stigma naked........................ 1. *Monotropa*
Flowers racemose; stigma hairy on the margin......... 2. *Hypopitys*

1. MONOTROPA L.

Succulent white, yellowish or reddish plants; flower nodding, but the capsule erect; sepals 2–4; petals 5 or 6; stamens 10–12; capsule 5-celled, 5-valved; seed with the testa prolonged at both ends. (Greek, once and turned.)

M. uniflora L.   Indian Pipe.

Stems white or reddish, turning blackish in drying, 10–25 cm. tall; flowers 15–20 mm. long; capsule obtusely angled, 10–15 × 8–10 mm.

Rich woods, Hyder—Newf.—Fla.—Mex.—Calif., and in Japan—India. Fig. 791.

2. HYPOPITYS (Dill.) Adans.

Yellowish or reddish plants with sessile scales and the flowers in a nodding one-sided raceme which soon becomes erect; terminal flower 5-merous, the lateral ones 3- to 4-merous; petals saccate at the base; stamens 6–10; anthers horizontal; ovary 3- to 5-celled; styles short; stigmas funnel-form with ciliate margins. (Greek, under a fir tree.)

H. latisquama Rydb. Pinesap.

Plant pinkish, slightly fragrant, pubescent above, 1–3 dm. tall; scales ovate, 10–15 mm. long; sepals spatulate or cuneate with acute tip, ciliate, 7–10 mm. long; petals obovate or cuneate, 10–15 mm. long, rounded and sinuate at the apex, ciliate and pubescent; capsule scaly, about 8 mm. long.

Southeast Alaska—Mont.—N. Mex.—B. C. Fig. 792.

38. ERICACEAE (Heath Family)

Ours all shrubs or subshrubs; leaves simple, often leathery and persistent; flowers perfect, usually gamopetalous; calyx of 4 or 5 sepals, usually partly united; corolla regular or nearly so; stamens as many or twice as many as the corolla lobes; anthers 2-celled, the sacs often prolonged into tubes; ovary 2- to 5-celled; fruit a capsule, berry or drupe.

1A. Fruit a septicidal capsule; corolla deciduous; anthers unappendaged. (RHODODENDREAE)
  1B. Corolla of separate petals, capsule dehiscent from the base.
    1C. Leaves wooly beneath.......................... 1. *Ledum*
    2C. Leaves glabrous and shiny..................... 2. *Cladothamnus*
  2B. Corolla gamopetalous, capsule dehiscent from the top.
    1C. Seed flat, winged.
      1D. Leaves evergreen........................... 3. *Rhododendron*
      2D. Leaves deciduous........................... 4. *Menziesia*
    2C. Seed angled or rounded.
      1D. Stamens 5; capsule 2- to 5-celled............. 5. *Loiseleuria*
      2D. Stamens 10; capsule 5-celled.
        1E. Corolla saucer-shaped..................... 6. *Kalmia*
        2E. Corolla ovoid............................. 7. *Phyllodoce*
2A. Fruit a loculicidal capsule; anthers often awned. (ANDROMEDAE)
  1B. Low heath-like shrubs with small thick imbricate leaves; corolla campanulate.............................. 8. *Cassiope*
  2B. Shrubs; corolla urceolate or ovate-cylindric.
    1C. Anther cells beaked but not awned............ 9. *Chamaedaphne*
    2C. Anther cells awned.......................... 10. *Andromeda*
3A. Fruit a drupe or the capsule enclosed by the fleshy accrescent calyx.
  1B. Fruit the fleshy calyx surrounding the ovary (GAULTHERIAE)................................ 11. *Gaultheria*
  2B. Fruit a drupe with 4 or 5 nutlets. (ARBUTAE)....... 12. *Arctostaphylos*

1. LEDUM L.

Resinous, branching, evergreen shrubs; leaves alternate, coriaceous, thick, with revolute margins; flowers white, from terminal scaly buds; calyx small, persistent, 5-lobed; corolla of 5 separate spreading petals;

stamens 5–10, exerted; anthers small, opening by terminal pores; ovary 5-celled, stigma 5-lobed; capsule 5-valved. (Greek, from a plant now placed in a different family; *Cistus ledon,* the Rock Rose.)

Leaves linear; stamens about 10.........................1. *L. decumbens*
Leaves oblong, stamens 5–10............................2. *L. groenlandicum*

1. *L. decumbens* (Ait.) Lodd.   Narrow-leaved Labrador Tea.
  L. palustris decumbens (Ait.) Hult.

Similar to *L. groenlandicum* but much smaller and more decumbent, 1–5 dm. tall; leaves linear, 10–25 mm. long, 0.5–3 mm. wide; pedicels very pubescent.

Common in muskegs and alpine situations, east Asia—Greenl.—Newf.—Skagway—Aleutians. Fig. 793.

2. *L. groenlandicum* Oeder.   Labrador Tea.
  L. pacificum Small.

3–10 dm. tall; leaves oblong to linear-oblong, obtuse, strongly revolute, densely red-wooly beneath, green and rugose above, 15–50 × 3–10 mm.; flowers numerous; petals about 5 mm. long; stamens slender; pedicels 15–25 mm. long, recurved in fruit; capsule 4–6 mm. long.

Muskegs and woods, central Alaska—Greenl.—Mass.—Pa.—Wash. Fig. 794.

2. CLADOTHAMNUS Bong.

Erect or ascending shrub; leaves alternate, deciduous, entire; flowers one or two, terminal; sepals and petals each 5, distinct or nearly so; stamens 10 with filaments dilated at the bases; style curving; stigma capitate; ovary and capsule 5-celled. (Greek, branch and bush.)

C. pyrolaeflorus Bong.   Copper Bush.

6–12 dm. tall, the bark exfoliating; leaves ovate to oblanceolate, 15–40 mm. long, shining above, paler beneath, mucronulate; sepals linear, acute; petals coppery pink, oblong, 10–12 mm. long; capsule flattened, about 6 mm. in diameter.

Alpine and subalpine, southeast Alaska—Ore. Fig. 795.

3. RHODODENDRON L.

Ours evergreen shrubs or subshrubs; leaves alternate, entire, short-petioled; calyx 5-parted, often small; corolla from rotate to campanulate, the limb 5-lobed, often somewhat irregular; stamens usually 10; anthers opening by pores at the apex; ovary 5- to 10-celled; capsule separating into 5–10 valves, seed numerous. (Greek, rose and wood.)

Flowers 1–3..1. *R. kamtschaticum*
Flowers several, borne in umbels.......................2. *R. lapponicum*

1. *R. kamtschaticum* Pall.   Kamchatka Rhododendron.
  Thororhodion kamtschaticum (Pall.) Small.

A subshrub a few centimeters tall; leaves spatulate to obovate, 15-35 mm. long, ciliate on the margins and on the veins beneath, cuneate at the base, rounded and mucronulate at the apex; flowers borne on the new growth; sepals ovate or elliptic, foliaceous and with ciliate margins, 12-15 mm. long; corolla rose-purple, 35-45 mm. across, the lobes ovate with finely ciliate margins; base of corolla densely pubescent. Ssp. *glandulosum* (Standl.) Hult. is a lower-growing form more or less glandular on the leaf margins and the corolla nonciliate.

An Asiatic species extending into western Alaska, the subspecies in Seward Peninsula and lower Yukon valley. Fig. 796.

2. *R. lapponicum* (L.) Wahl. Lapland Rose Bay.
R. parvifolium Adams.

A low, depressed, prostrate shrub 5-25 cm. tall; leaves oval or elliptic, obtuse, entire, 6-16 mm. long, more or less revolute on the margins, the lower surface brownish in age, both surfaces covered with scales as also the peduncles and capsule; corolla pinkish-purple, 15-20 mm. broad; capsule ovoid, about 5 mm. high.

Alpine, in Alaska north of 63 degrees; circumpolar, south to New York. Fig. 797.

4. MENZIESIA Smith.

Erect branching shrubs; leaves deciduous, alternate, membranous; flowers 4-merous, in corymbs from terminal buds; calyx 4-lobed; corolla urceolate; stamens usually 8, included; anthers linear-sagittate, opening by terminal chinks; ovary 4-celled; stigma 4-lobed or toothed. (Archibald Menzies was a surgeon and naturalist.)

M. ferruginea Smith. Rusty Menziesia.

An odorous shrub 15-30 dm. tall; leaves oblanceolate, mucronate, hirsute above, on the margins and on the veins beneath, 2-6 cm. long; calyx ciliate-margined; corolla coppery pink, 7-9 mm. long; capsule ovoid.

In woods, central Alaska—Wyo.—Ore. Fig. 798.

5. LOISELEURIA Desv.

A low, glabrous, depressed, caespitose, evergreen subshrub; leaves small, linear-oblong, coriaceous, entire, obtuse, petioled; calyx deeply 5-parted, the divisions ovate-lanceolate, reddish-purple, persistent; corolla campanulate, 5-lobed; stamens 5, opening by slits; ovary 2- or 3-celled; capsule 2- or 3-valved, the valves 2-cleft. (Loiseleur was a French botanist.)

L. procumbens (L.) Desv. Alpine or Trailing Azalea.

A diffusely-branched subshrub sometimes forming mats up to 2 or 3 dm. in diameter but usually much smaller; leaves crowded, 3-7 mm. long, pale with a prominent ridge beneath; flowers pink or white, about

4 mm. long, in small terminal clusters; capsules 2-2.5 mm. in diameter.

Mostly alpine, circumpolar, south to New Hampshire and the 49th parallel. Fig. 799.

6. KALMIA L.

Glabrous, evergreen shrubs; leaves in ours opposite; coriaceous; flowers in terminal or axillary corymbs with deciduous bracts; sepals 5, coriaceous, persistent; corolla rotate, 10-keeled, 5-lobed; stamens 10; anthers at first enclosed in pouches of the corolla, awnless, opening by terminal pores; capsule 5-valved. (Peter Kalm was a pupil of Linnaeus.)

K. polifolia Wang. Swamp Laurel.
  K. microphylla (Hook.) Heller.
  K. occidentalis Small.

A sparingly branched shrub 1-3 dm. tall; leaves 15-35 mm. long, dark green above, glaucous beneath, entire, revolute; sepals purplish, ovate, concave, about 3 mm. long; corolla rose, 12-18 mm. wide; capsules about 4 mm. thick and 5 mm. long.

Muskegs, southeast Alaska and Yukon—Newf.—Penn.—Mich.—Mont.—Calif. Fig. 800.

7. PHYLLODOCE Salisb.

Low, branching, evergreen shrubs; leaves narrow, coriaceous, crowded, linear, obtuse; flowers in terminal corymbs; sepals 5, persistent; stamens 10, included; anthers awnless, opening by pores; filaments glabrous; capsule 5-valved to the middle; seeds minute, with coriaceous testa. (Greek, a sea nymph.)

1A. Flowers blue..1. *P. coerulea*
2A. Flowers pink to red............................2. *P. empetriformis*
3A. Flowers yellowish.
  1B. Corolla glandular-puberulent...................3. *P. glanduliflora*
  2B. Corolla glabrous.............................4. *P. aleutica*

1. *P. coerulea* (L.) Babingt. Blue Mountain Heather.

8-15 cm. tall with ascending branches; leaves 4-10 mm. long, less than 2 mm. wide, their margins scabrous or serrulate; calyx teeth lanceolate, acute; pedicels glandular, elongating in fruit; corolla ovoid, 7-8 mm. long, glabrous; capsules about 4 mm. in diameter. Hybridizes with *P. aleutica*.

Attu Island and east North America, Eurasia. Fig. 801.

2. *P. empetriformis* (Smith) D. Don. Red or Purple Heather.

Plants up to 15 cm. tall, tufted or matted; leaves 6-15 mm. long, revolute; calyx lobes ovate; corolla campanulate, pink to red, 7-9 mm. long; capsules globular, about 3 mm. in diameter.

Southeast Alaska—Mack.—Alberta—Colo.—Calif. Fig. 802.

3. **P. glanduliflora** (Hook.) Cov. — Yellow Heather.

Stems 1-3 dm. tall; leaves subsessile, linear-oblong, serrulate, rugose, with a narrow furrow above and a light, minutely hairy line below, 6-10 mm. long; pedicels, calyx and corolla glandular-pubescent; sepals lanceolate, acute; corolla urceolate, about 8 mm. long.

South Alaska—Alta.—Wyo.—Ore. Fig. 803.

4. **P. aleutica** (Spreng.) A. Heller. — Aleutian Heather.

Plants up to 20 cm. tall; leaves linear, 5-11 mm. long, obtuse, serrulate; calyx lobes linear to lanceolate; corolla globose-urceolate, 6-8 mm. long.

East Asia, Aleutians—Bering Sea region and Pr. William Sd. Fig. 804.

8. CASSIOPE D. Don.

Low, branching, evergreen shrubs or subshrubs; leaves thick; flowers axillary or terminal, nodding on slender pedicels; sepals usually 5, thickened at the base; corolla campanulate, usually 5-lobed; stamens included; anthers attached near the apex, opening by large terminal pores and tipped by recurving awns; style thickened below; capsule 4- to 5-valved. (Cassiope of Greek mythology was mother of Andromeda.)

1A. Leaves spreading, flowers terminal..........................1. *C. stelleriana*
2A. Leaves 4-ranked, appressed.
  1B. Leaves furrowed on back................................2. *C. tetragona*
  2B. Leaves not furrowed on back.
    1C. Diameter of stem with appressed leaves 2.5 mm. or more. 3. *C. mertensiana*
    2C. Diameter of stem with appressed leaves 1.5-2 mm........4. *C. lycopodioides*

1. **C. stelleriana** (Pall.) DC. — Alaska Heather.
Harrimanella stelleriana (Pall.) Cov.

Stems 5-20 cm. long; leaves very numerous, oblanceolate, 3-5 mm. long, flattish above, slightly keeled beneath; calyx lobes oval with yellowish margins, about 3 mm. long; corolla white, lobed about half way to the base, about 6 mm. long; capsule erect.

Alpine, east Asia and the Aleutians along the coast to Wash. Fig. 805.

2. **C. tetragona** (L.) D. Don. — Four-angled Mountain Heather.

More or less decumbent at the base, the ascending branches 1-2 dm. tall; leaves very thick, ovate, 3-5 mm. long with a deep furrow on the back; peduncles 10-25 mm. long; sepals about 2.5 mm. long, acute; corolla white or pink, 5-6 mm. long; capsule much longer than the calyx; diameter of branches including leaves 4-5 mm.

Alpine, east Asia and the Aleutians along the coast to Wash. Fig. 806.

3. **C. mertensiana** (Bong.) D. Don. — Mertens Mountain Heather.

Similar to *C. tetragona;* leaves ovate-lanceolate, 2.5-4 mm. long, concave above, round-keeled on the back; peduncles 8-15 mm. long; sepals pinkish, ovate, acute, 2.5–3 mm. long; corolla white or pinkish, 6–8 mm. long; capsule a little longer than the calyx.

Alpine, southeastern Alaska—Alta.—Mont.—Calif. Fig. 807.

4. *C. lycopodioides* (Pall.) D. Don.   Club-moss Mountain Heather.

Stems more or less prostrate, 5–20 cm. long; leaves very closely and evenly appressed, 2–3 mm. long; sepals ovate, obtuse, the margins hyaline, about 2 mm. long; corolla white, about 6 mm. long, the ovate lobes nearly as long as the tube; capsule a little longer than the calyx.

Rocky alpine, east Asia—Aleutians—southeast Alaska. Fig. 808.

9. CHAMAEDAPHNE Moench.

An erect branched shrub with rather slender terete branches; leaves alternate, coriaceous, evergreen; flowers in terminal, leafy racemes; calyx of 5 distinct sepals bracted at the base; corolla oblong-cylindric with 5 recurved teeth; stamens 10, included; anthers-sacs tapering upward into tublar beaks, not awned; ovary 5-celled; capsule 5-valved. (Greek, ground or low Daphne.)

C. calyculata (L.) Moench.   Leather-leaf.
    Andromeda calyculata L.
    Cassandra calyculata (L.) D. Don.

6–12 dm. tall with pubescent twigs; leaves thick, rugose above and covered underneath with minute, roundish, scurfy scales which often occur on the upper surface also, the margins minutely wavy toward the tips, 12–40 mm. long; corolla about 6 mm. long; capsule about 4 mm. in diameter, a little longer than the calyx.

Swamps and wet woods, circumpolar, south to Ga.—Ill.—B. C. Fig. 809.

10. ANDROMEDA L.

A low, glabrous, evergreen shrub; leaves narrow, alternate, coriaceous, strongly revolute; flowers in terminal corymbs; sepals 5, persistent; corolla globose-urceolate with 5 recurved teeth; stamens 10, included; filaments bearded; anthers with ascending awns; ovary 5-celled; capsule subglobose, 5-valved; seed shining. (In mythology Andromeda was a daughter of Cassiope.)

A. polifolia L.   Bog Rosemary.

1–4 dm. tall; leaves oblong to linear, dark green above, glaucous beneath, 2–4 cm. long, mucronulate; pedicels 10–20 mm. long; sepals triangular, acute, about 1 mm. long; corolla pink, about 6 mm. long.

Bogs, circumpolar, south to N. J.—Idaho—Wash. Fig. 810.

11. GAULTHERIA (Kalm) L.

Shrubs with hairy twigs; leaves alternate, coriaceous, evergreen; calyx 5-cleft, persistent; corolla urceolate or campanulate, 5-toothed or lobed; stamens 10, included; filaments dilated above the base; anthers opening by terminal pores; capsule enclosed by the enlarged and fleshy calyx forming a berry-like fruit. (Named for Dr. Gaultier of Quebec.)

Racemes many-flowered..................................1. G. shallon
Racemes 1- to 6-flowered..................................2. G. miqueliana

1. G. shallon Pursh. Salal

Partially decumbent or erect, stout, 2–12 dm. tall; leaves oval or ovate, serrate, mucronate, cordate at the base, 3–8 cm. long; flowers in glandular-pubescent bracted racemes; calyx with prominent, stiff, reddish-brown, glandular hairs; corolla ovoid, pubescent, 6–8 mm. long; filaments hairy; anthers with 4 awns; fruit purple.
Southeast Alaska—Calif. Fig. 811.

2. G. miqueliana Takeda.

Stems up to 35 cm. tall, procumbent at the base, the branches ascending; leaves short-petioled, oval to oblong-oval, 15–35 × 8–16 mm.; calyx lobes triangular, glandular-pubescent on the back, the apex ciliolate; corolla ovoid-urceolate, about 5 mm. long; anthers 4-aristate at the apex; fruit globose, 10 mm. long.
An east Asiatic species found on Kiska Island.

12. ARCTOSTAPHYLOS Adans.

Flowers in small, terminal, bracteolate racemes; calyx small, 4- to 5-parted; corolla urceolate with 4 or 5 recurved lobes; stamens included; anthers with 2 recurved awns on the back; ovary 4- to 10-celled; fruit a drupe with 1–8 more or less coherent nutlets. (Greek, bear and bunch of grapes.)

Stems long-trailing; leaves evergreen..................1. A. uva-ursi
Stems short; leaves deciduous..........................2. A. alpina

1. A. uva-ursi (L.) Spreng. Bearberry. Kinnikinnick.

Depressed and spreading over ground and rocks, forming patches sometimes 1–2 m. in diameter; leaves spatulate, reticulate, the apex rounded, the base cuneate; corolla white, 4–5 mm. long; fruit red, globose, 6–10 mm. in diameter, usually containing 5 coalesced nutlets.
Circumpolar, south to Va.—Ill.—N. Mex.—Calif. Fig. 812.

2. A. alpina (L.) Spreng. Alpine Bearberry.
Arctous alpina (L.) Niedzu.
Mairania alpina (L.) Desv.

A depressed, prostrate subshrub, 3–10 cm. tall; leaves spatulate or ovate, finely crenate, reticulate-veined, 15–30 mm. long; corolla white or pink; fruit bluish-black when ripe, 6–8 mm. in diameter. The typical form is usually alpine. In woods at lower elevations is variety *rubra* (Fern.) Rhed. & Wils. (*A. rubra* Fern.) (*Arctous erythrocarpa* Small.) which is somewhat larger-growing and has red fruit.

Circumpolar, south to Newf., N. Hamp. and B. C., the var. in Alaska—Man.—B. C. Fig. 813.

39. VACCINIACEAE (Blueberry Family)

Ours all shrubs or trailing vines; leaves alternate, often coriaceous, simple, sometimes evergreen; flowers small, perfect, white, pink, or red, clustered or solitary; calyx tube adherent to the ovary, 4- or 5-toothed, -lobed or -parted; corolla gamopetalous with 4 or 5 lobes, or in *Oxycoccus* of nearly distinct free petals; stamens twice as many as the corolla-lobes; ovary 4- to 10-celled; fruit a berry.

Petals united..1. *Vaccinium*
Petals distinct and reflexed..........................2. *Oxycoccus*

1. VACCINIUM L.

Ours all branching shrubs; calyx lobes small; fruit a many-seeded berry with or without bloom. (Ancient Latin name for the blueberry.)

```
1A. Corolla open campanulate; leaves evergreen...............1. V. vitis-idea
2A. Corolla cylindric to urceolate; leaves deciduous.
   1B. Tall shrubs, 5 dm. or more tall.
      1C. Fruit red.............................................3. V. parvifolium
      2C. Fruit blue or black.
         1D. Leaves finely serrulate.
            1E. Leaves firm, strongly reticulate...................7. V. paludicola
            2E. Leaves thin, not reticulate.......................6. V. membranaceum
         2D. Leaf margins entire or nearly so.
            1E. Corolla ovoid; early-flowering....................4. V. ovalifolium
            2E. Corolla depressed urceolate, flowering later.......5. V. alaskensis
   2B. Low shrubs, less than 5 dm. tall.
      1C. Flowers arising from scaly buds on old wood.........2. V. uliginosum
      2C. Flowers borne on current seasons growth.
         1D. Twigs distinctly angled, usually more than
                25 cm. tall...................................7. V. paludicola
         2D. Twigs not distinctly angled, usually less than
                25 cm. tall..................................8. V. caespitosum
```

1. V. vitis-idea L.   Mountain Cranberry. Lingen Berry.

Low evergreen subshrub 5–15 cm. tall with a more or less creeping stem; leaves thick, obovate, green and shining above, pale and spotted beneath, 5–15 mm. long, the margins slightly revolute; corolla 4-lobed, light rose, about 5 mm. long; berry bright red, acid, 6–8 mm. in diameter. Our form is smaller than the European and has been separated as subspecies *minus* Lodd.

Circumpolar, the ssp. south to Mass., Minn., and Wash. Fig. 814.

2. V. uliginosum L.   Bog Blueberry.

A much branched shrub 1–6 dm. tall; leaves obovate, thickish, entire, glaucescent and paler beneath, 1–2 cm. long; calyx lobes rounded; corolla light pink; berries blue-black with bloom, from oblate to cylindrical, 5–15 mm. in diameter. This is the common blueberry of interior Alaska and used in large quantities. In southeast Alaska it is largely a bog or alpine dweller and not much used.

Circumpolar, south to Newf., Maine, N. Y., B. C. Fig. 815.

3. *V. parvifolium* Smith. Red Huckleberry.

5-15 dm. tall with green, sharply-angled branches; leaves oblong or oval, obtuse or rounded at both ends, mucronulate, 1-3 cm. long, entire except on basal shoots on which they are often serrulate and evergreen; flowers solitary and axillary; fruit red, translucent, pleasantly acid, 7-10 mm. in diameter.

South Alaska—Idaho—Calif. Fig. 816.

4. *V. ovalifolium* Smith. Early Blueberry.

5-15 dm. tall with slender twigs; leaves glabrous, entire, 15-50 mm. long, pale and glaucous beneath; flowers solitary, preceding the leaves; corolla light pink, 5-7 mm. long; fruit blue with bloom; globular or slightly oblate, 8-12 mm. in diameter.

This and the next species furnish most of the blueberries gathered in the Pacific coast region of Alaska. Japan—Aleutians—Oregon, and in eastern North America. Fig. 817.

5. *V. alaskensis* Howell. Alaska Blueberry.

A shrub 6-18 dm. tall with stout reddish twigs; leaves oval, paler beneath, acute, entire or irregularly serrulate, 2-7 cm. long; flowers borne singly, appearing with the leaves; corolla depressed urceolate, green shaded red; berry variable, from depressed-globose to pyriform, reddish-black to blue-black, with or without bloom, 10-15 mm. in diameter.

Woods, Prince William Sound—Ore. Fig. 818.

6. *V. membranaceum* Dougl. Thin-leaved Blueberry.

A widely spreading shrub with angled twigs, 3-12 dm. tall; leaves oval, thin, very finely serrate, only slightly paler beneath, 2-7 cm. long, mucronulate at the apex; corolla depressed-globose; fruit globose or slightly oblate, dark purple to black, 8-10 mm. in diameter.

Southeast Alaska—Mich.—Ore. Fig. 819.

7. *V. paludicola* Camp. Swamp Blueberry.

Stems 15-60 cm. tall, the branches angled and puberulent; leaves elliptic-obovate, subcoriaceous, green and shining, the margins minutely glandular serrulate, 20-35 $\times$ 10-15 mm.; corolla ovoid-urceolate, pink, about 6 mm. long; berry with bloom, about 10 mm. in diameter.

Swampy places, southeast Alaska and B. C. Fig. 820.

8. *V. caespitosa* Michx. Dwarf Blueberry.

Stems much branched, 6-25 cm. tall; leaves obovate, often cuneate at the base, serrulate, the teeth mucronulate, somewhat rugose above and net-veined beneath, 1-3 cm. long; flowers pink; berry blue with bloom, 6-8 mm. in diameter, quite sweet. Resembles a low-growing, small-fruited *V. paludicola*.

Central Alaska—Labr.—Maine—N. Y.—Wis.—Colo.

2. OXYCOCCUS (Tourn.) Hill.

Delicate trailing or creeping vines; leaves small, alternate, nearly sessile, persistent; flowers solitary or few, pendulous, slender peduncled, red or pink; petals 4, narrow, recurved; stamens 8; anther-sacs prolonged into slender tubes with terminal pores; fruit a globose or ellipsoid, acid, red berry. (Greek, sour berry.)

O. microcarpus Turcz.   Swamp Cranberry.

O. oxycoccus and *O. intermedia* of reports from Alaska.

Stems very slender, creeping through the moss and rooting at the nodes, 1-4 dm. long; leaves thick and leathery, ovate with rounded bases and revolute margins, acute at the apex, whitish underneath, 4-8 mm. long; flowers 1-4, terminal; petals 4-6 mm. long, berry globose to ellipsoid, 6-10 mm. in diameter.

Circumpolar, south to Manitoba and Alberta. Fig. 821.

40. DIAPENSIACEAE (Diapensia Family)

Ours a low, tufted subshrub; leaves simple, alternate or basal, persistent; flowers perfect, axillary, regular; calyx 5-parted, persistent; corolla 5-lobed, 5-cleft, or 5-parted, deciduous; stamens 5, in ours inserted on the corolla tube and alternate with its lobes; ovary superior, 3-celled; style persistent; stigma 3-lobed; capsule 3-celled, 3-valved; seed minute.

DIAPENSIA L.

Glabrous densely tufted subshrubs; leaves thick and firm; flowers on erect peduncles, white or pink; calyx bracted at the base, the sepals oval, obtuse, firm; corolla campanulate, 5-lobed; stamens inserted in the sinuses of the corolla; seed reticulated. (Greek, by fives.)

D. lapponica L. ssp. *obovata* (F. Schmidt) Hult.   Diapensia.

Stems usually much branched forming dense cushion-like tufts; leaves crowded, spatulate, sessile, rounded at the apex, usually curved, entire, 4-10 mm. long; peduncles becoming 2-4 cm. long in fruit; corolla whitish; 7-8 mm. long.

Alpine-arctic, east Asia and Alaska, the typical *D. lapponica* in east North America and Eurasia. Fig. 822.

41. PRIMULACEAE (Primrose Family)

Annual or perennial herbs; leaves simple; flowers perfect, regular; sepals 4-9, partially united; corolla lobes 4-9; stamens as many as the corolla lobes and opposite them, partly adnate to the tube; ovary

1-celled with free central placenta; fruit a 1-celled capsule opening by 2-8 valves.

1A. Leaves all basal; flowers borne on scapes.
  1B. Corolla lobes reflexed........................... 1. *Dodecatheon*
  2B. Corolla lobes erect or spreading.
    1C. Corolla tube shorter than the calyx........... 2. *Androsace*
    2C. Corolla tube equaling or exceeding the calyx.
      1D. Corolla open at the throat................... 3. *Primula*
      2D. Corolla crested at the throat............... 4. *Douglasia*
2A. Stems leafy.
  1B. Flowers in rather dense clusters............... 5. *Lysimachia*
  2B. Flowers sessile in the axils..................... 6. *Glaux*
  3B. Flowers on long axillary peduncles............. 7. *Trientalis*

1. DODECATHEON L.

Perennials with leaves in basal rosettes; flowers borne in an involucrate umbel on a naked scape; calyx 5-lobed, persistent, reflexed in flowering; corolla 4- or 5-parted, the lobes reflexed, the tube short; stamens 5, their filaments united, their anthers long and attached by their bases; ovary superior; style filiform; stigma capitate; ovules numerous. The various species of this genus are known as Shooting Stars or as Bird Bills. (Greek, twelve gods.)

1A. Anthers with a distinct filament tube.
  1B. Filament tube one-half as long as the anther or longer 1. *D. pauciflorum*
  2B. Filament tube less than one-half as long as the anther.. 2. *D. macrocarpum*
2A. Filament tube very short or none.
  1B. Leaves broad with rounded or truncate base............ 3. *D. frigidum*
  2B. Leaves gradually narrowed into a bordered petiole...... 4. *D. viviparum*

1. **D. pauciflorum** (Durand) Greene.

Leaves glabrous, 3–10 cm. long; blades oblanceolate, entire; scapes 1- to 10-flowered; corolla purple; anthers 4–5 mm. long. This and the following may be races of the same species. The form mostly reported under this name is in reality the next.

Yukon—Gt. Slave L.—Sask.—Nebr.—Utah.

2. **D. macrocarpum** (Gray) Knuth.
    D. superbum Pennell & Stair.

Rootstock usually short; leaves variable, oblanceolate, spatulate-oblong or ovate, up to 25 cm. long including the petiole, entire or sinuate-dentate; scapes up to 45 cm. tall in fruit, 3- to many-flowered, glabrous or slightly glandular in the inflorescence; corolla pale at the base, the violet or rose-purple lobes 10–18 mm. long; filament-tube yellow; capsule cylindric, 12–17 mm. long. The form with wide, ovate leaves has been described as var. *alaskanum* Hult.

Kodiak Isl.—Tanacross—Wash. Fig. 825.

3. **D. frigidum** C. & S.

Leaves ovate, obtuse, the margins usually wavy, 2–5 cm. long on petioles up to 7 cm. long; scapes 10–35 cm. tall, 1- to 7-flowered, glabrous below, glandular in the inflorescence; corolla lobes 5, 10–18 mm. long, bluish or rose-purple; filaments 1 mm. or less long; anthers acute, 4–6 mm. long, purple; capsule 1 cm. or less long.

East Siberia—arctic Alaska—lower Mackenzie R.—northeast B. C. Fig. 823.

4. *D. viviparum* Greene.
   D. integrifolium Michx. pro parte.

Rootstock stout; leaves oblanceolate, thick, up to 25 cm. long including petiole, occasionally denticulate; scapes up to 6 dm. long in fruit, few-flowered, glandular in the inflorescence; corolla lobes 4, up to 25 mm. long, purplish with a yellow ring around the base and purple at the base of the stamens; anthers purple, almost sessile, 6–10 mm. long.

Prince William Sound to Ore. Fig. 824.

2. ANDROSACE (Tourn.) L.

Low herbs with a dense basal tuft of leaves; flowers small, borne singly or in umbels on a scape; calyx 5-lobed or 5-parted; corolla salver- or funnel-form; the tube shorter than the calyx, the limb 5-lobed; stamens 5, included; style very short; stigma capitate; capsule 5-valved, 2- to many-seeded. (Greek, man's shield, from the shape of the leaf of some species.)

1A. Plant low, cushion-like..........................4. *A. ochotensis*
2A. Plant with rosulate basal leaves.
  1B. Stemmed, caespitose perennial..................1. *A. chamaejasme lehmanniana*
  2B. Plants acaulescent.
    1C. Umbels several- to many-flowered...........2. *A. septentrionalis*
    2C. Umbels 1- or 2-flowered.....................3. *A. alaskana*

1. *A. chamaejasme* Host. ssp. *lehmanniana* (Spreng.) Hult.
   A. carinata Torr.

Stems branched with the leaves in rosulate clusters at the ends of the branches; leaves oblanceolate, 4–10 mm. long; scapes usually less than 10 cm. tall; bracts narrow, acute; calyx about 2.5 mm. long, its lobes oval or oblong, obtuse; corolla cream-colored with yellow eye, the limb 8–10 mm. across. Var. *andersonii* Hult. has bracts more or less saccate at the base.

Eurasia—Victoria Isl.—Mackenzie delta—Kodiak—Aleutians. Fig. 826.

2. *A. septentrionalis* L.

A winter annual with a cushion of leaves at the base; leaves oblanceolate or oblong, acute, somewhat pubescent, denticulate or entire, 4–40 mm. long; scapes nearly glabrous, 6–40 cm. tall; bracts subulate, calyx about 3 mm. long, its lobes triangular with a rib to the base of the calyx; corolla white, small, slightly exceeding the calyx.

Eurasia—Victoria Isl.—Ellesmereland—N. Mex.—Calif. Fig. 827.

3. **A. alaskana** Cov. & Standl.

Leaves in a dense rosette at the top of a perennial caudex, ciliate on the margins and usually more or less pubescent on upper surface, usually 3-toothed at the apex, up to 25 mm. long; scapes several to many, 1- or 2-flowered, pubescent with simple and forked hairs when young, glabrate in age, up to 14 cm. long; flowers sessile with one lanceolate bract at the base of each flower; calyx about 5 mm. long; corolla slightly exceeding the calyx; seed dark brown, angular, 2-2.5 mm. long.

Alpine, Seward and Shumagin Islands. Fig. 828.

4. **A. ochotensis** Willd.

Cushions 1-3 cm. high; leaves 4-8 × 1-2 mm., obtuse, ciliate on the margins and more or less hirsute on the upper surface; peduncles 4-15 mm. long; calyx in fruit 2-2.5 mm. long, the teeth lanceolate; corolla rose-purple, its tube as long as or longer than the calyx.

East Siberia—Cape Lisburne—St. Matthew Isl. Fig. 829.

3. PRIMULA L.

Perennials with leaves in a basal rosette; flowers borne in an umbel at the top of a scape; calyx persistent, 5-toothed, usually angled; corolla funnel-form or salver-form, the tube equaling or exceeding the calyx; stamens 5, inserted on the tube or throat of the corolla; capsule 1-celled, 5-valved at the summit, many-seeded. (Latin, first, from the early blooming habits of some species.)

```
1A. Lobes of corolla entire.
   1B. Leafless sheaths at base lacking...................... 9. P. nivalis
   2B. Leafless sheaths at base present..................... 10. P. tschuktschorum
2A. Lobes of corolla emarginate or obcordate.
   1B. Lobes of corolla very deeply cordate................. 1. P. cuneifolia
   2B. Lobes of corolla less deeply cleft.
      1C. Bracts of the involucre oblong with saccate auricles
            at the base........................................ 8. P. sibirica
      2C. Bracts of the involucre tapering to a point.
         1D. Flowers small, leaves mostly entire............. 4. P. egalikensis
         2D. Flowers larger, leaves mostly toothed.
            1E. Scapes usually less than 1 dm. tall.
               1F. Flowers pale, the limb usually less than
                     10 mm. across.............................. 6. P. parvifolia
               2F. Flowers lilac or rose-purple.
                  1G. Limb of corolla 12-20 mm. across.......... 5. P. borealis
                  2G. Limb of corolla smaller................... 7. P. mistassinica
            2E. Plants usually more than 1 dm. tall.
               1F. Leaves copiously farinose beneath..............3. P. incana
               2F. Leaves green beneath or only slightly farinose.
                  1G. Limb of corolla 5-8 mm. broad..............2. P. stricta
                  2G. Limb of corolla 12-20 mm. broad............5. P. borealis
```

1. **P. cuneifolia** Ledeb. Wedge-leaved Primrose.

Leaf blades spatulate to oblanceolate, cuneate at the base, rounded and toothed at the apex, 10-25 mm. long; scapes elongating in fruit,

25 mm. to 25 cm. tall, 1- to 5-flowered; calyx 2–4 mm. long, its teeth lanceolate; corolla pink to purplish, the limb 12–25 mm. across, the lobes deeply cleft. The ssp. *saxifragaefolia* (Lehm.) Hult. is the smaller, more common form.

Typical form is Asiatic, extending into the western Aleutians and Seward Penin., the ssp., from the Aleutians to southeast Alaska. Fig. 830.

2. *P. stricta* Hornem.

Leaves green or sparingly farinose beneath, oblanceolate to narrowly ovate, 5–40 × 2–15 mm.; scapes 2–30 cm. tall, rather stout; bracts lance-subulate, usually somewhat saccate or gibbous at the base, 3–8 mm. long; umbel 2- to 8-flowered; calyx urceolate-campanulate, 4–6 mm. long at maturity, the lobes one-half as long as the tube; corolla lilac or violet, the lobes shallowly notched; capsule slightly longer than the calyx.

Of scattered circumpolar distribution, south to Que., Ont., Alta.

3. *P. incana* Jones.

Leaves farinose beneath, elliptic, oblong-ovate or spatulate, 15–80 × 5–20 mm., shallowly denticulate; scapes 5–45 cm. tall, 2- to 14-flowered; bracts lanceolate to linear-oblong, flat, broadly gibbous at the base, 5–10 mm. long, usually equaling or exceeding the flowering pedicels; fruiting pedicels up to 25 mm. long; calyx farinose, the lobes shorter than the tube; corolla lilac, the limb 6–10 mm. broad; capsule ellipsoid, only slightly exceeding the calyx.

Central Alaska—Mack.—Colo.—Utah.

4. *P. egalikensis* Wormskj.   Greenland Primrose.

Leaves oval or lance-ovate, narrowed into winged petioles, the whole 6–20 mm. long, the margins entire or wavy; scapes slender, 5–23 cm. tall; umbel 1- to 9-flowered; bracts lanceolate, acuminate, dilated and somewhat saccate at the base; calyx 3–6 mm. long, the teeth short and acute; corolla tube yellow, the limb white, 4–8 mm. broad, the lobes cleft one-third to one-half their length; capsules slender, more than twice as long as the calyx.

West Alaska—Greenl.—Newf.—Que.—Alta.—B. C.

5. *P. borealis* Duby.   Northern Primrose.

Leaves cuneate-obovate to rhombic-spatulate, 6–45 mm. long including the margined petiole, more or less dentate above; scapes usually 5–10 cm. tall, sometimes taller; 1- to 10-flowered; bracts lance-subulate, dilated and often slightly saccate at the base; calyx 5–6 mm. long in fruit, the lobes nearly equaling the tube; capsule cylindric, slightly exceeding the calyx. Var. *ajanensis* (Busch) Hult. is usually somewhat smaller and is farinose on the lower surface of the leaves and in the inflorescence.

Asia—west Alaska—Banks Land. Fig. 831.

6. *P. parvifolia* Duby.   Small-leaved Primrose.

Leaves cuneate-obovate, spatulate or rhombic, denticulate, the lower scarcely petioled; scapes 4–10 cm. tall, rarely taller, 2- to 9-flowered; bracts lance- or linear-subulate, 2–5 mm. long, dilated and thickened but not saccate at the base; calyx 3–5 mm. long, the lobes about equaling the tube. Closely related to the preceding.

Asia, the Bering Sea and Seward Peninsula region.

7. *P. mistassinica* Michx.   Lake Mistassini Primrose.

Variable, leaves oblanceolate, spatulate, or cuneate-obovate, 5–70 × 2–6 mm., many of them dentate; scapes 3–21 cm. tall, 1- to 10-flowered; bracts linear-subulate, usually not saccate at the base, 2–6 mm. long; pedicels filiform, much exceeding the bracts; calyx 3–6 mm. long, the lobes equaling the tube; corolla tube yellow, the limb pink to bluish purple, 8–20 mm. broad; capsule subcylindric, 2–3 mm. in diameter, one and one-half times as long as the calyx.

East Alaska and Yukon—Labr.—Newf.—Maine—Mich.—Wis.—Minn.—B. C.

8. *P. sibirica* Jacq.   Siberian Primrose.

Leaves oval or elliptic, the blade 6–25 mm. long, the petioles often longer than the blade, the margins entire or minutely denticulate; scape slender, 5–20 cm. tall, 1- to 4-flowered; bracts oblong or obovate, 4–11 mm. long; calyx at maturity 5–6 mm. long, the tube twice as long as the oblong-ovate lobes; corolla lilac, the limb 10–18 mm. broad; capsule narrow, usually twice as long as the calyx.

Arctic Eurasia, Alaska and Yukon. Fig. 832.

9. *P. nivalis* Pall.   Snow Primrose.

Leaves elliptic to oblanceolate, up to 12 cm. or more long including the margined petiole, evenly serrate, farinose below; scapes stout, 1–2 dm. tall, 2- to 10-flowered; pedicels up to 4 cm. long in fruit; corolla lilac-purple; capsule 12–15 mm. long.

An east Asia species found around Cape Prince of Wales.

10. *P. tschuktschorum* Kjellm.   Chukch Primrose.
P. eximia Greene and *P. macounii* Greene.

Rootstock thick and short; leaves oblanceolate to ovate, sometimes quite narrow, the margins entire to crenate-dentate or serrate, 3–9 cm. long; scapes stout, 4–24 cm. tall, farinose in the inflorescence, few- to many-flowered; calyx 5–8 mm. long, the lobes about twice as long as the tube; corolla violet with lavender eye, the limb spreading, 12–20 mm. broad; capsule up to 20 mm. long.

East Asia and west Alaska. Fig. 833.

4. DOUGLASIA Lindl.

Low, perennial, cushion plants, suffrutescent at the base; leaves linear, imbricated, persistent, the dried ones covering the branches; flowers in our species solitary; calyx 5-angled, lobed to about the middle; corolla pink or violet; ovary 1-celled, usually 2- or 3-seeded; seeds brown, pitted. (David Douglas of Scotland made botanical explorations in northwest America.)

Leaves glabrous with ciliate margins.................. 1. *D. arctica*
Leaves stellate pubescent............................. 2. *D. gormanii*

1. *D. arctica* Hook.

2-5 cm. tall; leaves closely imbricated; peduncles stellate-canescent; leaves narrowly oblanceolate, 4-8 × 1-2 mm., obtuse, thin, entire; calyx campanulate-turbinate, the lobes lanceolate, mucronate, 2 mm. long; corolla rose-pink, the tube 5-6 mm. long, the lobes cuneate, 3 mm. long, erose.

Arctic coast of Yukon to the Mackenzie R.

2. *D. gormanii* Const.

Leaves appressed, pubescent with simple and forked hairs, 4-10 × 1-2 mm., withering persistent and thickly investing the branches; peduncles stellate-pubescent, from very short to 3 cm. long in fruit; corolla rose-pink.

Central Alaska—Yukon. Fig. 834.

5! LYSIMACHIA (Tourn.) L.

Ours an erect, perennial, leafy marsh plant; leaves opposite, entire, rather narrow, the lower ones reduced; flowers yellow, in peduncled axillary spikes; sepals linear, 5-7; corolla deeply 5- to 7-parted with narrow lobes; stamens 5-7, exerted, alternating with small sterile staminodia. (Greek, release and strife.)

L. thyrsiflora L.   Tufted Loosestrife.
Naumburgia thyrsiflora (L.) Duby.

Stem simple, 3-7 dm. tall; leaves sessile, linear to lanceolate, 5-10 cm. × 8-24 mm., acute, spotted, the lower ones reduced to scales; spike head-like, on peduncles 1-3 cm. long; pedicels short; calyx about 3 mm. long, spotted; corolla about 7 mm. long, the divisions linear and spotted near the apex.

Circumpolar, south to Penn. and Colo. Fig. 835.

6. GLAUX (Tourn.) L.

A low succulent perennial; leaves opposite, entire; flowers small, axillary, white or pinkish; corolla none; calyx campanulate and fleshy, colored like a corolla; stamens 5, inserted at the base of the calyx; capsule 5-valved at the summit; seeds few. (Greek, sea green.)

G. maritima L.   Sea Milkwort.

Stems very leafy, usually simple but often branched, 5-25 cm. tall;

leaves sessile, oval to linear oblong, 5–20 mm. long; calyx 3–4 mm. long, the lobes oval.

Sea beaches and salt marshes, circumpolar, south to N. J. and Calif. Fig. 836.

7. TRIENTALIS L.

Low perennials with tuberous rootstocks; stems simple with one to several small leaves along the stem and a cluster of larger leaves at the top; flowers few, often solitary, borne on slender peduncles from the axils of the upper leaves; corolla rotate, white or pinkish, parted to near the base; capsule 5-valved, few-seeded. (Latin, one third of a foot, referring to the height of some of the plants.)

T. europea L.   Star Flower.

Stems usually 1 dm. or less tall but may reach 2 dm.; leaves obovate or oblanceolate, cuneate at the base, 1–8 cm. long; flowers 1–3; sepals usually 7, narrow; corolla 12–18 mm. broad, 5- to 7-lobed; stamens mostly 7, arising from a ring at the base of the corolla; seed covered by a fine white network. The typical form occurs in the interior but most of the collections from our area are of the ssp. *arctica* (Fisch.) Hult.

East Asia—Athabasca region—B. C.—Aleutians. Fig. 837.

42. PLUMBAGINACEAE (Plumbago Family)

Ours a perennial herb; leaves basal and tufted; flowers small, perfect, regular; calyx tubular or funnelform, 5-toothed, plaited at the sinuses; stamens 5, opposite the corolla segments; anthers 2-celled; ovule solitary; fruit a utricle or achene enclosed by the calyx.

ARMERIA Willd.

Tufted fleshy herb; leaves narrow, in dense tufts; flowers in dense heads on naked scapes, subtended by bracts, the outer ones forming a sort of involucre, the lower ones reflexed and more or less united into a sheath. (An old Latin name.)

A. maritima (Mill.) Willd.   Sea Pink.
    A. vulgaris arctica (Wallr.) Hult.
    Statice armeria L.

Scapes 1–4 dm. tall; heads densely glomerate, leaves narrowly linear; bracts wide with rounded apex, the inner scarious; calyx scarious with dark, thickened base and ribs, pubescent at the base and on the ribs; corolla pink, purple, or white. Occurs in two forms. Var. *sibirica* (Turcz.) Lawr. Outer bracts one-half as long as the inner or less; leaves 6 cm. or less long. Var. *purpurea* (Mert. & Koch) Lawr. Outer bracts more than half as long as the inner; leaves flat, recurved or slightly contorted and canaliculate, 3–18 cm. long.

Circumpolar, var. *sibirica* arctic coast; var. *purpurea* Kotzebue and Aleutians—southeast Alaska. Fig. 838.

PLATE XXXI

Scale in millimeters.
FIG.
731. *Geranium erianthum* DC. Sepal, petal and leaf.
732. *Geranium sanguineum* L. Sepal and leaf.
733. *Geranium bicknellii* Britt. Fruit and leaf.
734. *Linum perenne lewisii* (Pursh) Hult. Leaf and fruit.
735. *Impatiens noli-tangere* L. Flower, leaf and fruit.
736. *Callitriche verna* L. Fruit, emersed and immersed leaves.
737. *Empetrum nigrum* L. Pistillate and staminate flowers and leaf.
738. *Acer glabrum* var. *douglasii* (Hook.) Dipp. Samara and leaf.
739. *Viola glabella* Nutt. Flower, stipule and leaf.
740. *Viola biflora* L. Flower, stipule and leaf.
741. *Viola renifolia* var. *brainerdii* (Greene) Fern. Fruit, stipule and leaf.
742. *Viola epipsila repens* (Turcz.) W. Bckr. Flower, stipules and leaf.
743. *Viola selkirkii* Pursh. Flower and leaf.
744. *Viola langsdorfii* Fisch. Flower, stipule and leaf.
745. *Viola adunca* Smith. Flower, leaf and stipule.
746. *Elaeagnus commutata* Bernh. Leaf, flower, fruit and stone.
747. *Shepherdia canadensis* (L.) Nutt. Fruit, leaf and stone.
748. *Circaea alpina* L. Flower, fruit and leaf.
749. *Epilobium angustifolium* L. Leaves and flower.
750. *Epilobium latifolium* L. Leaf and flower.

FLORA OF ALASKA 383

PLATE XXXI

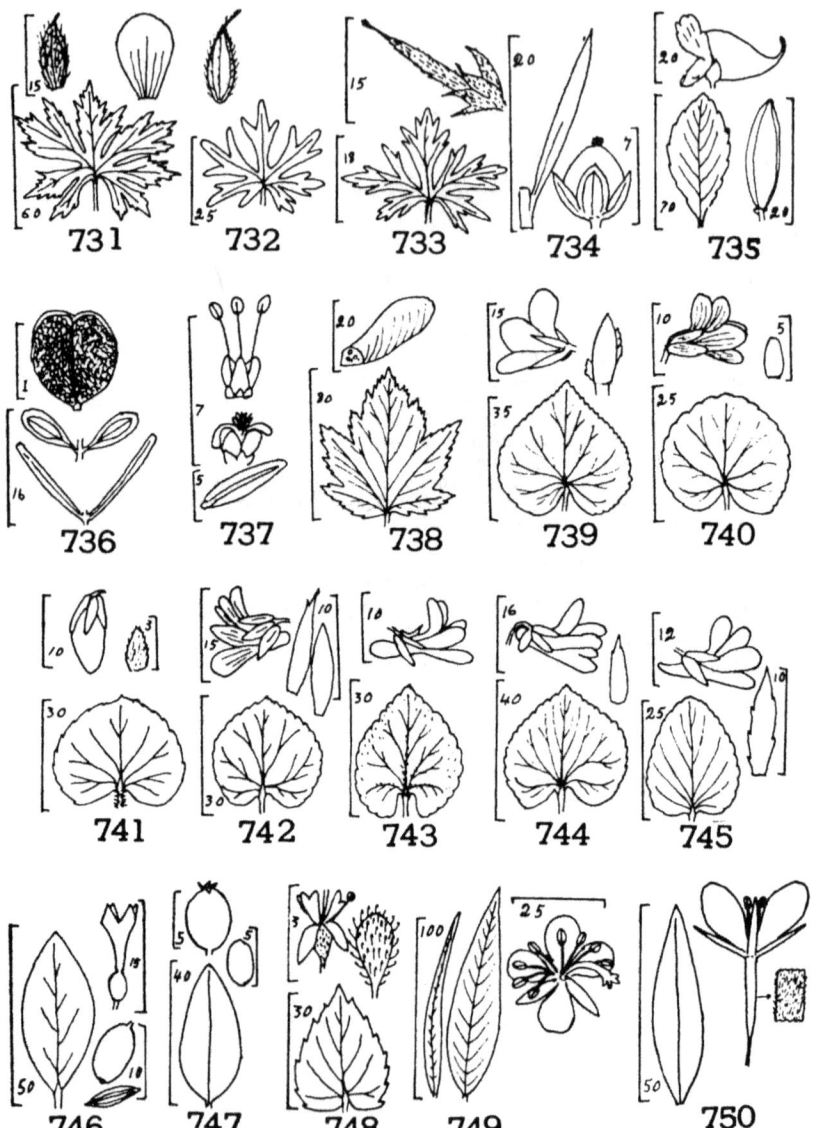

PLATE XXXII

Scale in millimeters.
Fig.
151. *Epilobium luteum* Pursh. Leaf and flower.
152. *Epilobium davuricum* Fisch. Leaves and flower.
153. *Epilobium anagallidifolium* Lam. Leaf and seeds.
154. *Epilobium leptocarpum* Hausskn. Leaf and flower.
155. *Epilobium palustre* L. Leaves and end of stolon.
156. *Epilobium adenocaulon* Hausskn. Leaf and flower.
157. *Epilobium glandulosum* Lehm. Leaf and flower.
158. *Epilobium behringianum* Hausskn. Leaf and flower.
159. *Epilobium hornemannii* Rchb. Leaf and flower.
160. *Hippuris montana* Ledeb. Whorl of leaves, fruit and anther.
161. *Hippuris vulgaris* L. Whorl of leaves, flower and fruit.
162. *Hippuris tetraphylla* L.f. Whorl of leaves and fruit.
163. *Myriophyllum spicatum* L. Leaves and fruit.
164. *Oplopanax horridus* (Sm.) Miq. Half of leaf, flower and fruit.
165. *Osmorrhiza obtusa* (Coult. & Rose) Fern. Fruit.
166. *Osmorrhiza purpurea* (Coult. & Rose) Suksd. Fruit.
167. *Osmorrhiza chilense* Hook. & Arn. Fruit.
168. *Bupleurum americanum* (Coult. & Rose.) Stem leaf, section of carpel and fruit.
169. *Conioselinum benthami* (Wats.) Fern. Part of leaf, fruit and section of carpel.
170. *Conioselinum cnidifolium* (Turcz.) Pors. Part of leaf, fruit and section of carpel.
171. *Heracleum lanatum* Michx. Leaflet, fruit and section of carpel.

FLORA OF ALASKA

PLATE XXXII

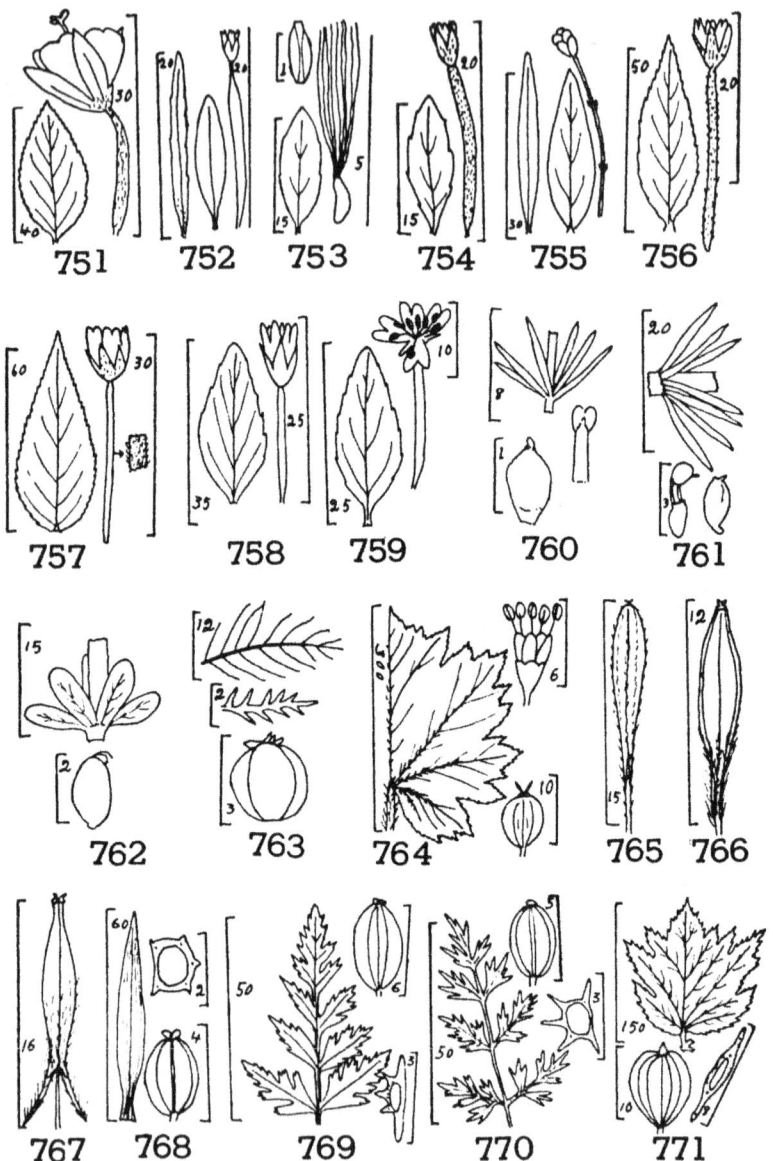

PLATE XXXIII

Scale in millimeters.
FIG.
772. *Angelica lucida* L. Fruit, part of leaf and section of carpel.
773. *Angelica genuflexa* Nutt. Fruit, part of leaf, section of carpel.
774. *Ligusticum mutellinoides alpinum* (Ledeb.) Thellung. Fruit, leaf and section of carpel.
775. *Ligusticum hultenii* Fern. Fruit, part of leaf, section of carpel.
776. *Oenanthe sarmentosa* Presl. Part of leaf, fruit, section of carpel.
777. *Cicuta maculata* L. Leaflet, fruit and section of carpel.
778. *Cicuta douglasii* (DC.) Coult. & Rose. Leaflet, fruit and section of carpel.
779. *Cicuta mackenzieana* Raup. Leaflets, fruit and section of carpel.
780. *Sium suave* Walt. Leaflet, fruit and section of carpel.
781. *Cornus canadensis* L. Fruiting plant and flower cluster.
782. *Cornus suecica* L. Fruiting plant and single flower.
783. *Cornus stolonifera* Leaf, flower, and stone.
784. *Chamaphila umbellata* (L.) Pursh. Leaf, flower, and stamen.
785. *Moneses uniflora* (L.) Gray. Leaf, flower, fruit, and stamen.
786. *Pyrola chlorantha* Swartz. Leaf, fruit, and stamen.
787. *Pyrola grandiflora* Radius. Leaf, fruit, and stamen.
788. *Pyrola asarifolia incarnata* (DC.) Fern. Leaf, fruit, and stamen.
789. *Pyrola minor* L. Leaf, fruit, and stamen.

FLORA OF ALASKA

PLATE XXXIII

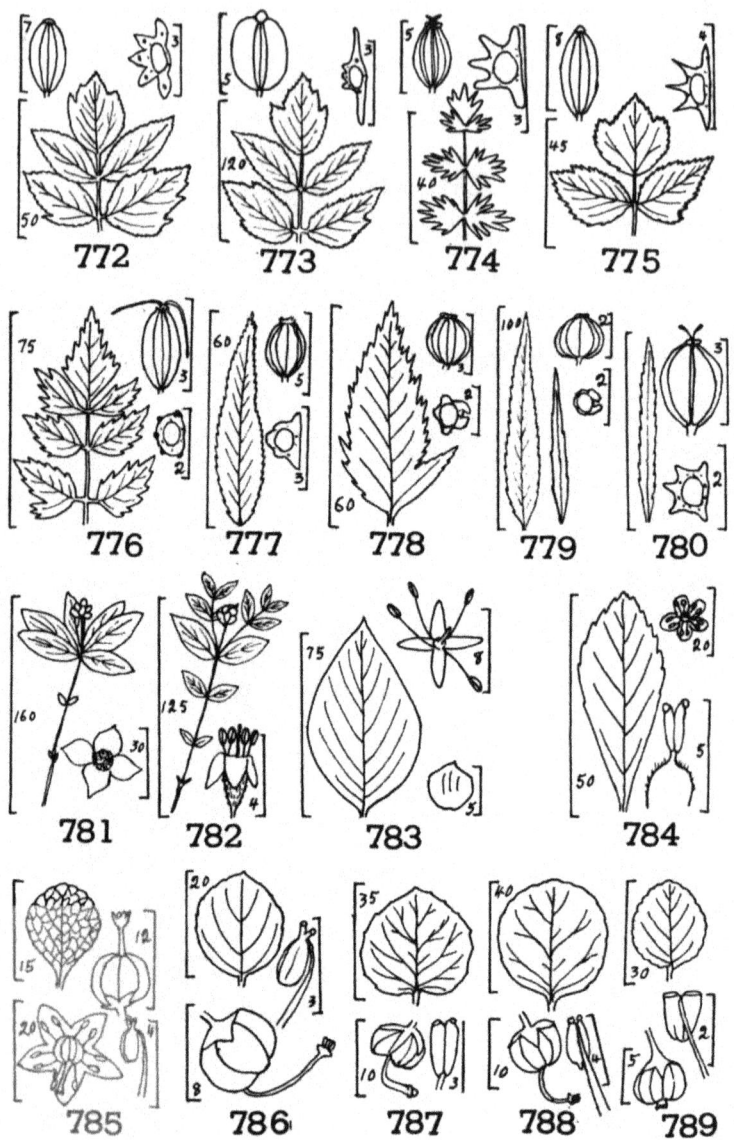

PLATE XXXIV

Scale in millimeters.
Fig.
790. *Pyrola secunda* L. Leaf, fruit, and anther.
791. *Monotropa uniflora* L. Flowering plant.
792. *Hypopitys latisquama* Rydb. Flower and fruit.
793. *Ledum decumbens* (Ait.) Lodd. Under surface of leaf, stamen, and fruit.
794. *Ledum groenlandicum* Oeder. Under surface of leaf, stamen, and fruit.
795. *Cladothamnus pyrolaeflorus* Bong. Leaf, flower, and fruit.
796. *Rhododendron kamtschaticum* Pall. Flower and leaf.
797. *Rhododendron lapponicum* L. Flower, leaf, and fruit.
798. *Menziesia ferruginea* Smith. Leaf, flower, and fruit.
799. *Loiselevria procumbens* (L.) Desv. Flower, leaf, and fruit.
800. *Kalmia polifolia* Wang. Flower, leaf, and fruit.
801. *Phyllodoce coerulea* (L.) Bab. Flower and leaf.
802. *Phyllodoce empetriformis* (Smith) D. Don. Flower and leaf.
803. *Phyllodoce glanduliflora* (Hook.) Cov. Flower and leaves.
804. *Phyllodoce aleutica* (Spreng.) Hill. Flower, pistil, and stamen.
805. *Cassiope stelleriana* (Pall.) DC. Flower, fruit, and leaf.
806. *Cassiope tetragona* (L.) D. Don. Part of stem with flower, back of leaf, and fruit.
807. *Cassiope mertensiana* (Bong.) D. Don. Flower and leaf.
808. *Cassiope lycopodioides* (Pall.) D. Don. Part of stem with flower, leaf, and fruit.
809. *Chamaedaphne calyculata* (L.) Moench. Flower, leaf, and fruit.
810. *Andromeda polifolia* L. Leaf, flower, and fruit.
811. *Gaultheria shallon* Pursh. Flower and leaf.
812. *Arctostaphylos uva-ursi* (L.) Spreng. Flower, leaf, and stamen.
813. *Arctostaphylos alpina* (L.) Spreng. Leaf, anther, and flower.
814. *Vaccinium vitis-idea* L. Flower, leaf, and stamen.

FLORA OF ALASKA

PLATE XXXIV

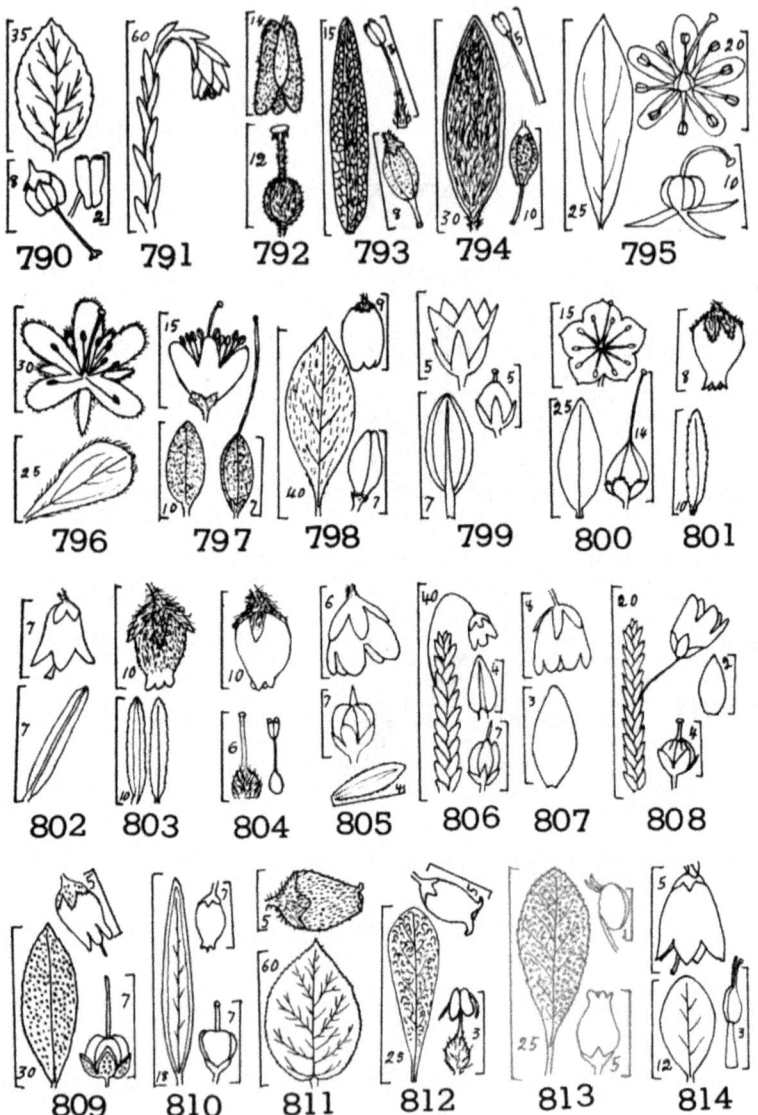

PLATE XXXV

Scale in millimeters.
FIG.
815. *Vaccinium uliginosum* L. Flower, leaf, and stamen.
816. *Vaccinium parvifolium* L. Flower, leaf, and anther.
817. *Vaccinium ovalifolium* Smith. Flower, leaf, and stamen.
818. *Vaccinium alaskensis* Howell. Flower, leaf, and anther.
819. *Vaccinium membranaceum* Dougl. Flower and leaf.
820. *Vaccinium paludicola* Camp. Flower, leaf, and stamen.
821. *Oxycoccus microcarpus* Turcz. Flower, stamen, and leaves.
822. *Diapensia lapponica obovata* (Fr.Schm.) Hult. Tip of stem with flower.
823. *Dodecatheon frigidum* C. & S. Flower, leaf, and dehiscing capsule.
824. *Dodecatheon viviparum* Greene. Flower, leaf, and capsule.
825. *Dodecatheon macrocarpum* (Gray) Kunth. Flower and capsule.
826. *Androsace chamaejasme lehmanniana* (Spreng.) Hult. Limb of corolla, leaf and calyx.
827. *Androsace septentrionalis* L. Leaves and flower.
828. *Androsace alaskana* Cov. & Standl. Leaf and flower.
829. *Androsace ochotensis* Willd. Fruit and base of stem.
830. *Primula cuneifolia saxifragifolia* (Lehm.) Hult. Flower, leaf, and dehisced capsule.
831. *Primula borealis* Duby. Limb of corolla, leaf, and capsule.
832. *Primula sibirica* Jacq. Limb of corolla, leaf, and capsule.
833. *Primula tschuktschorum* Kjellm. Limb of corolla, leaf, and capsule.
834. *Douglasia gormanii* Const. Dehisced capsule and top of stem.
835. *Lysimachia thrysiflora* L. Flower and leaf.
836. *Glaux maritima.* Flower, leaf, and capsule.
837. *Trientalis europea arctica* (Fisch.) Hult. Flower and leaf.
838. *Armeria maritima* (Mill.) Willd. Leaf and calyx.

FLORA OF ALASKA

PLATE XXXV

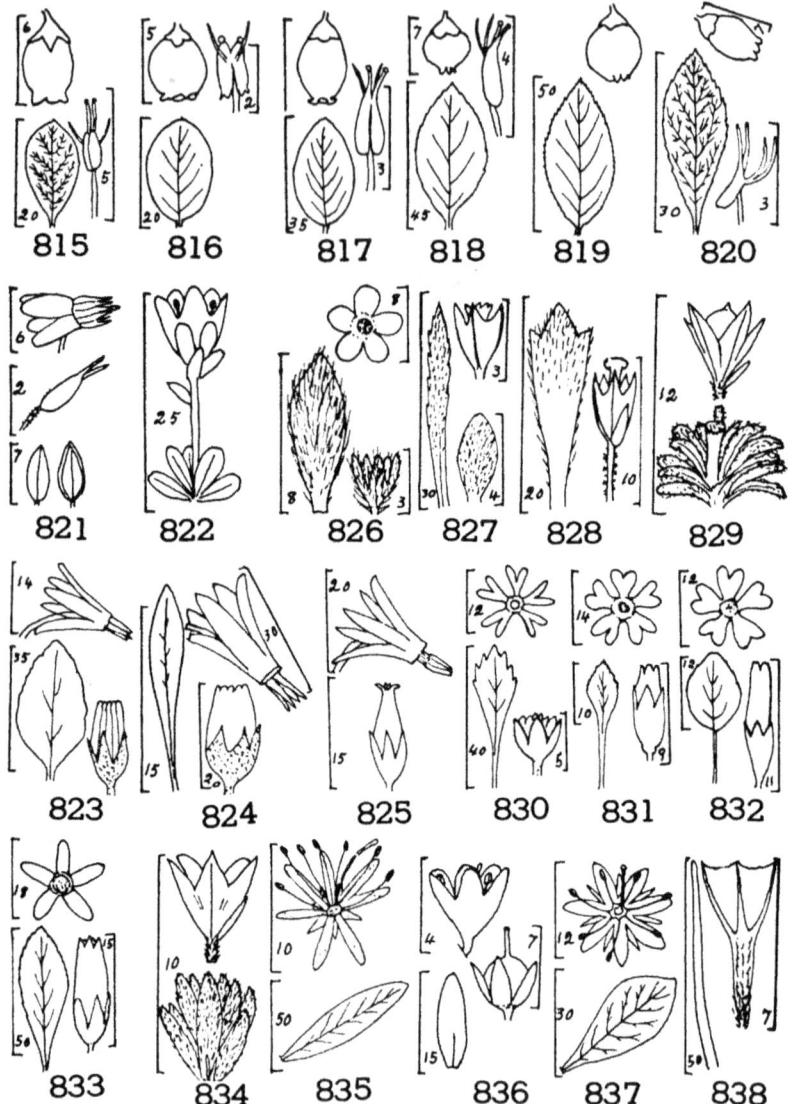

43. GENTIANACEAE (Gentian Family)

Annual or perennial, bitter, mostly glabrous herbs; leaves usually opposite; flowers perfect, regular; calyx persistent, 4- or 5-lobed or parted; corolla funnelform, campanulate, club-shaped, or rotate with 4 or 5 lobes and the stamens partly adnate and alternate with its lobes; ovary bicarpellary, 1-celled, superior, with 2 parital placentae; fruit a many-seeded capsule.

1A. Leaves simple and entire.
  1B. Corolla with 1 or 2 nectiferous pits at the base of
      each lobe .... 1. *Swertia*
  2B. Corolla usually without nectiferous pits.
    1C. Corolla rotate .... 2. *Lomatogonium*
    2C. Corolla funnelform to campanulate............. 3. *Gentiana*
2A. Leaves trifoliate or crenate. (MENYANTHACEAE)
  1B. Leaves trifoliate................................. 4. *Menyanthes*
  2B. Leaves simple, crenate.......................... 5. *Fauria*

1. SWERTIA L.

Simple-stemmed perennials; leaves alternate or opposite; flowers purple or blue; corolla rotate, usually 5-parted, each division bearing a pair of nectiferous pits; stamens inserted at the base of the corolla; style short or none; stigma 2-lobed; capsule ovate; seed margined. (Emanuel Swert was a German herbalist.)

S. perennis L.

Stems 1–6 dm. tall; lower leaves spatulate or oblanceolate, the blades 3–10 cm. long; stem leaves few, smaller, varying to lanceolate; calyx lobes lanceolate, 5–6 mm. long; corolla lobes 10–12 mm. long, often toothed at the apex; capsule a little longer than the calyx; seed strongly wing-margined.

Shumagin I.—Central Alaska—Colo.—Utah. Fig. 839.

2. LOMATOGONIUM Braun.

Slender, branched, glabrous annuals; calyx united at the base only and with 4 or 5 divisions; corolla 4- or 5-parted, the divisions acute and with a pair of narrow appendages at the bases; stamens inserted on the base of the corolla; anthers versatile; ovary with stigma decurrent along the sutures; capsule 2-valved; seed small, numerous.

L. rotatum (L.) Fries.   Marsh Felwort.
Pleurogyne rotata (L.) Griseb.

Stems occasionally simple but usually with erect or ascending branches, 1–4 dm. tall; leaves linear, 1–3 cm. long; sepals linear, 3-nerved, acute, about as long as the corolla; corolla white or blue, divided nearly to the base, the segments 10–12 mm. long.

In wet soil, circumpolar, south to N. H. and Colo. Fig. 840.

3. GENTIANA L.

Glabrous annual, biennial or perennial herbs, often with a basal rosette of leaves; Flowers often variable in size on the same plant; corolla 4- or 5-lobed, often with teeth or plaits in the sinuses; stamens included; style short or none; stigmas 2; capsules with numerous ovules. (Gentius was King of Illyria.)

```
1A. Corolla with plaits in the sinuses.
   1B. Annuals or biennials.
      1C. Dwarf plants with solitary terminal flowers...... 1. G. prostrata
      2C. Swamp plants with axillary flowers............. 2. G. douglasiana
   2B. Perennials.
      1C. Flowers whitish............................... 3. G. algida
      2C. Flowers normally blue.
         1D. Stem leaves 2-4 pairs, dwarf................. 4. G. glauca
         2D. Stem leaves 4-10 pairs, taller................ 5. G. platypetala
2A. Corolla without plaits in the sinuses.
   1B. Corolla with a fringe in the throat.
      1C. Stem very short, peduncles elongated............ 8. G. tenella
      2C. Stem longer, peduncles short.
         1D. Calyx lobes rounded, much shorter than
             the tube.................................... 9. G. auriculata
         2D. Calyx lobes acute, much longer than the tube.10. G. acuta
   2B. Corolla without fringed crown in throat.
      1C. Flowers small, usually less than 20 mm. long.
         1D. Leaves ovate, corolla little longer than the
             calyx........................................11. G. aleutica
         2D. Leaves narrower, corolla proportionately larger.
            1E. Low alpine-arctic plant....................13. G. arctophila
            2E. Taller plant of lower elevations.............12. G. propinqua
      2C. Flowers larger.
         1D. Corolla lobes fringed on the sides............ 6. G. barbata
         2D. Corolla lobes not fringed..................... 7. G. tonsa
```

1. G. prostrata Haenke.   Moss Gentian.
Chondrophylla americana (Engelm.) A. Nels.

Stems low, usually more or less procumbent and branched from the base, 2–10 cm. long; leaves numerous, small, closely ascending, faintly white-margined; calyx 8–10 mm. long with 4 scarious-margined lobes; corolla blue, 4-lobed, 12-20 mm. long; capsule oblong, 7-14 mm. long, long-stipitate, the stipe often projecting beyond the end of the persistent corolla.

Arctic slope—Alba.—Colo.—B. C. Also in Eurasia. Fig. 841.

2. *G. douglasiana* Bong.   Swamp Gentian.

Diffusely branched, 10-25 cm. tall, basal leaves elliptic-oblong to ovate-deltoid, up to 1 cm. in length; stem leaves shorter, ovate-deltoid; corolla white or blue, 8-12 mm. long, the plaits conspicuous, usually 2-forked at the apex; capsule obovate, flattened.

In muskegs near the coast, Kenai Pen.—Ore. Fig. 843.

3. *G. albida* Pall.   Whitish Gentian.
   G. frigida Haenke.
   G. romanzovii Ledeb.

Few-flowered, 5-10 cm. tall; leaves mostly basal, rather thick, linear to oblanceolate, 3-10 cm. long, the upper pair of stem leaves connate; calyx up to 2 cm. long, its uneven lobes with 2 teeth or lobes; corolla yellowish white, often tinged blue and purple spotted above, 3-5 cm. long, the lobes triangular, acute, the plaits broad.

East Asia—Seward Pen.—Colo.—Utah—B. C. Fig. 843.

4. *G. glauca* Pall.   Glaucous Gentian.

Stems simple, 2-4 cm. tall; basal leaves ovate or obovate, thick, 8-12 mm. long; stem leaves elliptic-ovate; calyx about 8 mm. long, its lobes lanceolate; corolla greenish blue, 15-18 mm. long, the lobes obtuse; plaits entire or with a small lobe.

Alpine-arctic, east Asia—arctic Alaska—Mont.—B. C. Fig. 844

5. *G. platypetala* Griseb.   Broad-petaled Gentian.
   G. calycosa, *G. gormani*, and *G. covillei* of reports.

Stems several from the base, 1-4 dm. tall, rather stout and leafy; leaves thick, sessile or partly clasping, oval, obtuse, 15-40 mm. long; calyx about 15 mm. long, parted at one or two sides, each part with 1-3 slender teeth; corolla bright blue, 3 cm. or more long. Our showiest Gentian.

Mostly alpine, Kodiak I. along the coast to Ketchikan. Fig. 845.

6. *G. barbata* Froel.   Smaller Fringed Gentian.
   G. procera Holm.
   G. macounii Holm.
   G. raupii Pors.

Stems erect, angled, 25-50 cm. tall; lowest leaves spatulate or oblong-lanceolate, obtuse, the upper stem leaves linear lanceolate and acute; branches 1- to 3-flowered with 2 or 3 pairs of leaves; calyx 15-30 mm. long, unequally cleft into acute and carinate lobes the lobes about as long as the tube; corolla deep blue, 2-4 cm. long, the 4 lobes with long fringes on the sides; smaller plants are usually unbranched.

Asia—central Alaska—Mackenzie district—N. Y.—Minn. Fig 846.

7. **G. detonsa** Rottb.
   G. serrata Gunn.

Stems erect, 10-25 cm. tall, simple or sparingly branched from near the base; basal leaves spatulate to oblanceolate; upper stem leaves lanceolate to linear; calyx lobes not carinate; corolla up to 4 cm. long, the lobes narrow, erose at the tips.

Bering Strait region — Mackenzie district — Greenland — Iceland — north Europe.

8. **G. tenella** Rottb. Slender Gentian.

Stems 2-10 cm. tall, usually branched; leaves oblong or the lowest spatulate, 4-10 mm. long; peduncles at maturity up to 8 cm. long; calyx deeply parted, its lobes foliaceous and somewhat unequal; corolla blue, up to 1 cm. long; capsule narrow, a little longer than the persistent corolla.

Eurasia—west Alaska—Greenland—Colo.—Calif. Fig. 847.

9. **G. auriculata** Pall. Auricled Gentian.

Stems 6-18 cm. tall; lower leaves oblong-lanceolate, the upper ovate; calyx lobes rounded, much shorter than the tube; corolla violet-blue, up to 25 mm. long, its lobes ovate; peduncles wing-angled.

An east Asian species found on Attu Island. Fig. 848.

10. **G. acuta** Michx. Northern Gentian.

Stem slightly wing-angled, 1-4 dm. tall; basal leaves spatulate or obovate, obtuse; upper leaves lanceolate, acute, 2-5 cm. long; flowers blue, numerous, up to 18 mm. long, pedicelled and with 2 sepals wider than the others; corolla lobes usually 5 but often 4 with a fimbriate crown at the base of the lobes. The var. *plebeja* (Cham.) Wittst. differs from the type in having few (up to 6) internodes and blunt lower stem leaves.

Asia—Aleutians—Dawson—Lab.—N. D.—Ariz.—Calif. Fig. 849.

11. **G. aleutica** C. & S. Aleutian Gentian.

Stems 4-8 cm. tall; calyx lobes ovate-lanceolate, 2-3 times as long as the tube; corolla up to 15 mm. long, white, yellowish or violet, the lobes nearly as long as the tube; capsule slightly surpassing the persistent corolla.

Attu I.—Sitka and Juneau.

12. **G. propinqua** Rich. Four-parted Gentian.

Slender annual, usually branched at the base and above, 1-4 dm.

tall, slightly wing-angled, often purplish; basal leaves spatulate, the upper lanceolate, 1-2 cm. long; flowers pedicelled, long and narrow, widely variable in size; the flowers near the top of the plant may be 18 mm. long with some near the base of same plant as small as 4 mm. long; calyx lobes very unequal; corolla normally blue, its 4 lobes acute and sometimes denticulate; capsule a little longer than the corolla. Our commonest Gentian.

Asia—Wiseman—Lab.—Alba.—B. C. Fig. 851.

13. *G. arctophila* Griseb. Arctic Gentian.

Low annual, usually branching at the base, 3-15 cm. tall; basal leaves obovate; stem leaves ovate-oblong and acute; calyx lobes somewhat scarious-margined; corolla up to 2 cm. long, the round-ovate lobes acuminate-cuspidate. Perhaps only an arctic race of *G. propinqua*.

Arctic coast—Alaska Range—Great Bear Lake—Coronation Gulf.

4. MENYANTHES (Tourn.) L.

Perennial bog plant; leaves trifoliate with long petioles; flowers perfect, racemose or paniculate, borne on long scapes; calyx deeply 5-parted, persistent; corolla short funnelform, bearded within, white, usually tinted with rose; stamens with filiform filaments and sagittate anthers; capsule ovoid. (Greek, month and flower.)

M. trifoliata L. Buckbean.

Rootstock stout, scaly; leaflets glabrous, 4-10 cm. long; scape 1-3 dm. long; corolla about 15 mm. long.

Circumpolar, south to Penn.—Iowa—Colo.—Calif. Fig. 852.

5. FAURIA Franch.

Leaves simple, glabrous, reniform, all basal on long petioles; flowers in a close, bracted panicle at the top of a long scape; pedicels enlarged below the calyx; stamens exerted; anthers strongly sagittate.

F. crista-galli (Menz.) Makino. Deer Cabbage.
Menyanthes crista-galli Menz.
Nephrophyllidium crista-galli (Menz.) Gilg.

Rootstock thick and scaly; leaves evenly crenate, 6-14 cm. broad; calyx lobes about 3 mm. long; corolla white, about 7 mm. long; capsule linear-ovoid, about 12 mm. long.

Asia and Prince William Sound—Wash. Fig. 853.

44. APOCYNACEAE (Dogbane Family)

Our species a perennial herb with acrid milky juice; leaves entire and opposite; flowers perfect and regular; sepals 5, persistent; corolla of 5 partly united petals; stamens 5, inserted on the corolla tube and alternate with its lobes; anthers 2-celled; ovary of 2 distinct carpels with united stigma; fruit of 2 distinct follicles.

APOCYNUM (Tourn.) L.

Stems branched; leaves mucronate; flowers rather small; corolla campanulate, the tube with 5 small appendages in the throat, alternating with the stamens; stamens attached at the base of the corolla, the anthers sagittate and adhering to the stigma; follicles slender; seed with a long coma. (Greek, against dog.)

A. androsaemifolium L.   Spreading Dogbane.

A glabrous perennial with spreading branches, 3–6 dm. tall; leaves pubescent on the veins beneath, otherwise glabrous, pale beneath, 3–7 cm. long; sepals lanceolate, 2–3 mm. long; corolla about 6 mm. long, its lobes finally reflexed.

Central Alaska—Que.—Ga.—Ariz.—B. C. Fig. 854.

45. POLEMONIACEAE (Phlox Family)

Flowers perfect, regular; calyx of 5 partly united sepals; corolla from rotate to salver-shaped and 5-lobed; stamens inserted on the corolla, often at different levels, and alternate with its lobes; ovary superior, mostly 3-celled; capsule 3-valved.

```
1A. Leaves pinnately compound......................... 1. Polemonium
2A. Leaves bipinnatifid................................5. Gilia
3A. Leaves simple.
   1B. Perennials, calyx not enlarging..................2. Phlox
   2B. Annuals.
      1C. Calyx distended and finally ruptured by the
             ripening capsule............................4. Microsteris
      2C. Calyx enlarging but not rupturing in fruit.......3. Collomia
```

1. POLEMONIUM (Tourn.) L.

Our species all perennial herbs; leaves alternate, simply pinnate; calyx campanulate, cleft to about the middle; corolla mostly campanulate but may be almost rotate; stamens inserted near the base of the corolla; ovules several to many, rarely only 1 or 2 to each cell. There has been much confusion regarding the species. (Derivation of name uncertain.) The species are often called Jacob's Ladder or Greek Velerian.

```
1A. Leaves glabrous or nearly so........................1. P. acutiflorum
2A. Leaves viscid-pubescent.
   1B. Corollas 10–15 mm. long..........................2. P. pulcherrimum
   2B. Corollas 15 mm. or more long.....................3. P. boreale
```

1. *P. acutiflorum* Willd.
   P. occidentale Greene

Stems glabrate below, glandular-pubescent above, 2–8 dm. tall; leaflets 15–27, acute at the apex, 8–35 mm. long; calyx 7–9 mm. long, its lobes lanceolate; corolla blue or purple, rarely white, 15–25 mm. long. A dwarf form with corolla fully 25 mm. long and with very wide corolla lobes occurs in the Bering Sea region. This may be a cross with *Polemonium boreale macranthemum*.

Eurasia—Arctic coast—Mackenzie district—Alba.—B. C. Fig. 855.

2. *P. pulcherrimum* Hook.
   P. fasciculatum Eastw.
   P. rotatum Eastw.

Stems 1-many from a woody rootstock, usually more or less spreading, often branched, 10–35 cm. tall; leaflets up to 31, orbicular to narrowly ovate, oblique, usually less than 8 mm. long in the common form; calyx 5–8 mm. long, the lobes lanceolate and usually obtuse at the apex; corolla blue with yellow tube and rounded lobes. Var. *lindleyi* (Wherry) nov. comb. (*P. lindleyi* Wherry; Am. Midl. Nat. 27; 748, 1942) is a robust form with leaflets up to 14 mm. long and corollas 12–18 mm. long.

Bering Strait—Yukon—Wyo.—Calif. Fig. 856.

3. *P. boreale* Adams.
   P. lanatum Pall.

Stems 1–2 dm. tall, occasionally taller, stout and often quite erect, glandular-pubescent at least near the inflorescence; leaves up to 12 cm. long and bearing up to 25 leaflets; leaflets oval, ovate, or lanceolate, up to 12 mm. long; calyx 7–10 mm. long, the lobes acute; corolla campanulate, 20–25 mm. long. Ssp. *macranthemum* (C. & S.) Wherry of the Bering Sea region has larger flowers and leaflets up to 18 mm. long. Ssp. *richardsonii* (Grah.) nov. comb. (*P. richardsonii* Grah. Edinb. N. Phil. J. 4:175, 1827) is a dwarf, far northern race with small leaves and white anthers.

Eurasia — Mackenzie district — Greenland and Arctic — southeast Alaska. Fig. 857.

2. PHLOX L.

Our species low, diffuse, spreading perennials; leaves mostly opposite and entire; flowers showy, in our species solitary; calyx narrow, of 5 partly united, scarious-margined sepals; corolla salverform with slender tube and spreading limb; seed usually only one in each cavity. (Greek, flame.)

Leaves subulate, corolla white..........................1. *P. hoodii*
Leaves broader, corolla pink to blue.....................2. *P. sibirica*

1. *P. hoodii* Rich.   Moss Phlox.

Very densely caespitose from a woody rootstock; leaves sparingly lanate, apiculate, 4–10 mm. long; flowers sessile at the end of the branches; calyx 5–7 mm. long; limb of corolla about 1 cm. across.

Northeast Alaska—Yukon—Mackenzie district—Neb.—Idaho.

2. *P. sibirica* L.   Siberian Phlox.

Depressed and loosely, or sometimes densely caespitose; leaves narrow, linear, apiculate, villous pubescent, especially along the margins;

calyx about 10 mm. long, the narrow sepals sharp-pointed; corolla tube about same length as the calyx, its lobes 6-8 mm. long; style almost equaling the corolla tube.

Northeast Asia—Arctic—central Alaska. Fig. 858.

3. COLLOMIA Nutt.

Leaves alternate, entire; flowers in subcapitate clusters at the end of the stem and in the axils of the upper leaves; Calyx with scarious sinuses; corolla funnelform with the 5 stamens unequally inserted on the tube; ovules 1-few in each cell; seeds developing mucilage and spirocles when wetted. (Greek, gluten, for the mucilage of the wet seed.)

C. linearis Nutt.

Stems puberulent, 7-40 cm. tall; lower leaves linear-lanceolate, the upper lanceolate 2-7 cm. long, 2-8 mm. wide; corolla tube yellowish, the limb tinted pink or purple; calyx lobes lanceolate, acuminate; capsule about 4 mm. long.

Central Alaska—Minn.—Colo.—Ariz.—Calif. Fig. 859.

4. MICROSTERIS Greene

Small, usually branched annuals; leaves narrow and entire, the lower ones opposite; flowers small, axillary; calyx 5-cleft, scarious between the lobes; corolla salverform with a slender tube and 5-lobed limb; capsule 3-celled with few large seed. (Greek, small Steris.)

M. gracilis (Dougl.) Greene

Stems 1-4 dm. tall, glandular and puberulent above; basal leaves spatulate, the leaves becoming linear or linear-lanceolate above; 2-6 cm. long; calyx 7-10 mm. long; corolla 10-14 mm. long, the tube yellowish, the limb purplish or violet. Probably introduced.

Haines, and B. C.—Mont.—Wyo.—Calif.

5. GILIA R. and P.

Calyx campanulate, the tube more or less hyaline in the sinuses and bursted by the mature capsule; corolla trumpet-shaped or salverform; capsule usually many-seeded. (Philip Gil was a Spanish botanist.)

G. capitata Dougl.

Stems erect, glabrous or nearly so, up to 6 dm. tall; leaves bi- or tripinnatifid with linear segments; flowers blue, in terminal capitate clusters; lobes of the corolla about equaling the tube.

Occasionally persisting from cultivation. Native of the Pacific states.

46. HYDROPHYLLACEAE (Water-leaf Family)

Annual or perennial, usually hirsute or pubescent herbs; flowers white or blue, regular or nearly so; calyx deeply cleft or divided:

corolla mostly campanulate or funnelform with 5 lobes; stamens 5, attached to the base of the corolla and alternate with its lobes; filaments often bearded; ovary 1- or 2-celled developing into a few- to many-seeded capsule.

Styles united to the apex............................ 1. *Romanzoffia*
Styles free at the apex.............................. 2. *Phacelia*

1. ROMANZOFFIA Cham.

Low perennials; leaves chiefly basal, roundish or reniform; flowering stems scapose; corolla campanulate, white or slightly tinted; stamens unequal; ovary 2-celled or nearly so; ovules many. The plants have very much the aspect of *Saxifraga*. (Romanzoff was a Russian who sent Kotzebue to Alaska.)

Calyx and pedicels glabrous.......................... 1. *R. sitchensis*
Calyx and pedicels pubescent......................... 2. *R. unalaschcensis*

1. *R. sitchensis* Bong.   Mist Maid.

Only slightly pubescent, 1–2 dm. tall; leaf blades mostly reniform, sometimes orbicular, the base cordate, glabrate, 10–35 mm. wide; corolla about 8 mm. long and nearly as broad; calyx lobes linear-lanceolate, nearly half as long as the corolla; capsule ovoid. *R. minima* Brand appears to be a very depauperate form of this species. It has been collected at Craig.

Kodiak I. east along the coast to Cailf. and Alba.—Mont. Fig. 860.

2. *R. unalaschcensis* Cham.

8–20 cm. tall; leaves similar to those of *R. sitchensis* but viscid-pubescent beneath and on the petioles, up to 33 mm. wide; calyx 5–7 mm. long, at least two-thirds the length of the corolla; capsule pubescent. A rare form with nearly glabrous leaves is the var. *glabriuscula* Hult.

East Aleutians — Kodiak Island group and Vancouver I. — Calif. Fig. 861.

2. PHACELIA Juss.

Leaves various in form, in our species alternate; flowers in scorpoid racemes or cymes, perfect; calyx 5-lobed, slightly enlarging in fruit; corolla white, blue or purple, 5-lobed, appendaged within; stamens 5, the filaments adnate to the tube of the corolla; ovules 2–many on each of the 2 placentae; seeds reticulate or roughened. (Greek, a cluster, referring to the flowers of some species.)

Plant annual or biennial.............................. 1. *P. franklinii*
Plant perennial...................................... 2. *P. mollis*

1. *P. franklinii* (R. Br.) Gray.   Franklin Phacelia.

Stems 2–5 dm. tall, softly hirsute, often much branched; leaves similarly pubescent, 3–7 cm. long, pinnately parted into linear-oblong, entire, toothed or incised acute lobes; inflorescence dense; calyx lobes

linear, acute, up to 8 mm. long in fruit; corolla bluish or nearly white, about 8 mm. long; stamens slightly exerted; ovules numerous.
Yukon—Great Slave Lake—Mich.—Wyo.—Idaho—B. C. Fig. 862.

2. *P. mollis* Macbr. Silky Phacelia.

Silky-pubescent throughout; stems 1–4 dm. tall from a branching caudex; leaves 3–10 cm. long, somewhat doubly pinnatifid; inflorescence dense and spike-like; corolla blue, violet or white, 5–6 mm. long; stamens long-exerted, more than twice the length of the corolla.
Haines—Yukon. Fig. 863.

An undetermined species of Phacelia in fruit was collected at Chicken in the Fortymile district in 1941. The stems arise from a thick rootstock, are up to 18 cm. tall, silky-canescent; basal leaves lanceolate, cut more than half way to the midrib into 3–6 pairs of rounded ovate lobes, silky above, more densely so beneath; inflorescence of 1–5 dense, head-like, very silky glomerules.

47. BORAGINACEAE (Borage Family)

Our species all annual, biennial or perennial herbs; leaves alternate, simple, entire and bristly; flowers perfect, mostly regular, in scorpoid racemes or spikes which often unroll like a fern frond; calyx mostly 5-lobed, -cleft or -parted and usually persistent; corolla from nearly rotate to salver-shaped, 5-lobed with the 5 stamens adnate to its tube; ovary 4-lobed, developing normally into four 1-seeded nutlets.

1A. Nutlets with hooked prickles, at least on the margins.
  1B Nutlets spreading or divergent on the low receptacle.... 1. *Cyanoglossum*
  2B. Nutlets erect on elevated receptacle.
    1C. Fruiting pedicels erect............................. 2. *Lappula*
    2C. Fruiting pedicels recurved or reflexed............. 3. *Hackelia*
2A. Nutlets unarmed, often roughened.
  1B. Receptacle flat or merely convex.
    1C. Nutlets obliquely attached......................... 4. *Mertensia*
    2C. Nutlets attached by the very base................. 5. *Myosotis*
  2B. Receptacle conic or elongated.
    1C. Calyx in fruit much enlarged, veiny-reticulate
        and folded...................................... 10. *Asperugo*
    2C. Calyx only moderately enlarged in fruit.
      1D. Corolla yellow or orange....................... 7. *Amsinckia*
      2D. Corolla blue or white.
        1E. Nutlets attached below the middle and with a
           margined, truncate back.................... 6. *Eretrichium*
        2E. Nutlets attached at about the middle, not as above.
          1F. Nutlets more or less keeled on the outer
            surface, the scar ovate or orbicular....... 8. *Plagiobotrys*
          2F. Nutlets not keeled, the scar linear or
            dilated at the base...................... 9. *Cryptanthe*

1. CYANOGLOSSUM (Tourn.) L.

Hirsute or hispid tall herbs; basal leaves with long margined petioles; calyx lobes spreading or reflexed; corolla funnelform or salverform, the tube short, the throat closed by 5 scales; stamens included;

ovary separating into 4 diverging nutlets in fruit; nutlets covered with short barbed prickles. (Greek, dog's tongue.)

C. *boreale* Fern.   Northern Wild Comfrey.

Perennial, leafless above, 4–8 dm. tall; upper stem-leaves clasping the stem; corolla blue, 6–8 mm. across; fruiting pedicels recurved; nutlets ovoid-pyriform, 4–5 mm. long.

Liard River Hot Springs—Què.—N. B.—N. Y.—Minn.—B. C.

2. LAPPULA (Riv.) Moench.

Rough-hairy herbs; leaves narrow and pubescent; flowers small, blue, borne in terminal scorpoid racemes; calyx lobes narrow; throat of the corolla closed by 5 scales; nutlets with barbed prickles along the edge and sometimes smaller ones on the dorsal surface.

Marginal prickles of nutlets in 2 rows, their bases distinct....1. *L. myosotis*
Marginal prickles in 1 row, their bases broad and often
  confluent...2. *L. redowski*

1. *L. myosotis* Moench.   European Stickseed.
   L. echinata Gilib.

Stems branched, 15–40 cm. tall; leaves narrow, all except the lowest sessile, 1–3 cm. long; pedicels short and not deflexed in fruit; corolla blue, about 2 mm. wide; prickles of the nutlets stout and hooked.

A roadside weed, native of Eurasia but widely naturalized. Fig. 864.

2. *L. redowski* (Hornem.) Greene.   Redowski Stickseed.
   L. occidentalis (Wats.) Greene.

Similar to *L. myosotis* but the stem unbranched below but with ascending branches above; nutlets papillose-tuberculate on the back with the marginal bristles flat and their bases more or less united.

Central Alaska—Sask.—Mo.—N. M.—Wash. Fig. 865.

3. HACKELIA Opiz.

Biennial or perennial; inflorescence naked or rarely sparsely bracteate; pedicels recurved or deflexed in fruit; style definitely surpassed by the nutlets; nutlets attached by a large oblique submedial ovate or deltoid areola; ventral keel extending over only the upper half of the nutlet.

H. leptophylla (Rydb.) Johnst.

Stems up to 8 dm. tall, finely pubescent with reflexed hairs, leafy, branched above; basal leaves oblanceolate; stem leaves lanceolate, very thin, 1–2 dm. long; inflorescence much branched and many-flowered; corolla blue, 2–3 mm. across; fruit about 5 mm. in diameter; margins of the nutlets with varying linear-lanceolate prickles up to 3 mm. long.

Along Glenn Highway—Mont. and Wyo.

4. MERTENSIA Roth.

Perennial herbs; flowers blue, rather large, borne in terminal racemes or panicles; calyx deeply 5-cleft, persistent; corolla tubular-funnelform or trumpet-shaped with 5 imbricated lobes; stamens with short, often flattened filaments. (C. F. Mertens was a German botanist.)

1A. Trailing seashore plants.
  1B. Corollas 6–7 mm. long..................................1. *M. maritima*
  2B. Corollas 10–11 mm. long.............................2. *M. asiatica*
2A. Not maritime, stems ascending.
  1B. Calyx lobes glabrous..................................3. *M. eastwoodae*
  2B. Calyx lobes pubescent................................4. *M. paniculata*

1. *M. maritima* (L.) S. F. Gray   Sea Lungwort.
Pneumaria maritima (L.) Hill.

Pale green, often glaucous, the stems usually forming a loose mat on beach gravel, 2–6 dm. long; leaves fleshy, the lower petioled, the uppermost sessile, 2–6 cm. long; calyx much enlarged in fruit, the lobes becoming broad and orbicular with a sharply pointed apex; nutlets smooth and shining.

Interrupted circumpolar, south to the Aleutians, Queen Charlotte Islands and Mass. Fig. 866.

2. *M. asiatica* (Takeda) Macbr.   Asiatic Lungwort.

Differs from *M. maritima* in the large corolla, long styles and broad filaments.

An asiatic species found on the western Aleutians.

3. *M. eastwoodae* Macbr.   Eastwood Lungwort.

Stems erect, 2–6 dm. tall; cauline leaves elliptic-lanceolate, 2–10 cm. long, 1–4 cm. broad, acuminate, both surfaces strigose with the hairs pointing toward the apex or the upper surface nearly glabrous; corolla 12–15 mm. long, 3–5 times as long as the calyx; nutlets more or less shiny.

Seward Pen.—Kokrines Mts.—Takotna—Lake Clerk.

4. *M. paniculata* (Ait.) Don.   Tall Lungwort.

Plant hirsute throughout, strigose in the inflorescence, 2–7 dm. tall; leaves on sterile branches cordate to oval, often 1–2 dm. long on petioles 15–30 cm. long; stem leaves lanceolate, 4–10 cm. long; calyx lobes from ½ to as long as the corolla tube, the tube about equaling the limb; corolla 12–16 mm. long; nutlets tuberculed and wrinkled. Var. *alaskana* (Britt.) L. O. Will. has narrow leaves, calyx lobes glabrous on the back but ciliate on the margins.

Common, most of Alaska—Que.—Mich.—Iowa—Mont.—Wash. Fig. 867.

5. MYOSOTIS (Rupp.) L.

Rather low herbs; the perfect regular flowers borne in 1-sided racemes; corolla usually blue, sometimes pink or white, often with an eye, the tube about the length of the calyx, the throat with appendages; nutlets small, smooth and shining. (Greek, mouse ear.)

Hairs of the calyx spreading..............................1. *M. alpestris*
Hairs of the calyx appressed..............................2. *M. palustris*

1. *M. alpestris* Schmidt subsp. *asiatica* Vesterg.   Forget-me-not.

Stems erect, 1–3 dm. tall; basal leaves spatulate or oblanceolate and petioled, 3–7 cm. long; stem leaves linear-lanceolate; corolla bright blue with yellow eye. This is Alaska's official territorial flower.

Arctic coast—Mackenzie district—Alba.—Colo.—B. C. Fig. 868.

2. *M. palustris* (L.)   Forget-me-not.

Stems decumbent and rooting at the nodes; flowers similar to *M. alpestris* but later and more continuously produced.

Sparingly escaped from cultivation in several places. Native of Eurasia.

Myosotis arvensis L. a native of Eurasia with flowers 2–3 mm. across has been reported from Mt. McKinley National Park.

6. ERETRICHIUM Schrader

Depressed, pulvinate-caespitose, arctic-alpine perennials; leaves crowded on the short branches; flowers blue with short funnelform corolla; calyx ascending in fruit; nutlets obliquely attached to a conic receptacle, smooth but with an obliquely truncate apex, the truncate portion surrounded by a margin, in ours consisting of teeth with bristly points. (Greek, wool and small hairs.)

1A. Limb of corolla 9–13 mm. across......................1. *E. splendens*
2A. Limb of corolla 5–7 mm. across.
  1B. Flowers raised on distinct sparingly leafy stems......2. *E. aretioides*
  2B. Flower clusters sessile..............................3. *E. chamissonis*

1. *E. splendens* Kearney.   Showy Eretrichium.

Caudex much branched and forming a mat of numerous short, sterile, leafy shoots and of fewer elongated flowering stems 4–13 cm. long; lower leaves closely appressed, 15–20 mm. by 2–3 mm., tapering to slender petioles, the upper leaves sessile; racemes few-flowered; corolla bright blue; teeth of the nutlets about two-thirds as long as the body.

Alpine, central and northwest Alaska.

2. *E. aretioides* (C. & S.) DC.

Densely villous with long and soft white hairs, often papillose-dilated at the base; leaves 4–10 mm. long; flowering stems 2–12 mm. tall; corolla sky blue; nutlets 1.5 mm. long, the teeth of the border about equaling the body, more or less connate at the base, and bearing minute bristles on margin and apex.

Siberia—arctic Alaska and Yukon—central Alaska—Pribilof Islands. Fig. 869.

3. *E. chamissonis* DC.

Dense and villous; flower clusters at the end of branches sometimes elongated in fruit to 1-2 cm.; flowers and nutlets much as in *E. aretioides* but the bristles at the apex of the teeth on the nutlets show a tendency to be divergent or reflexed.

East Asia, Pribilof Islands and Bering Sea Coast.

7. AMSINCKIA Lehm.

Coarse, rough-hispid biennials; leaves linear to oblong or ovate; calyx persistent; corolla yellow, salver-shaped, with long tube; nutlets rough, bony, attached below the middle. The species found in our region are weeds probably introduced from the Pacific Northwest. (Amsinck was a burgomaster of Hamburg.)

Stems erect, leaves less than 1 cm. wide....................2. *A. lycopsoides*
Stems decumbent with broader leaves......................1. *A. menziesii*

1. *A. menziesii* (Lehm.) Nels. & Macbr.   Menzies Fiddle-neck.

Stems branched, 3-8 dm. tall; leaves oblanceolate to long-ovate to lanceolate, strongly but sparsely setulose-hispid; corolla light yellow, about 8 mm. long; nutlets covered with small tubercules but without tessellated ridges.

Introduced, Nome—B. C.—Idaho—Calif.—Introduced further east. Fig. 870A.

2. *A. lycopsoides* Lehm.   Fiddle-neck.

Resembles *A. menziesii;* calyx lobes lanceolate, often 1 cm. long; corolla up to 1 cm. long, orange; nutlets with tessellated ridges, interposed with smaller tubercules.

Introduced, Alaska Range—Wash.—Calif. Fig. 870B.

8. PLAGIOBOTRYS F. & M.

Diffusely branched annuals, soft pubescent or hispid; leaves narrow and entire; sepals persistent and often enlarging in fruit; corolla white with yellow crested throat; nutlets rugose, keeled on both sides near the apex. (Greek, oblique and scar.)

1A. Corolla about 1 cm. in diameter......................4. *P. hirtus*
2A. Corolla much smaller.
 1B. Nutlets glossy...3. *P. cusickii*
 2B. Nutlets dull.
  1C. Calyx densely strigose...........................1. *P. orientalis*
  2C. Calyx stiffly hispid..............................2. *P. cognatus*

1. *P. orientalis* (L.) Johnst.

Stems usually branched and of spreading growth, 1-3 dm. long; leaves linear, 2-7 cm. by 1-6 mm., strigose-hispid; sepals 2 mm. long at anthesis, up to 5 mm. long in fruit; corolla 2.5 mm. long, 2 mm. wide; nutlets 1.5–2.25 mm. long, coarsely rugose or reticulated.

Kamchatka—Aleutians—Kodiak and Katmai.

2. P. cognatus (Greene) Johnst.
   Allocarya cognata Greene

Branching from the base, appressed strigose-pubescent, especially the inflorescence; leaves linear, the lower 2–4 cm. long; racemes loose; fruiting calyces spreading; nutlets acuminate with the scar just above the base.

Probably introduced, native of western America. Fig. 871.

3. P. cusickii (Greene) Johnst.
   Allocarya cusickii Greene

Diffusely branching, 1–2 dm. tall, canescent with appressed setose-hispid pubescence; nutlets ovate-oblong, vitrous-shining, 1 mm. long, carinate ventrally only, the back with depressed rugae and few tuberculations; scar almost basal, narrowly linear.

Reported from Fairbanks, probably introduced from further south.

4. P. hirtus (Greene) Johnst.
   Allocarya hirta Greene.

Stems branched, 15–40 cm. tall; lower leaves narrowly linear, 2–8 mm. long, the upper wider; calyx densely brown-villous, almost 3 mm. long in flower; corolla in appearance much like a white Forget-me-not.

Near Juneau, probably adventative from the Pacific Northwest.

9. CRYPTANTHE Lehm.

Hispid branched annuals; leaves narrow and entire; flowers white with 5 crests closing the throat of the corolla; calyx lobes connivent around the nutlets at maturity; nutlets in our species shining, rounded on the back, attached by fully half its length, the scar a groove forked at the base. (Greek, hidden flower.)

C. torreyana (Gray) Greene.

Branched, 1–2 dm. tall; base of many of the leaves pustulate; leaves 10–25 mm. long; corolla about 1.5 mm. wide; calyx in fruit 5–8 mm. long, sepals with a row of very stiff hairs up the center and abundant ascending hairs on the margins; nutlets 2 mm. long.

Introduced at Skagway, B. C.—Alba.—Colo.—Calif. Fig. 872.

10. ASPERUGO (Tourn.) L.

Low procumbent annual; leaves hispid; calyx foliaceous, strongly reticulate-veiny, enlarged in fruit; corolla shorter than the calyx, the limb spreading; nutlets ovoid, granular-tuberculed, keeled, attached by the middle. (Latin, very rough.)

A. procumbens L.　　　　　　　　　　German Madwort. Catchweed.

Leaves oblong or spatulate, up to 8 cm. long; corolla blue, about 2 mm. broad; fruiting calyx 8–12 mm. broad.

A weed, native of Europe.

48. LAMIACEAE (Mint Family)

Our species all aromatic herbs with 4-angled stems; leaves simple, opposite or whorled; flowers in axillary clusters or spikes; corolla with a short or long tube, the limb mostly 2-lipped with 2 lobes on the upper lip and 3 lobes on the lower; stamens 4 or one pair abortive; anthers 2-celled; ovary 4-lobed, 4-celled, each cell developing into a 1-seeded nutlet and included in the persistent calyx. This family is often known as *Labiatae*.

```
1A. Corolla nearly regular, 4- or 5-toothed.
    1B. Anther-bearing stamens 2.......................1. Lypcopus
    2B. Anther-bearing stamens 4.......................2. Mentha
2A. Corolla bilabiate.
    1B. Calyx with a protruberence on the upper side......3. Scutellaria
    2B. Calyx not gibbous on upper side.
        1C. Stamens 4, the upper pair longer than the lower.
            1D. Calyx 5-toothed............................4. Glecoma
            2D. Calyx 2-lipped.............................5. Dracocephalum
        2C. Lower stamens longer than the upper.
            1D. Calyx 2-lipped, closed in fruit................6. Prunella
            2D. Calyx 5-toothed.
                1E. Anther-sacs transversely 2-valved..........9. Galeopsis
                2E. Anther-sacs not transversely 2-valved.
                    1F. Nutlets 3-sided, truncate above..........7. Lamium
                    2F. Nutlets nearly terete rounded above......8. Stachys
```

1. LYCOPUS (Tourn.) L.

Mint-like herbs, slightly aromatic, perennial by slender stolons or suckers; leaves lanceolate or oblanceolate with serrate margins; flowers small, in dense, verticillate, bracted clusters; calyx regular or nearly so, 4- or 5-toothed; corolla funnelform or campanulate, nearly equally 5-lobed; upper pair of stamens rudimentary; nutlets 3-angled, truncate, smooth. (Greek, wolf's foot.)

Calyx teeth obtuse, the edges smooth....................1. *L. uniflorus*
Calyx teeth acuminate, finely ciliate on the edges.......2. *L. lucidus*

1. *L. uniflorus* Michx.　　　　　　　　Northern Bugleweed.

Stem slender, finely puberulent, tuberous thickened at the base. 1–6 dm. tall; leaves glabrous, 25–70 mm. long, distinctly petioled; calyx teeth 4, ovate-lanceolate, not subulate; corolla much longer than the calyx, its lobes spreading.

Wet soil, southeast Alaska—Newf.—N. C.—Neb.—Ore. Fig. 873.

2. *L. lucidus* Turcz.　　　　　　　　Western Water Horehound.

Stems nearly glabrous or pubescent, especially at the nodes, rather stout, leafy, 3–9 dm. tall; leaves nearly sessile, 4–12 cm. long; calyx about 3 mm. long with 5 subulate-lanceolate teeth about as long as the tube, ciliate on the margins; corolla scarcely exceeding the calyx.

Circle Hot Springs, B. C.—Neb.—Kan.—Ariz.—Calif. and in east Asia. Fig. 874.

2. MENTHA (Tourn.) L.

Strongly aromatic perennials; flowers perfect, small, purple, pink or white, borne in dense, axillary clusters, often appearing spicate; calyx campanulate, 10-ribbed, 5-lobed, regular or nearly so; corolla nearly regular, 4-lobed, the upper lobe larger than the others; stamens 4; anther-sacs 2, parallel; nutlets ovoid, smooth. (Minthe was a fabled Greek nymph.)

1A. Whorls of flowers all axillary......................1. *M. arvensis*
2A. Whorls of flowers mostly in spikes.
  1B. Spikes slim, usually interrupted, leaves sessile or
    nearly so..2. *M. spicata*
  2B. Spikes thicker, leaves petioled....................3. *M. piperita*

1. *M. arvensis* L.   Wild Mint.
    M. canadensis L.

Perennial by suckers, pubescent or glabrate; stems erect, usually branched and pubescent, at least along the angles, up to 8 dm. tall; leaves oval or ovate to lanceolate, with the margins crenate to sharply serrate; corolla white to pink. A very variable and widespread species.

Circumboreal, south to Va.—Neb.—N. M. Fig. 875.

2. *M. spicata* L.   Spearmint.

Stems erect, glabrous, 3–7 dm. tall; leaves lanceolate, sessile or nearly so, sharply serrate, up to 7 cm. long; flowers in bracted whorls in an interrupted spike; bracts subulate-lanceolate, ciliate; calyx teeth subulate, about as long as the tube.

Has become established at a few places in Alaska. Native of Europe but widely naturalized.

3. *M. piperita* L.   Peppermint.

Perennial by subterranean suckers; stems glabrous, usually erect, 3–8 dm. tall; leaves lanceolate, dark green, sharply serrate; bracts lanceolate, acuminate; calyx teeth subulate, shorter than the tube.

Found in a few places, native of Europe but widely naturalized.

3. SCUTELLARIA L.

Annual or perennial herbs (some species shrubby); flowers perfect; calyx 2-lipped, the upper with a crest; corolla violet with a 2-lipped limb, the upper lip arched; stamens 4, the anthers ciliate, those of the upper pair 2-celled, those of the lower 1-celled. (Latin, a dish, from the appendaged calyx.)

S. galericulata L.   Marsh Skullcap.
S. epilobifolia Hamilton

Perennial by stolons, puberulent, 2-7 dm. tall; leaves short-petioled, sessile near the top of the stem, oblong-lanceolate, crenate, 2-5 cm. long; flowers solitary in the axils; corolla blue, pubescent, 15-20 mm. long.

Swamps and edge of lakes, central Alaska—Mackenzie district—Newf.—N. C.—Neb.—Ariz.—Calif. Fig. 876.

Marrubium vulgare L. Horehound, was once collected at Juneau but has not become established. It is a woolly, usually much-branched plant; leaves oval to nearly orbicular, rugose-veined; flowers in dense axillary clusters. Native of Eurasia.

4. GLECOMA L.

Our species a low, creeping perennial; flowers in axillary verticels; corolla 2-lipped, the tube exerted and enlarged above; upper lip erect and 2-lobed or emarginate; lower lip spreading and 3-lobed; nutlets ovoid, smooth. (Greek name for Thyme or Pennyroyal.)

G. hederacea L.   Ground Ivy.
Nepeta hederacea (L.) B.S.P.

Stems puberulent, up to 5 dm. long, the branches ascending; leaves orbicular or reniform, crenate, 1-4 cm. broad; calyx teeth unequal, lanceolate, acuminate; clusters few-flowered; corolla light blue, 14-20 mm. long, the tube 2 or 3 times as long as the calyx.

Southeast Alaska, native of Eurasia.

Nepeta cataria L. Catnip was once collected at Sitka. It is a densely canescent perennial 5-10 dm. tall; leaves coarsely crenate-dentate; flowers in spiked clusters. It is native to Europe and widely naturalized in temperate climates.

5. DRACOCEPHALUM L.

Herbs with blue or purple flowers in axillary or terminal clusters; calyx tubular, 15-nerved, 5-toothed, the upper tooth the largest; corolla 2-lobed and erect, the lower 3-lobed and spreading; anther-sacs diverging; nutlets ovoid, smooth. (Greek, dragon head.)

D. parviflorum Nutt.   Dragon Head.

Annual or biennial somewhat branched herb 2-5 dm. tall; leaves 3-8 cm. long, coarsely serrate; bracts pectinate with awl-pointed teeth; corolla light blue, scarcely longer than the calyx.

Central Alaska—Mackenzie district—Que.—N. Y.—Mo.—Ariz. Fig. 877.

6. PRUNELLA L.

Perennial pubescent herbs with petioled toothed leaves; flowers borne in terminal or axillary bracted spikes; calyx 2-lipped, the tube

10-ribbed; stamens 4 but 2 sterile, the fertile stamens with forked filaments, the 2-celled anthers borne on one prong; nutlets smooth. (Derivation of name doubtful.)

P. *vulgaris* L. subsp. *lanceolata* (Barton) Hult.   Heal-all.

Stems procumbent or ascending, sometimes nearly erect, 8–40 cm. tall; leaves oval, ovate or lanceolate, from almost entire to dentate, 2–10 cm. long; spikes dense; bracts broadly ovate-orbicular, cuspidate, with ciliate margins, more or less purplish on the edges; corolla violet, 8–12 mm. long. Var. *aleutica* Fern. with bracts tomentose or lanate on the back and the calyx dark purple occurs on the Aleutian Islands.

Whole species circumboreal, south to Fla.—N. M.—Calif. Fig. 878.

7. LAMIUM L.

Annual or perennial herbs with petioled, usually broad, toothed or incised leaves; flowers strongly 2-lipped, borne in axillary clusters; calyx campanulate, 5-lobed, the upper tooth slightly the larger; corolla slightly inflated in the throat; upper lip concave, entire; lower lip 3-lobed, the lateral lobes small, the middle lobe notched; stamens all fertile. (Greek, throat, from the ringent corolla.)

L. *album* L.   White Dead Nettle.

Perennial, pubescent, rather stout, 3–6 dm. tall; leaves 3–8 cm. long; calyx teeth subulate, spreading, the upper one a little wider; corolla white, 22–25 mm. long, tube about same length as the calyx, contracted at the base, an oblique ring of hairs within.

Established around Juneau. Native of Europe.

8. STACHYS (Tourn.) L.

Our species perennial herbs; leaves toothed or incised; flowers verticillate in the upper axils and an interrupted spike; calyx campanulate with 5 nearly equal teeth; corolla purplish, its tube not exceeding the calyx, the upper lip concave, the lower lip spreading and 3-lobed; anther-sacs divergent; nutlets ovoid or oblong. (Greek, spike, from the inflorescence.)

Upper and lower leaves sessile, middle ones short-petioled.. 1. *S. palustris*
Upper leaves sessile, petioles increasing toward the base....2. *S. emersonii*

1. S. *palustris* L. subsp. *pilosa* (Nutt.) Epling.   Hedge Nettle.

Stems erect, often branched, 3–10 dm. tall; leaves lanceolate or oblong-lanceolate, 4–10 cm. by 1–3 cm., dentate; calyx pubescent and with subulate teeth; corolla 12–16 mm. long, its upper lip pubescent.

Probably introduced, central Alaska—Newf.—N. Y.—Ill.—N. M. and in Eurasia. Fig. 879.

2. S. *emersonii* Piper.   Emerson Hedge Nettle.

About 1 m. tall; leaves about 6 pairs, ovate, cordate or subcordate at the base, coarsely crenate, sparingly pilose-pubescent on both surfaces, 6–7 cm. by about 4 cm.; petioles 2–4 cm. long; internodes exceeding the leaves; flowers 1 or 2 in the axils of the upper leaves, the upper contracted into a leafy-bracted spike; corolla 12 mm. long, purplish, puberulent on the upper lip, the lower lip white-spotted.

Along the coast, Anette I. to Calif.

9. GALEOPSIS L.

Erect branching annuals; flowers borne in verticillate axillary clusters; calyx 5-ribbed with 5 subequal lobes; corolla 2-lobed, dilated at the throat, the upper lip arched and entire, the lower lip 3-cleft, the middle lobe obcordate; anthers 2-celled; nutlets ovoid, slightly flattened, smooth. (Greek, weasel-like.)

G. bifida Boenn.   Hemp Nettle.
G. tetrahit Auct.

Stems retrorsely rough-hispid, 4–9 dm. tall, swollen below the joints; leaves ovate-lanceolate, coarsely serrate, 3–10 cm. long; corolla purplish or variegated with white, 15–20 mm. long, about twice as long as the calyx.

An introduced weed, native of Europe and Asia. Fig. 880.

49. SCROPHULARIACEAE (Figwort Family)

Our species all herbs; flowers perfect, sometimes nearly regular but usually distinctly 2-lipped; stamens usually 4, sometimes also a rudimentary fifth, or only 2 fertile, didymous, inserted on the corolla; ovary 2-celled with axial placentae; fruit a 2-celled, 2-valved, usually many-seeded capsule.

```
1A. Corolla spurred..................................... 1. Linaria
2A. Corolla not spurred.
  1B. Stamens 2.
    1C. Corolla elongated and deeply cleft.
      1D. Ovules many................................ 6. Syntheris
      2D. Ovules 2.................................... 8. Lagotis
    2C. Corolla rotate................................. 7. Veronica
  2B. Anther-bearing stamens 4, a fifth filament present.
    1C. Corolla tubular, 2-lipped...................... 2. Pentstemon
    2C. Corolla 2-cleft, declined...................... 3. Collinsia
  3B. Stamens 4, all anther-bearing.
    1C. Corolla nearly regular, flowers on scapes........ 5. Limosella
    2C. Corolla long-campanulate....................... 9. Digitalis
    3C. Corolla 2-lipped.
      1D. Stamens not enclosed in upper lip of corolla... 4. Mimulus
      2D. Stamens enclosed in upper lip of corolla.
        1E. Anther-sacs dissimilar, the inner one pendulous by its apex.
          1F. Upper lip of corolla much longer than
              the lower............................10. Castilleja
          2F. Upper lip of corolla scarcely longer
              than lower...........................11. Orthocarpus
        2E. Anther-sacs similar and parallel.
          1F. Upper lip of corolla with recurved
              margins..............................12. Euphrasia
          2F. Upper lip of corolla not recurved. ..
```

1G. Calyx scarcely or not inflated in fruit....13. *Pedicularis*
2G. Calyx much inflated and veiny in fruit..14. *Rhinanthis*

1. LINARIA (Tourn.) L.

Stems erect, flowers in terminal racemes or spikes; sepals partly united; corolla decidedly 2-lipped, the tube spurred at the base, the throat partly closed by a convex fold; stamens enclosed; capsules short, opening by 3-toothed pores at the apex. (Latin, linum, flax, which some species resemble.)

L. vulgaris Hill.   Butter and Eggs.

Perennial by short rootstocks, 2–10 dm. tall, glabrous or pubescent above; leaves linear, entire, sessile, 2–7 cm. long; corolla yellow with orange throat, 2–3 cm. long.

Naturalized in a few places in Alaska and widely elsewhere. Native of Europe.

2. PENTSTEMON Mitchell.

Perennials, mostly branched from the base; leaves opposite; flowers irregular, in terminal racemes or panicles; calyx 5-parted; corolla with elongated tube, the limb 2-lipped; upper lip 2-lobed, the lower 3-lobed; stamens 4, the fifth sterile filament usually bearded; capsule ovoid, 2-valved; seeds numerous. (Greek, five stamens.)

1A. Leaves wide, ovate-lanceolate, serrate..............1. *P. diffusus*
2A. Leaves narrower and entire.
  1B. Flowers 18–25 mm. long.........................2. *P. gormanii*
  2B. Flowers 8–12 mm. long..........................3. *P. procerus*

1. *P. diffusus* Dougl.   Diffuse Beard-tongue

Stem glabrous or puberulent, 2–6 dm. tall; leaves glabrate, serrate; inflorescence interrupted; calyx ciliate, 6–8 mm. long, the sepals lanceolate, acuminate; corolla blue or purple, about 2 cm. long.

Hyder—B. C.—Ore. Fig. 881.

2. *P. gormanii* Greene   Gorman Beard-tongue.

Stems clustered, decumbent at the base, glandular-pubescent above, 1–5 dm. tall; lower leaves petioled, narrowly spatulate or linear; upper leaves sessile, linear to narrowly lanceolate, 3–8 cm. long; calyx densely pubescent, nearly 1 cm. long, the lobes attenuate; corolla rose-purple; capsule about 1 cm. long.

Central Alaska—Yukon—B. C. Fig. 882.

3. *P. procerus* Dougl.

Stems decumbent at the base, glabrous or slightly pubescent, 10–35 cm. tall; basal leaves oblanceolate or linear-oblanceolate, petioled, glabrous, 4–6 cm. long; inflorescence compact but interrupted below; calyx glabrous, about 5 mm. long, the teeth cuspidate; corolla purplish blue.

Nome and southeast Alaska—Yukon—Sask.—Colo.—Calif. Fig. 883.

3. COLLINSIA Nutt.

Winter annual or biennial herbs; leaves opposite or verticillate; flowers axillary; calyx campanulate, 5-cleft; corolla tube short, the limb 2-lipped; upper lip 2-cleft, the lower lip larger and 3-lobed, the middle lobe keeled and enclosing the stamens. (Zaccheus Collins was a botanist of Philadelphia.)

C. parviflora Dougl. Blue Chickweed.

Stems weak, the branches spreading, 5–30 cm. tall; leaves oblong or lanceolate, 1–4 cm. long, sometimes with a few teeth, the upper often whorled; corolla 5–7 mm. long, blue or whitish; seeds concave.

Haines and Hyder—Ont.—Mich.—Colo.—Ariz. Fig. 884.

4. MIMULUS L.

Annual or perennial herbs with opposite, mostly toothed leaves; flowers axillary and peduncled; calyx angled, unequally 5-lobed; corolla with a reflexed, 2-lobed upper lip and a spreading 3-lobed lower lip; capsule many-seeded, enclosed by the calyx. (Latin, a buffoon, from the grinning corolla.)

Flowers yellow..1. *M. guttatus*
Flowers rose-red..2. *M. lewisii*

1. *M. guttatus* DC. Yellow Monkey-flower.
   M. langsdorfii Donn.

Stems glabrous below, pubescent above, 1–9 dm. tall; leaves variable, the lower petioled, the upper sessile or clasping, glabrous; calyx 10–15 mm. long, puberulent; corolla 2–4 cm. long, spotted on the lower lip. A variable species. At Craig the author collected dwarf plants less than 1 dm. tall with flowers nearly 4 cm. long growing alongside plants 4–5 dm. tall with flowers 3 cm. long.

Wet places, Aleutians—Talkeetna—B. C.—Mont.—Mexico—Calif. Fig. 885.

2. *M. lewisii* Pursh. Lewis Monkey-flower.

Stems 3–8 dm. tall, more or less viscid-pilose; leaves oblong to lanceolate, dentate, pubescent; flowers on long peduncles; calyx glandular-pubescent, up to 2 cm. long, the teeth triangular and acuminate; corolla 35–50 mm. long.

Hyder—Minn.—Colo.—Ariz.—Calif. Fig. 886.

5. LIMOSELLA L.

Low, glabrous, floating or creeping annuals, or perennial by stolons; leaves basal, entire, slender-petioled; flowers small, white, pink or purplish, borne singly on scape-like peduncles; corolla nearly regular. (Greek, seated in mud.)

L. aquatica L. Mudweed.

Leaves narrowly spatulate or with no blade distinct from the petiole, 2–7 cm. long, the blade ¼ to ⅓ as long as the petiole; peduncles shorter than the leaves; corolla about 2 mm. broad; capsule about 3 mm. long.

Imuruk Basin and Atka I., of wide geographic distribution. Fig. 887.

6. SYNTHERIS Benth.

Low perennials with mostly basal leaves; flowers blue or pink in terminal spikes or racemes; calyx of 4 slightly united sepals; corolla irregularly 2-lipped or wanting; filaments exerted, the anther-cells parallel; capsule short, emarginate; seeds several, flat. (Greek, together and a door, in allusion to the valves of the pod.)

S. borealis Pennell. Kitten tails.

Stems woolly with brown hairs, 5–15 cm. tall; basal leaves cordate in outline, doubly serrate, woolly, especially along the margins; stem leaves few, reduced; flowers in a head-like spike; calyx lobes acute, woolly; capsule emarginate with woolly margins.

Alaska Range and south Yukon. Fig. 888.

7. VERONICA (Tourn.) L.

Annual or perennial herbs; leaves usually opposite but sometimes alternate or verticillate; flowers blue or whitish, axillary, racemose or spicate; calyx mostly 4-parted; corolla rotate, 4-lobed; stamens 2, divergent, inserted at the base of the upper corolla lobe; styles united with a capitate stigma; capsule flat, usually notched or 2-lobed at the apex; seed flat or concave on one side. (Named for St. Veronica.)

```
1A. Flowers in axillary racemes (Euveronica).
  1B. Capsules pubescent.
    1C. Stem less than 5 cm. tall...?.................. 1. V. grandiflora
    2C. Stems 1–3 dm. long............................ 2. V. chamaedrys
  2B. Capsule glabrous or nearly so.
    1C. Capsule much wider than long................. 3. V. scutellata
    2C. Capsule nearly as long as wide................ 4. V. americana
2A. Flowers in terminal spikes or racemes (Veronicella)
  1B. Perennials.
    1C. Capsule wider than long.
      1D. Corolla pale violet with darker lines.......... 5. V. tenella
      2D. Corolla whitish with violet lines.............. 6. V. serpyllifolia
    2C. Capsule as long as or longer than wide, not or only
        slightly notched.
      1D. Fruiting pedicels 8–11 mm. long.............. 7. V. stelleri
      2D. Fruiting pedicels 2–5 mm. long.............. 8. V. wormskjoldii
  2B. Annuals.
    1C. Pedicels longer than the ovate sepals........... 9. V. persica
    2C. Pedicels shorter than the lanceolate to linear sepals.
      1D. Leaves narrow, nearly entire.................10 V. peregrina
      2D. Leaves wider, crenate-serrate................11. V. arvensis
```

1. *V. grandiflora* Gaertn. Large-flowered Speedwell.

Pubescent with flat, many-celled hairs; stems decumbent at the

base and with short internodes; leaves 3-5 pairs, broadly oval, obscurely serrate, 15-35 mm. long, contracted into a short petiole; peduncles 1-3, surpassing the leaves, 3- to 8-flowered; corolla blue, 10-15 mm. across.
Kamchatka—Unalaska. Fig. 889.

2. *V. chamaedrys* L. Germander Speedwell.

Stems ascending, slender, pubescent in 2 lines, 1-3 dm. tall; leaves ovate, sessile or nearly so, pubescent, incised-dentate, 12-30 mm. long; corolla light blue, 5-8 mm. across.
Sparingly adventative, native of Europe.

3. *V. scutellata* L. Skullcap Speedwell.

Glabrous or sparingly pubescent; stems slender, weak, 1-5 dm. tall; leaves linear or linear-lanceolate, nearly entire, sessile and slightly clasping, 25-75 mm. by 2-6 mm.; corolla blue, 4-6 mm. across; capsule emarginate at base and apex.
Yukon—Newf.—Va.—Colo.—Calif.

4. *V. americana* (Rof.) Schwein. Brooklime.

Stems glabrous, 2-6 dm. long, usually decumbent and rooting at the base; leaves short-petioled, oblong-lanceolate, 2-8 cm. long, serrate or sometimes almost entire; flowers in long, slender, bracted racemes; corolla blue or nearly white, rarely pink, 4-6 mm. wide; capsule thick, orbicular, slightly notched at the apex.
Growing in water or mud, Aleutians—central Alaska—Newf.—S. C.—Mex.—Calif. and west shore of Bering Sea. Fig. 890.

5. *V. tenella* All. Low Speedwell.
V. humifusa Dickson

Lower portion of stem decumbent and rooting at the nodes, ascending portion 5-30 cm. tall; leaves short-petioled or sessile, suborbicular to ovate, entire or denticulate, 5-18 mm. long, the upper reduced; inflorescence pubescent and often glandular; corolla 3-4 mm. wide; capsule retuse at the apex.
Pacific coast regions of Alaska, circumboreal, south to Maine—N. Y.—Wis.—Colo.—Mex. Fig. 891.

6. *V. serpyllifolia* L. Thyme-leaved Speedwell.

Similar to *V. tenella* but the flowers are smaller, the upper part of the stem is less pubescent and with shorter hairs, and the lower decumbent part of the stem does not root.
Hyder and Haines—Lab.—Ga.—N. M.—Calif.

7. *V. stelleri* Pall. Steller Speedwell.
V. alpina unalaschkensis C. & S.

Stems ascending from a usually decumbent base, 8-35 cm. tall, hirsute, leafy to the base of the inflorescence; leaves ovate, sharply serrate, up to 4 cm. long but usually much smaller; corolla blue, 8–11 mm. wide; capsules ovate, 7–8 mm. long. Some forms described as *V. stelleri* var. *glabrescens* Hult. and *V. wormskjoldii nutans* (Bong.) Pennell may be hybrids of the two species.

East Asia—Aleutians and Pribilof I.—southeast Alaska. Fig. 892.

8. *V. wormskjoldii* R. & S.   Alpine Speedwell.
    V. alpina var. *wormskjoldii* (R. & S.)

Stems 1–3 dm. tall, usually simple, pubescent, glandular above; leaves oval or ovate, entire or crenulate, sessile, 1–3 cm. long; corolla blue, campanulate, about 5 mm. wide; capsule emarginate, 4–5 mm. long; a form with wider, distinctly toothed leaves is the var. *nutans* (Bong.) Pennell.

Central Aleutians—Nome—Lab.—N. H.—Ariz. Fig. 893.

9. *V. persica* Poir.   Persian Speedwell.
    V. buxbaumii Tenore

Stems pubescent, diffusely branched, spreading or ascending, 1–3 dm. tall; leaves ovate or oval, deeply crenate-dentate, 10–25 mm. long; flowers blue, about 1 cm. broad, borne on slender pedicels from the axils of the alternate leaves; calyx lobes spreading; capsule nearly 2 times as wide as long.

Sparingly adventative; native of Europe.

10. *V. peregrina* L. var. *xalapensis* (H.B.K.) Pennell.   Neckweed.

Annual, 1–3 dm. tall, more or less pubescent; leaves thick, the lower petioled and opposite, the upper flower-bearing ones alternate; racemes spike-like; corolla whitish, 2–3 mm. wide; capsule orbicular, cordate at the apex, of nearly same length as the calyx lobes.

Probably introduced, widespread in the Americas. Fig. 894.

11. *V. arvensis* L.   Corn Speedwell.

Stems pubescent, 5–25 cm. tall; lower leaves petioled and opposite, ovate, crenate; upper leaves sessile, the floral ones reduced to bracts and alternate; corolla blue or whitish, 2 mm. wide, shorter than the calyx; capsule shorter than the calyx, 2 mm. long, obcordate.

Sparingly adventative, native of Europe.

8. LAGOTIS Gaertn.

Perennial glabrous herbs; rootstocks from nearly upright to horizontal; stems scapiform with reduced leaves on the upper part; flowers bluish, in a dense terminal spike, each solitary and sessile in the axis of a bract; corolla 2-lipped, the upper usually crenulate, the lower divided into 2 widely diverging lobes; ovary 2-celled, 2-ovuled.

L. glauca Gaertn.

Stems up to 35 cm. tall, lower leaves ovate to reniform, crenate, up to 15 cm. long; spikes 13–20 mm. thick; stamens shorter than the upper lip of the corolla. Var. *stelleri* (C. & S.) Trautv. has ovate or lanceolate leaves, often with sharp-pointed teeth, the blades seldom more than 6 cm. long and spikes 10–15 mm. thick.

An asiatic species, the head form extending to the Talkeetna Mts. and Bering Sea regions, the variety to the Arctic Coast and Yukon. Fig. 895.

9. DIGITALIS (Tourn.) L.

Tall biennial or perennial herbs; leaves large, alternate; flowers in terminal spikes or racemes, showy; calyx 5-parted; corolla declined, somewhat 2-lipped; stamens ascending, mostly included; seeds numerous, rugose. (Latin, finger of a glove, from the shape of the corolla.)

D. purpurea L.   Foxglove.

Stems erect, 6–20 dm. tall, pubescent; basal and lower leaves ovate or ovate-lanceolate, slender-petioled, dentate; upper leaves smaller, becoming sessile; corolla purple to white, spotted within, up to 45 mm. long.

Sparingly escaped from cultivation, native of Europe.

10. CASTILLEJA Mutis.

Herbs, partially parasitic on the roots of other plants; leaves alternate; flowers red, yellow, purple or white, in dense, leafy-bracted spikes, the bracts usually colored and more conspicuous than the flowers; calyx flattened, 4-lobed and more deeply cleft above and below than on the sides; corolla flattened, 2-lobed, the upper lip arched and entire, the lower lip short and 3-lobed; stamens inclosed in the upper lip; capsule many-seeded.

```
1A. Lower lip of corolla at least one third as long as the upper (galea).
   1B. Leaves ovate-lanceolate with 2 or 3 pairs of
         lateral lobes.................................... 1. C. parviflora
   2B. Leaves linear to lanceolate, the lobes if present, linear.
      1C. Annual......................................... 2. C. annua
      2C. Perennial.
         1D. Calyx lobes distinct 3–8 mm. from apex,
                longer than the united part.............. 3. C. pallida
         2D. Calyx lobes 0.5–2.5 mm. long, shorter than
                the united part.
            1E. Bracts violet-purple....................... 4. C. raupii
            2E. Bracts yellow or yellowish.
               1F. Corolla 15–20 mm. long.
                  1G. Stems 20–35 cm. tall.................... 5. C. yukonis
                  2G. Stems 7–12 cm. tall..................... 6. C. hyperborea
               2F. Corolla 10–13 mm. long
                  1G. Stem and inflorescence heavily villous... 7. C. villosissima
                  2G. Stems puberulent or finely pubescent.... 8. C. muelleri
2A. Lower lip of corolla less than one-fifth the length of the galea.
   1B. Calyx lobes obtuse or rounded, bracts yellowish... 9. C. unalaschcensis
```

2B. Calyx lobes acute to acuminate, bracts red or dull yellowish.
  1C. Bracts all acute or acuminate....................10. *C. miniata*
  2C. Bracts partly or wholly obtuse or rounded.
    1D. Corolla 18–25 mm. long; inflorescence
      elongating ....11. *C. hyetophila*
    2D. Corolla 25–30 mm. long; inflorescence short
      and dense...............................12. *C. chryomactis*

1. *C. parviflora* Bong.

Stems glabrous, 1–4 dm. tall; leaves 2–4 cm. long, quite variable, but with from 2–6 dm. long linear or subulate teeth and a long pointed apex; bracts similar, reddish; flowers about 15 mm. long; calyx about 1 cm. long, pubescent, the lanceolate lobes about one-third as long as the tube; corolla tube about as long as the calyx; galea about 5 mm. long with a small tooth at the base of the apex and a pubescent ridge on the back.

Prince William Sound—Queen Charlotte I. Fig. 896.

2. *C. annua* Pennell.

Stems solitary, much branched, finely appressed-pubescent, villous in the inflorescence, about 5 dm. tall; leaves lanceolate, 3-ribbed, finely pubescent; bracts greenish-yellow, proximately becoming purple; calyx 12–13 mm. long; corolla 13–16 mm. long; galea 5–6 mm. long; lower lip 3–4 mm. long.

Tanana valley near Fairbanks.

3. *C. pallida* (L.) Spreng.

Leaves caudate; anterior lip of corolla about two-thirds the length of the galea; spikes relatively dense, the bracts overlapping and appressed. A circumpolar, polymorphic species represented in our region by 5 races as follows:

1A. Bracts yellowish, inflorescence merely hirsute.
  1B. Stems with spreading hairs, usually 2–5 dm. tall,
    the leaves pubescent........................ Subsp. *typica*
  2B. Stems usually appressed-pubescent, 1–3 dm. tall,
    leaves glabrate.............................Subsp. *caudata*
2A. Inflorescence villous, bracts usually violet-purple
  (except in 1B.)
  1B. Corolla 20 mm. long; bracts and villous hairs of the
    inflorescence yellow........................Subsp. *auricoma*
  2B. Corolla 14–18 mm. long; bracts purplish or ochroleuceous,
    hairs white.
    1C. Stems 1–3 dm. tall; leaves entire................Subsp. *mexiae*
    2C. Stems 5–15–25 cm. tall; leaves entire or some of
      them lobed................................Subsp. *elegans*

The typical form is asiatic and occurs on the Bering Sea coast. Subsp. *caudata* Pennell has linear-attenuate or -caudate leaves 5–9 cm. long; corolla 15–20 mm. long. It ranges from Seward Pen. and Nunivak Island.—Mackenzie. Fig. 897. Subsp. *auricoma* Pennell has stems 15–20 cm. tall; leaves linear-lanceolate. 2–3 cm. long. Known from the Chandalar River. Subsp. *mexiae* (Eastw.) Pennell has linear-lanceolate leaves

4-6 cm. long, the bracts violet-purple. It is found from the Alaska Range and Matanuska to the Wrangell Mts. Subsp. *elegans* (Ostenf.) Pennell has linear or linear-lanceolate leaves 3-6 cm. long and violet-purple bracts. It occurs on Seward Pen. and the Arctic Coast to Hudson Bay.

4. *C. raupii* Pennell.

Stems several, 3-5 dm. tall, finely retrorse-pubescent; leaves linear, attenuate or caudate, 3-6 cm. long; bracts oval, becoming lanceolate, with a pair of lateral lobes; inflorescence villous; calyx 13-16 mm. long, cleft one-half its length, violet-purple; corolla 15-18 mm. long; galea 5-6 mm. long, acute, green, hirsute, with wide, glaborus, purplish margins; lower lip 2.5-3 mm. long.

Tanana Valley—Keewatin—James Bay—Peace River. Fig. 898.

5. *C. yukonis* Pennell.

Stems 20-35 cm. tall, purplish, pubescent with spreading or retrorse white hairs; inflorescence hirsute or villous with yellowish hairs; leaves linear, attenuate, 2-6 cm. long; bracts lanceolate with 1 or 2 pairs of short lateral lobes, obtuse or rounded, yellowish; calyx 13-18 mm. long, cleft three-fifths to two-thirds its length; galea 7-8 mm. long; lower lip 4-5 mm. long.

Yukon Territory.

6. *C. hyperborea* Pennell.

Stems from a much-branched crown, hirsute-pubescent; leaves lance-linear, attenuate, 2-4 cm. long, the lowest entire but most with 1 or 2 pairs of narrow lateral lobes; calyx 13-17 mm. long, cleft one-half its length; galea 5-8 mm. long, green with pale yellow margins; lower lip 3-4 mm. long.

Seward Pen.—central Yukon. Fig. 899.

7. *C. villosissima* Pennell.

Stems 5-14 cm. tall; leaves linear lanceolate, 1-4 cm. long, some of the upper ones with a pair of divaricate lobes; bracts ovate, with 1 or 2 pairs of lobes; obtuse or rounded, yellowish; calyx 10-13 mm. long, cleft about one-half length, its lobes very short; galea 5-6 mm. long; lower lip 3-4 mm. long.

Southwest Yukon. Fig. 900.

8. *C. muelleri* Pennell.

Stems several, 15-30 cm. tall, puberulent or finely pubescent, hirsute with yellowish hairs in the inflorescence; leaves linear, attenuate, entire, finely pubescent, 25-35 mm. long; bracts lanceolate or ovate-lanceolate, the lowest entire, the upper with a pair of slender lobes; calyx cleft two-fifths of its length, the lobes cleft only about 1.5 mm.; galea 5-6 mm. long; the lower lip 4-5 mm. long.

Southwest Yukon.

9. *C. unalaschcenis* (C. & S.) Malte.

Stems lanate-pubescent to glabrate, villous-hirsute in the inflorescence, 3–6 dm. tall; leaves entire, 5–10 cm. long, strongly 3-ribbed; bracts yellowish to orange, oval, 2–3 cm. long, entire or the upper ones with 1 or 2 pairs of teeth; calyx 18–22 mm. long, cleft one-half its length; corolla 2–3 cm. long. Subsp. *transnivalis* Pennell of northwest B. C. and Yukon is a smaller form with leaves lanceolate, 4–6 cm. long; corolla 15–20 mm. long.

Along the coast, Aleutian and Pribilof I.—southeast Alaska. Fig. 901.

10. *C. miniata* Dougl.

Stems 2–6 dm. tall glabrous nearly to the inflorescence; leaves lanceolate or linear, 3–6 cm. long, 3-nerved, glabrous; bracts crimson, often more or less cleft; calyx teeth lanceolate, acute, about 5 mm. long; corolla up to 30 mm. long; galea up to 15 mm. long; lower lip small.

Southeast Alaska—Alba.—Colo.—Utah—Ore. Fig. 902.

11. *C. hyetophila* Pennell.

Stems several, 3–6 dm. tall, glabrous or slightly pilose, villous-hirsute in the inflorescence; leaves linear-lanceolate or narrowly lanceolate, 3–10 cm. long, usually entire, 3-ribbed; bracts elliptic or oval the upper with a pair of lateral lobes, distally red; calyx 15–25 mm. long the teeth 3–7 mm. long; corolla 18–35 mm. long; galea 9–14 mm. long; lower lip 1–1.5 mm. long.

Southeast Alaska. Fig. 903.

12. *C. chrymactis* Pennell.

Stems several, erect, 3–5 dm. tall, glabrous or with sparse appressed hairs, villous in the inflorescence; leaves lanceolate, acuminate, 6–10 cm. long, entire, 3-ribbed; bracts mostly with 1–3 pairs of slender lobes; calyx 2–3 cm. long, the teeth 3–7 mm. long, red; corolla 25–30 mm. long; galea 11–15 mm. long; lower lip 1–2 mm. long.

Glacier and Yakutat Bays.

11. ORTHOCARPUS Nutt.

Alternate-leaved annuals related to *Castilleja;* leaves sessile, pectinately cleft or entire, those of the inflorescence sometimes highly colored; flowers perfect, in terminal spikes; calyx tubular or tubular-campanulate, 4-cleft; corolla very irregular, the upper lip erect and not exceeding the saccate, 3-lobed lower lip; capsula oblong; seeds many, reticulate. (Greek, erect fruit.)

O. hispidus Benth.   Lesser Paintbrush.

Stems usually simple, 10–25 cm. tall, pubescent throughout; leaves 2–4 cm. long with linear-lanceolate lobes, the flowering leaves similar but shorter with more and stiffer lobes; flowers about 15 mm. long; calyx lobes nearly as long as the tube, linear; corolla whitish or cream-colored, the upper lip sharp-pointed and seemingly longer than the 3-lobed lower lip.

Skagway, probably introduced from western U. S.

12. EUPHRASIA (Tourn.) L.

Erect, usually branching herbs partially parasitic on other plants; leaves opposite, dentate or incised; flowers in leafy, terminal spikes; calyx tubular, 4-cleft; corolla 2-lipped, the upper lip 2-lobed, the lower lip larger, with 3 spreading lobes; capsule oblong; seeds many, oblong, longitudinally ribbed. The species are known as Eyebright. (Greek, delight.)

Inflorescence nearly capitate..........................1. *E. mollis*
Inflorescence more elongate...........................2. *E. subarctica*

1. *E. mollis* (Ledeb.) Wettst.

Stems pubescent, 4–12 cm. tall; leaves 4–10 mm. long; inflorescence compact; calyx densely pilose, its triangular teeth barely acute. Closely related to *E. subarctica*.

East Asia, the Aleutians and southwest Alaska. Fig. 904.

2. *E. subarctica* Raup.

E. disjuncta of reports from Alaska and Yukon.

Stems finely puberulent, 6–30 cm. tall, often branched below; leaves 8–18 mm. long, ovate or orbicular, crenate with 7–11 teeth; bracts large and resembling the leaves but with more pointed teeth; corolla 4–5.5 mm. long and with a yellow eye; capsule 4–5 mm. long, about equaling the very acute calyx teeth.

Central and southwest Alaska—Lab.—Newf.—Maine—Alba. Fig. 905.

13. PEDICULARIS (Tourn.) L.

Annual, biennial or perennial herbs; leaves pinnate or pinnatifid; flowers perfect in terminal spikes or racemes; calyx cleft on the lower side, 2- to 5-lobed; corolla strongly 2-lipped; upper lip (galea) compressed, often beaked or toothed; lower lip 3-lobed, the lobes usually spreading; stamens ascending under the upper lip; capsule compressed and obliquely beaked. (Latin, louse.) The species are often known as Lousewort, but the Eskimo call the arctic species Bumble-bee Plant.

1A. Leaves verticillate.
  1B. Leaves deeply pinnatifid.......................12. *P. chamissonis*
  2B. Leaves 1- to 2-pinnately parted..................13. *P. verticillata*
2A. Leaves alternate (occasionally opposite).
  1B. Galea with a conical or thick-subulate beak.
    1C. Stem low more or less leafy.................... 7. *P. lapponica*
    2C. Stem scapiform (or with 1 pair of leaves)........ 8. *P. ornithorhyncha*
  2B. Galea with apex more or less incurved.
    1C. Annuals or biennials with branching stems.
      1D. Flowers yellowish............................ 9. *P. labradorica*

         2D. Flowers purplish-red.
            1E. Stems 25–75 cm. tall..........................10. *P. parviflora*
            2E. Stems less than 2 dm. tall..................11. *P. pennellii*
      2C. Stems simple from perennial roots.
         1D. Stems scapiform.
            1E. Corolla yellow............................... 1. *P. capitata*
            2E. Corolla purplish........................... 2. *P. sudetica*
         2D. Stems leafy.
            1E. Corolla yellowish.
               1F. Corollas about 12 mm. long................. 3. *P. flammea*
               2F. Corolla 15–20 mm. long................... 4. *P. oederi*
            2E. Corolla rose to purplish.
               1F. Spike densely lanate..................... 5. *P. lanata*
               2F. Spike pubescent but not densely lanate.
                  1G. Stems scape-like with 1–3 leaves........ 2. *P. sudetica*
                  2G. Stems more leafy....................... 6. *P. langsdorfii*

1. *P. capitata* Adams.

Stems usually pubescent, 3–12 cm. tall; leaves few, slender-petioled, the pinnate divisions deeply cut or toothed; flowers 2–6 in a capitata cluster; calyx 5-lobed, the lobes crenate; corolla up to 35 mm. long.

Bering Sea—Ellesmereland—Hudson Bay—Aleutians. Fig. 906.

2. *P. sudetica* Willd.

Stems solitary or few, glabrate but villous in the inflorescence, 15–40 cm. tall, scape-like but with a few leaves; base leaves lanceolate in outline and long-pointed; flowers in dense spikes which become elongated in fruit; calyx villous; corolla 15–22 mm. long, the galea recurved, 6–7 mm. long.

Eurasia — all of Alaska — Ellesmereland — James Bay — Mackenzie. Fig. 907.

3. *P. flammaea* L.

Stems glabrous or slightly woolly, 4–10 cm. tall; leaves few, 2–6 cm. long, pinnately divided into oblong or oval crenate divisions; calyx with 5 lanceolate teeth; corolla tube and lower lip yellowish, the galea tinged purple or crimson and about 6 mm. long.

A specimen from Goodnews Bay seems to belong here but the main range is east of our area. Fig. 908.

4. *P. oederi* Vahl.

Stems 6–20 cm. tall; leaves 3–7 cm. long, pinnately divided into dentate segments 3–5 mm. long; spikes 3–10 cm. long; calyx lobes lanceolate and more or less ciliate on the margins; corolla 18–22 mm. long, yellowish with purple-tinged galea about 8 mm. long and boat-shaped; lower lip deeply cleft with rounded lobes.

Eurasia—Kotzebue—Yukon—Mont.—Aleutians. Fig. 909.

5. *P. lanata* Willd.

Whole plant woolly except the lower leaves which are glabrous; leaves 2–6 cm. long, the divisions up to 6 mm. long with crenate to pinnatifid margins; spikes in fruit usually much longer than the remainder of the stem, often more than 2 dm. long; corolla rose-purple, about 2 cm. long.

Eurasia—Arctic Alaska—Ellesmereland—Greenland—Lab.—B. C.—Aleutians. Fig. 910.

6. *P. langsdorfii* Fisch.
    P. arctica R.Br.

Stems 4–18 cm. tall; leaves pinnatifid, the segments ovate with crenate margins, 1–3 mm. long; spikes dense, 3–5 cm. long; calyx about 8 mm. long, woolly, the lobes lanceolate with hairy margins; corolla 20–25 mm. long; galea 10–14 mm. long with a small tooth near the apex. The plant of the arctic is more pubescent in the inflorescence and has somewhat smaller flowers.

East Asia—Arctic Coast and central Alaska—Kodiak—Aleutians. Fig. 911.

7. *P. lapponica* L.

Stems usually simple, leafy, 10–25 cm. tall; leaves lanceolate, up to 35 mm. long, pinnately incised into oblong, serrate lobes; spikes short and dense, almost capitate; flowers light yellow, 12–14 mm. long, the galea erect and arched.

Rare in our area, Eurasia—Baffin Land—Greenland—Mackenzie. Fig. 912.

8. *P. ornithorhyncha* Benth.
    P. pedicellata Bunge.

Stems 1–2 dm. tall, appearing scapose but usually with one pair of leaves; basal leaves long-petioled, pinnatifid almost to the midrib, the pinnae again pinnatifid, the teeth acute; galea deeply bent, the outside measuring almost 1 cm. long.

Southeast Alaska—Wash. Fig. 913.

9. *P. labradorica* Panzer.
    P. euphrasioides Steph.

Stems hirsute, usually much branched, 1–4 dm. tall; lower leaves pinnatifid, the upper merely crenate, 2–4 cm. long; flowers in axils of upper leaves or spicate, about 15 mm. long, yellow or sometimes the galea tinged reddish-purple; galea short with a very short beak and 2 lanceolate teeth at lower side of apex; pod 2 times as long as the calyx.

Asia—nearly all of Alaska—Greenland—Lab.—B. C. Fig. 914.

10. *P. parviflora* Smith.

Stems usually glabrous and branched, 3–9 dm. tall; stem leaves deeply pinnatifid, the segments crenately toothed, the uppermost reduced; flowers solitary in the upper axils or in loose terminal spikes,

10-13 mm. long; galea boat-shaped, 3-5 mm. long; calyx 2-cleft.
Central and south Alaska—Hudson Bay—Lake Mistassini—Ore. Fig. 915.

11. *P. pennellii* Hult.

Stem glabrous, widely branched from the base, 10-15 cm. tall; leaves pinnatisect, the segments toothed; calyx glabrous, 2-parted, the teeth dentate; corolla rose-purple, 10-14 mm. long; galea erect with 2 acute teeth well back from the apex.
Alaska Pen.—Kotzebue—Lake Illiamna. Fig. 916.

12. *P. chamissonis* Stev.

Robust perennial 2-6 dm. tall; leaves usually in whorls of 3 or 4, up to 9 cm. long, deeply pinnatifid, the lanceolate divisions serrate or incised; corolla reddish, up to 25 mm. long; galea boat-shaped, scarcely as long as the lower lip.
East Asia—Aleutians—Alaska Pen.—Pribilof I. Fig. 917.

13. *P. verticillata* L.

Stems somewhat pubescent, usually clustered, 1-4 dm. tall; basal leaves long-petioled, those of the stem verticillate and short-petioled or sessile; flowers in spikes, a few in the axils of the upper leaves; galea about 5 mm. long.
Eurasia through Alaska and Yukon. Fig. 918.

Pedicularis groenlandica Retz. The Little Elephants has been reported from Alaska. It is characterized by the galea which is much prolonged into a narrow recurved beak.

14. RHINANTHUS L.

Annual, erect, mostly branching herbs with opposite leaves; flowers perfect, solitary in the axils of the upper leaves, becoming 1-sided spikes; calyx compressed, 4-toothed, becoming inflated in fruit, reticulate; corolla 2-lipped, the upper compressed, with 2 minute teeth below the apex, the lower lip shorter with 3 lobes; anthers hairy, the sacs distinct; capsule orbicular, flat, dehiscent, containing several winged seeds. (Greek, nose-flower, from the beaked corolla.)

R. *minor* L. subsp. *groenlandicus* (Chab.) Neum. Rattlebox.
R. *borealis* Sternb. R. *crista-galli* C. & S.

Stems glabrous or pubescent above, 1-7 dm. tall; leaves lanceolate or oblong-lanceolate, becoming broader at the base and more pointed on the upper part of the stem, scabrous; calyx short-hairy, ciliate on the margins; corolla yellow; fruiting calyx 1 cm. or more broad.
Aleutians—Talkeetna—Yukon—Greenland—N. H.—Conn.—N. M.—Ore. Fig. 919.

50. LENTIBULARIACEAE (Bladder-wort Family)

Small scapose herbs growing in water or wet places; leaves, when submerged, dissected into filiform segments and in our species bladder-bearing; flowers perfect; calyx of 2 or 5 sepals; corolla 2-lipped, the tube spurred or saccate; stamens 2; ovary 1-celled with central placenta; style short or none.

Plants of wet places with entire leaves.................. 1. *Pinguicula*
Submerged plants with dissected leaves................ 2. *Utricularia*

1. PINGUICULA (Tourn.) L.

Perennials of wet places with 1-flowered scapes; leaves in rosettes, thick, producing a musilagenous secretion to which insects adhere; corolla in our species blue to purple, the tube produced into a nectar-bearing spur. (Pinguis, fat, in allusion to the greasy leaves.)

Scape villous; corolla less than 10 mm. long.............1. *P. villosa*
Scape smooth, corolla 12-25 mm. long....................2. *P. vulgaris*

1. *P. villosa* L.   Hairy Butterwort.

Scapes finely villous, 3–8 cm. tall; leaves 3–5, 6–12 mm. long; flowers 3–5 mm. broad; corolla with the upper lip 2-parted, the lower lip 3-parted, the tube contracted into a straight spur 3–5 mm. long. Grows in sphagnum bogs and is very hard to detect except when in flower.

Circumpolar, south to Unalaska, Prince William Sound, southeast Alaska, Lab. Fig. 920.

2. *P. vulgaris* L.   Bog Violet. Common Butterwort.

Scapes glabrous or nearly so, 3–20 cm. tall; leaves 3–7 ovate to elliptic, 15–40 mm. long; lips of the corolla equally spreading, the upper 2-lobed, the lower 3-lobed; spur subulate, acute.

Circumpolar, south to Aleutians—Wash.—Mont.—Mich.—N. Y. Fig. 921.

2. UTRICULARIA L.

Aquatic plants with immersed, finely-dissected, bladder-bearing leaves; bladders small, urn-shaped and provided with valvular lids, small aquatic animals thus being entrapped and digested; sepals 2; corolla yellow or yellowish with spur at the base. (Utriculus, a little bladder.)

1A. Leaf segments flat......................................2. *U. intermedia*
2A. Leaf segments filiform.
  1B. Leaves 2–5 cm. long..............................1. *U. macrorhiza*
  2B. Leaves less than 1 cm. long......................3. *U. minor*

1. *U. macrorhiza* LeConte.   Common Bladderwort.
    U. vulgaris Auct.

Stems submerged and very leafy; leaves 2- to 3-pinnately dissected; bladders 2–4 mm. long; scapes 1–3 dm. long, 5- to 10-flowered; spur horn-like, slightly curved; corolla bright yellow; pedicels recurved in fruit. Often included in the European *U. vulgaris* but differs in several respects.

All our area except the Arctic—Fla.—Mo.—Okla.—Lower Calif. Fig. 922.

2. *U. intermedia* Hayne. Flat-leaved Bladderwort.

Leaves 5–15 mm. long, trichotomous at the base; bladders on separate, leaflets branches; scapes 1- to 4-flowered; corolla yellow, the spur acute and appressed to the lower lip and nearly as long; fruiting pedicels erect. Commonly propagates by velvety winter buds.

Infrequent, circumboreal south to N. J.—Ind.—Calif. Fig. 923.

3. *U. minor* L. Lesser Bladderwort.

Stems slender with scattered alternate leaves; bladders not abundant, 2 mm. long; scapes 5–15 cm. long, 1- to 10-flowered; corolla pale yellowish, the upper lip very small; spur short and blunt; pedicels recurved in fruit.

Infrequent, circumpolar south to Conn.—Penn.—Ind.—Colo.—Calif. Fig. 924.

51. OROBANCHACEAE (Broom-rape Family)

A family of root-parasites without green foliage; leaves reduced to appressed scales; flowers perfect, sessile in the axils of the scales or solitary on peduncles in the axils of the scales; calyx 4- to 5-toothed, corolla much as in *Scrophulariaceae;* ovary 1-celled with 2–4 parietal placentae; seeds numerous, reticulated, wrinkled or striate.

Glabrous, thick, fleshy, brownish-red plant .... 1. *Boschniakia*
Plants glandular-pubescent and more slender .... 2. *Orobanche*

1. BOSCHNIAKIA C. A. Mey.

Stems thick and fleshy with numerous flowers in a cone-like spike; the whole plant of a reddish color; base of the anthers rounded. (Boschniak was a Russian botanist.)

B. rossica (C. & S.) B. Fetsch. Poque.
B. glabra C. A. Mey.

Stems 10–35 cm. tall from tuber-like formations parasitically attached to the roots of *Alnus;* lower scales triangular and sharp-pointed, broadest at the base, the upper blunt and broadest near the middle, often ciliate on the edges, otherwise glabrous; corolla 10–15 mm. long.

Asia—Seward Pen.—Mackenzie—Vancouver I. Fig. 925.

2. OROBANCHE (Tourn.) L.

Glandular or viscid-pubescent herbs parasitic on the roots of various plants; flowers long-peduncled; calyx campanulate with acute or acuminate lobes; corolla oblique, the tube elongated and curved, the upper lip 2-lobed, the lower lip 3-lobed. (Greek, choke-vetch.)

Stems 1–2 cm. long......................................1. *O. uniflora*
Stems 4–10 cm. long..................................2. *O. fasciculata*

1. **O. uniflora** L. One-flowered Cancer-root.
    Aphyllon uniflorum T. & G.   *Thalesia uniflora* (L.) Britt.

Stems very short and nearly subterranean, bearing 1–4 peduncles 5–20 cm. tall; corolla tinged violet, 15–20 cm. long, puberulent.
Sand Point and Kodiak; B. C.—Newf.—S. C.—Texas. Fig. 926.

2. **O. fasciculata** Nutt. Clustered Cancer-root.

Aphyllon fasciculatum Gray.   *Thalesia fasciculata* (Nutt.) Britt.
Stems rising 2–8 cm. above the surface, densely glandular-pubescent; peduncles few to several, 2–10 cm. long; corolla yellowish or purplish, 15–25 mm. long.
Yukon—Ind.—Neb.—Calif. Fig. 927.

52. PLANTAGINACEAE (Plantain Family)

Annual or perennial herbs, mostly with basal leaves; flowers subtended by bracts, usually in dense spikes; calyx 4-parted, inferior; corolla campanulate or tubular with 4 lobes, scarious, nerveless, persistent; pistils 1; ovary superior, 1- to 4-celled; stamens 2 or 4; fruit usually a circumsissle capsule.

PLANTAGO (Tourn.) L.

Our species are acaulescent herbs with strongly ribbed or fleshy leaves and flowers in dense spikes on rather long peduncles. (The Latin name.)

1A. Leaves linear.
  1B. Bracts linear, much longer than the calyx.........2. *P. aristata*
  2B. Bracts ovate or orbicular, about the same length
      as the calyx.....................................1. *P. maritima*
2A. Leaves wider.
  1B. Leaves ovate, abruptly contracted at base.........3. *P. major*
  2B. Leaves lanceolate to ovate, sometimes narrowly so.
    1C. Capsule indehiscent..............................4. *P. macrocarpa*
    2C. Capsule a circumsissle pyxis.
      1D. Seed concave on the inner surface............5. *P. lanceolata*
      2D. Seed nearly flat on the inner surface.
        1E. Plant sparingly pubescent with brown wool
          at the base...........................6. *P. eriopoda*
        2E. Plant somewhat villous with little or no
          wool at base..........................7. *P. canescens*

1. **P. maritima** L. subsp. *juncoides* (Lam.) Hult.   Goose-tongue.

A seaside perennial with fleshy, linear leaves; scapes either longer or shorter than the leaves, pubescent, especially just below the spike; spikes dense, blunt, 3–10 cm. long; corolla pubescent within, the lobes spreading; capsule 2–3 mm. long, 2- to 3-seeded.

Along the coasts this species is widespread in temperate climates. Fig. 928.

2. *P. aristata* Michx.　　　　　　　　　　　Large-bracted Plantain.

Annual, dark green; scapes erect, 12-35 cm. tall; leaves linear, acuminate, entire, narrowed into slender petioles, 3-8 mm. wide; spikes dense, cylindric, 2-15 cm. long; bracts linear, up to 3 cm. long; pyxis 2-seeded, the seed concave on the face.

Dawson, probably adventative from central U.S.A.

3. *P. major* L.　　　　　　　　　　　Common Plantain.

A weed with somewhat pubescent, oval or ovate leaves 5-15 cm. long on petioles of same length or less, 5- to 7-ribbed; scapes 1-5 (-7) dm. tall; spikes 4-20 cm. long; pyxis ovoid, about 3 mm. long. Reports of *P. asiatica* from Alaska refer to a form of this species.

Generally distributed in settled parts of our region and of wide geographic distribution. Fig. 929.

4. *P. macrocarpa* C. & S.　　　　　　　　　　　Seashore Plantain.

Leaves mostly 5- to 7-nerved on dilated petioles; scapes equaling or exceeding the leaves; spikes 2-5 cm. long; bracts fleshy and very dark-colored and with scarious margins; capsule ovoid-oblong, 6-8 mm. long; seeds 2, hollowed on the face.

Commander and Aleutian I. along the coast to Wash. Fig. 930.

5. *P. lanceolata* L.　　　　　　　　　　　Ribgrass. Buckhorn.

More or less pubescent; leaves 3- to 5-nerved, usually on long petioles and with a tuft of hairs at the base; scapes much exceeding the leaves, up to 6 dm. long; spikes dense, cylindrical, 2-8 cm. long; sepals broadly scarious-margined with green midrib.

Native of Europe but widely introduced in temperate regions.

6. *P. eriopoda* Torr.　　　　　　　　　　　Saline Plantain.

Perennial; leaves narrowly lanceolate or oblanceolate, entire somewhat pubescent, up to 2 dm. long; scapes 15-40 cm. tall; spikes up to 15 cm. long at maturity, sparse below but dense above; sepals oblong-ovate, with wide scarious margins.

Yukon—Mackenzie—Keewatin—Neb.—N. M.—Calif.

7. *P. canescens* Adams
P. septata Morris.

Perennial; leaves lanceolate to oblanceolate, 5-ribbed, with long or short petioles, entire or remotely dentate, the blade up to 15 cm. long and 3 cm. wide but often quite small, more or less pubescent; scapes 1-5 dm. tall; spikes up to 9 cm. long, dense; pyxis very finely reticulated.

Asia, central Alaska—Mackenzie and Mont. Fig. 931.

53. RUBIACEAE (Madder Family)

Our species all herbaceous plants; leaves opposite or whorled, entire; flowers perfect but often dimorphous or trimorphous; ovary inferior, 2- to 4-celled; stamens as many as the lobes of the corolla and alternate with them; fruit in our species of two 1-seeded carpels.

GALIUM L.

Annual or perennial herbs with 4-angled stems and whorled leaves; flowers small, mostly white, in cymes or panicles; calyx obsolete; corolla rotate; stamens mostly 4; styles 2; fruit separating at maturity into 2 indehiscent carpels. (Greek, milk, which some species were used to curdle.)

```
1A. Fruit bristly.
   1B. Flowers in terminal panicles...................1. G. boreale
   2B. Flowers solitary or in 3's.
       1C. Stem leaves in 4's, wide.......................2. G. kamtschaticum
       2C. Stem leaves in 6's to 8's.
           1D. Annual, leaves narrow, 3-7 cm. long...........3. G. aparine
           2D. Perennial, leaves wider, 1-3 cm. long.........4. G. triflorum
2A. Fruit smooth.
   1B. Leaves mostly in 5's (or 6's).......................6. G. trifidum columbianum
   2B. Leaves in 4's.
       1C. Pedicels long, thin, retrorsely scabrous.........6. G. trifidum
       2C. Pedicels short, thick, glabrous..................5. G. brandgei
```

1. *G. boreale* L.   Northern Bedstraw.

Erect perennial, 2-6 dm. tall; leaves 3-nerved, lanceolate, in whorls of 4; sometimes the leaves may be almost linear or there may be fascicles of leaves in the axils; flowers in a large terminal panicle and very ornamental; fruit about 2 mm. in diameter.

Common except in the arctic; circumboreal, south to N. J.—Ind.—Mo.—N. M.—Calif. Fig. 932.

2. *G. kamtschaticum* Steller.   Northern Wild Liquorice

Stems weak, 1-3 dm. tall; leaves broadly oval, orbicular or obovate, 3-nerved, obtuse, mucronulate, 10-30 mm. by 7-20 mm., sometimes ciliate; flowers terminal in 3's.

Asia—Aleutians—southeast Alaska—Wash. and in Que.—N. Y. Fig. 933.

3. *G. aparine* L.   Cleavers.

Stems weak, prostrate or scrambling over other vegetation, 3-15 dm. long, hispid on the angles; leaves linear-oblanceolate, 30-70 mm. by 2-5 mm., rough on the midrib and margins; fruit about 4 mm. in diameter.

Aleutians—southeast Alaska; of very wide distribution. Fig. 934.

4. *G. triflorum* Michx.   Sweet-scented Bedstraw.

Stems diffuse, glabrous or nearly so, shining, 3-10 dm. long; leaves in 6's, 2-6 cm. long, 3-12 mm. wide; peduncles often exceeding the leaves, 1- to 5-flowered but mostly 3-flowered; fruit long-hispid with hooked hairs, about 3 mm. broad.

Aleutians—central Alaska—Greenland and circumboreal south to Fla.—La.—Texas—Calif. Fig. 935.

5. *G. brandegei* Gray.   Brandegee Bedstraw.

Forming dense, low, leafy mats, mostly glabrous; leaves small, broadly spatulate, less than 10 mm. long and equaling or exceeding the internodes; flowers lateral, solitary; pedicels short, stout, glabrous, often not exceeding the fruit in length. Closely related to the next species.

Seward Pen.—Lab.—Iceland—Maine—Great Lakes—N. M.—Calif.

6. *G. trifidum* L.   Small Bedstraw.

Stems slender, diffuse and weak, retrorsely hispid; leaves narrowly linear to broadly spatulate, 4–10 mm. long; flowers minute; pedicels slender, solitary or terminal in 3's, scabrous, long and curved; fruit glabrous, its carpels about 1.5 mm. thick. Subsp. *columbianum* (Rydb.) Hult. is less diffuse; the stems ascending, 2–4 dm. long; leaves of the stem usually in 5's, those of the branches in 4's, 5–15 mm. long.

Aleutians—Mackenzie—Maine—N. Y.—Ind.—Colo. the subsp. in the coastal sections. Whole species circumboreal. Fig. 936.

54. CAPRIFOLIACEAE (Honeysuckle Family)

Shrubs, trees, vines or perennial herbs; leaves opposite; flowers perfect; calyx adnate to the ovary, its limb 3- to 5-parted; corolla gamopetalous, from rotate to tubular, often gibbous at the base, its limb 5-lobed and often 2-lipped; fruit a 1-seeded pod or more often a berry.

1A. Stamens 4; herbaceous trailing evergreen............1. *Linnaea*
2A. Stamens 5, adnate to the corolla and alternate with its
    lobes; shrubs.
  1B. Corolla tubular; stigma capitate.
    1C. Corolla irregular..............................2. *Lonicera*
    2C. Corolla regular................................3. *Symphoricarpus*
  2B. Corolla rotate or nearly so.
    1C. Leaves pinnate................................4. *Sambucus*
    2C. Leaves simple.................................5. *Viburnum*

1. LINNAEA (Gronov.) L.

Slender, trailing evergreens, somewhat woody, with ascending branches; flowers fragrant, pinkish, borne on slender, drooping pedicels at the forked top of the erect peduncles; calyx 5-lobed; corolla bell-shaped to funnelform, 5-lobed; fruit 1-seeded. (Named for Linneus, the father of modern botany.)

L. borealis L.   Twin-flower.

Leaves somewhat coriaceous, the blades oval or orbicular, generally crenate above the middle; peduncles at the fork and just below each flower with 2 glandular scales. Represented in our area by 3 forms.

The typical form with rounded leaf-apex and campanulate corolla in central Alaska and westward. Subsp. *americana* (Forbes) Hult. found from the Bering Sea across the continent has a longer, more trumpet-shaped corolla. Subsp. *longiflora* (Torr.) Hult. found in southeast Alaska—Idaho—Calif. has long, narrow sepals; a long narrow corolla; elliptical leaves which are acutish at the tip.

Whole species is circumboreal. Fig. 937.

2. LONICERA L.

Erect or climbing shrubs; leaves opposite, entire; corolla tubular, funnelform or campanulate, often gibbous at the base; fruit a few-seeded berry. (Adam Lonitzer was a German botanist.)

A vine with flowers in heads.............................1. *L. glaucescens*
A shrub with flowers in pairs on axillary peduncles......2. *L. involucrata*

1. *L. glaucescens* Rydb.   Glaucous Honeysuckle.

Stems twining, glabrous; leaves glabrous above; glaucous and more or less pubescent below, ovate 2-8 cm. long, the upper pair connate-perfoliate; corolla yellow, changing to reddish, 20-25 mm. long, gibbous at the base.

Liard River Hot Springs—Mackenzie—Penn.—N. C.—Ohio.—Okla.

2. *L. involucrata* (Rich.) Banks.   Black Twinberry.

A shrub with 4-angled twigs, 1-3 m. tall; leaves pubescent, at least on the veins and margins; corolla nearly regular, 10-12 mm. long, strongly gibbous at the base; fruit black with dark red involucre.

Southeast Alaska—Que.—Penn.—Ky.—Mont.—Calif. Fig. 938.

3. SYMPHORICARPOS (Dill.) Ludwig.

Shrubs; leaves simple, opposite, short-petioled; flowers perfect, white or pink; corolla campanulate; stamens 4 or 5, inserted on the corolla; ovary 4-celled; fruit a 2-seeded berry. (Greek, borne together, from the clustered fruit.)

S. rivularis Suksd.   Snowberry.

Erect or diffuse shrub 3-12 dm. tall; leaves ovate to oblong, obtuse at each end, often whitish or pubescent beneath, 15-50 mm. long; flowers in small axillary clusters; corolla bearded within, about 6 mm. long; berry snow-white, 5-10 mm. in diameter. Probably only a subspecies of *S. albus* (L.) Blake.

Southeast Alaska—Que.—Va.—Colo. Fig. 939.

4. SAMBUCUS (Tourn.) L.

Shrubs growing in clumps; leaves odd-pinnate, the leaflets finely serrate; flowers small, whitish, in compound cymes; corolla rotate or saucer-shaped with the 5 stamens inserted at its base; berry-like drupe 3- to 5-celled and -seeded. (The Latin name of the elder.)

S. racemosa L. subsp. *pubens* (Michx.) Hult. Red-berried Elder.

A shrub 1–4 m. tall with pyramidal inflorescence of whitish flowers which turn brown in drying; cyme 5–8 cm. by 4–6 cm.; drupe scarlet, occasionally orange.

Common in the Pacific coast regions, central Alaska—Newf.—Ga.—Colo.—Calif. Fig. 940.

5. VIBURNUM (Tourn.) L.

Shrubs or small trees; leaves simple; flowers white with spreading 5-lobed corolla; stamens 5, the style short and 3-cleft; ovary 1- to 3-celled but the fruit with a single compressed seed. (Ancient Latin name.)

V. edule (Michx.) Raf. Few-flowered Highbush Cranberry.
V. pauciflorum Pylaie

10–25 dm. tall; leaves variable, more or less pubescent beneath and usually 2 glands at the base of the blade; cymes rather few-flowered, short-rayed; drupe 8–10 mm. long, the stone flat.

Seward Pen.—Mackenzie—Newf.—Penn.—Colo.—Wash. Fig. 941.

55. ADOXACEAE (Moschatel Family)

A glabrous perennial with scaly or tuberiferous rootstock; basal leaves ternately divided; flowers small, greenish, borne in a terminal capitate cluster; corolla rotate, 4- to 6-lobed; stamens twice as many as the lobes of the corolla, borne in pairs on its tube; anthers peltate, 1-celled; ovary 3- to 5-celled; ovules 1 in each cell; fruit a small drupe with 3–5 nutlets.

ADOXA L.

The only genus. (Greek, without glory, i.e. insignificant.)

A. moschatellina L. Moschatel. Musk Root.

Stems simple, weak, 6–15 cm. tall, bearing a pair of ternate leaves; heads few-flowered, 6–8 mm. in diameter; drupes green, bearing the persistent calyx-lobes.

Circumpolar, south to Wis.—Iowa—Colo. Fig. 942.

56. VALERIANACEAE (Valerian Family)

Herbs with opposite, entire or pinnately divided leaves and no stipules; flowers small, in cymes; calyx tube adnate to the ovary, its limb inconspicuous in flower but often pappus-like in fruit; corolla tubular, funnelform or salver-shaped with 3–5 lobes, sometimes gibbous or spurred at the base; stamens 1–4, exerted; ovary 3-celled, only one maturing, giving rise to a 1-seeded fruit.

VALERIANA (Tourn.) L.

Heavy-scented perennials with small whitish or pinkish flowers in close cymes; calyx-limb at first inrolled, developing into 5–15 plumose bristles in fruit; corolla funnelform or salver-shaped, 5-lobed, often saccate at the base; stamens 3; achene flattened, 1-nerved on one side, 3-nerved on the other side. (Latin, valere, to be strong.)

1A. Corolla of pistillate flowers 2–3 mm. long............1. *V. septentrionalis*
2A. Corolla of pistillate flowers 5–8 mm. long.
  1B. Upper stem leaves all 3- to 7-foliate.............3. *V. sitchensis*
  2B. Upper stem leaves simple or 3-foliate.............2. *V. capitata*

1. *V. septentrionalis* Rydb.   Northern Valerian.

Stems erect, 2–4 dm. tall, glabrous or the inflorescence minutely pubescent; stem leaves usually 3-pairs with 5–7 segments, the segments oval to linear-lanceolate; flowers white, about 3 mm. wide.

Atlin—Great Bear Lake—Hudson Bay—Newf.—Wyo.

2. *V. capitata* Pall.   Capitate Valerian.

Stems glabrous, slender, 2–6 dm. tall; flowers in a capitate cluster which elongates in fruit; lower leaves simple, the upper trifoliate or 3-lobed, the center part much wider than the lateral.

Common, Eurasia across Alaska and Yukon to Mackenzie. Fig. 943.

3. *V. sitchensis* Bong.   Sitka Valerian.

Stems glabrous except in the inflorescence, 4–8 dm. tall; stem leaves 3- to 5-foliate, the lower less divided or simple; leaflets coarsely toothed; inflorescence rather dense; corolla white or pinkish, 6 mm. long.

Mountain meadows, Talkeetna Mts.—southeast Alaska—Mont.—Idaho—Ore. Fig. 944.

57. CAMPANULACEAE (Bellflower Family)

Our species all perennial (or biennial) caulescent herbs; leaves simple, alternate; calyx entirely enclosing the 2- to 5-celled ovary; flowers perfect; corolla 5-lobed, blue or rarely white; stamens 5, inserted with the corolla at the line where the calyx becomes free from the ovary; fruit a 2- to 5-celled capsule; seeds numerous.

Corolla regular..1. *Campanula*
Corolla 2-lipped..2. *Lobelia*

1. CAMPANULA (Tourn.) L.

Calyx 5-cleft; capsules opening by pores usually formed by the uplifting of small lids. (Diminutive of Latin, campana, a bell.)

1A. Calyx with deflexed appendages in the sinuses.......1. *C. dasyantha*
2A. Calyx without appendages at the sinuses.
  1B. Corolla rotate, lobed almost to the base..........2. *C. aurita*
  2B. Corolla campanulate, lobes not longer than the tube.
    1C. Calyx pubescent.
      1D. Calyx lobes laciniate.......................4. *C. lasiocarpa*
      2D. Calyx lobes entire..........................3. *C. uniflora*
    2C. Calyx glabrous.
      1D. Style much exceeding the corolla............5. *C. scouleri*
      2D. Style shorter than the corolla...............6. *C. rotundifolia*

1. *C. dasyantha* Bieb.

Stems 3–10 cm. tall, 1-flowered; leaves mostly basal, ovate to ovate-spatulate, crenate with gland-tipped teeth, the stem leaves varying to lanceolate or linear; calyx lobes triangular-lanceolate, about 1 cm. long; corolla deep blue, 25–35 mm. long, the tube longer than the lobes.

Japan—Aleutians. Fig. 945.

2. *C. aurita* Greene.

Stems erect or ascending, often tufted, 1- to 3-flowered, 7–25 cm. tall; leaves oblong to linear, sessile by a narrow base, the lower ciliate on the margin near the base, entire or with a few teeth; calyx teeth lanceolate, usually with a pair of acute lobes near the base; corolla spreading, 10–15 mm. long.

Yukon valley—Mackenzie. Fig. 946.

3. *C. uniflora* L. Arctic Harebell.

Stems 5–15 cm. tall, glabrous or nearly so; lower and basal leaves spatulate, narrowed into a petiole; upper leaves narrower; flowers erect or ascending, pubescent at the base of the calyx; loves of the corolla about equaling the tube; capsule ascending, opening by pores near the summit.

Aleutians—Arctic, circumpolar, to Colo. Fig. 947.

4. *C. lasiocarpa* Cham. Mountain Harebell.

Stems 3–15 cm. tall, the alpine form usually 1-flowered, the lowland form branched; lower leaves glabrous or somewhat ciliate, upper leaves more ciliate; hypanthium villous; sepals 6–10 mm. long and lobed; corolla 15–25 mm. long. This is our most widely distributed and common species.

Throughout our areas except the high Arctic. Asia—Alba.—B. C. Fig. 948.

5. *C. scouleri* Hook. Scouler Harebell.

Glabrous or slightly pubescent; stems slender, clustered from branching rootstocks, 1–3 dm. long; lower leaves ovate, acute, remotely serrate, 2–3 cm. long; upper leaves narrowed to linear bracts; flowers nodding; calyx lobes twice as long as the tube; corolla pale or white, the acute reflexed lobes nearly equaling the tube.

Wrangel—Calif.

6. *C. rotundifolia* L. Harebell. Blue Bells of Scotland.

Perennial by slender rootstocks; stems 15–50 cm. tall; basal leaves nearly orbicular to cordate, usually dentate, 6–25 mm. long, often wanting at flowering time; upper leaves narrower, often linear; flowers drooping or spreading; corolla 15–25 mm. long; capsule pendulous, ribbed, opening by valves at the base. This is a very diverse group from which several species have been described; the typical form is near to or identical with the form known by some as *C. petiolata* DC. which has linear or subulate lobes of the corolla; narrow linear upper leaves and usually branching stem. The latest to be described is *C. latisepala* Hult. which has usually only 1 but may have up to 3 flowers with triangular calyx lobes 2-3 mm. wide at the base and up to 12 mm. long; the upper leaves relatively wide. This form connects through the variety *dubia* Hult. which has narrower upper leaves and sepals and numerous other variations with the typical form of the species. Var. *alaskana* Gray and *C. heterodoxa* are other names used in this group.

Circumboreal, south to N. J.—Ind.—Neb.—N. M.—Calif. Fig. 949.

2. LOBELIA (Plum.) L.

Corolla split nearly to the base on the upper side; upper 2 lobes narrow, spreading or reflexed; lower 3 lobes united into a broad lip; stamens 5, the filaments monodelphous, the anthers united, two or all bearded at the apex; capsule 2-valved.

L. *kalmii* L.

Biennial; stems slender, 15-35 cm. tall, simple or with a few branches; basal leaves spatulate or oblanceolate, the stem leaves narrower, varying to linear, up to 4 cm. long; sepals acute, about 3 mm. long; corolla 7-8 mm. long.

Liard River Hot Springs — Great Slave Lake — Man. — N. D. — Mont.—Wash. Fig. 950.

PLATE XXXVI

Scale in millimeters.

Fig.
839. *Swertia perennis* L. Flower, leaf and nectiferous pit.
840. *Lomatigonium rotatum* (L.) Fries. Flower and pair of leaves.
841. *Gentiana prostrata* Haenke. Two joints of stem and flower.
842. *Gentiana douglasiana* Bong. Portion of corolla laid open and leaf.
843. *Gentiana algida* Pall. Leaf and flower.
844. *Gentiana glauca* Pall. Portion of corolla, leaf and seeds.
845. *Gentiana platypetala* Griseb. Flower and pair of leaves.
846. *Gentiana barbata* Froel. Flower and basal leaf.
847. *Gentiana tenella* Rottb. Flower and two lobes of corolla.
848. *Gentiana auriculata* Pall. Flower and lower stem leaf.
849. *Gentiana acuta* var. *plebeja* Corolla laid open and lower stem leaf
850. *Gentiana aleutica* C. & S. Flower and leaf.
851. *Gentiana propinqua* Rich. Flower and leaf.
852. *Menyanthes trifoliata* L. Leaf and flower.
853. *Fauria cristi-galli* (Menz.) Mikano. Leaf, flower and fruit.
854. *Apocynum androsaemifolium* L. Flower, leaf and fruit.
855. *Polemonium acutiflorum* Willd. Flower and section of leaf.
856. *Polemonium pulcherrimum* Hook. Flower and section of leaf.
857. *Polemonium boreale* Adams. Flower and tip of leaf.
858. *Phlox sibirica* L. Flower and leaf.
859. *Collomia linearis* Nutt. Flower and leaves.
860. *Romanzoffia sitchensis* Bong. Flower, fruit and leaf.
861. *Romanzoffia unalaschcensis* Cham. Flower and fruit.
862. *Phacelia franklinii* (R.Br.) Gray. Leaf, flower and fruit.
863. *Phacelia mollis* Macbr. Leaf and flower.

PLATE XXXVI

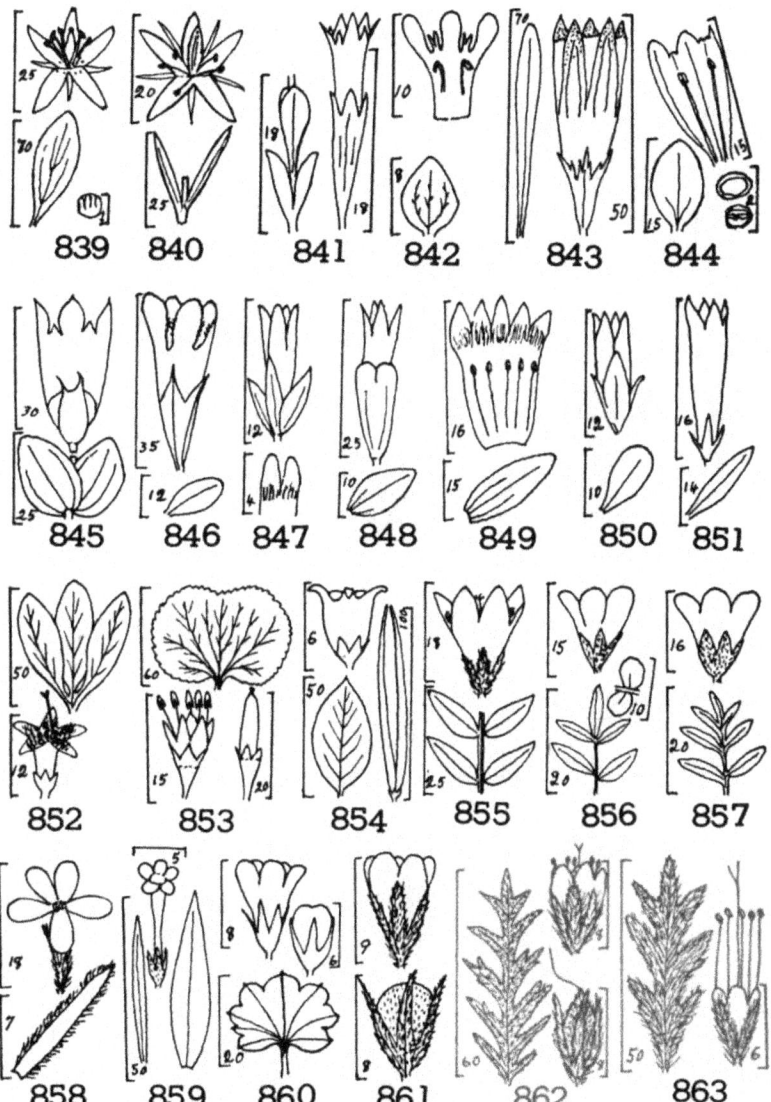

PLATE XXXVII

Scale in millimeters.
FIG.
864. *Lappula myosotis* Moench. Nutlet.
865. *Lappula redowskii* (Hornem.) Greene. Nutlet.
866. *Mertensia maritima* (L.) S. F. Gray. Leaf, flower and nutlet.
867. *Mertensia paniculata* (Ait.) Don. Leaf, flower and nutlet.
868. *Myosotis alpestris asiatica* Vesterg. Leaf, limb of corolla, nutlet and fruiting calyx.
869. *Eritrichium aretioides* (C. & S.) DC. Leaf, flower and nutlet.
870A. *Amsinckia menziesii* (Lehm.) Nels & Macbr. Nutlet.
870B. *Amsinckia lycopsoides* Lehm. Nutlet.
871. *Plagiobotrys cognatus* (Greene) Johnst. Leaf, fruiting calyx and nutlets.
872. *Cryptanthe torreyana* (Gray) Greene. Leaf, calyx and nutlets.
873. *Lycopus uniflorus* Michx. Leaf and calyx.
874. *Lycopus lucidus* Turcz. Leaf, calyx and group of nutlets.
875. *Mentha arvensis* L. Leaf, calyx and nutlet.
876. *Scutellaria galericulata* L. Leaf and flower.
877. *Dracocephalum parviflorum* Nutt. Leaf, fruiting calyx and nutlet.
878. *Prunella vulgaris lanceolata* (Barton) Hult. Leaf, calyx and nutlet.
879. *Stachys palustris pilosa* (Nutt.) Epling. Leaf and calyx.
880. *Galeopsis bifida* Boenn. Leaf, calyx and nutlet.
881. *Pentstemon diffusus* Dougl. Calyx and leaf.
882. *Pentstemon gormanii* Greene. Leaves and fruit.
883. *Pentstemon procerus* Dougl. Leaf, flower and fruit.
884. *Collinsia parviflora* Dougl. Flower and leaf.
885. *Mimulus guttatus* DC. Leaves and flower.
886. *Mimulus lewisii* Pursh. Leaf and flower.

FLORA OF ALASKA 439

PLATE XXXVII

PLATE XXXVIII

Scale in millimeters.
FIG.
887. *Limosela aquatica* Leaf, flower and fruit.
888. *Syntheris borealis* Pennell. Fruit and leaf.
889. *Veronica grandiflora* Gaertn. Flower and leaf.
890. *Veronica americana* (Raf.) Schwein. Leaf and fruit.
891. *Veronica tenella* All. Fruit and leaf.
892. *Veronica stelleri* Pall. Fruit and leaf.
893. *Veronica wormskjoldii* R. & S. Fruit and leaf.
894. *Veronica peregrina* var. *xalapensis* (H.B.K.) Pennell. Fruit and leaves.
895. *Lagotis glauca* Gaertn. Leaves and flower.
896. *Castilleja parviflora* Bong. Fruit and leaf.
897. *Castilleja pallida caudata* Pennell. Leaves and flower.
898. *Castilleja raupii* Pennell. Lower and upper leaves, and flower.
899. *Castilleja hyperborea* Pennell. Leaf and flower.
900. *Castilleja villosissima* Pennell. Leaves, corolla and capsule.
901. *Castilleja unalaschcensis* (C. & S.) Malte. Leaves and corolla.
902. *Castilleja miniata* Dougl. Leaf and flower.
903. *Castilleja hyetophila* Pennell. Leaf and flower.
904. *Euphrasia mollis* (Ledeb.) Wettst. Fruit and leaf.
905. *Euphrasia subarctica* Raup. Fruit, flower and leaf.
906. *Pedicularis capitata* Adams. Leaf and flower.
907. *Pedicularis sudetica* Willd. Leaf and flower.
908. *Pedicularis flammea* L. Flower and section of leaf.
909. *Pedicularis oederi* Vahl. Flower and section of leaf.
910. *Pedicularis lanata* Willd. Flower and section of leaf.
911. *Pedicularis langsdorfii* Fisch. Flower and section of leaf.

FLORA OF ALASKA

PLATE XXXVIII

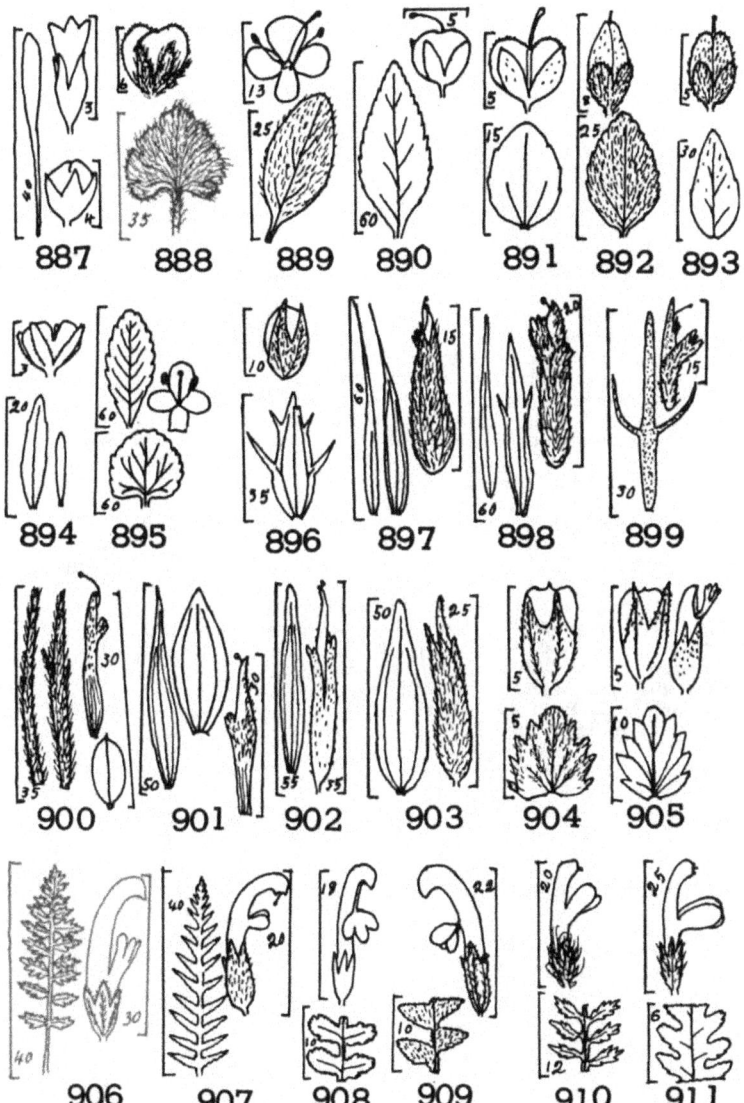

PLATE XXXIX

Scale in millimeters.

FIG.
912. *Pedicularis lapponica* L. Flower and leaf.
913. *Pedicularis ornithorhyncha* Benth. Corolla and section of leaf.
914. *Pedicularis labradorica* Panzer. Flower and sections of leaves.
915. *Pedicularis parviflora* Smith. Flower and section of leaf.
916. *Pedicularis pennellii* Hult. Flower and section of leaf.
917. *Pedicularis chamissonis* Stev. Flower and section of leaf.
918. *Pedicularis verticillata* L. Flower and section of leaf.
919. *Rhinanthus minor groenlandicus* (Chab.) Neum. Leaf, seed and flower.
920. *Pinguicula villosa* L. Whole plant.
921. *Pinguicula vulgaris* L. Flower and basal rosette of leaves.
922. *Utricularia macrorhiza* LeConte. Flower and leaf.
923. *Utricularia intermedia* Hayne. Winter bud, bladders and leaf.
924. *Utricularia minor* L. Floyer and section of stem.
925. *Boschniakia rossica* (C. & S.) B. Fedisch. Flower with bract and capsule.
926. *Orobanche uniflora* (C. & S.) B. Fedisch. Flower and stem.
927. *Orobanche fasciculata* Nutt. Flower and part of stem.
928. *Plantago maritima juncoides* (Lam.) Hult. Leaf, fruit and seed.
929. *Plantago major* L. Leaf, seed and fruit.
930. *Plantago macrocarpa* C. & S. Leaf, fruit and seed.
931. *Plantago canescens* Adams. Leaf, fruit and seed.
932. *Galium boreale* L. Fruit and whorls of leaves.
933. *Galium kamtschaticum* Steller. Fruit and leaf.
934. *Galium aparine* L. Fruit and whorl of leaves.
935. *Galium triflorum* Michx. Fruit and leaf.
936. *Galium trifidum* L. Fruit with pedicel and whorl of leaves.
937. *Linnaea borealis americana* (Forbes) Hult. Flower and leaf.

PLATE XIL

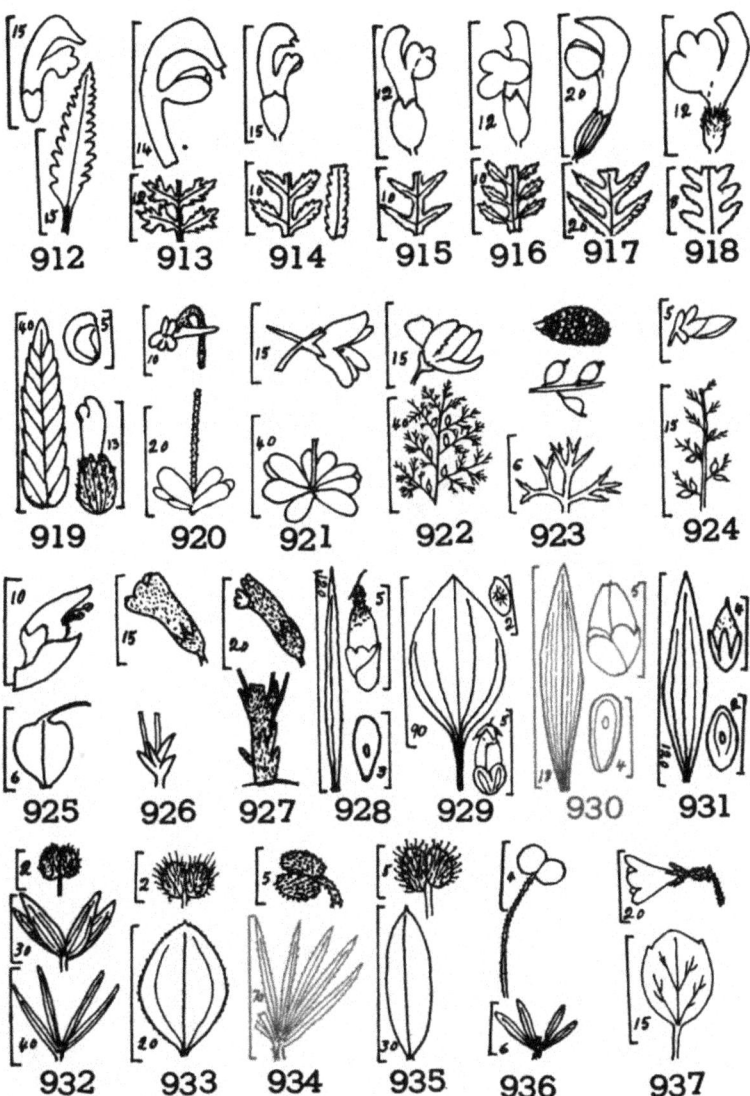

PLATE XL

Scale in millimeters.

FIG.
938. *Lonicera involucrata* (Rich.) Banks. Fruit and leaf.
939. *Symphoricarpus rivularis* Suksd. Flower and leaf.
940. *Sambucus racemosa pubens* (Michx.) Hult. Flower and leaflet.
941. *Viburnum edule* (Michx.) Raf. Flower, stone and fruit.
942. *Adoxa moschatellina* L. Flower, fruit and leaf.
943. *Valeriana capitata* Pall. Flower, seed and upper leaf.
944. *Valeriana sitchensis* Bong. Flower, seed and upper leaf.
945. *Campanula dasyantha* Bieb. Flower and leaf.
946. *Campanula aurita* Greene. Flower and leaves.
947. *Campanula uniflora* L. Flower, leaf and fruit.
948. *Campanula lasiocarpa* Cham. Flower and leaves.
949. *Campanula rotundifolia* L. Flower and leaves.
950. *Lobelia kalmii* Flower and leaves.

FLORA OF ALASKA

PLATE XL

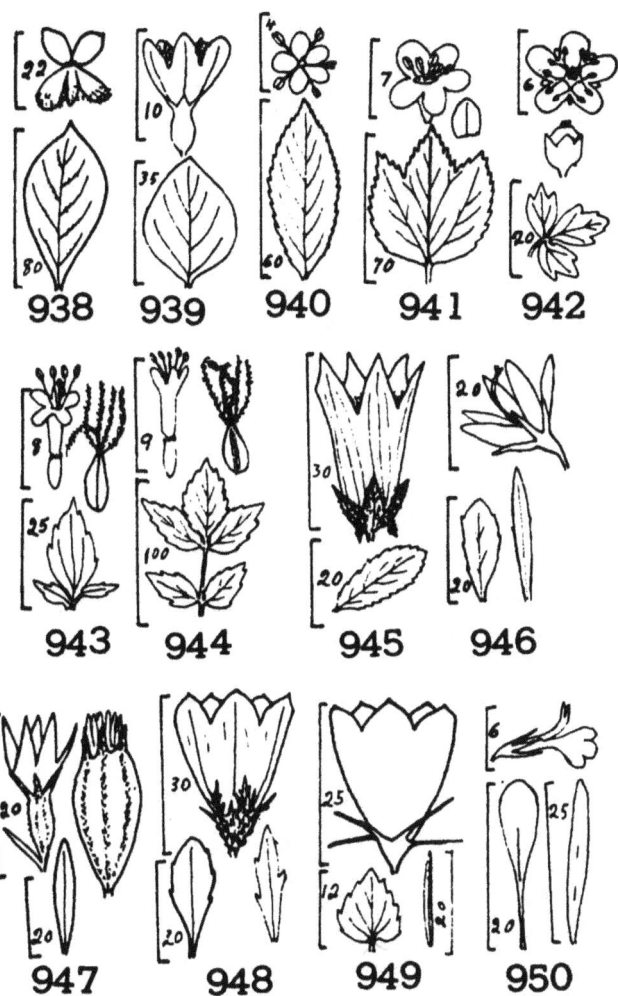

58. CICHORIACEAE (Chicory Family)

Herbs (in ours) with bitter or milky sap; leaves alternate or basal; flowers all alike, perfect and fertile, in heads with bracts (phyllaries) in 1 to several series, and often with smaller ones at the base; corolla of united petals forming a tube which is split on one side giving rise to a straplike ligule usually 5-toothed at the apex; stamens 5, united by their anthers into a tube around the pistil; style 2-cleft, filiform; ovary 1-celled, becoming an achene. This family is often combined with the Asteraceae and known as the *Compositae*.

```
1A. Pappus none. ..................................... 1. Lapsana
2A. Pappus of plumose bristles.
    1B. Receptacle chaffy. ........................... 2. Hypochaeris
    2B. Receptacle naked. ............................ 3. Picris
3A. Pappus of simple bristles.
    1B. Heads solitary on scapes.
        1C. Pappus tawny. ............................ 4. Apargidium
        2C. Pappus white.
            1D. Achenes muricate at apex. ............ 5. Taraxacum
            2D. Achenes smooth (see also Taraxacum
                kamtschaticum). ..................... 6. Agoseris
    2B. Heads several to many; stems usually leafy.
        1C. Achenes flattened.
            1D. Achenes beakless. .................... 7. Sonchus
            2D. Achenes beaked. ...................... 8. Lactuca
        2C. Achenes not flattened.
            1D. Flowers whitish. ..................... 9. Prenanthes
            2D. Flowers yellow, rarely pinkish.
                1E. Pappus white. ....................10. Crepis
                2E. Pappus sordid or tawny. ..........11. Hieracium
```

1. LAPSANA L.

Erect branching annuals; leaves dentate or pinnatifid; heads small, yellow, slender-peduncled; phyllaries 8 with a short outer series; receptacle flat, naked. (Greek, Lampsana, the name of a crucifer.)

L. communis L.   Nipplewort.

Stem paniculately branched, 3–10 dm. tall, pubescent below, glabrous above; heads very numerous, the involucres in fruit 5–6 mm. long.

Introduced in southeast Alaska. Native of Europe.

2. HYPOCHAERIS (Vaill.) L.

Perennials with scapose, often branched stems; leaves mostly basal, those of the stem few and scale-like; heads large, long-peduncled; flowers yellow; receptacle chaffy; achenes 10-ribbed. (Greek, for pigs, which are fond of its roots.)

H. radicata L.   Cat's Ears.

Stems 2–4 dm. tall; leaves spreading, oblanceolate to obovate, pinnatifid to dentate, hirsute; heads 25–30 mm. broad; achenes beaked.

Introduced weed, native of Europe.

3. PICRIS L.

Erect, hispid herbs; flowers yellow in rather large heads; principal phyllaries in 1 series, nearly equal, with 2 or 3 series of exterior spreading ones; receptacle flat, short-fibrillate; achenes 5- to 10-ribbed and transversely wrinkled, narrowed at base and summit; pappus of slender plumose bristles. (Greek, bitter.)

P. hieracoides L. ssp. kamtschatica (Ledeb.) Hult.

Biennial, up to 1 m. tall, quite densely hispid; leaves lanceolate to oblanceolate, up to 15 cm. long, the lower narrowed into petioles; involucre 12–15 mm. high; phyllaries narrow, strongly setose; achenes bright brownish red. Fig. 951.

Attu Island and East Asia.

4. APARGIDIUM T. and G.

Practically acaulescent plants with fusiform roots; leaves narrow; heads turbinate with scales in 2- or 3-series, borne on scapes; pappus tawny, of barbellate bristles. (Likeness to Apargia.)

A. boreale (Bong.) T. and G.

Scorzonella borealis (Bong.) Greene.

Leaves linear-lanceolate, 1–2 dm. long, entire or with a few teeth; scapes 15–30 cm. tall; flowers yellow; involucres about 12 mm. high, composed of lanceolate scales with prominent midrib; achenes about 6 mm. long with 10 prominent longitudinal ribs.

Prince William Sound to Humboldt Co., Calif. Fig. 952.

5. TARAXACUM Zinn.

Acaulescent biennial or perennial herbs; heads many-flowered, large, borne on slender, hollow scapes; involucres double, the outer of short phyllaries, the inner of long, linear, erect phyllaries in a single row; flowers yellow (flesh-colored in one species); achenes 4- to 5-ribbed, the ribs usually roughened, the apex prolonged into a slender beak bearing the pappus of capillary bristles. (Derivation from supposed medicinal properties.)

The dandelions are a difficult group. There is much apomixis, i.e., the achenes develop without fertilization. Many of the forms produce no pollen. This apomixis gives rise to numerous distinguishable constant forms. Modern writers on the genus treat these as species. These correspond with cultivated garden varieties rather than with species in the true sense as recognized in most groups. It will be noted that most of these forms are of local or limited distribution. As collections increase more will be found. Dr. Gustaf E. Haglund of the Riksmuseet at Stockholm, Sweden, has done much work on the group as represented in Alaska and Yukon and his treatment is followed here as a matter of convenience rather than approval.

1A. All phyllaries lacking appendages below the apex.
  1B. Low-growing native species with small heads.
    1C. Involucres dilute-green, broad, short; outer
      phyllaries whitish-green, up to 9 mm. long. . 54. *T. collinum*
    2C. Involucre usually blackish-green, narrower, outer
      phyllaries dark.
      1D. Lateral lobes of the leaves more or less
        retrorse, short, more or less broad, acute. .52. *T. alaskanum*
      2D. Lateral lobes of the middle leaves patent or
        somewhat attenuate, often with blunt apex
        or more or less claw-like with rounded
        corners.
        1E. Petioles purple, terminal lobes of leaves
          short; achenes red, smooth or nearly
          so. .... 3. *T. kamtschaticum*
        2E. Petioles usually pale, terminal lobes of leaves
          hastate to hastate-triangular; achenes
          brownish-black with small sharp spines
          on top. ....55. *T. sibiricum*
  2B. Taller (15–30 cm. or more) robust introduced
    species with large heads.
    1C. Outer phyllaries broad (4–6 mm.) usually more
      or less patent; petioles wing-margined, pale
      or slightly rose-colored. ....50. *T. undulatum*
    2C. Outer phyllaries narrower (2–5 mm. broad) most
      of them reflexed petioles more or less red-colored.
      1D. Outer phyllaries strongly reflexed, more or less
        whitish green. .... 49. *T. retroflexum*
      2D. Outer phyllaries obliquely reflexed, grayish-
        green or more or less red-colored.
        1E. Lateral lobes of leaves deltoid with more or
          less straight upper margin.
          1F. Lateral lobes of leaves short; outer
            phyllaries narrow (about 2 mm.
            broad), strongly radiating. ....51. *T. vagans*
          2F. Apex of lateral lobes longer, outer phyl-
            laries broader.
            1G. Terminal lobe of inner leaves long,
              sagittate, outer phyllaries rather
              short. .... 48. *T. decorifolium*
            2G. Terminal lobe of inner leaves shorter,
              broad; outer phyllaries longer. ......46. *T. cinericolor*

2E. Lateral lobes of leaves more or less claw-like
with more or less convex upper margin. . 47. T. dahlstedtii
2A. At least some of the phyllaries with large or small
appendages below the apex.
1B. Achenes small (about 3 mm. long) with narrow
cylindrical beak, brownish brick-red. ....45. T. scanicum
2B. Achenes larger, usually with broad beak, only in a
few species more or less red.
1C. Achenes blackish or blackish-green with short
conical beak; low-growing species.
1D. Involucre rather small (9-14 mm. long, 5-8
mm. broad); achenes totally spinose. .... 2. T. phymatocarpum
2D. Involucre larger, achenes more or less smooth
at the base. .... 1. T. hyperarcticum
2C. Achenes mostly not blackish, beak usually
longer; rostrum 2-3 times longer than the
achene.
1D. Petioles more or less intensely red-colored.
1E. Low-growing species (about 10 cm. tall); involucres small, at most about 12 mm. in
diameter.
1F. Involucres about as long as broad, outer
phyllaries narrowly scarious-mar-
gined. .... 5. T. angulatum
2F. Involucres longer than broad, outer phyllaries lacking scarious margins. ......18. T. festivum
2.E. Larger species, 15-45 cm. tall; involucres
larger.
1F. Achenes more or less red-colored.
1G. Achenes with conic-cylindrical to cylindrical beak about 1.25 mm. long;
lateral lobes of leaves more or less
dentate ....24. T. lateritium
2G. Achenes with broader, shorter beak;
lateral lobes of leaves entire. .... 9. T. callorhinorum
2F. Achenes olivaceous to straw-colored to
more or less brown.
1G. Robust species, phyllaries without scarious margins. ....11. T. chlorostephum
2G. Medium-sized species; phyllaries with at
least narrow scarious margins.
1H. Outer phyllaries cordate to ovate (3-5
x 5-8 mm.). ....39. T. pribylofense
2H. Outer phyllaries ovate to ovate-lanceolate, narrower. ....23. T. lacerum
2D. Petioles only slightly or not at all red-colored.
1E. Small species, less than 10 cm. tall. Scapes of
some may elongate in fruit.
1F. Leaves entire with a few small teeth
only. ....42. T. speirodon
2F. Leaves more or less lobed.
1G. Lateral lobes narrow, tapering to a
patent or forward-turning point. ....13. T. demissum
2G. Lateral lobes short and broad.
1H. Petioles wing-margined. ....26. T. leptoglossum
2H. Petiole narrow, not wing-margined.
1J. Outer phyllaries lacking scarious
margins. ....29. T. microceras
2J. Outer phyllaries with distinct although sometimes narrow scarious margins.
1K. Terminal lobe of leaves elongate
with ligulate apex. .... 4. T. andersonii
2K. Terminal lobe of leaves short.

1L. Involucres about 14 mm. high,
    blackish-green. ....27. *T. leptopholis*
2L. Involucres about 11 mm. high,
    olivaceous-green. ....31. *T. multesimum*
2E. Larger species with broader heads.
  1F. Leaves not lobed.
    1G. Petioles more or less broadly wing-margined.
      1H. Outer phyllaries narrowly ovate-lanceolate, about as long as the inner ones. ....20. *T. hypochoeropsis*
      2H. Outer phyllaries ovate to ovate-lanceolate, about half as long as the inner ones. ....19. *T. flavovirens*
    2G. Petioles narrow or narrowly wing-margined.
      1H. Appendages of outer phyllaries 0.4-0.8 mm. long. ....21. *T. integratum*
      2H. Appendages of outer phyllaries 0.7-1.5 mm. long. ....41. *T. signatum*
  2F. Leaves more or less lobed, or some leaves with tooth-like lobes.
    1G. All or most of the ligules involute or more or less canaliculate.
      1H. Outer phyllaries 7-9 mm. long, ovate-lanceolate. ....33. *T. ochraceum*
      2H. Outer phyllaries shorter, ovate. ...... 8. *T. caligans*
    2G. Ligule more or less flat.
      1H. Lateral lobes of leaves narrow to linear.
        1J. Outer phyllaries membranous, whitish-green. ....40. *T. scotostigma*
        2J. Outer phyllaries more or less herbaceous, pale green. ....43. *T. sublacerum*
      2H. Lateral lobes of leaves broader.
        1J. Lateral lobes of leaves more or less claw-like.
          1K. Anthers with pollen, lateral lobes of leaves approximate, entire or very sparsely dentate. ...22. *T. kodiakense*
          2K. Anthers lacking pollen, lateral lobes of leaves less approximate, with larger teeth. ....35. *T. paralium*
        2J. Lateral lobes of leaves deltoid or replaced by short tooth-like lobes.
          1K. Outer phyllaries nearly as 'ong as the inner ones, more or less lanceolate.
            1L. Leaves narrow, 10-15 mm. wide, lateral lobes few and poorly developed. ....38. *T. phalolepis*
            2L. Leaves broader, lateral lobes more numerous, short, broad. ....14. *T. dumetorum*
          2K. Outer phyllaries much shorter than the inner ones, ovate to ovate-lanceolate.
            1L. Outer phyllaries more or less membranous, whitish-green, partly more or less reddish.
              1M. Leaves broad, densely rosulate, petioles wing-margined. ....34. *T. ovinum*

2M. Leaves narrow, not rosulate,
   petioles narrow. ....53. *T. carneocoloratum*
2L. Outer phyllaries herbaceous.
 1M. Outer phyllaries with broad (1 mm.) scarious margins.
  1N. Appendages of the outer phyllaries about 1.5 mm. long. ....25. *T. latilimbatum*
  2N. Appendages of the outer phyllaries about 0.5 mm. long. ....36. *T. patagiatum*
 2M. Outer phyllaries narrowly scarious-margined.
  1N. Involucres dark, blackish-olivaceous to blackish-green.
   1P. Leaves narrow (5 mm. broad) with weakly developed lateral lobes. ....15. *T. eurylepium*
   2P. Leaves broader, lateral lobes short and broad.
    1Q. Lateral lobes of the leaves with a somewhat contracted, short apex. .... 6. *T. arietinum*
    2Q. Lateral lobes tapering to a short, more or less retrorse apex.
     1R. Involucres about 17 mm. high, broad 30. *T. mitratum*
     2R. Involucre shorter, fairly narrow. . 28. *T. maurolepium*
  2N. Involucres lighter colored, more or less olive-green.
   1P. Petioles narrow, somewhat reddish. ..... 7. *T. aureum*
   2P. Petioles more or less wing-margined, pale.
    1Q. Terminal lobes of leaves small, triangular-rhomboid, often with short acumen.
     1R. Leaves dark green 16. *T. eyerdamii*
     2R. Leaves light or yellowish green. .44. *T. trigonolobium*
    2Q. Terminal lobes of leaves of medium size, usually more or less sagittate, tapering to the apex or blunt.
     1R. Achenes smooth with scale-like spines toward the apex only.33. *T. oncophorum*
     2R. Achenes more spiny.

1S. Achenes large, about 6 mm. long including beak. ..17. *T. fabbeanum*
2S. Achenes smaller, 4.6-4.8 mm. long. ....10. *T. chamissonis*

1. *T. hyperarcticum* Dahlst.

Leaves long-petioled, entire to few sinuate-lobed, 10–20 cm. long; scapes many, rose-violet at the base; involucre 12–15 mm. high; outer phyllaries ovate to ovate-lanceolate, dark green; achenes olivaceous, muricate at top, tuberculate below, about 5 mm. long.

Cape Thompson, Nova Zemla, Greenland.

2. *T. phymatocarpum* J. Vahl.

Leaves small, subentire, sparsely and minutely dentate to briefly lobed; heads 1–3, rarely more; involucre blackish, 9–14 mm. high, 5–8 mm. wide; outer phyllaries ovate; achenes 5 mm. long, about 1.45 mm. wide, muricate at top to tuberculate at base.

Elim and Teller—Greenl.

3. *T. kamtschaticum* Dahlst.

Leaves 1 dm. or less long, lobed, the lobes rounded or obtuse, the petiole reddish-purple; scapes in anthesis low, in fruit up to 15 cm. tall, smooth; involucre olive-green to black-green; outer phyllaries without conspicuous nerves, unappendaged, with entire, greenish, scarious margins. This is easily distinguished from the other species by the chestnut-red achene which is smooth or very nearly so.

From Kamchatka to most of Alaska. Fig. 953.

4. *T. andersonii* Hagl.

Leaves narrow, ligulate lanceolate, dark green, with short, deltoid, lateral lobes, up to 7 cm. long; scapes short at flowering but elongating in fruit to 25 cm.; phyllaries with appendages below the apex, dark green; achenes dark brown, rugulose below and spiny at the top, 4.5–4.9 mm. long.

Skagway and Popoff Isl. of the Shumagin Isls. Fig. 954.

5. *T. angulatum* Hagl.

Low-growing, about 10 cm. tall; leaves 3–7 cm. long, about 1 cm. wide, shallowly lobed, the lobes deltoid; scapes longer than the leaves, subglabrous; involucre about 12 mm. long and broad; outer phyllaries ovate, 4.5–7 mm. long, 2–3 mm. wide, only one or a few provided with appendages; achenes 4.3–4.5 mm. long, tawny-olive, spinulose above.

St. Matthew Isl.

6. T. arietinum Hagl.

Plants about 15 cm. tall; leaves about 10 cm. long and 1 cm. wide, sinuate-lobed; scapes slender, nearly glabrous; involucre about 17 mm. high; outer phyllaries loosely appressed, 5–8 mm. long, 2–3.5 mm. wide with appendages 0.5–1 mm. long; achenes 4.7–5 mm. long, with acute spines above and tuberculate.

Alatna River and Richardson Highway between Summit and McCarty.

7. T. aureum Hagl.

Plants 2–3 dm. tall; leaves ascending, obovate-oblanceolate, toothed to shallowly lobed, about 12 cm. long and 2 cm. wide; scapes exceeding the leaves; involucre 12–17 mm. high; outer phyllaries ovate, 5–7 mm. long, 2.5–4 mm. wide, with wide scarious margins and appendages up to 1 mm. long.

Teller.

8. T. caligans Hagl.

Plant about 1 dm. tall; leaves up to 2 cm. wide, with subtriangular lateral lobes; scapes about equaling the leaves, more or less red-colored; involucres 10–12 mm. high; outer phyllaries ovate, 2–3 mm. wide, 5.5–8 mm. long, with appendages 0.5 mm. long; achenes 5 mm. long, brown.

Tonsina Lodge on Richardson Highway.

9. T. callorhinorum Hagl.

Medium size; outer leaves spatulate, sparsely toothed; inner leaves lobed with few triangular lobes; petioles purple; scapes exceeding the leaves; heads dark, almost black, 15–18 mm. high; outer phyllaries ovate to ovate-lanceolate, the margins light rose to purplish; appendages about 0.5 mm. long; achenes red, spiny at top, tuberculate or smooth at base.

St. Paul and Unalaska Isls.

10. T. chamissonis Greene, emend Hagl.

Plant with many leaves, about 3 dm. tall; leaves 12–20 cm. long, 2–4 cm. wide, not lobed but deeply and irregularly toothed with backward-pointed teeth; involucre 15–17 mm. high; outer phyllaries cordate to wide-ovate, (2.5–) 4.5 mm. wide, 6–8 mm. long; inner phyllaries with prominent appendages; achenes brown, 4.6–4.8 mm. long, somewhat spinose at top, smooth below.

St. Paul, St. Matthew and Hall Isls.

11. T. chlorostephum Hagl.

Plants robust, 25–50 cm. tall; leaves sinuate-lobed, the lobes subtriangular and toothed; outer phyllaries narrowly ovate-lanceolate to lanceolate, 3–4.5 mm. by 12–15 mm., narrowly white-margined, with appendages 0.5–1.5 mm. long; inner phyllaries sublinear; anthers with scanty pollen; achenes light olive-brown, 5 mm. long, short spinose above.

Eklutna, Kodiak and vicinities.

12. *T. chromocarpum* Hagl.

Leaves blue-green, lobes triangular, entire or minutely denticulate and with rose-colored petioles; outer phyllaries narrowly white-margined; achenes mahogany-red, about 4 mm. long, densely and acutely spinulose at the top.

Unalaska.

13. *T. demissum* Hagl.

Plants small; leaves lanceolate, deeply lobed with 3 or 4 pairs of deltoid lateral lobes, the terminal lobe hastate-sagittate; involucre about 13 mm. high; outer phyllaries ovate or ovate-lanceolate, 2–3 mm. wide, 3–8 mm. long, the appendages 1–2 mm. long.

Hooper Bay.

14. *T. dumetorum* Greene.

Large; leaves up to more than 3 dm. long, oblanceolate, often broadly so, acutish, the margin not deeply, but very unevenly and laciniately cut, the teeth spreading; scapes mostly 3 dm. or more tall; outer phyllaries large, pale and thin, before flowering almost as long as the inner; achenes olive-green, spinulose at the summit, otherwise smooth or the ribs tuberculate.

Ranch Valley, Yukon—Alta.—Wyo.

15. *T. eurylepium* Dahlst.

A small form resembling *T. phymatocarpum* J. Vahl and probably identical with it.

Bering Sea and Arctic Coast regions.

16. *T. eyerdamii* Hagl.

Plant medium in size; leaves slightly to rather deeply lobed, the terminal lobe triangular, the petioles pale red; involucre about 15 mm. high, the scape pubescent below the head; phyllaries light, striate, prominently appendaged; achenes buff, 4.5–5 mm. long, spiny above.

East Aleutians. Fig. 955.

17. *T. fabbeanum* Hagl.

Plant medium, 7–25 cm. tall; leaves rather deeply lobed, the lobes deltoid and usually sharply toothed; involucre 13–19 mm. high; outer phyllaries 3–4 mm. by 4.5–9.5 mm.; inner phyllaries conspicuously white-margined; achenes olive-brown, 6 mm. long.

St. Paul and St. Lawrence Isls.

18. *T. festivum* Hagl.

Plants small, 5–10 cm. tall; leaves 6–8 cm. long, 5–10 mm. wide, shallowly lobed; scapes more or less curved, sparsely pilose below the head; involucre 10–14 mm. high, black-green; outer phyllaries with small but well-developed appendages.

Point Barrow.

19. *T. flavovirens* Hagl.

Plants 15–35 cm. tall, not robust; leaves thin, yellow-green, obovate-oblong or oblong, 5–10 (–20) cm. long, the margins toothed; scapes 1 or few; involucre olive-green, about 16 mm. high; outer phyllaries lighter green than the inner ones, with broad white scarious margins, one or a few with appendages; achenes about 3.5 mm. long, sharply muricate.

Haines, Whitehorse and B.C.

20. *T. hypochoeropsis* Hagl.

Plant tall; leaves obovate-oblanceolate, broad, not lobed, some with prominent teeth; involucre about 25 mm. broad, light green; outer phyllaries long, some with appendages; anthers without pollen; achenes light brownish-olive, about 4 mm. long, short-spinose at top, rugulose below.

Anchorage.

21. *T. integratum* Hagl.

Plants up to 25 cm. tall; leaves oblong-oblanceolate, 4–13 cm. long, 5–15 mm. wide, sinuate-dentate; involucre 12–15 mm. high; outer phyllaries ovate or ovate-lanceolate, 2–3.5 mm. wide, 6–7 mm. long, white-margined; anthers without pollen; achenes umber, 4.7–5 mm. long, short spinulose above, rugulose below.

St. Michael, Teller, Pt. Lay and Camden Bay. Fig. 956.

22. *T. kodiakense* Hagl.

Plants 5–20 cm. tall; leaves light green, up to 3 cm. wide with few short lateral lobes; terminal lobe ovate-triangular, short-acuminate; petiole pale, wide-winged; involucre medium-sized, fleshy, olive-green; outer phyllaries ovate to ovate-lanceolate, 1.5–4 mm. wide, 6–10 mm. long with appendages 1–1.5 mm. long; anthers bearing pollen.

Kodiak Island.

23. *T. lacerum* Greene.

Medium size, the scapes in fruit up to 40 cm. tall; leaves deeply pinnatifid, the narrow acute lobes usually sharply deflexed backward; heads several to many, large; nearly all phyllaries with corniculate appendages; achenes about 4 mm. long, spinulose at apex to nearly smooth at base.

Widely distributed, Bering Sea—Arctic Isls.—Labr.—Newf.—Alta. Fig. 957.

24. *T. lateritium* Dahlst.

Outer leaves merely dentate, the inner sinuate-lobed, the lobes toothed; scapes 1-3, pale below, copper-colored above; involucre 13-15 mm. high, fuscous or blackish-green; outer phyllaries wide-ovate-lanceolate, the appendages small; achenes 3.5-4 mm. long, spinulose toward the apex, the pyramid prominent.

Bering Sea and Arctic Coast districts. Fig. 958.

25. *T. latilimbatum* Hagl.

Plants about 35 cm. tall; leaves long, about 2 cm. wide, remotely sinuate-dentate to sinuate-lobed with long pale petioles; involucre about 15 mm. high, pale olive-green; outer phyllaries ovate to ovate-lanceolate with appendages 1.5 mm. long; anthers without pollen; achenes honey yellow, about 4.5 mm. long, tuberculous-rugulose with a few spines on top.

Black Hill Creek, Yukon.

26. *T. leptoglossum* Hagl.

Plant small, 10-15 cm. tall; leaves many, up to 5 or 6 cm. long with a few broad lateral lobes and ligulate terminal one; involucre 10-12 mm. high, light green; outer phyllaries ovate-lanceolate, some with appendages; achenes dark straw-yellow, the top minutely but sharply spiny, more or less tuberculate below.

Karluk, Kodiak Isl.

27. *T. leptopholis* Hagl.

Plant small, about 10 cm. tall; leaves more or less prostrate, with 4-6 approximate and short lateral lobes; scapes fairly villous; involucre about 14 mm. high; outer phyllaries margined, the appendages small, black; achenes spiny at top, otherwise nearly smooth.

Glacier and Yakutat bays.

28. *T. murolepium* Hagl.

Plants 10-25 cm. tall; leaves narrowly lanceolate, long, 1-2 cm. wide, light green, sinuate-lobed, the lobes small, subtriangular; involucre about 14 mm. high, dark olive-green; outer phyllaries ovate, 3-5 mm. wide, 8-10 mm. long, white-margined, the appendages 0.5-2.5 mm. long; achenes 4-4.5 mm. long, short-spiny on top, rugulose below.

Umiat and Shaktolik.

29. *T. microceras* Hagl.

Plant small; leaves 3–5 cm. long, 5–15 mm. wide, the outer not lobed, the inner with a few short deltoid lobes; involucre about 13 mm. high, olive-green; outer phyllaries ovate-lanceolate, 1–2 mm. wide, 4–6 mm. long, the appendages small; achenes umber-brown, 4–4.5 mm. long with short, fine spines at top, tuberculate or the base smooth.

Worthington Glacier near Valdez.

30. *T. mitratum* Hagl.

Medium size; leaves lanceolate, the terminal lobe large, hastate-sagittate or sagittate, the lateral lobes deltoid, broad; involucre about 17 mm. high, blackish-green; outer phyllaries with broad white margins, provided with 1–2 mm. long appendages; achenes olive-ochre, 4.9 mm. long, spines small, tuberculate-rugulose, smooth toward the base.

Katmai region and Hope on Kenai Penin.

31. *T. multesimum* Hagl.

Plant low, 5–8 cm. tall; leaves about 5 cm. long, 5–15 mm. wide with lateral lobes approximate, deltoid; terminal lobe triangular-hastate; involucre about 11 mm. high; outer phyllaries 1.5 mm. wide, 4–5 mm. long, white-margined, one or a few with small appendages; anthers without pollen.

Moose Pass in Kenai Penin.

32. *T. ochraceum* Hagl.

Plants medium, 10–20 cm. tall; leaves many, long, narrow, with claw-like or deltoid lobes and prolonged terminal lobe; involucre olive-green, 15–18 mm. high; outer phyllaries ovate-lanceolate, scarious-margined, with appendages below the apex; achenes brownish, short-spiny on top, tuberculate or smoothish at the base, 4.5 mm. long.

Tonsina Lodge and Gulkana on Richardson Highway.

33. *T. oncophorum* Hagl.

Plant medium; outer leaves toothed, inner leaves with deltoid lateral lobes usually not opposite; terminal lobe small; involucre olive-green, about 13 mm. high; outer phyllaries ovate to ovate-cordate, 3–4 mm. wide, 5–8 mm. long; anthers without pollen; achenes light-colored, 4.1–4.5 mm. long, shortly and sparsely muricate at top, otherwise quite smooth.

Attu Island.

34. *T. ovinum* Greene emend Hagl.

Plants small, up to 20 cm. tall; leaves light yellowish green, somewhat decumbent, often only sinuate-dentate or sinuate-lobed, the lobes triangular; involucre medium, 12–15 mm. high; outer phyllaries ovate-lanceolate with fairly small appendages; anthers with pollen; achenes brown, spinulose at top, smooth at the base.

Ranch Valley, Yukon to B. C. and Alta.

35. T. paralium Hagl.

Plants 1–2 dm. tall; leaves 5–10 cm. long, about 2 cm. wide; lateral lobes 3–4 on each side, deltoid, short claw-like; involucre about 1 cm. high; outer phyllaries ovate-lanceolate, 1–2 mm. wide, 6–7 mm. long, pale yellow-green, conspicuously margined, with appendages about 1 mm. long; achenes brown, 5 mm. long, short-spinulose at top.

Anchorage and Kenai.

36. T. patagiatum Hagl.

Plants 7–15 cm. tall; leaves light green with fairly triangular, short lateral lobes and hastate terminal lobe; petioles pale; involucre about 15 mm. high, light green; outer phyllaries ovate, 2–4.5 mm. wide, 5–8 mm. long, with appendages up to 1 mm. long; achenes 5 mm. long, tuberculate with spinose tip.

Seward—Fairbanks—Coal Cr. in east Alaska.

37. T. pellianum Porsild.

Plant small; leaves glabrous, firm, bright green excepting the purplish midrib, lanceolate, subentire, 6–8 cm. long when mature, 8–10 mm. wide; scapes 8–10 cm. tall; involucre about 15 mm. high, olive-green; tips of the phyllaries distinctly corniculate; anthers dark yellow, containing pollen; achenes dark straw-colored, the tips spiny, 4.24 mm. long.

Pelly Mts.

38. T. phalolepis Hagl.

Plants about 15 cm. tall; leaves about 10 cm. long, 10–15 mm. wide, firm, glabrous, entire to sinuately lobed, the lobes short and few; involucre olive-green, about 16 mm. high; outer phyllaries long with rather large appendages; anthers without pollen.

Chitina River Glacier.

39. T. pribilofense Hagl.

Plants 2–3 dm. tall; leaves from sharply toothed to moderately deeply lobed, the lobes subtriangular, acute, tooth on upper edge; petioles reddish-purple; involucre about 16 mm. high, light olive-green; outer phyllaries 3–5 mm. wide, 5–8 mm. long; appendages up to 1.5 mm. long; anthers with little pollen; achenes brown, 5–5.5 mm. long, spinulose at top, slightly tuberculate.

St. Paul Isl.

40. T. scotostigma Hagl.

Plants about 20 cm. tall; leaves light green, more or less lobed, the lateral lobes with fairly long patent points; involucre 11–13 mm. high; phyllaries dirty yellowish-green, membranous, with scarious margins and acute corniculate appendages; anthers and stigmas black-green; achenes cinnamon-buff, 4.5 mm. long, squamulose.

Gakona, Tanacross and Fairbanks. Fig. 959.

41. *T. signatum* Hagl.

Plant medium; leaves narrowly oblong-lanceolate, about 1 cm. wide, dentate but not lobed; involucre about 13 mm. high, olive-green; outer phyllaries 1.5–2.5 mm. wide, 5–7.5 mm. long, narrowly but distinctly white-margined and with wide appendages; achenes isabella color, 5 mm. long, short-spiny on top, tuberculate.

Skagway.

42. *T. speirodon* Hagl.

Plant low, 5–15 cm. tall; leaves about 5 mm. rarely up to 1 cm. wide, entire or with a few teeth; involucre about 10 mm. high, black-green; outer phyllaries 5–8 mm. long, 1–3 mm. wide with rather small appendages; achenes olive-buff, 5–5.5 mm. long, squamulate to tuberculate entire length.

Point Hope and Port Clarence region.

43. *T. sublacerum* Hagl.

Plants 1–3 dm. tall; leaves light green, lobulate-dentate to short-lobed; involucre 10–14 mm. high. Resembles *T. lacerum* but has lighter heads with lighter, conspicuously scarious-margined outer phyllaries; the achenes are narrower and the lateral lobes of the leaves are shorter and broader.

Fairbanks—Big Delta—Glenallen—Tacotna.

44. *T. trigonolobium* Dahlst.

Leaves light green with approximate short, wide, more or less triangular lobes, the terminal lobe small, mucronate; scapes many; involucre 10–14 mm. high, blackish-green; outer phyllaries ovate-triangular to ovate-lanceolate, somewhat calloused, the inner phyllaries corniculate; anthers without pollen.

Kamchatka—Aleutians—Kenai. Fig. 960.

45. *T. scanicum* Dahlst.

Leaves glabrescent-green, many-lobed, the lobes spreading, toothed; outer phyllaries long, narrowly lanceolate to linear-lanceolate, callose below the apex or a few corniculate; achenes acute spinulose at apex, short-spinulose to tuberculate below, about 3 mm. long, 1 mm. wide, the pyramid 1 mm. long.

Introduced, Anchorage and Fairbanks.

46. *T. cinericolor* Hagl.

Medium, about 2 dm. tall; leaves gray-green, with long, patent lateral lobes separated by narrow interlobes, the terminal lobes small; involucre 16–19 mm. high, olive-green; outer phyllaries spreading or reflexed, lanceolate, often purple; achenes olive-ochre, 3.8–4 mm. long, spiny at top, somewhat tuberculate to smooth at the base.

Introduced, Skagway, Juneau, Lake Bennett, Unalaska.

47. *T. dahlstedtii* Lindb. f.

A vigorous weed; lobes of the leaves more or less claw-like with convex upper margin; outer phyllaries obliquely reflexed. This is one of the group of European dandelions introduced in America.

Orca and King Cove.

48. *T. decoraifolium* Hagl.

Medium to about 3 dm. tall; leaves cobwebby pilose on midrib, the lateral lobes deltoid, spreading, toothed; outer phyllaries spreading to subreflexed, (2.5–) 4 mm. wide, about 11 mm. long; anthers polliniferous; achenes 3.8–4.1 mm. long, minutely spiny at top, tuberculate, smooth at base.

Introduced type but known only from Juneau.

49. *T. retroflexum* Lindb. f.

Plant vigorous; leaves with petioles up to 3 dm. long, deeply lobed, the lobes triangular, acute, sharply toothed; petioles rose-purple at base; involucres up to 20 mm. high; outer phyllaries strongly reflexed; achenes 3.5 mm. long.

Introduced weed in many parts of Alaska.

50. *T. undulatum* Lindb. f. & Markl.

This is another one of the weedy European species. The broad outer phyllaries are spreading, not strongly reflexed.

In America known only from Juneau.

51. *T. vagans* Hagl.

Plants up to about 35 cm. tall; leaves light green, up to 4 cm. wide and 30 cm. long, sparsely cobwebby, mostly deeply lobed, the lobes dentate; involucre 17–20 mm. high; outer phyllaries subreflexed, about 2 mm. wide, 10–15 mm. long, acuminate; achenes 3.5 mm. long, spinulose on top, tuberculate, smooth at base.

The commonest of our introduced dandelions, Nome and Unalaska east and south.

52. *T. alaskanum* Rydb.

Leaves 3–5 cm. long, deeply runcinate-pinnatifid with triangular retrorse lobes; scapes 4–5 cm. tall; phyllaries fuscous, not corniculate, the inner linear-lanceolate, long-acuminate, the outer scarcely half as long, lanceolate, spreading or somewhat reflexed; achenes brownish, spinulose-muricate above, 4 mm. long.

Alaska Range, Bering Sea and Arctic. Fig. 961.

53. *T. carneocoloratum* A. Nels.

Taproot short with 1 or more crowns each with several leaves and 1 or 2 scapes; leaf blades 3–6 cm. long, 10–15 mm. wide with 3–6 pairs of subacute and more or less triangular teeth; scapes 10–15 cm. tall; heads rather large; outer phyllaries in 2 series, at first erect, later spreading; inner phyllaries margined, slender-corniculate, green becoming pink; flowers definitely flesh-colored; achenes 4 mm. long, spinulose-muricate at top only.

Mt. McKinley Park.

54. *T. collinum* DC.

Leaves spatulate-oblong, runcinate-dentate; scapes exceeding the leaves; outer phyllaries ovate-lanceolate, not corniculate, subvillous on the margins; achenes spinulose-muricate at apex.

Asia and Unalaska.

55. *T. sibiricum* Dahlst.

Leaves 5–10 cm. long, pale green, deeply runcinate-lobed, the lobes toothed; involucres 12–15 mm. high, more or less blackish-green; outer phyllaries ovate, somewhat acuminate; anthers polliniferous; achenes 3.5–4 mm. long, sparingly short-spinulose at top.

Canol Road, Chickaloon, Nome, Teller and in Asia.

6. AGOSERIS Raf.

Acaulescent perennials with strong taproot; involucral bracts imbricated in a few series, the outer broader and shorter; achenes 10-ribbed, narrowed above into a beak; pappus of numerous white capillary bristles. (Greek, goat and chicory.)

1A. Beak equaling or exceeding the body of the achene. . .1. *A. gracilens*
2A. Beak shorter than the body of the achene.
  1B. Ligulae yellow. ....2. *A. scorzoneraefolia*
  2B. Ligulae orange, turning purplish. ....3. *A. aurantiaca*

1. *A. gracilens* (Gray) Ktze.

Leaves oblanceolate to nearly linear, usually entire or some with a few short lobes, 1–2 dm. long; scapes slender, 1–4 dm. tall, villous below the head; heads 18–20 mm. high, the phyllaries linear-lanceolate; flowers orange, turning purple.

Skagway, B. C.—Alta.—Ida.—Utah—Calif.

2. *A. scorzoneraefolia* (Schrad.) Greene.

Leaves oblanceolate or linear-oblanceolate, glabrous, 1-3 dm. long, entire or rarely denticulate; scapes 1-2 (-3) dm. tall, villous under the head; involucres 2-3 cm. high, the phyllaries broadly lanceolate, acute, villous-ciliate; flowers light yellow turning pinkish in age.
Yukon boundary—Alta.—S. Dak.—Colo.—Nev.—Ore. Fig. 962.

3. *A. aurantiaca* (Hook.) Greene.

Leaves oblanceolate, entire, dentate or lobed; scapes 2-6 dm. tall, villous under the head; involucres 15-20 mm. high; inner phyllaries lanceolate and acute, the outer phyllaries villous; flowers orange turning purple.
Southeast Alaska—Alta.—Colo.—Utah—B.C. Fig. 963.

7. SONCHUS (Tourn.) L.

Introduced weeds; stems leafy; flowers yellow; involucre campanulate with long inner and short outer phyllaries; achenes flattened, ribbed, not beaked; pappus of numerous white capillary bristles. (Greek name of the Sow-Thistle.) The species of *Sonchus* are native of Europe and introduced in many parts of the world but not yet common in Alaska.

1A. Perennial; expanded heads 4-5 cm. wide .... 1. *S. arvensis*
2A. Annual; heads smaller.
  1B. Leaves lyrate-pinnatifid, achenes transversely wrinkled. .... 2. *S. oleraceus*
  2B. Leaves not deeply pinnatifid; achenes not wrinkled. 3. *S. asper*

1. *S. arvensis* L.   Field Sow-Thistle.

Stems 5-11 dm. tall; lower leaves runcinate-pinnatifid, 15-25 cm. long, the base auricled and clasping; heads 2 cm. high, 3-5 cm. broad when expanded; achenes oblong, slightly flattened and with thick ribs.

2. *S. oleraceus* L.   Common Sow-Thistle.

Tall and glabrous; lower leaves petioled, the upper clasping and with pointed auricles, lyrate-pinnatifid, the lobes spinulose-dentate; achenes 3-ribbed and transversely roughened. Fig. 964.

3. *S. asper* (L.) All.   Spiny Sow-Thistle.

Stems tall and glabrous; lower leaves spatulate or obovate, the upper lanceolate with auricled clasping base, spinulose-denticulate on the margins; achenes flattened, 3 mm. long, smooth. Fig. 965.

8. LACTUCA (Tourn.) L.

Tall leafy-stemmed herbs with alternate leaves; heads in panicles, cylindric, becoming conical in fruit; phyllaries imbricated in about 3 series; achenes flattened, contracted into a beak at the apex; pappus of numerous capillary bristles. (From lac, milk, on account of the milky juice.)

1A. Pappus brown. .... 1. *L. spicata*
2A. Pappus white.
  1B. Flowers blue. .... 2. *L. tartarica*
  2B. Flowers yellow. .... 3. *L. virosa*

1. *L. spicata* (Lam.) Hitchc.   Tall Blue Lettuce.

Tall, glabrous annual or biennial; leaves deeply pinnatifid, hispid on the veins beneath; heads numerous; flowers blue; achenes flat with short, stout beak.

Southeast Alaska—Newf.—N. Car.—Colo.—Ore. Fig. 966.

2. *L. tartarica* (L.) C. A. Mey.   Large-flowered Blue Lettuce.
   L. pulchella (Pursh) DC.

Glabrous, leafy perennial, 3–10 dm. tall; leaves linear-lanceolate, lanceolate or oblong, acute, entire to dentate or pinnatifid, 5–20 cm. long; panicle usually narrow; heads 15–20 mm. high.

Yukon Valley—Gt. Bear Lake—Mich.—Mo.—N. Mex.—Calif.

3. *L. virosa* L.   Prickly Lettuce.

Biennial with erect stem 5–20 dm. tall; leaves oblong or oblanceolate, spinulose-margined, denticulate or somewhat pinnatifid, 1–3 dm. long, those of the stem auriculate-clasping; heads numerous, the involucres about 1 cm. long.

Introduced at Manly Hot Springs, native of Europe.

9. PRENANTHES L.

Perennial herbs; leaves alternate, dentate, lobed or pinnatifid; heads rather small; flowers white, yellowish or purplish; involucre cylindric, the phyllaries in 1 or 2 series with smaller ones at the base. (Greek, drooping blossom.)

P. alata (Hook.) Dietrich.   Rattlesnake Root.
   P. lessingii Hult. *Nabalus hastatus* (Less.) Heller.

Somewhat pubescent, 2–4 dm. tall; leaves hastate-deltoid, sharply toothed, with margined petiole; main phyllaries lanceolate, about 1 cm. long; pappus of numerous rather stiff light reddish-brown bristles.

Aleutians—Prince William Sd.—Ida.—Ore. Fig. 967.

10. CREPIS L.

Annual or perennial herbs; leaves entire, toothed or pinnatifid; heads small or medium; flowers yellow or orange; phyllaries in 1 series with smaller ones at the base; pappus of white slender capillary bristles. (Greek, sandal, application not explained.)

1A. Achenes slightly dilated at the insertion of the pappus;
   low plants with numerous heads.
   1B. Achenes conspicuously beaked; stems 1 - 2 dm. tall.1. *C. elegans*
   2B. Achenes scarcely beaked; stems less than 5 cm. tall.2. *C. nana*
2A. Achenes not dilated at the insertion of the pappus;
   plants taller. ....3. *C. capillaris*

1. *C. elegans* Hook.
   Youngia elegans (Hook.) Rydb.

Perennial; stem branched, glabrous, 8–20 cm. tall; leaves entire to sinuately pinnatifid with triangular lobes, the cauline lance-linear; heads about 8 mm. high; achenes about 4 mm. long.

Central Alaska—Mack.—Sask.—Wyo.—B.C. Fig. 968.

2. *C. nana* Rich.
   Youngia nana (Rich.) Rydb.

In low, dense, sometimes almost stemless tufts; leaves mostly basal, ovate or spatulate, entire, repand-dentate or lyrate; involucre of 8–10 phyllaries 8–10 mm. high and thickened on the backs at the base; achenes cylindric, 5 mm. long, slightly narrowed above, minutely roughened.

Asia—Alaska—Victoria Isl.—Baffin Isl. — Newf. — Colo. — Nev.— Calif. Fig. 969.

3. *C. capillaris* (L.) Wallr.

Annual introduced weed, 3–8 dm. tall; leaves oblanceolate, mostly more or less lacinate-pinnatifid, those of the stem clasping with auriculate base; heads numerous, 8–10 mm. high.

Native of Europe.

11. HIERACIUM (Tourn.) L.

Hairy perennial herbs; leaves entire or dentate, abundant at the base but usually few on the stems; involucre cylindric or campanulate with the bracts in 2 or 3 series and a few small ones at the base; achenes cylindric, 10- to 15-ribbed; pappus of 1 or 2 series of sordid or brownish capillary bristles. (Greek, hawk, from the supposition that hawks used the plants to strengthen their eyesight.)

1A. Involucre bracts of the rather large heads more or
   less imbricate. ....4. *H. canadense*
2A. Involucral bracts of the small heads of almost a
   single series with small calyculate ones below.
   1B. Flowers white; heads several to many. ....1. *H. albiflorum*
   2B. Flowers yellow, heads few, about 8 mm. high.
      1C. Heads somewhat glandular and short-hairy.......2. *H. gracile*
      2C. Heads densely wooly. ....3. *H. triste*

1. **H. albiflorum** Hook. White Hawkweed.

Stems 3-6 dm. tall, long hairy below, glabrate above; upper part of stem nearly naked; lower leaves narrowing into a winged petiole, the upper much reduced; phyllaries long and narrow.

Southeast Alaska and Yukon—Sask.—Colo.—Calif. Fig. 970.

2. **H. gracile** Hook. Slender Hawkweed.

Stems 1-4 dm. tall; leaves spatulate or oblong, glabrate, repand-denticulate or entire; peduncles and involucre black-hairy but hairs much shorter than in *H. triste*. Our form is less glandular than the type and has been described as var. *alaskanum* Zahn. A vigorous form with stems 3-4 dm. tall and bearing 10-15 or more subumbellate heads on long, slender black-villous peduncles is the var. *yukonense* Porsild.

Alaska—Gt. Bear Lake—Alta.—N. Mex.—Calif. Fig. 971.

3. **H. triste** Cham. Wooly Hawkweed.

Stems 1-3 dm. tall; leaves obovate or spatulate, entire, glabrate or sparsely hairy; involucres and peduncles densely covered with long, dark gray or brownish wool. The var. *tristeforme* Zahn approaches *H. gracile* and may be a hybrid with that species.

East Asia—Aleutians—southeast Alaska. Fig. 972.

4. **H. canadense** Michx.

Stems erect, leafy, simple or branched above, 3-10 dm. tall; lower leaves oblanceolate, the upper lanceolate, 3-10 cm. long, sessile, distinctly dentate; involucre 10-12 mm. high; phyllaries glabrous and somewhat blackish.

Near Yukon boundary—Labr.—Penn.—Iowa—Ore.

59. ASTERACEAE (Aster Family)

Herbs (in ours, further south some are shrubs or trees) with various leaves but without stipules; flowers in heads subtended by an involucre of few to many bracts (phyllaries) arranged in one to many series; calyx tube completely adnate to the ovary, its limb when present forming the pappus; corolla tubular, usually 5-lobed or 5-cleft, or that of some marginal flowers expanded into a ligule and forming the ray flowers, which when present make the heads radiate as distinguished discoid heads where the rays are absent; stamens usually 5, borne on the corolla and alternate with its lobes; ovary 1-celled, developing into an achene; style of fertile flowers 2-cleft. An immense family usually divided into tribes only about half of which are represented in our territory. This group is also known as *Carduaceae* and is often combined with the *Cichoriaceae* and known as *Compositae*. The name *Asteraceae* can also be applied to the combined group.

KEY TO THE TRIBES

1A. Anthers caudate at the base; rays none.
   1B. Anthers unappendaged at the tip; heads small. .... 2. *Inuleae*
   2B. Anthers with elongated appendages at tip; heads large. .... 6. *Cynareae*
2A. Anthers not caudate at base; heads usually with ray flowers.
   1B. Receptacle naked.
      1C. Bracts of the involucre usually well imbricated. ... 1. *Astereae*
      2C. Bracts little or not at all imbricated. .... 5. *Senecioneae*
   2B. Receptacle chaffy.
      1C. Bracts of the involucre dry and scarious. .... 4. *Anthemideae*
      2C. Involucral bracts herbaceous. .... 3. *Heliantheae*

1. ASTEREAE

1A. Pappus wanting or of a few capillary bristles. .... 1. *Bellis*
2A. Pappus of capillary bristles.
   1B. Ray flowers yellow.
      1C. Leaves mostly basal. .... 2. *Haplopappus*
      2C. Stems leafy. .... 3. *Solidago*
   2B. Ray flowers not yellow.
      1C. Phyllaries in 2-5 series. .... 4. *Aster*
      2C. Phyllaries in 1 or 2 series. .... 5. *Erigeron*

2. INULEAE

1A. Heads all fertile. .... 8. *Gnaphalium*
2A. Heads dioecious or nearly so.
   1B. Phyllaries in 2-5 series. .... 6. *Antennaria*
   2B. Phyllaries in many series. .... 7. *Anaphalis*

3. HELIANTHEAE

1A. Heads with ray flowers. ....10. *Helianthus*
2A. Heads without rays or rays inconspicuous.
   1B. Pappus none. .... 9. *Madia*
   2B. Pappus of 2 or 4 barbed awns. ....11. *Bidens*

4. ANTHEMIDEAE

1A. Heads radiate (rays usually white).
   1B. Receptacle chaffy.
      1C. Rays short, 2-5 mm. long. ....12. *Achillea*
      2C. Rays 1 cm. or more long. ....13. *Anthemis*
   2B. Receptacle naked.
      1C. Phyllaries in several series. ....14. *Chrysanthemum*
      2C. Phyllaries in few series. ....15. *Matricaria*
2A. Heads discoid.
   1B. Heads small in spike-like or racemose-paniculate inflorescences. ....16. *Artemisia*
   2B. Heads solitary or corymbose.
      1C. Receptacle conical. ....15. *Matricaria*
      2C. Receptacle flat or convex.
         1D. Achenes sessile. ....17. *Tanacetum*
         2D. Achenes raised on pedicels which remain attached to the receptacle. ....18. *Cotula*

5. SENECIONEAE

1A. Leaves all basal; flower heads on scapes. ....19. *Petasites*
2A. Leaves mostly opposite; flowers yellow. ....20. *Arnica*
3A. Leaves alternate.
   1B. Heads discoid. ....21. *Cacalia*
   2B. Heads usually with ray flowers. ....22. *Senecio*

6. CYNAREAE

Leaves prickly. ....24. *Cirsium*
Leaves not prickly. ....23. *Saussurea*

ARTIFICIAL KEY

1A. Heads radiate.
  1B. Rays yellow.
    1C. Pappus of capillary bristles.
      1D. Leaves opposite. ....20. *Arnica*
      2D. Leaves alternate.
        1E. Phyllaries in 2-4 series. .... 3. *Solidago*
        2E. Phyllaries in 1 series with a few smaller basal ones.
          1F. Phyllaries broad. .... 2. *Haplopappus*
          2F. Phyllaries narrow. ....22. *Senecio*
    2C. Pappus not of capillary bristles.
      1D. Pappus of 2 scales. ....10. *Helianthus*
      2D. Pappus of 4 retrorsely barbed awns. ....11. *Bidens*
      3D. Pappus none or a mere crown.
        1E. Leaves linear and simple. .... 9. *Madia*
        2E. Leaves 1- to 3-pinnately compound. ....17. *Tanacetum*
  2B. rays not yellow.
    1C. Pappus of capillary bristles.
      1D. Rays inconspicuous, stems scapose ....19. *Petasites*
      2D. Rays conspicuous.
        1E. Rays in more than 1 row; phyllaries in 1 row or series. .... 5. *Erigeron*
        2E. Rays in 1 row, phyllaries in several series. ... 4. *Aster*
    2C. Pappus none or a mere crown.
      1D. Leaves entire.
        1E. Acaulescent. .... 1. *Bellis*
        2E. Stem leafy. ....14. *Chrysanthemum*
      2D. Leaves lobed or divided.
        1E. Rays 2-4 mm. long. ....12. *Achillea*
        2E. Rays longer.
          1F. Leaves toothed or lobed. ....14. *Chrysanthemum*
          2F. Leaves dissected.
            1G. Receptacle chaffy. ....13. *Anthemis*
            2G. Receptacle naked. ....15. *Matricaria*
2A. Heads discoid.
  1B. Pappus of capillary bristles.
    1C. Leaves prickly. ....24. *Cirsium*
    2C. Leaves not prickly.
      1D. Receptacle bristly or chaffy. ....23. *Saussurea*
      2D. Receptacle naked.
        1E. Stem scape-like. ....19. *Petasites*
        2E. Stem leafy.
          1F. Wooly plants.
            1G. Heads all fertile. .... 8. *Gnaphalium*
            2G. Heads dioecious or nearly so.
              1H. Phyllaries in 2-5 series. .... 6. *Antennaria*
              2H. Phyllaries in many series. .... 7. *Anaphalis*
          2F. Plants not wooly.
            1G. Flowers white. ....21. *Cacalia*
            2G. Flowers yellow. ....22. *Senecio*
  2B. Pappus not of capillary bristles.
    1C. Receptacle bristly.
      1D. Pappus of 4 retrorsely barbed awns. ....11. *Bidens*
      2D. Pappus none or a mere crown. ....16. *Artemisia*
    2C. Receptacle naked.
      1D. Achenes borne on persistent pedicels. ....18. *Cotula*
      2D. Achenes not on pedicels.

1E. Heads small in spike-like or racemosely-
paniculate inflorescence. .... 16. *Artemisia*
2E. Heads larger. .... 15. *Matricaria*

1. BELLIS (Tourn.) L.

Tufted; leaves all basal; phyllaries imbricated in 1 or 2 series, nearly equal; receptacle convex (or conic), naked; ray flowers pistillate, white, pink or purplish; disk flowers yellow, perfect, the corollas tubular with a 4- to 5-toothed limb; achenes obovate, flattened; pappus none or a ring of minute bristles. (Latin, pretty.)

B. perennis L.   European Daisy.

Leaves obovate, obtuse, slightly toothed, narrowed into a margined petiole; scapes 5–25 cm. tall bearing a single head; ray flowers numerous, linear.

Occasionally escaped from gardens; native of Eurasia.

2. HAPLOPAPPUS Endl.

Low perennials with caespitose, woody caudices; leaves narrow, firm; phyllaries appressed, thin, ovate to lanceolate; receptacle naked, alveolate; ray flowers fertile; disk flowers perfect, the corollas somewhat enlarged upward and with 5-toothed margin; achene white-villous; pappus of white capillary bristles. (Greek, simple pappus.)

H. macleanii Brandegee.
Stenotus borealis Rydb.

Leaves mostly basal, linear, ciliate on the margins, about 1 cm. long and 1 mm. wide; heads on scapes 2–6 cm. long; ray and disk flowers yellow.

Known from several widely scattered localities in Yukon.

3. SOLIDAGO L.

Caulescent perennials; leaves alternate, entire or toothed; heads small, several-flowered, with small yellow rays; phyllaries well imbricated in several series; receptacle small, alveolate; pappus of capillary bristles; achenes usually ribbed. (Latin, to make whole.)

1A. Heads about 8 mm. high. .... 1. *S. multiradiata*
2A. Heads smaller.
  1B. Heads very numerous, the inflorescence usually
     more or less spreading. .... 2. *S. elongata*
  2B. Heads less numerous, in narrow, spike-like inflorescence.
    1C. Phyllaries lanceolate and acute at the summit... 3. *S. lepida*
    2C. Phyllaries linear-elliptic, rounded at the summit. 4. *S. decumbens*

1. *S. multiradiata* Ait.   Northern Golden Rod.

Stems often several, glabrate below, pubescent above, up to 6 dm. tall but often very dwarf; leaves nearly glabrous but ciliate on the margins, at least below, the lower spatulate or oblanceolate and narrowed into a margined petiole, the upper sessile; heads several and glomerate in a terminal cluster; phyllaries narrowly lanceolate and thin-edged; rays prominent and linear; achenes pubescent, about 3 mm. long. Var. *arctica* (DC.) Fern. has elongated upper leaves.

Common. Alaska—Labr.—Newf.—Colo.—B.C. Fig. 973.

2. *S. elongata* Nutt.

Stems leafy above, 3-8 dm. tall; leaves narrowly lanceolate, acuminate, up to 1 dm. long, somewhat serrate; heads 3-4 mm. long, very numerous, the branches of the inflorescence sometimes ascending but usually spreading; phyllaries linear-lanceolate, acute. By some authors considered to be a variety of *S. lepida*.

Central Alaska—Gt. Slave L.—Mont.—Nev.—Calif. Fig. 974.

3. *S. lepida* DC.

Stems 3–10 dm. tall, leafy; leaves oblong to lanceolate, coarsely and sharply serrate, up to 1 dm. long; inflorescence rather compact and spike-like, up to 1 dm. long, sometimes but little exceeding the leaves; heads 5–6 mm. long; phyllaries linear-lanceolate, attenuate-acute.

Unalaska along the coast to Calif. and Sask.—Que.—Mich.—Utah. Fig. 975.

4. *S. decumbens* Green var. *oreophila* (Rydb.) Fern.

Stems usually clustered, 2–6 dm. tall; leaves glabrous, spatulate or oblanceolate, crenate-serrate toward the apex, the stem leaves reduced and few; phyllaries linear or oblong; heads 4–6 mm. high; achenes hirsute.

Central Alaska—Mack.—Colo.—N. Mex.—B.C. Fig. 976.

4. ASTER (Tourn.) L.

Perennials with alternate leaves; heads 1 to many with purple, pink, or white rays, never yellow; phyllaries in several series, herbaceous or herbaceous-tipped; receptacle flat or convex, alveolate; disk flowers perfect, yellow changing to red brown or purplish; achenes mostly flattened and nerved; pappus of numerous slender capillary bristles. (Greek, a star.)

```
1A. Pappus white or lightly tinged brown.
   1B. Heads solitary. .................................. 1. A. alpinus
   2B. Heads usually more than one.
      1C. Leaves very narrow. .......................... 3. A. junceus
      2C. Leaves wider.
         1D. Lower leaves cordate or subcordate at the
             base. ..................................... 7. A. ciliolatus
         2D. Lower leaves glaucous, narrowing into a
             winged petiole. ........................... 8. A. laevis
```

2A. Pappus medium to dark brown.
  1B. Stems 3 - 10 dm. tall.
    1C. Involucre and pedicels glandular. .... 6. *A. modestus*
    2C. Involucres not glandular. .... 4. *A. subspicatus*
  2B. Stems 8 - 40 cm. tall.
    1C. Leaves linear and entire. .... 2. *A. yukonensis*
    2C. Leaves wider and usually toothed. .... 5. *A. sibiricus*

1. *A. alpinus* L.   Alpine Aster.
    A. alpinus ssp. *vierhapperi* Onno.

Stems 1 to several from a thickish caudex, pilose, 15–30 cm. tall; leaves numerous, entire, pilose, those of the stem reduced; heads large, the involucre about 1 cm. high, 15–25 mm. broad; phyllaries of nearly equal length; rays violet, 1 cm. or more long.
    Yukon—Mack.—Colo. and in Eurasia. Fig. 977.

2. *A. yukonensis* Cronq.   Yukon Aster.

Stems somewhat pubescent, at least near the top, slender, 1 to several arising from a woody caudex, 1–2 dm. tall; leaves linear, acute, up to 6 cm. long and 3.5 mm. wide; heads usually one but occasionally 2, the involucre about 1 cm. high and about as broad; phyllaries pubescent, some of them mucronate; rays bluish violet, up to 1 cm. long.
    Lake Kluane. Fig. 978.

3. *A. junceus* Ait.   Rush-like Aster.
    A. junciformis Rydb.

Stems 2–6 dm. tall, glabrate below, hirsute near the top; leaves linear-lanceolate or linear-oblanceolate, entire, 4–10 cm. long, 2–6 mm. wide; involucres about 7 mm. high by 1 cm. wide; inner phyllaries longer than the outer; rays with narrow ligules which are white to purple.
    Southwest and central Alaska—Gt. Slave L.—N.S.—Penn.—Colo.—Wash. Fig. 979.

4. *A. subspicatus* Nees.
    A. foliaceus Lindl.

Stems 3–9 dm. tall, smooth below, pubescent above; leaves entire or with a few short teeth, oblanceolate or the upper lanceolate or linear-lanceolate; involucres about 1 cm. high, 12–15 mm. broad; ray flowers purple, 12–20 mm. long; phyllaries green and nearly glabrous except on the margins.
    Aleutians—along coast to Calif.—Ida.—Colo.—N. Mex. and in Labr. and Que. Fig. 980.

5. *A. sibiricus* L.   Siberian Aster.
    A. richardsonii Hook.

More or less pubescent throughout, 1–4 dm. tall; lower leaves petioled, the upper sessile, 2–7 cm. long; heads 1–4 on each stem, the stems usually clustered; involucres about 12 mm. high and 15 mm. broad; phyllaries lanceolate, pubescent, the inner purplish; rays purple; pappus purplish brown.

All of Alaska—Mack.—Alta.—B.C. Fig. 981.

6. *A. modestus* Lindl.   Great Northern Aster.
    A. unalaskensis var. *major* Hook.

Stem stout, leafy, branched above, pubescent and glandular near the inflorescence, up to 15 dm. tall; leaves lanceolate, partly clasping, acuminate at the apex with a few sharp, distant teeth, up to 12 cm. long, pubescent; heads at the end of short branches; phyllaries linear-subulate, little imbricated; rays purple to violet, 1 cm. or more long.

Alaska Penin.—Ont.—Minn.—Ore. Fig. 982.

7. *A. ciliolatus* Lindl.   Lindley Aster.
    A. lindleyanus T. and G.

Stems glabrous or sparingly pubescent with crisp hairy lines above, 3–10 dm. tall; lower leaves cordate to obovate, serrate, up to 15 cm. long; upper leaves with winged petioles or sessile, the margins sometimes entire; involucres 7–8 mm. high, about 1 cm. broad; phyllaries linear-lanceolate with a green oblanceolate midrib; rays blue or violet, 10–12 mm. long; pappus nearly white.

Liard Hot Springs—Labr.—N. Hamp.—Ohio—Wyo.—B.C.

8. *A. laevis* L.   Smooth Aster.

Stem upright, 3–12 dm. tall, glabrous and glaucous; leaves thick, entire or somewhat serrate, the basal ones tapering to winged petioles, the upper ones sessile and clasping; involucres 8–9 mm. high, about 1 cm. broad; rays blue or violet.

Near Yukon boundary—Maine—Ala.—La.—N. Mex.—B. C.

5. ERIGERON L.

Biennials or perennials; leaves alternate; heads 1 to many, radiate or discoid; phyllaries in 1 or 2 series, not much imbricated, usually narrow and not herbaceous; receptacle naked, flat, punctuate; rays usually narrow, in more than 1 series, pistillate; disk flowers yellow; pappus of a single series of rough capillary bristles, often with an outer whorl of short ones. Differs from *Aster* chiefly in the numerous narrow rays and in the involucre. (Greek, early old, in allusion to the pappus.)

1A. Rays inconspicuous or short.
  1B. Tubular-filiform pistillate flowers present between
      the hermaphrodite flowers and the outer ligu-
      late flowers. .... 14. *E. acris*
  2B. Tubular-filiform pistillate flowers absent. .... 13. *E. lonchophyllus*
2A. Ray flowers more or less conspicuous.
  1B. Leaves deeply divided. .... 11. *E. compositus*

2B. Leaves entire or nearly so.
  1C. Stems usually 1 dm. or less tall.
    1D. Plant densely caespitose. ....10. *E. purpuratus*
    2D. Stems usually only one.
      1E. Plant densely gnaphaloid-lanate. .... 7. *E. muirii*
      2E. Plant pubescent but not densely lanate.
        1F. Rays 1 mm. or more wide. .... 6. *E. hyperboreus*
        2F. Rays very narrow. .... 9. *E. humilis*
  2C. Stems usually more than 1 dm. tall.
    1D. Stems 5 – 30 cm. tall.
      1E. Rays 3 – 6 mm. long. .... 8. *E. uniflorus*
      2E. Rays longer.
        1F. Heads often more than 1. .... 3. *E. caespitosus*
        2F. Heads usually solitary.
          1G. Basal leaves narrow, elongate, acuminate
            or attenuate. .... 4. *E. yukonensis*
          2G. Basal leaves oblanceolate, tapering to the
            petiole. .... 5. *E. grandiflorus*
    2D. Stems usually 3 dm. or more tall.
      1E. Head usually solitary, large. .... 1. *E. peregrinus*
      2E. Heads usually several.
        1F. Rays more than 0.5 mm. wide. .... 2. *E. glabellus*
        2F. Rays less than 0.5 mm. wide. ....12. *E. philadelphicus*

1. *E. peregrinus* (Pursh) Greene.

Aster peregrinus Pursh.

Perennial, 2–5 dm. tall, densely villous on upper part, glabrate near the base; involucre about 15 mm. broad; rays purplish, 1 cm. or more long. Some collections from Douglas Island seem to be ssp. *callianthemus* (Greene) Cronq. which has glandular involucres. This species combines some characteristics of both *Aster* and *Erigeron*.

Common in the coast regions, Commander and Aleutian Islands—Colo.—Utah—Calif. Fig. 983.

2. *E. glabellus* Nutt. ssp. *pubescens* (Hook.) Cronq.

Hispid-bristly throughout, 3–6 dm. tall, often much branched from the base; heads on rather long peduncles; involucre about 8 mm. high and 15 mm. broad; rays long, white to pinkish-purple; when spread gives the head a diameter of 3–4 cm.

Central Alaska—Mack.—Wisc.—Colo.—Mont.—B.C. Fig. 984.

3. *E. caespitosus* Nutt.

Perennial with stout taproot and usually branched caudex; stems several, densely pubescent with short spreading hair, 5–35 cm. tall; leaves pubescent, the basal ones narrowly oblanceolate or spatulate, tapering to a petiole and up to 12 cm. long, the cauline smaller and sessile; heads solitary or few, 9–12 mm. wide; phyllaries appressed, thickened on the back; rays blue, white or pink, 5–15 mm. long.

Central Alaska and Yukon—Sask.—N. Dak.—Colo.—Ariz.—Wash. Fig. 985.

4. *E. yukonensis* Rydb.

Perennial with branched caudex; stems 6–40 cm. tall, villous-hirsute; leaves narrow, acuminate, hirsute-ciliate, at least along the margins, heads 1–4, mostly solitary, up to 17 mm. wide; involucre 7–10 mm. high, the phyllaries wooly-villous, narrow, with purplish tips; rays 45–75, pink to bluish purple, 10–15 mm. long; achenes 2-nerved, hairy.

Dawson—Lake Kluane—Whitehorse—Mack. Fig. 986.

5. *E. grandiflorus* Hook.

Stems decumbent at the base, 4–25 cm. tall; basal leaves oblanceolate, tapering to a petiole, hirsute-pilose, 1–9 cm. long, 4–8 mm. wide; stem leaves several, lanceolate to ovate; heads large, solitary; involucres 8–10 mm. high; phyllaries long villous or pilose, greenish below, reddish purple on the margins and the nearly naked tips; achene copiously hirsute.

Central Alaska—Alta.—B. C. Fig. 987.

6. *E. hyperboreus* Greene.

E. alaskanus Cronq.

Stems 5–10 cm. tall, spreading-hirsute; basal leaves oblanceolate, tapering to a short petiole or subsessile, 1–5 cm. long; stem leaves linear, few and reduced; heads solitary, 9–15 mm. wide; involucre 5–8 mm. high, somewhat viscid or glandular; phyllaries slender, attenuate, green to purplish black; rays 40–60, 2-toothed at the apex, 9–12 mm. long, 1–2 mm. wide, blue; achenes villous-hirsute.

Seward Penin.—Kivelina—Porcupine River. Fig. 988.

7. *E. muirii* Gray.

Perennial with stout caudex, the whole plant densely gnaphaloid-lanate; basal leaves oblanceolate or spatulate, 15–30 mm. long, 5–7 mm. wide; stem leaves several; involucre 8–9 mm. high; phyllaries purple under the dense tomentum; rays 75–100, 10–13 mm. long; achenese villous-hirsute.

Cape Thompson and Anaktuvuk Pass region.

8. *E. uniflorus* L. ssp. *eriocephalus* (J. Vahl) Cronq.

E. eriocephalus J. Vahl.

Stems 3–35 cm. tall, sparingly to densely villous with crinkled hairs; leaves villous when young, approaching glabrate with age, the basal oblanceolate, up to 9 cm. long and 8 mm. wide; stem leaves few; heads usually solitary, the disk 15–30 mm. wide; involucre usually densely wooly-villous; phyllaries tinted deep purple; rays 100 or more, white to pink or purple, 3–6 mm. long.

Arctic Alaska—Greenl.—Que.—Central Alaska. Fig. 989.

9. *E. humilis* Grah.
   E. unalaskensis (DC.) Ledeb.

   Stems 2-20 cm. tall, villous throughout; basal leaves oblanceolate, up to 8 cm. long and 11 mm. wide; heads solitary; involucre 6-9 mm. high; phyllaries black-purple, rays 50-150, 3-6 mm. long, erect or ascending.
   Circumpolar, south to Que. and Mont. Fig. 990.

10. *E. purpuratus* Greene.

    Caespitose; stems 1-10 cm. tall, leafy at the base, more or less villous; leaves oblanceolate or spatulate, up to 3 cm. long and 5 mm. wide; heads solitary, the disk 10-15 mm. wide; involucre 7-10 mm. high, villous; rays 60-90, white or pink, 4-8 mm. long.
    Central Alaska—Yukon—B.C. Fig. 991.

11. *E. compositus* Pursh.

    Hispid-hirsute throughout; leaves mostly basal, 1-4 times ternately lobed or dissected; stem leaves few and reduced; stems scapiform, up to 25 cm. tall; heads solitary, the disk 8-20 mm. wide; rays 20-60, white, pink or blue, up to 12 mm. long but usually shorter and often inconspicuous. The typical form has leaves 2-4 times ternate with long linear divisions. Var. *glabratus* Macoun has 2-3 times ternate leaves with shorter divisions. Var. *discoideus* Gray has leaves simply ternate.
    Central Alaska—Greenl.—Que.—S. Dak.—Ariz.—Calif. Fig. 992.

12. *E. philadelphicus* L.

    Biennial or short-lived perennial, more or less pubescent with long spreading hairs, 2-8 dm. tall, branched above; basal leaves oblanceolate to obovate, tapering into a short petiole, up to 15 cm. long and 3 cm. wide; stem leaves clasping, ample; heads few to many in an open inflorescence; phyllaries green with hyaline margins; rays more than 100, very narrow, pink or rose-purple.
    Liard Hot Springs—Mack.—Labr.—Maine—Texas—Calif.

13. *E. lonchophyllus* Hook.

    Biennial or short-lived perennial; stems 1 to many, 1-3 dm. tall, nearly glabrate at the base but increasingly bristly-hairy toward the top, the same type of hairs on the lower edge of the upper leaves and on the phyllaries; lower leaves narrowly oblanceolate, up to 15 cm. long; heads few or several, the involucre 4-9 mm. high; rays numerous, white or pinkish, 2-3 mm. long.
    Central Alaska—Yukon—Sask.—N. Dak.—N. Mex.—Calif. and in Ont. and Que. Fig. 993.

14. *E. acris* L.

Biennial or perennial; stems 1-8 dm. tall, subglabrous to spreading-hirsute; leaves subglabrous to hirsute, the basal one oblanceolate, entire or remotely serrulate, up to 10 cm. long and 15 mm. wide; heads one to many; involucres 5-12 mm. high; rays white to pink or purplish, 2.5-4.5 mm. long. This is our commonest *Erigeron* and occurs in 3 varieties. Var. *asteroides* (Andriz.) DC. 3-8 dm. tall, erect, heads several to many; peduncles and involucre glandular, rays pinkish. Var. *debilis* Gray plant 2-25 cm. tall; heads solitary or few, peduncles and involucre glandular; rays pinkish. Var. *elatus* (Hook.) Cronq. plant 1-4 dm. tall, erect; heads solitary or few; peduncles and involucre more or less hirsute, not glandular; rays white or pinkish, short.

Circumpolar, south to Maine, Mich., Colo., Calif. Fig. 994.

6. ANTENNARIA Gaertn.

Wooly, dioecious perennials; leaves basal and alternate; heads small, discoid, many-flowered; inflorescence dry, scarious, white, brown or rose; pistillate corollas with filiform corollas, the staminate flowers with tubular, 5-lobed corollas and rudimentary styles and ovaries; pappus of capillary bristles. There is much apomixis in this group which makes it very difficult to determine true species. Many of the forms here described should perhaps be regarded as subspecies or varieties. The treatment here largely follows the treatment of the genus by Dr. A. E. Porsild of the National Museum of Canada. (White pappus of sterile flowers suggests the antenna of certain insects.)

```
1A. Basal leaves prominently 3-nerved, 4-16 cm. long. ... 1. A. pulcherrima
2A. Basal leaves single-nerved, lateral nerves if any ob-
       scure.
  1B. Tall, rather broad-leaved plants with large heads. 2. A. howellii
  2B. Tall, medium or dwarf species with small heads.
    1C. Bracts of the involucre (phyllaries) with pale
           greenish-brown to olivaceous or dark brown
           with usually acuminate and erose tips.
      1D. Plant normally monocephalous.
        1E. Plants with stolons. ........................ 3. A. philonipha
        2E. Plants densely tufted, lacking stolons.
          1F. Male and female usually present. ......... 4. A. monocephala
          2F. Only the female plants known.
            1G. Inner phyllaries normally with blunt,
                  pale, straw-colored tips. ............ 24. A. pygmaea
            2G. Inner phyllaries with attenuate, olive-
                  brown tips. ........................ 5. A. angustata
      2D. Plant normally with more than one head.
        1E. Male and female plants usually present.
          1F. Involucre 4-5 mm. high. ................ 6. A. alaskana
          2F. Involucre about 6 mm. high.
            1G. Basal leaves spatulate, 10-15 mm. long. 7. A. neoalaskana
            2G. Basal leaves about 6 mm. long. ........ 8. A. densifolia
        2E. Only the female plant known.
          1F. Plant caespitose with numerous sessile
                 sterile rosettes.
            1G. Basal leaves narrowly oblanceolate, ta-
                  pering to the apex, pappus rufidu-
```

   lous .... 9. *A. ekmaniana*
  2G. Basal leaves spatulate-obovate, rounded
    at the apex; pappus white.
   1H. Plant pulvinate, basal leaves short,
     obovate. .... 10. *A. compacta*
   2H. Plant caespitose but not pulvinate;
     leaves longer. .... 11. *A. subcanescens*
 2F. Plants loosely caespitose with basal rosettes
   borne on well-developed prostrate or
   ascending stolons.
  1G. Achenes papillose.
   1H. Inflorescence rather compact; inner
     phyllaries dark brown. .... 12. *A. stolonifera*
   2H. Inflorescence open; inner phyllaries
     pale brown. .... 13. *A. pedunculata*
  2G. Achenes glabrous.
   1H. Heads 1-3, 7-9 mm. high. .... 14. *A. megacephala*
   2H. Heads 3-5, 6-7 mm. high. .... 15. *A. pallida*
2C. At least the inner phyllaries with papery, white,
 straw-colored or pink and usually ligulate
 non-attenuate tips.
 1D. Inner phyllaries pink.
  1E. Male plant as common as the female. .... 16. *A. dioica*
  2E. Only female plants known in our region.
   1F. Fruiting stems usually 20 cm. or more
    tall.
    1G. Plants with well developed stolons.
     1H. Basal leaves glabrate above; cauline
      leaves ample. .... 17. *A. alborosea*
     2H. Basal leaves not glabrate above, stem
      leaves reduced.
      1J. Basal leaves narrowly spatulate;
       phyllaries pale pink, soon be-
       coming pale gray or straw-
       colored. .... 18. *A. elegans*
      2J. Basal leaves oblanceolate; phyllaries
       pink even in age. .... 19. *A. rosea*
    2G. Plant with short, sessile offsets; leaves
     spatulate-obovate. .... 20. *A. oxyphylla*
   2F. Fruiting stems less than 20 cm. tall.
    1G. Heads nodding in youth; plants with mat-
     ted growth. .... 21. *A. breitungii*
    2G. Heads not nodding in youth; basal leaves
     erect. .... 22. *A. incarnata*
 2D. Inner phyllaries papery white or straw-colored;
  never pink.
  1E. Monocephalous or rarely with 2 or 3 heads.
   1F. Basal leaves 1-3 cm. long. .... 23. *A. shumaginensis*
   2F. Basal leaves rarely over 1 cm. long. ...... 24. *A. pygmaea*
  2E. Heads normally more than 1.
   1F. Tall plants, usually 20 cm. or more tall. ... 27. *A. leuchippi*
   2F. Dwarf to medium plants.
    1G. Heads nodding when young; inflores-
     cence glomerate; phyllaries papery
     white.
     1H. Cauline leaves without scarious tips. ... 25. *A. laingii*
     2H. Upper 1-3 cauline leaves with slender
      scarious tips. .... 26. *A. nitida*
    2G. Heads not nodding in youth; inflores-
     cence open; phyllaries thin and soft.
     1H. Upper 1-3 cauline leaves with slender
      scarious tips. .... 23. *A shumaginensis*
     2H. Upper 5-8 cauline leaves with broad,
      flat and very prominent scarious
      appendages. .... 29. *A. isolepis*

1. *A. pulcherrima* (Hook.) Greene.

Stems 2–5 dm. tall; basal leaves 4–12 cm. long; stem leaves narrow and much reduced; heads 4–20, the pistillate with involucres 7–8 mm. high; phyllaries in 3 series, lanate at the base, the tips scarious and pale brown; pappus white; achenes glabrous. Var. *angutisquama* Porsild of the Pelly Range has phyllaries long-attenuate and glabrate attenuate leaves.

Central Alaska—Gt. Bear L.—Newf.—Colo.—Wash. Fig. 995.

2. *A. howellii* Greene.

Young stolons flagellate, up to 1 dm. long; rosette leaves 25–50 mm. long, 5–20 mm. wide, glabrous above; mature stems 20–35 cm. tall, greenish purple with thin lanate tomentum; heads 4–8, the lateral ones on peduncles up to 1 cm. long; involucres about 1 cm. high; phyllaries linear-lanceolate, greenish brown and lanate below, with long-attenuate, tawny and papery tips.

Southeast Yukon—Sask.—Mont.—Ore. Fig. 996.

3. *A. philonipha* Porsild.

Plant matted with offsets 5–10 cm. long; basal leaves spatulate-obovate, grabrous above, about 15 mm. long and 4 mm. wide; stems slender and weak, the pistillate plant 8–14 cm. tall, the staminate shorter; pistillate involucres 6–7 mm. high; phyllaries of equal length, thin, hyaline, acuminate; pappus rufidulous; style exerted; achenes glabrous.

Seward Penin.—Artic Coast—Mack.—B. C. Fig. 997.

4. *A. monocephala* DC.

Often forming small mats with the offsets only a few cm. long; basal leaves spatulate-obcuneate, about 1 cm. long, mucronate, glabrous above; stems 2–10 cm. tall, rarely taller, the stem leaves with prominent scarious tips; pistillate involucres 4–5 mm. high; phyllaries dark brown to almost black in the middle; style long-exerted. Var. *exilis* (Greene) Hult. has silvery-appressed upper leaf surfaces and up to 5 cm. long runners.

East Asia—Mackenzie Mts.—B.C.—Aleutians. Fig. 998.

5. *A. angustata* Greene.

Caespitose with sessile or subsessile offsets; basal leaves narrowly oblanceolate, 8–13 mm. long, about 2 mm. wide; stems 5–14 cm. tall bearing 7–11 linear leaves with flat scarious tips; involucres 8–10 mm. high, thinly lanate at the base; phyllaries long-attenuate, olivaceous; style included or short-exerted; pappus white; achene glabrous.

Wainwright, Mack.—Greenl.—Canadian Rockies. Fig. 999.

6. *A. alaskana* Malte.

Caespitose with branched caudex; basal leaves narrowly spatulate-oblanceolate, 1–3 cm. long, 2–4 mm. wide, cinereous-tomentose on both surfaces; cauline leaves linear with short scarious tips; stems 3–17 cm. tall; heads 3–5; phyllaries densely imbricated and with olivaceous tips; pappus tawny.

Bering Sea region—central Alaska. Fig. 1000.

7. *A. neoalaskana* Porsild.

Caespitose, the offsets sessile; basal leaves spatulate, 10–15 mm. long, 2–3 mm. wide, appressed-tomentose on both surfaces, glabrate in age; stems 5–12 cm. tall, stiff, with 4–7 scarious-tipped leaves; heads mostly three; pappus white; tips of corolla pale yellow; style much exerted; achenes minutely papillose.

Known from Richardson Mts. and Sadlerochit River.

8. *A. densifolia* Porsild.

Densely caespitose, the offsets short and crowded; basal leaves densely congested, cuneate-obovate or broadly oblanceolate, obtuse and not mucronate, 5–6 mm. long, 3 mm. broad, densely and yellowishly tomentose on both surfaces; stems 6–9 cm. tall; cauline leaves 5–7 with subulate tips; heads 2–4 on 5 mm. long peduncles; corolla lobes purplish; style exerted, bifid; pappus white; achenes glabrous.

Mackenzie Mts.

9. *A. ekmaniana* Porsild.

Densely caespitose; basal leaves linear-oblanceolate, densely appressed-tomentose on both surfaces, 10–22 mm. long, 2–3 mm. wide; stems 1–2 dm. tall, purplish-tinged; heads 1–7; involucres about 7 mm. high; outer phyllaries lanceolate, the inner long-attenuate, light chestnut brown; pappus subrufescent; style exerted; achenes glabrous or minutely papillose.

East Asia—Alaska—Yukon—Greenl.—Labr. Fig. 1001.

10. *A. compacta* Malte.

Densely caespitose; basal leaves broadly oblanceolate-obovate, 4–8 mm. long, 2.5–4 mm. wide, densely appressed canescent-tomentose on both surfaces; stems 5–10 cm. tall, often arched; cauline leaves 5–9 with long scarious appendages; heads (1)2–4; involucres 6–7 mm. high; phyllaries oblong, dusky brown, in age tawny, the inner narrower, olivaceous, with attenuate, erose tips; style barely exerted; pappus white; achenes glabrous.

Seward Penin.—Victoria Isl.—Yukon—Mack. Fig. 1002.

11. *A. subcanescens* Ostf.

Caespitose, forming dense cushions; basal leaves oblanceolate, 15–25 mm. long including petiole, 4–5 mm. wide, thinly appressed-tomentose on both surfaces; stems 5–10 cm. tall, dark purplish, glandular-papillose under the indument; cauline leaves 5–7, with long scarious appendages; heads usually 3; involucre about 7 mm. high; phyllaries dark brown with greenish-brown tips; style exerted; pappus dirty white.

Cape Lisburne—Coronation Gulf—south Yukon—Alaska Range.

12. *A. stolonifera* Porsild.

Stolons leafy, freely rooting, 5–10 cm. long; basal leaves sericeous-tomentose, spatulate, obtuse, mucronate, 15–25 mm. long, 3–5 mm. wide; stems elongating in fruit to 14–18 cm. tall; uppermost stem leaves with scarious tips; heads 3–5; phyllaries dark; styles scarcely exerted; achenes small, papillose.

Southeast Yukon. Fig. 1003.

13. *A. pedunculata* Porsild.

Stolons procumbent, 5–10 cm. long; basal leaves oblanceolate, mucronate, 2 cm. long, 5 mm. wide, sericeous, becoming glabrescent in age; stems 15–22 cm. tall; heads 1–5, the lower with peduncles 3–6 cm. long; involucres 7–10 mm. high; corolla purple; styles strongly exerted, bifid; achenes strongly papillose, 1 mm. long.

Pelly Range and Umiat.

14. *A. megacephala* Fern.

Stolons densely leafy, short and suberect; basal leaves spatulate-obovate or broadly oblanceolate, mucronate, 8–12 mm. long, 3–4 mm. wide, the upper surface glabrescent in age; stems 5–12 cm. tall; cauline leaves 5–9, linear with prominent scarious tips; heads 1–3; involucres 8–10 mm. high; phyllaries dark green or olivaceous; style barely exerted, bifid; pappus white; achenes glabrous.

Southeast Yukon and north B. C.

15. *A. pallida* E. Nels.

Stolons well developed; basal leaves spatulate-oblanceolate, 10–15 mm. long, appressed wooly on both surfaces; stems 6–15 cm. tall; cauline leaves 7–9, almost lacking scarious appendages; heads 3–6; involucres 6–7 mm. high; inner phyllaries with dirty white erose tips; pappus white; achenes glabrous.

Aleutians—Alaska Range—southeast Alaska. Fig. 1004.

16. *A. dioica* (L.) Gaertn.

Offsets short and ascending; basal leaves obovate, mucronate, glabrous above, 1–2 cm. long, 3–6 mm. wide; heads 3–6; involucre 7–9 mm. high; inner phyllaries scarious, white or tinted rose; corollas rose-purple; styles exerted, bifid; stems 10–15 cm. tall; cauline leaves 7–10, the uppermost with slightly scarious margin.

A Eurasiatic species occurring in the west Aleutians. Fig. 1005.

17. *A. alborosea* Porsild.

Stolons creeping, branching, 5–10 cm. long; basal leaves oblanceolate-cuneate, mucronate, glabrous above 15–30 mm. long, 4–6 mm. wide; stems 20–35 cm. tall; stem leaves 14–20, linear-lanceolate, glabrate; heads 5–10; involucres 6–7 mm. high; inner phyllaries roseate, later straw-colored; style scarcely exerted; pappus white.

Central Alaska—Gt. Bear L.—Alta.—B.C. Fig. 1006.

18. *A. elegans* Porsild.

Humifuse; basal leaves spreading, linear-oblanceolate, acute, 8–20 mm. long, 2–3 mm. wide, appressed-sericeous on both surfaces; stems 12–20 cm. tall; upper stem leaves with scarious tips; heads 1–8. on elongated peduncles; involucres 5–6 mm. high; inner phyllaries oblong-lanceolate, acuminate, erose, pale rose when young, becoming gray or straw-colored.

Southeast Yukon and Gt. Bear Lake. Fig. 1007.

19. *A. rosea* (Eaton) Greene.

Stolons long, ligneous, branching; basal leaves oblanceolate, 10–25 mm. long, 2–4 mm. wide with densely appressed pale tomentum; stems 12–20 cm. tall; stem leaves 8–10, without scarious tips; heads 4–10; involucres 4–5 mm. high; inner phyllaries dark rose or pink, fading in age; pappus dirty white; style not exerted; achenes glabrous.

Pacific coast of Alaska—central Yukon—S. Dak.—Alta. Fig. 1008.

20. *A. oxyphylla* Greene.

Stolons short, leafy, ascending; basal leaves 10–20 mm. long, 4–7 mm. wide, obovate-oblanceolate, mucronate, silvery-gray on both surfaces; stems 16–30 cm. tall, slender; stem leaves 9–12, the upper with scarious tips; heads 3–10; involucres 6–7 mm. high; inner phyllaries pale pink turning straw-color; pappus white; corollas reddish-purple; style barely exerted; achenes glabrous.

Central Alaska—Yukon—Lake Athabaska. Fig. 1009.

21. *A. breitungii* Porsild.

Humifuse; stolons up to 8 cm. long; basal leaves narrowly spatulate-obovate, 5–7 mm. long, 2–3 mm. wide, cinereous-tomentose on upper surface but becoming glabrate in age; stems 8–15 cm. tall; stem leaves 8–10, linear-oblong with scarious tips; heads 4–8; involucre 5–6 mm. high; phyllaries rose pink; achenes papillose.

Alaska and south Yukon.

22. **A. incarnata** Porsild.

Stolons suberect, 3–5 cm. long; basal leaves oblanceolate or spatulate, acute, 1 cm. long, 2–3 mm. wide; stems 8–12 cm. tall; stem leaves 7 or 8, with acute or attenuate scarious tips; heads 4–10; involucres about 5 mm. high; inner phyllaries pale rose.
Pelly Range. Fig. 1010.

23. **A. shumaginensis** Porsild.

Stolons short and erect; basal leaves 1–3 cm. long, 3–6 mm. wide, spatulate-obovate, mucronate, glabrate above; stems 8–15 cm. tall, with 5–8 leaves; heads 1–3; involucres 6–7 mm. high; inner phyllaries pale; corolla purplish; style exerted.
Shumagin Isls., Naknek and Robertson River.

24. **A. pygmaea** Fern.

Stolons short, erect-ascending, forming tufts; basal leaves oblanceolate, mucronate, 8–14 mm. long, 3–4 mm. wide, glabrate above; stems 4–14 cm. tall, bearing about 9 glabrate leaves; heads solitary or occasionally 1 or 2 smaller heads below the terminal one; involucre about 7 mm. high; inner phyllaries with stramineous tips; pappus silky, white; style barely exerted.
Pelly Range—Labr.

25. **A. laingii** Porsild.

Offsets stolon-like; basal leaves 10–15 mm. long, about 3 mm. wide, oblanceolate-spatulate, acuminate, densely canescent-tomentose on both surfaces; stems 8–14 cm. tall with leaves not scarious-tipped; heads 3–8; involucres 5–6 mm. high; inner phyllaries ivory white, obtuse; pappus white; achenes glabrous.
Central Alaska—south Yukon—Rocky Mts. Fig. 1011.

26. **A. nitida** Greene.

Densely matted with freely branching stolons; basal leaves obovate-oblanceolate, silvery white on both surfaces, 5–15 mm. long, 3–5 mm. wide; stems 5–25 cm. tall bearing 8–20 linear leaves, the uppermost with attenuate scarious tips; inflorescence glomerate when young, open and branched in age; involucres 6–7 mm. high; phyllaries with papery white tips; styles barely exerted.
Central Alaska—Gt. Bear L.—James Bay—Sask.—B. C. and in Rocky Mts. to N. Mex. Fig. 1012.

27. **A. leuchippi** M. P. Porsild.

Leafy stolons up to 9 cm. long; basal leaves 20–25 mm. long, about 4 mm. wide, oblanceolate-spatulate, mucronate, white-tomentose on both surfaces; stems greenish-purple, 20–30 cm. tall with about 15 evenly spaced leaves; heads 6–10; involucres 5–6 mm. high; phyllaries mostly with white tips, in youth dotted with pink spots; corolla purple; style not exerted.

South Alaska and Yukon. Fig. 1013.

28. *A. subviscosa* Fern.

Humifuse; basal leaves oblanceolate-spatulate, obtuse, 8–15 mm. long, 2–5 mm. wide, densely white-tomentose on both surfaces; stems 8–14 cm. tall with about 10 leaves; heads 3–6; involucres 5–6 mm. high; inner phyllaries nearly white; pappus white; style barely exerted; achenes glabrous.

Southeast Yukon, Lake Athabasca, and Gaspe, Que. Fig. 1014.

29. *A. isolepis* Greene.

Humifuse with leafy stolons up to 5 cm. long; basal leaves oblanceolate, 1–2 cm. long, 2–5 mm. wide, appressed white-tomentose on both surfaces; stems 10–15 cm. tall, the upper cauline leaves with broad, flat, scarious tips; heads 3–6, rarely more; involucres 6–7 mm. high; phyllaries with erose papery white tips; pappus white.

Central Alaska—Gt. Bear L.—Labr.—Que.—B. C. Fig. 1015.

7. ANAPHALIS DC.

White-tomentose or wooly perennials; leaves alternate, entire; heads discoid, with polygamo-dioecious flowers; involucres hemispheric, the imbricated phyllaries in several series and pearly white; pistillate flowers with filiform corollas, the perfect but sterile central flowers with 5-toothed tubular corollas; pappus of capillary bristles. (Greek, name of some similar plant.)

A. margaritacea (L.) Benth. & Hook. Pearly Everlasting.

Stems 2–6 dm. tall; leaves 5–10 cm. long, 3–15 mm. wide, densely white-tomentose below, less so above, the upper surface glabrate in age; heads numerous, in a compound corymb, 6–7 mm. high, about 8 mm. broad; phyllaries pearly white. A very variable group some forms of which have been described as species.

East Asia—Aleutians—Labr.—Penn.—Kans. Fig. 1016.

8. GNAPHALIUM L.

Annual, biennial, or perennial herbs; leaves alternate, entire, narrow, wooly; heads discoid, of outer pistillate flowers with filiform corollas and a few perfect flowers with tubular corollas; phyllaries dry, scarious; pappus a row of capillary bristles. (Greek, referring to the wool.)

G. uliginosum L.

A wooly annual, 4–20 cm. tall, often diffusely branched; heads very small, in dense, terminal, leafy-bracted clusters; phyllaries linear, acute, brownish; achenes about two thirds of a mm. long.

Introduced as a weed in several places, native of Eurasia. Fig. 1017.

9. MADIA Molina.

Glandular-viscid, heavy-scented annuals; leaves entire, narrow and at least some of them alternate; heads radiate, 1- to many-flowered; rays yellow, small and inconspicuous; phyllaries in a single series, strongly inflexed, and each inclosing an achene; achenes angled, those of the ray flowers flattened and very oblique; pappus none. (Madi, the Chilian name.)

M. glomerata Hook.   Tarweed.

Plant 3–8 dm. tall, leafy, hirsute throughout, glandular in the inflorescence; leaves linear; heads glomerate, about 6 x 4 mm.; achenes from the ray flowers somewhat curved, those from the disk flowers 4- to 5-angled.

Matanuska—Yukon—Sask.—Colo.—Calif. Probably introduced in our area. Fig. 1018.

10. HELIANTHUS L.

Coarse annuals or perennials; leaves large, simple; heads large, 1 to many; rays yellow, neutral, spreading; phyllaries in several series; receptacle chaffy, the chaff subtending the disk flowers; achenes 4-angled or flattened. (Greek, sun and flower.)

H. annuus L.   Common Sunflower.

Stem hispid or scabrous, 3–25 dm. tall, usually branched; leaves broadly ovate, 3-ribbed, coarsely dentate; phyllaries usually long-acuminate. In cultivation the heads are often very large.

Adventitive at Fairbanks and Manly Hot Springs. Native of the central and southwestern states.

11. BIDENS L.

Herbs; leaves opposite, serrate, lobed or dissected; heads rather large, mostly with rays; phyllaries in 2 series, the outer often foliaceous; receptacle chaffy, the chaff subtending the disk flowers; achenes flat or quadangular, cuneate to linear; pappus of two or four teeth or subulate barbed awns. (Latin, 2-toothed, from the achene.)

B. cernua L.   Nodding Bur-Marigold.

Annual; 2–7 dm. tall; leaves sessile, lanceolate, distantly serrate, sometimes connate at the base; heads several to many, 15–25 mm. broad; achenes 4-angled with 4 awns, 5–6 mm. long.

Galena and Manly Hot Springs. Probably introduced. Fig. 1019.

12. ACHILLEA (Vail.) L.

Erect perennial leafy plants; leaves alternate, varying from serrate to tripinnatifid; heads corymbose, small, with white or rose-colored rays; phyllaries scarious-margined, in several unequal imbricated series; achenes oblong or obovate, flattened, margined; pappus none. (Named for Achilles of mythology.)

1A. Leaves pinnatifid. .... 1. *A. sibirica*
2A. Leaves bi- or tripinnate.
  1B. Involucres 5–7 mm. high. .... 2. *A. borealis*
  2B. Involucres 4–4.5 mm. high.
    1C. Ultimate leaf segments linear; rachis merely margined. .... 3. *A. lanulosa*
    2C. Ultimate leaf segments lanceolate; rachis winged. 4. *A. millefolium*

1. *A. sibirica* Ledeb.   Siberian Yarrow.
A. multiflora Hook.

Stems 3–12 dm. tall, villous; heads numerous, 4–5 mm. high and 4–6 mm. broad; phyllaries villous, elliptic with brown margins; ray flowers with much smaller ligules than in the following species.
    Asia—Bethel—Gt. Bear L.—Man.—B. C. and Gaspe Penin. Fig 1020.

2. *A. borealis* Bong.   Northern Yarrow.

Stems 2–5 dm. tall, more or less silky-villous; leaves 5–15 cm. long; heads larger than in the two following species; phyllaries lanceolate with prominent dark margins; ligules of the ray flowers white or pinkish.
    Common and widespread; Alaska—Newf.—Que.—N. Mex.—Calif. Fig. 1021.

3. *A. lanulosa* Nutt.

Copiously villous with long silky hairs, 2–6 dm. tall; leaves 5–10 cm. long, villous; phyllaries elliptic, obtuse, with greenish midrib and straw-colored or brownish margins; achenes margined.
    Alaska and Yukon—Sask.—Minn.—Kans.—Calif. Fig. 1022.

4. *A. millefolium* L.

Stems erect, 3–10 dm. tall, more or less villous; leaves 5–10 cm. long, finely villous to glabrate; primary segments spreading and more or less decurrent on the wing-margined rachis; achene scarcely margined.
    Klondyke Valley; native of Eurasia and common in the states. *A. ptarmica* L. sometimes persists from cultivation. The leaves are simply serrate and the ligules are 4–5 mm. long.

13. ANTHEMIS L.

Herbs with pinnatifid or dissected leaves; heads peduncled, rather large, radiate; involucre saucer-shaped or hemispheric; phyllaries scarious, in several series; receptacle conic or hemispheric, chaffy; pappus none or a small crown; achenes glabrous, terete or ribbed. (Ancient name of the Chamomile.)

A. cotula L.   Mayweed, Dog-Fennel.
Maruta cotula (L.) DC.

An ill-scented annual; rays white, disk flowers yellow; involucre 8-12 mm. broad; phyllaries oblong, obtuse, pubescent; achenes 10-ribbed. An introduced weed and not common.

A. tinctoria L. a yellow-flowered perennial sometimes persists after cultivation.

14. CHRYSANTHEMUM L.

Annual or perennial herbs; leaves alternate, usually dentate, incised or dissected; heads large, peduncled, usually radiate; phyllaries in 2 or 3 series, scarious-margined; receptacle flat or convex, naked; achenes angled or terete, 5- to 10-ribbed, those from the ray flowers commonly 3-angled; pappus none. (Greek, golden flower.)

```
1A. Leaves small, narrow, entire. ....................... 1. C. integrifolium
2A. Leaves larger, toothed or lobed.
    1B. Stem leaves cuneate-spatulate, toothed or lobed
        above. ........................................ 2. C. arcticum
    2B. Stem leaves linear-spatulate, pinnately incised. ... 3. C. leucanthemum
```

1. *C. integrifolium* Rich.

Perennial; stems 2-18 cm. tall, pubescent, scape-like with 1-4 leaves; leaves mostly basal, linear, 1-4 cm. long; heads solitary, the disk 8-15 mm. broad; involucre 5-8 mm. high; phyllaries rounded at the apex, green with wide brownish-black scarious margins; rays white, 5-10 mm. long.

Bering Sea—Arctic—central Alaska—north B. C. and in Asia. Fig. 1023.

2. *C. arcticum* L.   Arctic Daisy.

Stems usually simple, 1-6 dm. tall; leaves somewhat fleshy; heads solitary, the disk 15-25 mm. broad; rays white, 10-25 mm. long; phyllaries oblong, obtuse, with broad purplish-brown margins. A low form of the northern Bering Sea and Artic Coast with glabrous basal parts and cuneate, not pinnatifid leaves has been described as ssp. *polaris* Hult.

Circumpolar, south to south Hudson Bay and southeast Alaska. Fig. 1024.

3. *C. leucanthemum* L.   Ox-Eye Daisy.

Stems 3-9 dm. tall, simple or forking; basal leaves obovate or spatulate, dentate; upper leaves more or less incised; heads on long peduncles; involucres 12-15 mm. broad; phyllaries oblong-lanceolate, obtuse, with narrow band of purplish-brown and scarious margins; rays white, 12-15 mm. long.

Sparingly introduced; native of Eurasia.

15. MATRICARIA L.

Leaves alternate, 1- to 3-pinnatifid into narrow divisions; involucre saucer-shaped to hemispheric; phyllaries in 2-4 series, somewhat imbricate, obtuse, with scarious margins; receptacle conic or hemispheric, naked; achenes 3- to 5-ribbed. (Latin, mother and dear, from medicinal virtues of some species.)

1A. Annual; heads discoid.	1. *M. suaveolens*
2A. Heads radiate.	
1B. Plant 1-3 dm. tall.	2. *M. ambigua*
2B. Plant 3-6 dm. tall.	3. *M. inodora*

1. *M. suaveolens* (Pursh) Buch.   Pineapple Weed.

Chamomilla suaveolens (Pursh) Rydb.

Glabrous, leafy, much branched weed, 1-4 dm. tall; heads 8-10 mm. in diameter; phyllaries with broad, scarious margins; pappus an obscure crown. The odor is very distinctive.

Circumpolar as a native or introduced plant south to Mass. and Mo. Fig. 1025.

2. *M. ambigua* (Ledeb.) Kryl.   Arctic Chamomile.

M. grandiflora (Hook.) Britt.

Perennial; stems glabrous, branched above or simple and monocephalus; leaves 1- to 2-pinnately dissected, 2-7 cm. long; disks 12-20 mm. wide; phyllaries obtuse, glabrous, brown or with wide, dark brown, scarious margins; rays white, 15-25 mm. long.

Seward Penin.—Arctic—Hudson Bay—Baffin Land—Greenl. and in Eurasia. Fig. 1026.

3. *M. inodora* L.   Scentless Chamomile.

Annual or biennial; stems usually much branched and glabrous or nearly so; leaves 2- to 3-pinnately dissected into filiform lobes, up to 15 cm. long; phyllaries with brown scarious margins; rays white. Cultivated forms are usually double.

Adventive at Fairbanks. Native of Europe.

16. ARTEMISIA (Tourn.) L.

Odorous perennial herbs or shrubs with alternate leaves; heads usually small, discoid, many-flowered, usually nodding when young, with greenish or yellowish flowers; involucres campanulate or hemispheric; phyllaries in 2-4 series; achenes ellipsoid; pappus none. (Named for Artemisia, wife of Mausolus.)

1A. All leaves entire or the lower 3-toothed or lobed at
    the apex.
  1B. Plant glabrous. .... 1. *A. dracunculus*
  2B. Plants white-tomentose. .... 15. *A. gnaphalodes*
2A. Lower leaves deeply lobed to pinnate.
  1B. Low caespitose pulvinate species
    1C. Corolla pilose. .... 5. *A. glomerata*

2C. Corolla glabrous.
　　　1D. Leaves completely covered with long white
　　　　　hairs. .... 6. *A. senjavinensis*
　　　2D. Leaves sparingly silky, green above. .... 7. *A. globularia*
　2B. Plants not pulvinate-caespitose.
　　1C. Receptacle hairy. .... 4. *A. frigida*
　　2C. Receptacle glabrous.
　　　1D. Stem leaves large, often more than 5 cm. long,
　　　　　usually not divided to the midrib.
　　　　1E. Flowers reddish. ....16. *A. unalaskensis*
　　　　2E. Flowers yellowish-brown. ....17. *A. tilesii*
　　　2D. Stem leaves divided usually to the midrib.
　　　　1E. Phyllaries white-tomentose on the back.
　　　　　1F. Ultimate divisions of the leaves oblanceolate
　　　　　　　or spatulate. ....14. *A. kruhsiana*
　　　　　2F. Ultimate divisions of the leaves linear.
　　　　　　1G. Phyllaries with blackish-brown scarious
　　　　　　　　margins. ....12. *A. trifurcata*
　　　　　　2G. Phyllaries with light-colored scarious
　　　　　　　　margins. .... 13. *A. alaskana*
　　　　2E. Phyllaries lacking tomentum on the back.
　　　　　1F. Stem leaves 2-to 3-pinnatifid.
　　　　　　1G. Leaf segments ascending, all acute. ..... 8. *A. arctica*
　　　　　　2G. Leaf segments spreading.
　　　　　　　1H. Heads about 4 mm. in diameter. .... 9. *A. laciniata*
　　　　　　　2H. Heads about 9 mm. in diameter. ......10. *A. macrobotrys*
　　　　　2F. Stem leaves entire or lobed, few.
　　　　　　1G. Plant 1 dm. or less tall; densely white-
　　　　　　　villous. ....11. *A. aleutica*
　　　　　　2G. Plants usually taller.
　　　　　　　1H. Heads numerous, the phyllaries green. 2. *A. canadensis*
　　　　　　　2H. Heads fewer; phyllaries gray pubescent
　　　　　　　　on the back. .... 3. *A. borealis*

1. **A. dracunculus** L.　　　　　　　　　Linear-leaved Wormwood.
　　A. dracunculoides Pursh.

　　Glabrous; stems woody, branched, 4–9 dm. tall; leaves linear, the lower often 3- or more-cleft, the others entire, up to 4 cm. long; heads numerous, nodding, about 3 mm. long and broad; phyllaries ovate, green with wide scarious margins.
　　South Alaska and Yukon southward and in Eurasia. Fig. 1027.

2. **A. canadensis** Michx.　　　　　　　　　Canada Wormwood.

　　Plants with long taproot; stems 1 or few, 2–10 dm. tall; leaves glabrous to silky, divided into very narrow linear segments 0.5–2 mm. wide; heads numerous, suberect to nodding, 3–4 mm. long and at least as wide; phyllaries round-elliptic, green, with broad hyaline margins.
　　Central Alaska—Newf.—Vt.—Minn.—Colo.—Wash. Fig. 1028.

3. **A. borealis** Pall.　　　　　　　　　　Northern Wormwood.

　　Leaves mostly basal, minutely silky or glabrate, the ultimate segments linear-lanceolate; stems 1–several, 1–3 dm. tall; inflorescence a raceme or spike-like; heads 4–6 mm. broad, purplish to green; phyllaries pubescent to glabrate, with narrow scarious margins. The var. *purshii*

Bess. has smaller heads, is more permanently villous and the upper leaves mostly entire. Bering Strait—west Greenl.—Newf.—Gt. Lakes—Colo. also in north Asia. Fig. 1029.

4. *A. frigida* Willd. Prairie Sagewort.

Woody at the base, 2-5 dm. tall, whole plant silky-canescent, brownish in age; leaves 1-2 cm. long with linear-filiform divisions; heads numerous, nodding, racemose or racemose-paniculate, about 4 mm. broad.

Dry plains and hillsides, Alaska—L. Athabasca—Wisc.—Texas—Ariz. and in Eurasia. Fig. 1030.

5. *A. glomerata* Ledeb.

Caespitose and silky-villous; basal leaves 1-3 cm. long, 2- to 3-ternate; stem leaves few; stems 5-15 cm. tall; heads several in a capitate cluster, 5-6 mm. broad; phyllaries silky-villous, elliptical or oval with brown margins; disk corollas yellow, pubescent.

East Asia and Bering Sea and Arctic Coast districts of Alaska. Fig. 1031.

6. *A. senjavinensis* Bess.

Caespitose and silky-villous; leaves mostly basal, once or twice 3- to 5-fid, 5-15 mm. long, the lobes acutish; stems about 1 dm. tall; heads few or several in a capitate cluster, 4-5 mm. broad; phyllaries densely hirsute-villous on the back with dark margins; disk flower corollas yellow, glandular-glanduliferous. A beautiful species.

St. Lawrence Bay in Asia to Bering Strait region of Alaska. Fig. 1032.

7. *A. globularia* Cham.

Caespitose and silky-villous; leaves mostly basal, once or twice ternate, 1-3 cm. long; stems 5-12 cm. tall; heads several in a dense head, 5-8 mm. broad; phyllaries silky-hirsute with black or dark margins; corolla glabrous, yellow or pinkish.

East shore of Chuch Penin. in Asia—central Alaska. Fig. 1033.

8. *A. arctica* Less. Arctic Wormwood.

Glabrous or sparingly pubescent in the typical form; basal leaves petioled, 5-20 cm. long, twice or thrice pinnatifid; stems 2-6 dm. tall; heads several to many in a raceme, nodding, the lower on long peduncles, 7-10 mm. wide; phyllaries with green center and dark margins; corolla villous. A variable group. Var. *beringensis* Hult. of the Aleutians and Bering Sea region is lanate, the hairs of the inflorescence being rust-colored. Ssp. *comata* (Rydb.) Hult. of the Arctic Coast region is usually more or less lanate; stems 1-2 dm. tall; leaves with broad rachis and few short divaricate lobes.

East Asia—Yukon—Wash. Fig. 1034.

9. *A. laciniata* Willd.

More or less hirsute; lower leaves bipinnatifid; stem leafy, reddish-purple, 25–60 cm. tall; heads nodding, about 4 mm. broad; phyllaries light green with darker center and translucent scarious margins.

Globe on Livengood Highway and in Eurasia.

10. *A. macrobotrys* Ledeb.

Rootstock creeping; plant more or less hirsute-pilose; leaves mostly basal, petioled, bi- or tripinnatifid; stems 2–4 dm. tall, greenish or straw-colored; heads several to many, nodding, 5–6 mm. broad; phyllaries with light, scarious, erose margins.

Fairbanks and Ft. Selkirk and in Siberia. Fig. 1035.

11. *A. aleutica* Hult.

Caespitose and densely villous; leaves short-petioled, tripartite or pinnatisect; stems 2–5 cm. tall; heads few, about 5 mm. broad; outer phyllaries linear.

Middle Aleutians.

12. *A. trifurcata* Steph. var. *heterophylla* (Bess.) Kudo.

Caespitose and silky-villous or sericeous; basal leaves twice dissected into linear divisions; stems 1–2 dm. tall; heads in a spike-like raceme, 6–7 mm. broad; phyllaries densely villous; disk flowers yellow with glabrous or slightly pilose corollas.

East Asia—Coronation Gulf—Gt. Bear L.—Mt. McKinley Park. Fig. 1036.

13. *A. alaskana* Rydb.

Caespitose with woody rootstock; lower leaves pinnate with 5 divisions, white-tomentose on both surfaces; stems 2–5 dm. tall; inflorescence racemiform; heads nodding, 6–7 mm. broad; outer phyllaries villous-tomentose, the inner oval with scarious, erose margins.

Lake Kluane and Alaska. Fig. 1037.

14. *A. kruhsiana* Bess.
A. tyrrellii Rydb.

Silky-canescent, woody at the base; lower leaves twice pinnatifid with 3–5 primary divisions, the ultimate divisions spatulate or oblanceolate; stem leaves more simple, the uppermost simple; stems 2–4 dm. tall, branched; heads nodding, often on long peduncles, about 7 mm. broad; corollas glandular-glanduliferous and somewhat hairy.

East Asia to Yukon. Fig. 1038.

15. *A. gnaphaloides* Nutt.

Caespitose, white-tomentose throughout; stems 3–10 dm. tall; leaves

numerous, the lower oblanceolate, 5-10 cm. long, the upper linear; heads numerous in leafy panicles, densely tomentose, 2-3 mm. broad.
Bennett, probably introduced. Sask.—Ont.—Mo.—Colo.

16. *A. unalaskensis* Rydb.

Stems leafy, angled, striate, 3-12 dm. tall; leaves numerous, green and glabrate above, white-tomentose beneath, primary divisions usually 5, these again lobed and toothed, up to 1 dm. long and about as broad; heads numerous, in a leafy panicle, about 5 mm. long and wide; phyllaries ovate or oval, light green with scarious margins; corollas reddish-purple. The var. *aleutica* Hult. has narrower divisions of the leaves and the upper surface subtomentose-lanuginose.
Japan—Unalaska. Fig. 1039.

17. *A. tilesii* Ledeb.

Leafy perennial, striate; leaves sessile, 5-10 cm. long, acuminate, pinnatifid, soon glabrate above, white-tomentose beneath, the divisions 3-5, often again cleft or toothed; heads nodding in spike-like panicles; flowers yellowish-brown. A large and diverse group separable into 4 races as follows.

1A. Heads large, 6 - 8 mm. broad.
  1B. Unbranched with few heads; upper leaves lobed;
     leaves with narrow median lobe. .... *A. tilesii*
  2B. Taller and often branched with numerous heads;
     upper leaves entire; leaves with broad median
     lobe. .... Ssp. *unalaskensis*
                                                                        (Bess.) Hult.
2A. Heads smaller, 4 - 6 mm. broad.
  1B. Leaves strongly dissected into narrow, acute divi-
     sions; inflorescence often branched. .... Ssp. *gormanii* (Rydb.)
                                                                                Hult.
  2B. Upper leaves entire, inflorescence narrow. .... Ssp. *elatior* T. & G.

The typical form Eurasia—Alaska—Hudson Bay. Ssp. *elatior* Alaska and Yukon—Mont.—Ore. Ssp. *gormanii* central Alaska south and west to Naknek. Ssp. *unalaskensis* Cordova and Mt. McKinley Park—Nome—Aleutians. Fig. 1040.

18. *Artemisia* sp.

What may be an undescribed species occurs on the sandy shores of Lake Kluane. It is subfruticose, 2-4 dm. tall, silvery-canescent throughout; lower leaves 2-pinnatisect or often only deeply 3-fid at the apex; upper leaves simple; heads about 5 mm. broad; phyllaries densely wooly.

17. TANACETUM (Tourn.) L.

Strongly aromatic leafy perennials; leaves alternate, 1- to 3-pinnatifid; heads radiate but the rays inconspicuous; involucres hemispheric or depressed; receptacle convex, naked; achenes ribbed and with a flat top; pappus a short crown. (Name of uncertain derivation.)

1A. Heads numerous; introduced. .... 3. *T. vulgare*
2A. Heads solitary or few; native.
  1B. Heads 1 – 3, usually solitary. .... 1. *T. bipinnatum*
  2B. Heads usually more than 1. .... 2. *T. huronense*

1. *T. bipinnatum* (L.) Schultz-Bip.

Stems 2–5 dm. tall, striate, hirsute; leaves up to 2 dm. long, the primary divisions up to 5 cm. long; involucres 15–20 mm. broad; phyllaries hirsute, with brown scarious margins; corollas 4 mm. long; achenes 3 mm. long; pappus a 3- to 5-lobed crown.
Central Alaska west through Asia. Fig. 1041.

2. *T. huronense* Nutt.

Very similar to *T. bipinnatum* but with a larger number of smaller heads and with shorter ray flowers.
Central Alaska—Que.—Newf.—Maine—Mich.

3. *T. vulgare* L.   Common Tansy.

Stems stout, glabrous, 5–9 dm. tall; leaves pinnately divided, 1–3 dm. long; leaflets pinnatifid; heads numerous, 1 cm. or less broad.
Adventive in southeast Alaska. Native of Europe and widely naturalized. Fig. 1042.

18. COTULA L.

Low marsh plants with opposite leaves; heads discoid with a narrow row of marginal pistillate flowers; receptacle with short pedicels from which the achenes are deciduous; pappus not evident; achenes glabrous, compressed. (Greek, small cup, in allusion to the bases of the clasping leaves.)

C. coronopifolia L.   Mud-Disk.

Decumbent and slightly fleshy; stems about 1 dm. tall; lower leaves toothed, the upper leaves reduced and lanceolate with entire margins; heads about 8 mm. broad, borne on slender peduncles; achenes smooth on convex side, white-papillose on slightly concave surface.
Tidal flats, southeast Alaska. Widely distributed along the shores of both hemispheres.

19. PETASITES L.

Perennial herbs with thick creeping rootstocks; leaves basal, petioled, broad, reniform, cordate, triangular or sagittate, tomentose beneath; flowering stems scaly, scape-like, preceding the leaves; heads many-flowered, white or purplish, corymbose, some heads with fertile ray flowers and sterile tubular ones, others with all pistillate and fertile flowers; achenes narrow, 5- to 10-ribbed; pappus of soft white capillary bristles. (Greek, a broad brimmed hat in allusion to the large leaves.)

1A. Leaves lobed ¾ or more of the way to the base. ..... 5. *P. palmatus*
2A. Leaves sagittate, the margins merely serrate. .... 1. *P. sagittata*
3A. Leaves somewhat lobed but not more than about
    ½ way to the base.
  1B. Leaves lobed about ½ way to the base.
    1C. Leaves thin. .... 4. *P. vitifolius*
    2C. Leaves thick. .... 3. *P. hyperboreus*
  2B. Leaves scarcely lobed to lobed ¼ way to the base. 2. *P. frigidus*

1. *P. sagittatus* (Banks) Gray.   Arrow-leaf Sweet Coltsfoot.

Scapes 2-4 dm. tall, floccose; scales lanceolate, attenuate, 5-8 cm. long; petioles of leaves 1-4 dm. long, white tomentose; leaves up to 2 dm. long, the under surface white-tomentose, the upper surface becoming glabrate in age. Easily determined by the mature leaf.
Central Alaska—Labr.—Minn.—Colo.—B.C. Fig. 1043.

2. *P. frigidus* (L.) Fries.   Arctic Sweet Coltsfoot.

Stems 2-4 dm. tall, floccose, with lanceolate scales 5-8 cm. long, often bearing a small blade; leaves 5-13 by 5-15 cm., sometimes slightly longer than broad, on petioles 1-2 dm. long; involucres about 1 cm. high; flowers nearly white, fragrant. Seems to hybridize with *P. hyperboreus* and *P. sagittatus*.
Throughout Alaska and Yukon—Mack.—B.C. and in Eurasia. Fig. 1044.

3. *P. hyperboreus* Rydb.

Stems 1-3 dm. tall, floccose; scales lanceolate, 3-6 cm. long, sometimes with the suggestion of a blade at the tip; leaves reniform to deltoid, 5-15 cm. long, 8-20 cm. wide, on petioles 5-15 cm. long; involucres about 1 cm. long in the pistillate flowers; achenes 2 mm. long; pappus in flower 3-5 mm. long, in fruit 14-18 mm. long.
Alaska and Yukon—Hudson Bay—Alta.—Wash. Fig. 1045.

4. *P. vitifolius* Greene.

Leaves reniform to cordate-deltoid, 5-25 cm. broad, cut about half way to the base into divergent lobes, white-tomentose beneath, on petioles 1-3 dm. long; stems 12-60 cm. tall; achenes about 1 mm. long.
Yukon—Labr.—Que.—Minn.—Alta.

5. *P. palmatus* (Ait.) Gray.

Stems 15-60 cm. tall; leaves palmately 5- to 9-cleft from two thirds to almost the base, 7-20 cm. long and wide; involucres about 1 cm. high; flowers creamy white.
South Yukon—Newf.—N. Y.—Minn.—B. C. Fig. 1046.

20. ARNICA L.

Perennial herbs with opposite leaves and peduncled heads; ray flowers usually present and fertile; involucre campanulate or turbinate with the phyllaries in one or two subequal series; corollas yellow; receptacle flat, villous or fibrillate; disk flowers perfect, fertile, the corollas 5-lobed, the style with reflexed branches; achenes 5- to 10-ribbed; pappus of one whorl of rather rigid, usually barbellate bristles. (Derivation uncertain.)

1A. Anthers yellow.
  1B. Pappus white.
    1C. Leaves lanceolate or oblanceolate.
      1D. Achenes lanate-pilose. .... 1. *A. alpina*
      2D. Achenes glabrous or subglabrous below. ...... 2. *A. louiseana*
    2C. Leaves ovate, obovate or orbicular.
      1D. Lower leaves distinctly cordate. .... 4. *A. cordifolia*
      2D. Lower leaves truncate or wide-cuneate at the base. .... 3. *A. latifolia*
  2B. Pappus tawny or stramineous.
    1C. Phyllaries usually obtuse, the tips pilose within. 5. *A. chamissonis*
    2C. Phyllaries acute, the tips not pilose within.
      1D. Stem leaves 4-10 pairs.
      2D. Stem leaves 3 or 4 pairs. .... 6. *A. amplexicaulis*
        1E. Basal leaf blades broadly ovate or subcordate. 7. *A. diversifolia*
        2E. Lower leaves oblanceolate, short petioled. ... 8. *A. mollis*
        3E. Lower leaves petioled, ovate or ovate-oblong. 9. *A. parrui*
2A. Anthers purple.
  1B. Heads nodding, leaves mostly basal. .... 10. *A. lessingii*
  2B. Heads erect; stem leafy. .... 11. *A. unalaskensis*

1. *A. alpina* (L.) Olin.

Stems single or seldom several, arising from a loose crown, 10–45 cm. tall, thinly to densely villous or hirsute; lower leaves 3- to 5-nerved, oblanceolate with winged petioles; stem leaves 1–4 pairs, lanceolate; heads 15–22 mm. high, usually somewhat broader than high; phyllaries 10–20 in 2 series; achenes densely hirsute. Represented in our area by 2 subspecies. Ssp. *angustifolia* (Vahl) Mag. (*A. angustifolia* Vahl) of the Arctic is 5–15 cm. tall with narrow leaves and bearing a single head. Ssp. *attenuata* (Greene) Mag. (*A. attenuata* Greene) is 15–45 cm. tall with 3–7 heads. Var. *linearis* Hult. has linear leaves. Var. *vestita* Hult. has the whole plant cinereous-pubescent.

Alaska—Ellesmereland—Greenl.—Mont.—B. C. and in Eurasia. Fig. 1047.

2. *A. louiseana* Farr. ssp. *frigida* (Meyer) Mag.

A. nutans Rydb. *A. sancti-laurentii* Rydb. *A. brevifolia* Rydb. *A. mendenhallii* Rydb. *A. illiamnae* Rydb.

Stems 5–35 cm. tall, leafy to the middle, usually reddish at the base; leaves oblanceolate or elliptic to elliptic-lanceolate, glabrate to sparsely hispidulous-puberulent, the margins entire or usually some of them toothed; heads 1, rarely 2 or 3, nodding in anthesis, erect in fruit;

phyllaries 10–18, 9–14 mm. long, tinged reddish-purple; achenes usually glabrous below, sparsely hispid at the summit. Var. *pilosa* Mag. is a pilose form.

East Siberia, most of Alaska and Yukon to north B. C. The type is from Lake Louise in the Canadian Rockies. Fig. 1048.

3. *A. latifolia* Bong.

A. betonicaefolia Greene.

Rootstock horizontal; stems 2–6 dm. tall, sparingly hairy below, more densely so in the inflorescence; stem leaves 2–5 pairs, the lower ovate to elliptic-lanceolate and petioled, the upper sessile; heads 1–5; involucre 12–15 mm. long and wide, finely villous; achenes striate, from glabrous to glandular and a few hairs at the apex.

Southwest coast of Alaska—central Alaska—Colo.—Calif. Fig. 1049.

4. *A. cordifolia* Hook.

Stems usually simple, 15–45 cm. tall, glandular-pubescent; cauline leaves 2 or 3 pairs, the lower ovate to lanceolate, cordate at the base, glandular-puberulent, the upper reduced; heads 1–3, rarely more, large, 18–25 mm. high; phyllaries 14–18 mm. long, more or less ciliate; achenes 6.5–8 mm. long, uniformly but not densely hirsute.

Southeast Alaska—Yukon—N. Mex.—Calif. Fig. 1050.

5. *A. chamissonis* Less.

A. kodiakense Rydb.

Stems solitary, usually unbranched, 2–8 dm. tall, striate, variously pubescent, red tinged toward the base; cauline leaves 4–10 pairs, but little reduced above, all sessile or the lowermost petioled, lanceolate to oblanceolate, often connate at the base; heads 3–15, 12–18 mm. high; phyllaries conspicuously pilose at the tip; pappus subplumose; achenes 4.5–6 mm. long, tapering to the base, sparsely hirsute. The subsp. *foliosa* (Nutt.) Mag. is seldom red tinged; the pubescence is not moniliform; pappus barbellate, lower cauline leaves long petioled.

Aleutians—Mack.—Mont.—Utah—N. Mex.—Calif. Fig. 1051

6. *A. amplexicaulis* Nutt.

Stems 20–75 cm. tall, simple, branched only in the inflorescence, subglabrous to scabrid-glandular, the pubescence dense in the inflorescence; cauline leaves 4–10 pairs, elliptic-lanceolate, inconspicuously to strongly and sharply serrate-dentate, acute, all but the lowermost sessile; heads 10–15 mm. high; phyllaries 10–12 mm. long; achenes 4.5–6 mm. long, sparingly hirsute.

South Alaska—Mont.—Calif. Fig. 1052.

7. *A. diversifolia* Greene.

Stems simple, 15–40 cm. tall, the herbage usually pale green;

cauline leaves usually 3 pairs, the middle pair the largest, ovate to elliptic, more or less irregularly serrate-dentate, 4-8 cm. long, 25-60 mm. wide, mostly with winged petioles shorter than the blades; heads usually 3-5; achenes 5.5-6.5 mm. long, strongly angled, scantily short-hispidulous or hispid.

South Alaska—Alta.—Mont.—Calif.

8. *A. mollis* Hook.

Stems 3-6 dm. tall, sparingly crisp-hairy, glandular-hirsute in the inflorescence; lower leaves oblanceolate, the upper lanceolate, usually denticulate, pubescent on both surfaces; heads 1-3; phyllaries acuminate; achenes 5 mm. long, sparingly hirsute; pappus about 6 mm. long, light brown.

Yukon—Alta.—Colo.—Calif.

9. *A. parryi* Gray.

Stems 3-6 dm. tall, somewhat villous, glandular above; basal leaves petioled, the blades ovate or lanceolate, 3-10 cm. long, somewhat villous on both surfaces, the uppermost reduced; heads 3-20, usually nodding in anthesis; ray flowers usually wanting; pappus about 1 cm. long.

Yukon—Alta.—N. Mex.—Ore.

10. *A. lessingii* (T. & G.) Greene.

Rootstock horizontal; stems 1-3 dm. tall, villous with brown hairs; basal leaves small; stem leaves 2-4 pairs, mostly near the base, oblong, elliptic or oblanceolate, nearly smooth below, ciliate on the margins, 4-7 cm. long, 5-20 mm. wide; head solitary, nodding; involucre densely villous; rays about 2 cm. long; achenes nearly glabrous and striate; bristles light brown and barbellate. The subsp. *norbergii* Hult. & Mag. is taller growing and has 5-6 pairs of stem leaves.

Kamchatka—Arctic Alaska—Yukon—north B. C. Fig. 1053.

11. *A. unalaskensis* Less.

Rootstock covered by the fuscous fibrous remains of old leaves; stems 1-3 dm. tall, striate, villous and glandular-puberulent; cauline leaves about 3 pairs; lower leaves oblanceolate changing in the upper to lanceolate, 3- to 5-ribbed, hairy on both surfaces; heads solitary; involucre about 12 x 20 mm. phyllaries 3-nerved; achenes hirsute; bristles light brown and strongly barbellate.

Aleutians and the islands of Bering Sea to Japan. Fig. 1054.

21. CACALIA L.

Tall glabrous perennials; leaves alternate and petioled; heads rather small, discoid; involucre of 5 nearly equal bracts, usually with a few shorter outer ones; pappus of white bristles; achenes glabrous. (Ancient Greek name.)

C. auriculata DC.

Plant tall, the stem curved at each node; leaves reniform, the base cordate, the margins very unequally serrate, up to 25 cm. wide; inflorescence spike-like; heads about 1 cm. long.
East Asia and west Aleutians. Fig. 1055.

22. SENECIO (Tourn.) L.

Annual or perennial herbs with alternate or basal leaves; heads several or numerous, occasionally solitary, radiate or discoid, yellow; involucre cylindric or campanulate; phyllaries in one series often with smaller ones at the base; receptacle flat, naked, often pitted; achenes 5- to 10-ribbed; pappus of copious soft capillary bristles. (Senex, an old man, in allusion to the white pappus.)

```
1A. Involucre scales in a single row.
  1B. Annual or biennial. ............................. 1. S. congestus
  2B. Perennials.
    1C. Phyllaries pubescent.
      1D.Heads solitary, phyllaries with purplish or
          brown pubescence. ........................ 2. S. atropurpureus
      2D. Heads usually more than 1; phyllaries with gray
          or yellowish pubescence.
        1E. Leaves floccose on both surfaces. ............ 3. S. fuscatus
        2E. Leaves pubescent below only. .............. 4. S. yukonensis
    2C. Phyllaries glabrous.
      1D. Heads solitary or stem branched, each branch
          with 1 head.
        1E. Achenes hirtellous. ........................ 5. S. hyperborealis
        2E. Achenes glabrous.
          1F. Stem usually simple, glabrous or slightly
              tomentose at the base. ................. 6. S. resedifolius
          2F. Stem usually branched with markedly
              tomentose base. ...................... 7. S. conterminus
      2D. Heads in corymbose cymes.
        1E. Heads discoid.
          1F. Disk flowers red. ......................... 8. S. pauciflorus
          2F. Disk flowers yellow. ...................... 9. S. indecorus
        2E. Heads radiate.
          1F. Ligules short. ............................10. S. cymbalarioides
          2F. Ligules long. .............................11. S. pauperculus
2A. Involucral scales in 2 or more series or with outer
    scales at the base of the involucre.
  1B. Introduced annual weed. ..........................17. S. vulgaris
  2B. Native perennials.
    1C. Leaves pinnately or subpalmately divided. ......12. S. palmatus
    2C. Leaves not divided.
      1D. Phyllaries black-tipped. ....................13. S. lugens
      2D. Phyllaries not black-tipped.
        1E. Heads including rays 2-3 cm. in diameter.
          1F. Leaves elongated-deltoid. ................14. S. triangularis
          2F. Leaves ovate. ...........................15. S. sheldonensis
        2E. Heads 5-6 cm. in diameter. ................16. S. pseudo-arnica
```

1. *S. congestus* (R. Br.) DC. var. *palustris* (L.) Fern. Marsh-Fleabane.

Stems simple, stout, hollow, 2-7 dm. tall; leaves linear to oblong-lanceolate, dentate to shallowly pinnatifid; corymb dense and villous-lanate; rays yellow, short; mature pappus four to five times as long as the smooth achene.

Circumpolar, south to Alta. and the north shore of the Great Lakes. Fig. 1056.

2. *S. atropurpureus* (Ledeb.) B. Feditsch.

Stems 1-2 dm. tall, tomentose when young, often becoming glabrate in age; lower leaves ovate to obovate. A varied group giving rise to well-marked races. The typical form found in the Bering Sea region to Point Hope has comparatively wide leaves and large head with long rays. Var. *tomentosus* (Kjellm.) Hult. (*S. kjellmanii* Porsild) has pointed lower leaves and densely black-wooly involucre. Var. *dentatus* Gray has all the leaves conspicuously dentate, the lower leaves being lanceolate to oblong. Subsp. *frigidus* (Rich.) Hult. is the most common and widespread form; the lower leaves are less developed than in the type; the heads rather smaller and either radiate or discoid.

Circumpolar, south to Labr., Mack., Yukon, and Alaska. Fig. 1057.

3. *S. fuscatus* (Jord. & Fourr.) Hayek.

S. lindstroemii (Ostf.) Porsild. *S. denali* A. Nels.

Stems 10-25 cm. tall, more or less floccose; basal leaves obovate, petioled, floccose-tomentose beneath, glabrate in age; heads 1-5; phyllaries narrow, acuminate, purple; rays orange, sometimes with purplish tinge; achenes sparingly strigose-hirsute.

Bering Sea and Arctic coast to Mack. and in Eurasia. Fig. 1058.

4. *S. yukonensis* Porsild.

S. alaskanus Hult.

Stems 1-3 dm. tall, floccose-pilose, yellowish lanate in the inflorescence; basal leaves elliptic to lanceolate, entire or remotely sinuate-dentate, green above, white-tomentose beneath; heads 2-6, densely aggregated; phyllaries narrowly lanceolate, purplish with profuse yellowish indument; achenes glabrous.

Bering Strait—Arctic Coast—Yukon—Alaska Range. Fig. 1059.

5. *S. hyperborealis* Greenm.

Stems 1 to several, 1-2 dm. tall, simple or branched; lower leaves obovate and crenately margined to pinnately divided, 4-10 cm. x 10-25 mm.; stem leaves pinnatisect; heads radiate, the involucre glabrous;

achenes hispid on the margins, often puberulent on the sides.
Alaska, Yukon and Mack. Fig. 1060.

6. *S. resedifolius* Less.

Stems simple or branching from the base, 5–15 cm. tall, smooth, striate; lower leaves with broad lobe at the top and usually one pair of small lobes below; stem leaves much reduced; heads solitary, about 1 cm. high; phyllaries acute, decidedly purple; achenes glabrous.
Circumpolar, south to Newf., Gaspe Penin., Mont., Colo. Fig. 1061.

7. *S. conterminus* Greenm.

Quite variable; stems usually branched, 1–4 dm. tall, sometimes caespitose, more or less white-floccose; lower leaves ovate to spatulate, crenate-dentate to lobed; heads solitary at the ends of the branches, radiate; involucre floccose at the base, glabrous above; achenes glabrous.
Alaska—Yukon—Alta.—B.C. Fig. 1062.

8. *S. pauciflorus* Pursh.

Stems glabrous, 1–6 dm. tall; leaves thick and fleshy, the basal one long-petioled, elliptical to reniform, coarsely dentate, 15–40 mm. long; cauline leaves sessile with mostly obtuse pinnatifid lobing; heads 1–6, rarely more; phyllaries usually purple; disk corollas with red or red-orange lobes; achenes plump, glabrous.
Yukon—Gt. Bear L.—Sask.—Labr.—Newf.—Wyo.—Calif.

9. *S. indecorus* Greene.

Stems glabrous or glabrescent, 2–10 dm. tall; leaves membranous, the basal oblong, elliptical or rounded, dentate to lacerate; blades of basal leaves 2–7 cm. long on slender petioles; cauline leaves becoming lacerate-pinnatifid upward; heads 5–20; phyllaries green or with purple tips; achenes strongly ribbed, 2–3 mm. long.
Central Alaska—Gt. Slave L.—Que.—Mich.—Mont.—Ida.—Calif. Fig. 1063.

10. *S. cymbalarioides* Nutt.

Stems clustered, glabrous except in the axils and base of the petioles, 1–4 dm. tall; lower leaves entire or dentate toward the apex, glabrous, on petioles 1–8 cm. long; heads few to many, radiate; phyllaries 5–8 mm. long; achenes glabrous. A low-growing northern form with narrow leaves and small stem leaves is var. *borealis* (T. & G.) Greenm.
Yukon—Gt. Slave L.—Utah—N. Mex.

11. *S. pauperculus* Michx.

Stems 1–6 dm. tall, glabrous or glabrescent and frequently with flocculent tufts of white wool; basal leaves oblanceolate, spatulate or oblong-elliptic, crenate or crenate-dentate; upper stem leaves pinnatifid; heads 2–40, 5–9 mm. high; phyllaries greenish, rarely purple-tipped, glabrous or glabrescent.

Alaska—Gt. Bear L.—Labr.—Newf.—Va.—Mo.—Colo.—Ida.—B.C. Fig. 1064.

12. *S. palmatus* (Pall.) Ledeb.

Stems glabrous, from horizontal rhizomes, up to 15 dm. tall, leafy; leaves up to 20 cm. long, very deeply palmately lobed into 5 lanceolate lobes prominently and irregularly toothed; heads numerous in a dense corymb; pappus brown.

An Asiatic species found on Attu Isl. Fig. 1065.

13. *S. lugens* Rich.

Stem rather stout, 2–7 dm. tall, wooly when young becoming glabrate in age; basal leaves narrowly oblanceolate, sinuate-dentate; upper stem leaves much reduced and becoming linear or linear-lanceolate; heads several or numerous, in a close corymb, about 8 x 8 mm. phyllaries conspicuously black-tipped.

Alaska—Coronation Gulf—Man.—Wash. Fig. 1066.

14. *S. triangularis* Hook.

Stems several from the same clump, 5–15 dm. tall, leafy to the summit; leaves elongate-triangular, 5–20 cm. long, dentate with triangular teeth; heads several to many; involucres about 8 x 8 mm.; phyllaries linear.

South Alaska—Yukon—Sask.—Mont.—N. Mex.—Calif. Fig. 1067.

15. *S. sheldonensis* Porsild.

Stems glabrous, slender, 3–4 dm. tall, bearing about 10 leaves; leaves broadly lanceolate, glabrous, repand-denticulate, the lower petioled, the upper sessile or clasping and reduced; heads 3 or 4, long-peduncled, turbinate; phyllaries with hyaline margins and attenuate, dark-colored, pubescent tips; achenes glabrous.

Mt. Sheldon in Yukon.

16. *S. pseudo-arnica* Less.

Stems stout, 1–10 dm. tall, very leafy; leaves spatulate to oblanceolate, 6–15 cm. long, densely fine-wooly beneath, glabrous and rugose above; heads 1 to several, large, on short peduncles; achenes smooth; pappus dull.

Beaches, except the Arctic and south to Vancouver Isl. and in Asia and east America. Fig. 1068.

17. *S. vulgaris* L. Common Groundsel.

Stems 1-4 dm. tall, branched; lower leaves petioled, the upper ones sessile or clasping, usually wooly in the axils; leaves undulate to pinnatifid-lobed, rather fleshy; heads several or numerous, about 8 mm. high; phyllaries, especially the smaller ones at the base, black-tipped.

An introduced weed; native of Europe.

23. SAUSSUREA DC.

Perennial herbs with heads of purplish flowers which are all perfect; involucre of several series of imbricated bracts; anther tails ciliate; pappus double, the outer of short, rigid bristles, the inner of stout plumose bristles united at the base. (de Saussure was a Swiss botanist.)

1A. Leaves broad, regularly serrate. .... 1. *S. americana*
2A. Leaves lanceolate to linear, entire or with a few teeth.
  1B. Phyllaries of different lengths and regularly imbricate. .... 2. *S. angustifolia*
  2B. Most of the phyllaries of nearly equal length with shorter ones at the base.
    1C. Plant usually more than 1 dm. tall. .... 3. *S. nuda*
    2C. Plant less than 1 dm. tall. .... 4. *S. vicida*

1. *S. americana* D. C. Eaton.

Stems 4-10 dm. tall; lower leaves petioled and cordate or ovate; upper leaves lanceolate and nearly sessile; heads in a dense panicle; involucres about 12 x 8 mm., pubescent; phyllaries deltoid-ovate, the inner with dark margins.

Alpine meadows, southeast Alaska—Ida.—Ore. Fig. 1069.

2. *S. angustifolia* (Willd.) DC.

Stems 1-4 dm. tall, leafy, sometimes purple-tinged; leaves narrowly lanceolate to linear, entire or remotely dentate, glabrous to floccose; phyllaries acute, in 3 or 4 rows.

East Asia—Alaska—Yukon—Kewatin—Sask. Fig. 1070.

3. *S. nuda* Ledeb.

S. subsinuata Ledeb.

Stems 1-3 dm. tall; leaves wider than in *S. angustifolis* and usually repand-denticulate; receptacle naked; inflorescence dense.

East Asia and west Alaska. Fig. 1071.

4. *S. vicida* Hult.

Plants low, 2-15 cm. tall; lower leaves elliptic-lanceolate, entire or remotely denticulate, sessile or short-petioled, viscid pubescent; heads densely aggregated; receptacle squamate; phyllaries attenuate-triangular. The typical form is found in the Bering Sea district only. Var. *yukonensis* (Porsild) Hult. is less markedly viscid-pubescent.

East Asia—Alaksa—Canadian Rockies. Fig. 1072.

24. CIRSIUM (Tourn.) Mill.

Stout biennial or perennial herbs; leaves alternate with lobes or teeth ending in spines; heads discoid, the flowers usually purple; phyllaries in many series, prickly-tipped; pappus one series of plumose bristles united at the base and falling away together. (Greek, referring to the use of the thistle as a remedy for swollen joints.)

1A. Perennial; heads small, 2 cm. high or less. .... 1. *C. arvense*
2A. Biennials; heads larger.
  1B. All phyllaries spine-tipped. .... 2. *C. vulgare*
  2B. Inner phyllaries unarmed.
    1C. Tips of inner phyllaries dilated or twisted. ..... 3. *C. foliosum*
    2C. Tips of inner phyllaries not dilated or twisted.
      1D. Leaves arachnoid-pubescent below. .... 4. *C. edule*
      2D. Leaves pilose on the nerves below not arachnoid-pubescent. .... 5. *C. kamtschaticum*

1. *C. arvense* (L.) Scop.   Canada Thistle.

Perennial from creeping rootstocks; stems 3–10 dm. tall, branched above; heads numerous, campanulate; flowers purple, rarely white.

Has become established as a weed in several places, native of Europe.

2. *C. vulgare* (Savi) Tenore.   Common or Bull Thistle.
C. lanceolatum Auct.

Stems stout, 1–2 m. tall; leaves dark green, pinnatifid, the apex and triangular-lanceolate lobes tipped with long, stout prickles; phyllaries cottony, lanceolate, all tipped with prickles.

Introduced weed, native of Eurasia.

3. *C. foliosum* (Hook.) DC.

Stems 2–6 dm. tall, more or less arachnoid-hairy; leaves light green, from rather deeply pinnatifid to almost entire, the spines rather weak and yellowish; inner phyllaries with erose, scarious tips; corollas pale.

Yukon—S. Dak.—Colo.—Utah—B. C. Fig. 1073.

4. *C. edule* Nutt.   Edible Thistle.

Stems lightly pubescent-arachnoid, 1–2 m. tall; leaves pinnately cleft, the divisions 2- to 3-lobed, weakly spiny; heads solitary or 2 or 3;

phyllaries lanceolate, acuminate; corollas usually purple, sometimes pale. Hyder—Nev.—Calif.

5. *C. kamtschaticum* Ledeb.   Kamchatka Thistle.

Plant tall, up to 2 m. or more; leaves oblong-ovate or oval, deeply dentate to incisely pinnatifid, 8–25 cm. long, weakly prickly, the lower decurrent on the stem with prickly wings; heads one or few; phyllaries all attenuate-subulate from a narrow base.

An Asiatic species found in the western Aleutians. Fig. 1074.

PLATE XLI

Scale in millimeters.

FIG.
951. *Picris hieracoides kamtschatica* (Ledeb.) Hult. Leaf and fruit.
952. *Apargidium boreale* (Bong.) T. & G. Leaf and fruit.
953. *Taraxacum kamtschaticum* Dahlst. Leaf, achene and phyllaries.
954. *Taraxacum andersonii* Hagl. Leaf, phyllaries and achene.
955. *Taraxacum eyerdamii* Hagl. Leaves and phyllaries.
956. *Taraxacum integratum* Hagl. Leaf, phyllaries and achene.
957. *Taraxacum lacerum* Greene. Leaf, achene and outer phyllary.
958. *Taraxacum lateritium* Dahlst. Leaf, achene and outer phyllary.
959. *Taraxacum scotostigma* Hagl. Leaf, achene and outer phyllary.
960. *Taraxacum trigonolobium* Dahlst. Leaf, achene and outer phyllary.
961. *Taraxacum alaskanum* Rydb. Leaf, achene and outer phyllary.
962. *Agoseris scorzoneraefolia* (Schrad.) Greene. Leaf and achene.
963. *Agoseris aurantiaca* (Hook.) Greene. Leaf and achene.
964. *Sonchus oleraceus* L. Leaf and achene.
965. *Sonchus asper* (L.) All. Leaf and achene.
966. *Lactuca spicata* (Lam.) Hitchc. Leaf and achene.
967. *Prenanthes alata* (Hook.) Dietrich. Leaf and achene.
968. *Crepis elegans* Hook. Leaf and achene.
969. *Crepis nana* Rich. Leaf and achene.
970. *Hieracium albiflorum* Hook. Leaf and phyllary.
971. *Hieracium gracile* Hook. Leaf and phyllary.
972. *Hieracium triste* Cham. Leaf and phyllary.
973. *Solidago multiradiata* Ait. Leaves, flower and phyllary.
974. *Solidago elongata* Nutt. Leaf, flower and phyllary.

505

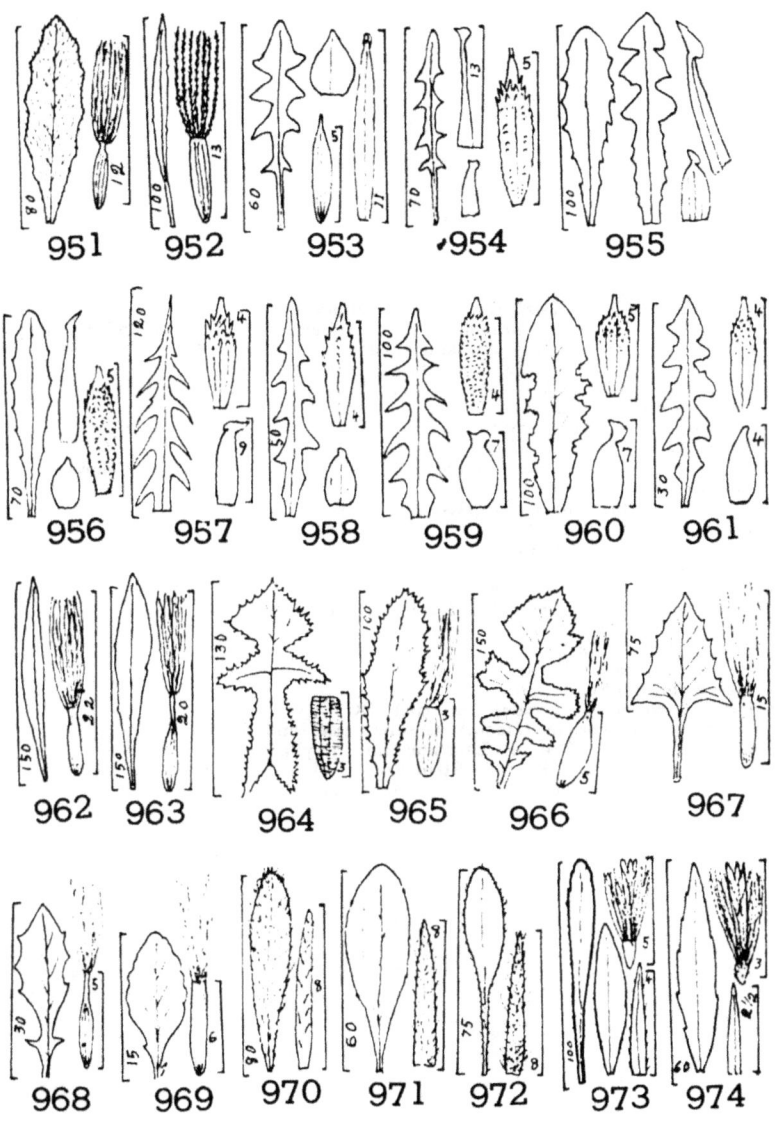

PLATE XLII

Scale in millimeters.

FIG.
975. *Solidago lepida* Nutt. Leaf, flower and phyllary.
976. *Solidago decumbens* var. *oreophila* (Rydb.) Fern. Leaf, flower and phyllary.
977. *Aster alpinus* L. Leaf, phyllary and achene.
978. *Aster yukonensis* Cronq. Leaf, ray flower and phyllary.
979. *Aster junceus* Ait. Leaf, ray flower and phyllary.
980. *Aster subspicatus* Nees. Leaf, ray flower and phyllary.
981. *Aster sibiricus* L. Leaf, ray flower and phyllary.
982. *Aster modestus* Lindl. Leaf, ray flower and phyllary.
983. *Erigeron peregrinus* (Pursh) Greene. All *Erigerons* are leaf, ray flower and phyllary.
984. *Erigeron glabellus pubescens* (Hook.) Cronq.
985. *Erigeron caespitosus* Nutt.
986. *Erigeron yukonensis* Rydb.
987. *Erigeron grandiflorus* Hook.
988. *Erigeron hyperboreus* Greene.
989. *Erigeron uniflorus eriocephalus* (J. Vahl) Cronq.
990. *Erigeron humilis* Grah.
991. *Erigeron purpuratus* Greene.
992. *Erigeron compositus* Pursh.
993. *Erigeron lonchophyllus* Hook.
994. *Erigeron acris* var. *asteroides* (Andriz.) DC.
995. *Antennaria pulcherrima* (Hook.) Greene. Leaf and flowers.
996. *Antennaria howellii* Greene. Leaf and inner phyllary.
997. *Antennaria philonipha* Porsild. Stem and basal leaves and phyllaries.
998. *Antennaria monocephala* DC. Leaves and flower.

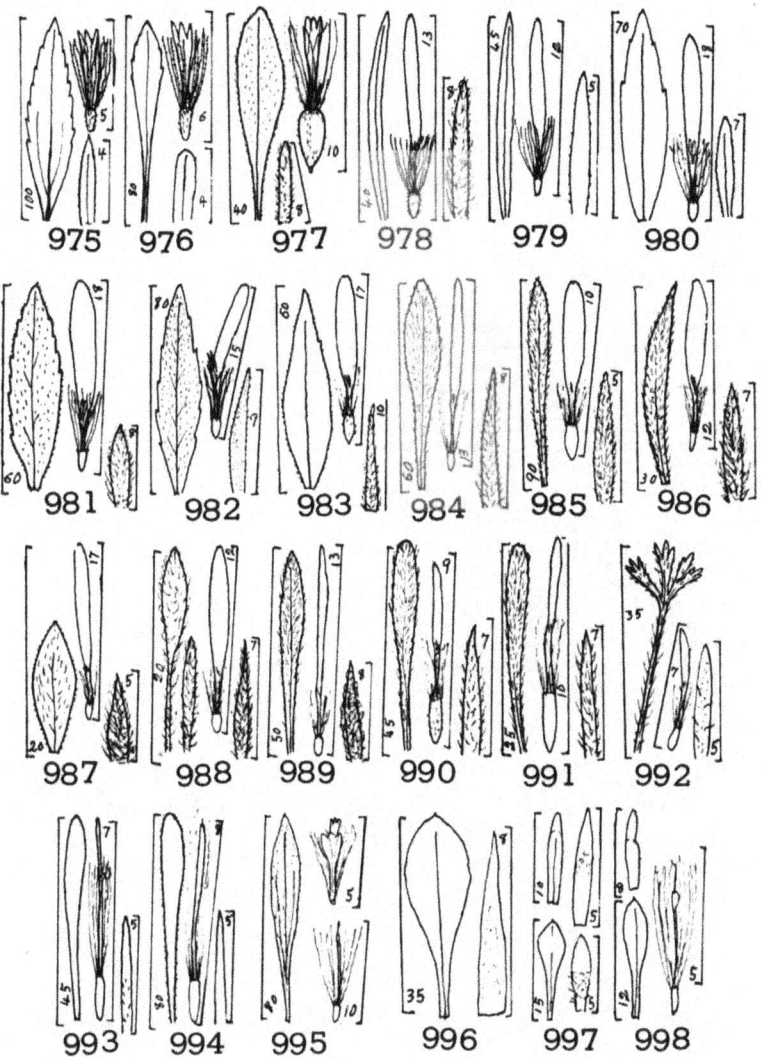

PLATE XLIII

Scale in millimeters.

FIG.
999. *Antennaria angustata* Greene. Leaves and inner phyllary.
1000. *Antennaria alaskana* Malte. Leaves, corolla and achene.
1001. *Antennaria ekmaniana* Porsild. Leaves and phyllaries.
1002. *Antennaria compacta* Malte. Leaves and phyllaries.
1003. *Antennaria stolonifera* Porsild. Leaves and achene.
1004. *Antennaria pallida* E. Nels. Leaves and phyllary.
1005. *Antennaria dioica* (L.) Gaertn. Leaves and phyllary.
1006. *Antennaria alborosea* Porsild. Phyllary and leaves.
1007. *Antennaria elegans* Porsild. Leaves and phyllary.
1008. *Antennaria rosea* (Eaton) Greene. Leaves and phyllary.
1009. *Antennaria oxyphylla* Greene. Leaves and phyllary.
1010. *Antennaria incarnata* Porsild. Leaves and phyllary.
1011. *Antennaria laingii* Porsild. Leaves and phyllary.
1012. *Antennaria nitida* Greene. Leaves and Phyllary.
1013. *Antennaria leuchippi* M. P. Porsild. Leaves and phyllary.
1014. *Antennaria subviscosa* Fern. Leaves and phyllary.
1015. *Antennaria isolepis* Greene. Leaves and phyllary.
1016. *Anaphalis margaritacea* (L.) Benth. & Hook. Leaf, flower and phyllary.
1017. *Gnaphalium uliginosum* L. Leaf, achene and phyllary.
1018. *Madia glomerata* Hook. Leaf, achene and outer phyllary.
1019. *Bidens cernua* L. Leaf, achene and phyllary.
1020. *Achillea sibirica* Ledeb. Ray flower, phyllary and section of leaf.
1021. *Achillea borealis* Bong. Ray flower, phyllary and section of leaf.
1022. *Achillea lanulosa* Nutt. Ray flower, phyllary and section of leaf.
1023. *Chrysanthemum integrifolium* Rich. Leaf, ray flower and phyllary.
1024. *Chrysanthemum arcticum* L. Leaf, ray flower.
1025. *Matricaria suaveolens* (Pursh) Rydb. Leaf, achene with corolla and phyllary.
1026. *Matricara ambigua* (Ledeb.) Kryl. Leaf, ray and disk flowers.

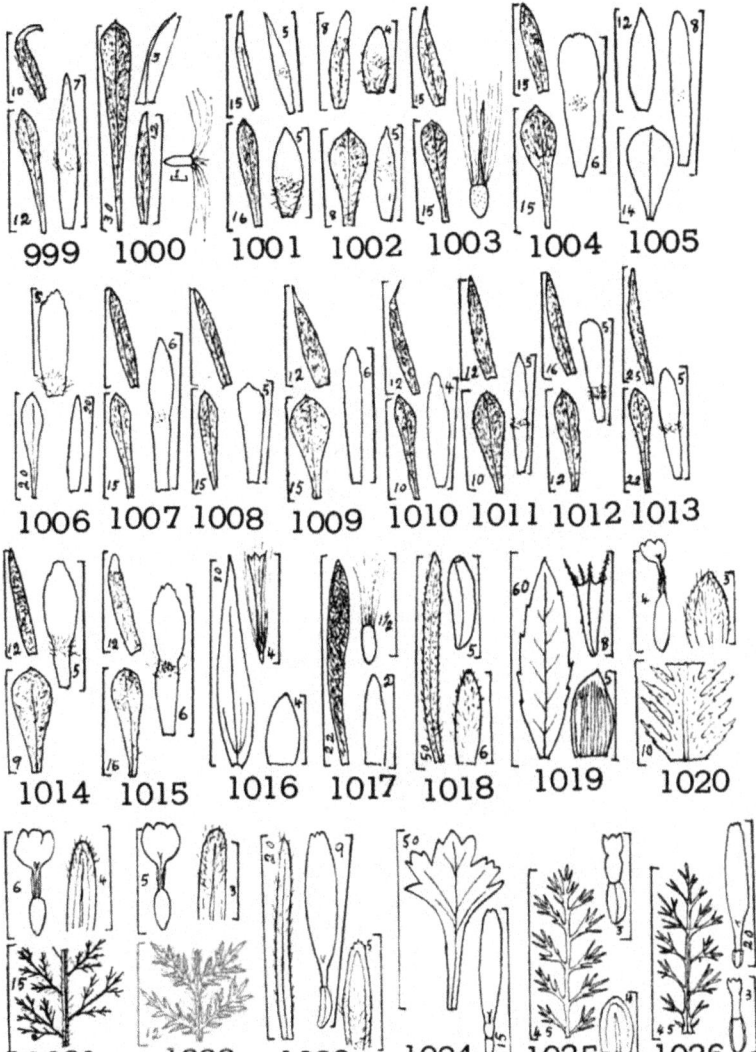

PLATE XLIV

Scale in millimeters.

FIG.
1027. *Artemisia darcunculus* L. Leaf, head and marginal flower.
1028. *Artemisia canadensis* Michx. Lower leaf, head and marginal flower.
1029. *Artemisia borealis* Pall. Lower leaf, head and disk flower.
1030. *Artemisia frigida* Willd. Head, achene and leaf.
1031. *Artemisia glomerata* Ledeb. Flower, phyllary and leaf.
1032. *Artemisia senjavinensis* Bess. Flower, phyllary and leaf.
1033. *Artemisia globularia* Cham. Head, leaf and flower.
1034. *Artemisia arctica* Less. Flower, phyllary and leaf.
1035. *Artemisia macrobotrys* Ledeb. Flower and leaf.
1036. *Artemisia trifurcata*.var. *heterophylla* (Bess.) Kudo. Phyllary and leaf.
1037. *Artemisia alaskana* Rydb. Phyllary, flower and leaf.
1038. *Artemisia kruhsiana* Bess. Head and leaf.
1039. *Artemisia unalaskensis* Rydb. Flower and leaf.
1040. *Artemisia tilesii elatior* T. & G. Leaf and flower.
1041. *Tanacetum bipinnatum* (L.) Schultz-Bip. Pinna.
1042. *Tanacetum vulgare* L. Pinna.
1043. *Petasites sagittatus* (Banks) Gray. Leaf.
1044. *Petasites frigidus* (L.) Fries. Leaf.
1045. *Petasites hyperboreus* Rydb. Leaf.
1046. *Petasites palmatus* (Ait.) Gray. Leaf.
1047. *Arnica alpina attenuata* (Greene) Mag. Leaf, achene and phyllary.
1048. *Arnica louiseana frigida* (Meyer) Mag. Achene, leaf and phyllary.
1049. *Arnica latifolia* Bong. Leaf and achene.

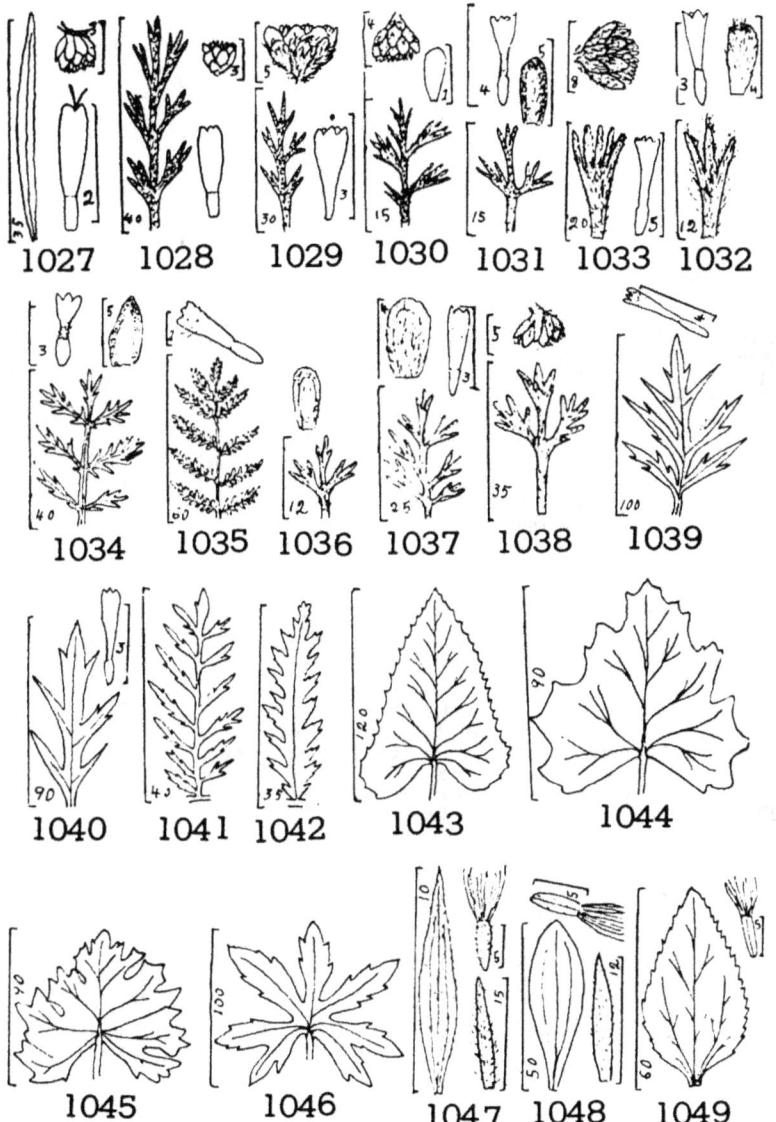

PLATE XLV

Scale in millimeters.

FIG.
1050. *Arnica cordifolia* Hook. Phyllary, leaf and achene.
1051. *Arnica chamissonis* Less. Leaf and flower.
1052. *Arnica amplexicaulis* Nutt. Leaf, achene and phyllary.
1053. *Arnica lessingii* (T. & G.) Greene. Leaf and flower.
1054. *Arnica unalaskensis* Less. Leaf and flower.
1055. *Cacalia auriculata* DC. Leaf, head and flower.
1056. *Senecio congestus* var. *palustris* (L.) Fern. Leaf and achene.
1057. *Senecio atropurpureus frigidus* (Rich.) Hult. Leaf and achene.
1058. *Senecio fuscatus* (Jord. & Fourr.) Hayak. Leaf and flower.
1059. *Senecio yukonensis* Porsild. Leaf, phyllary and flower.
1060. *Senecio hyperborealis* Greenm. Leaf and achene.
1061. *Senecio resedifolius* Less. Leaf and achene.
1062. *Senecio conterminus* Greene. Leaves and achene.
1063. *Senecio indecorus* Greene. Leaves and achene.
1064. *Senecio pauperculus* Michx. Leaves and achene.
1065. *Senecio palmatus* (Pall.) Ledeb. Leaf.
1066. *Senecio lugens* Rich. Leaf and achene.
1067. *Senecio triangularis* Hook. Leaf and flower.
1068. *Senecio pseudo-arnica* Lessing. Leaf and flower.
1069. *Saussurea americana* D. C. Eat. Leaf and achene.
1070. *Saussurea angustifolia* (Willd.) DC. Leaves and achene.
1071. *Saussurea nuda* Ledeb. Leaf and flower.
1072. *Saussurea vicida* var. *yukonensis* (Porsild) Hult. Leaf and flower.
1073. *Cirsium foliosum* (Hook.) DC. Leaf.
1074. *Cirsium kamtschaticum* Ledeb.

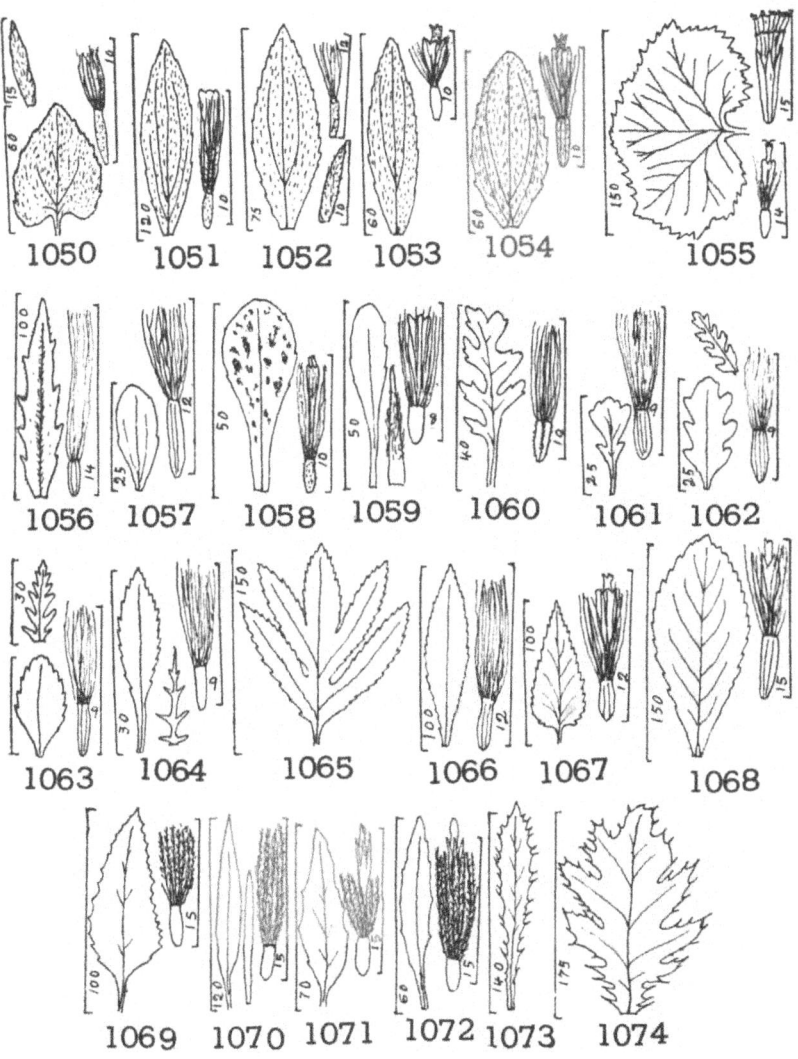

1050 1051 1052 1053 1054 1055
1056 1057 1058 1059 1060 1061 1062
1063 1064 1065 1066 1067 1068
1069 1070 1071 1072 1073 1074

ADDENDUM. Winter and Summer Keys to Willows (*Salix*) of Interior Alaska

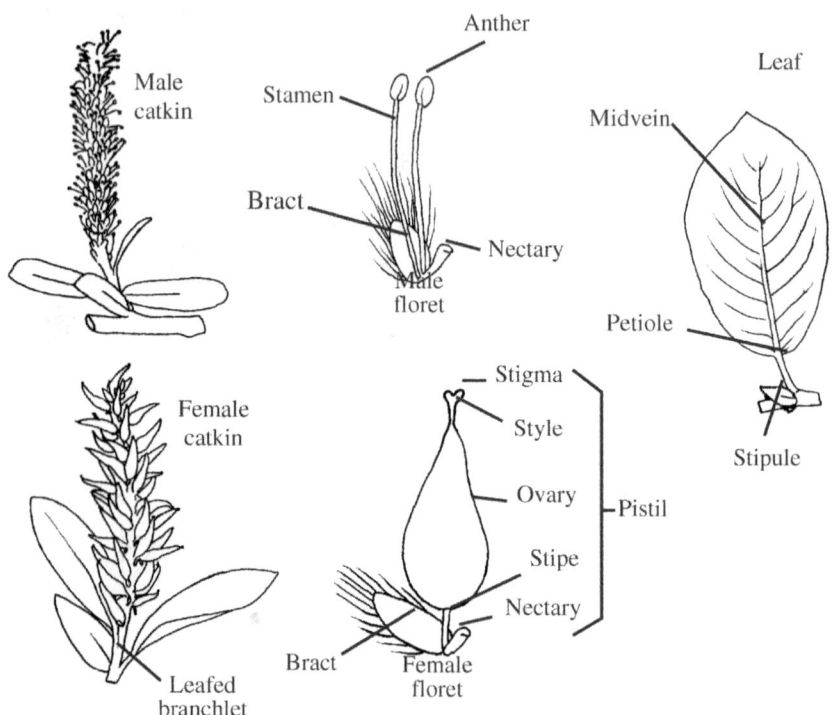

Winter key to the willow shrubs and trees of interior Alaska

This key is designed for the field identification of willows for wildlife browse surveys and harvest of dormant stems used in revegetation projects. Only willows reaching at least 1 meter in height are included in this key. Leaves and catkins from the previous season often remain on dwarf shrub willows protected under the snow cover and can often be identified with the summer or vegetative keys.

The best technique for identifying willows at an unfamiliar site is to first look for the obvious: dried catkins, leaves, petioles, or galls still attached to the plants. Dried leaves or catkins found on the ground under a willow can be used but with caution as they could come from other shrubs. Once a willow shrub is positively identified, a voucher specimen should be collected, dried between newspapers and labeled. A tag with the plant name should be left on the willow from which the identified cutting was harvested and left in the field for future reference. Confirmation of the identity can be done by rechecking the tagged plant left in the field later in the season when the leaves or catkins are developed, using the summer identification keys in this guide.

1.a Shrubs or trees with leaves shed or brown in winter; leaf scars arranged alternately on the stems, winter bud covered by a single scale (see p.36)........**2**
1.b Not as above..**not a willow**

2.a Dried leaves, stipules, and/or catkins remaining on the plant.....................**3**
2.b No leaf, stipule or catkin remaining on the plant...**15**

3.a Persistent stipules (sometimes only a few) on either side of the previous year's leaf scar...**4**
3.b No persistent stipule..**8**

4.a Stipules large, broad, leaflike, often numerous, giving the shrub a scraggly appearance...***S. richardsonii***
4.b Stipules elongated, not leaflike..**5**

5.a Leaves underside woolly, young stems either covered with yellowish dense hairs or coated with a bluish bloom such as found on grapes or plums............
....***S. alaxensis***
5.b Leaves underside not woolly..**6**

6.a Catkin buds large, black and shiny with elongated tip, opening early in the spring; leaf buds small, not shiny, more or less hairy; when present, rosette gall with scales pointed at the tip and long white silk toward the center, "pinecone gall" scales hairless.................................***S. pulchra***
6.b Catkins and leaf buds small, obscured under dense hairs, dried catkins often present; "pinecone gall" scales white hairy **7**

7.a Catkins remaining on the plant...8
7.b No catkins, but dried leaves on the plant 9

8.a Stipes long, catkins loose, pale yellow or light brown, curled.............................
....*S. bebbiana*
8.b Stipe absent or minute, catkins compact,...10

9.a Shrubs usually less than 2 m, female catkins to 50 mm long, capsules to 6 mm long, male catkins often remain on the shrub overwinter. Rosette galls up to 18 mm long.. *S. niphoclada*
9.b Shrubs to 6 m, catkins to 80 mm long, capsules to 9 mm long, male catkin rarely present on the shrub in winter. Rosette galls 10-40 mm long............
....*S. glauca*

10.a Stem tips thin, whiplike, leaves long and narrow, often curled lengthwise, margin finely toothed, underside covered with short stiff reddish hair
....*S. arbusculoides*
10.b Stem tip thicker, leaves broader, no short red hairs covering the leaf under sides...11

11.a Leaf underside woolly, petioles sometimes inflated, young stems either covered with yellowish dense hairs or coated with a bluish bloom such as found on grapes or plums...................................*S. alaxensis*
11.b Leaf underside not woolly, stems different than above................................12

12.a Leaf margin toothed, leaves pale underneath and dark above......................
....*S. pseudomonticola*
12.b Leaf margin untoothed, leaf color variable...13

13.b Buds scale visible, basal sections of some of leaf petioles remain attached to the stem,..*S. hastata*
13.a Bud scale obscured by dense hairs...14

14.a Shrubs usually less than 2 m. Leaves narrow, rounded at the tip, petiole to 3 mm long. Rosette galls up to 18 mm long.................*S. niphoclada*

14.b Shrubs to 6 m, leaves broader, tip often pointed, petioles 3-10 mm long. Rosette galls 10-40 mm long... *S. glauca*

15.a Stem tips thin, whiplike...*S. arbusculoides*
15.b Stem tips thicker...16

16.a	Young stems either covered with yellowish dense hairs or coated with a bluish bloom such as found on grapes or plums............*S. alaxensis*	
16.b	Not as above...17	

17.a	Stem erect, newly established sandbars or disturbed habitat near major stream or river. Early pioneer...*S. interior*	
17.b	Stem branched. Various habitats..18	

18.a	Tree with distinct trunk and canopy; when still young, shoots with long annual growth...19	
18.b	Shrub branching from the base..20	

19.a	Catkin buds black, not shiny, roundish and pointed at the tip, much larger than leaf buds, developing in late winter, exposing the white catkin silk; stems often branching at right angle, trunk bark grey, smooth ....*S. scouleriana*	
19.b	Catkins and leaf buds alike, waxy to the touch, yellow or brown, oval, shiny, developing in the spring; young stem waxy, branching at sharp angles, trunk bark rough, deeply furrowed ....*S. lasiandra*	

20.a	Young stems reddish-brown...21	
20.b	Young stems not reddish brown. Mostly south of the Alaska Range............ ....*S. barclayi*	

21.a	Catkin buds flattened dorsoventrally (duckbill shaped) in the winter, fuller but with a lateral pleat in the spring. Rosette gall with few deformed leaves. Only south of the Alaska Range...*S. sitchensis*	
21.b	Not as above... 22	

22.a	Catkin buds much larger than leaf buds...............*S. pseudomonticola*	
22.b	Catkin buds almost the same size as leaf buds..23	

23.a	Stems branching often at wide angle, common in disturbed sites, heavily browsed by moose, catkin bud minute, pointed at the tip, minute rosette..*S. bebbiana*	
23.b	Stem branching at sharper angles, catkin buds not as above.....................24	

24.a	Basal sections of some of leaf petioles remain attached to the stem;........... stem densely hairy..*S. hastata*	
24.b	Basal section of petiole does not remain attached to the stem................. ....*S. pseudomyrsinites*	

Flowering key to willows of interior Alaska. Adapted from Argus (2001).

Willows differ from other trees and shrubs by their leaves arranged alternately on the stems, their winter bud covered by a single scale, and their flowers bunched in dense catkins.

Use this key only if catkins are present; otherwise use the vegetative key (p. 39). This summer key relies heavily on female catkin characters. Individual willows are either male or female and the female catkins themselves often remain on the plant only for a short period. Occasionally, dried female catkins or dried leaves remain attached for an extended period and can be useful for identification, but care must be taken to make sure that they are still connected to the plant. Willows of several species frequently grow side by side with their branches entangled. Using dried leaves or catkins collected from the duff under the shrubs may mislead identification.

Poplar winter bud covered by several scales

Catkins that appear before the leaves are usually directly attached to the stems (see p.104) while those that develop simultaneously with the leaves or later are borne on more or less developed leafy branchlets (see p.106). The best way to obtain a full set of characteristics for willows whose catkins and leaves are not present at the same time is to tag a branch from which samples are collected at various times of the year. Make sure that only cuttings from branches observed forking above ground are associated. The branch samples labeled, pressed, and dried between newspapers can be kept indefinitely.

Willow winter bud covered by a single scale

1.a	Dwarf willow with branches lying flat on the ground, under 20 cm tall...	2
1.b	Upright shrub more than 20 cm tall or tree...	12
2.a	Roundish leaves strongly veined, dark green above and pale beneath; reddish petiole long, at least half as long as the leaf blade..........................	*S. reticulata*
2.b	Leaf shape variable, veins not so conspicuous; petiole short...................	3
3.a	Ovary hairy, sometimes only at the beak...	4
3.b	Ovary hairless...	10
4.a	Leaves finely toothed around the whole margin.........	*S. chamissonis*
4.b	Leaves margin untoothed or toothed only at the basal half.....................	5
5.a	Leaves green beneath...	6
5.b	Leaves pale beneath (pale waxy layer can be scraped with the fingernail revealing green plant tissue beneath)...	7
6.a	Leaf margin with a fringe of hairs; dried skeletonized leaves at the base of the plants; plant with central root, stems spreading............................	*S. phlebophylla*

6.b	Leaf margin hairless; dried leaves at the base of the plant not skeletonized, plant spreading with roots at internodes..............................*S. polaris*	
7.a	Style 0-0.5 mm...	8
7.b	Style longer than 0.5 mm..	9
8.a	Ovary red, pear-shaped with short stiff reddish hairs; leaves hairless, dark green glossy above, broadest near the tip; margin of the leaves toothed at the base; flexible branches trailing in the vegetation.....*S. fuscescens*	
8.b	Ovary grayish-green, barrel-shaped, densely white hairy; leaves elongated, hairy above and beneath, margin not toothed; shrub erect, branches stiff, grayish black hairy, dull (not shiny)...........................*S. niphoclada*	
9.a	Ovaries sparsely hairy with short flat crinkled hairs, leaf margin finely toothed-glandular, leaves hairless, nectary shorter than the stipe........... ....*S. arctophila*	
9.b	Ovary densely hairy with straight hairs, leaf margin untoothed or only finely toothed near the base; branchlets without roots; some of the leaves with long silky hair forming a "beard" at their tip, nectary longer than the stipe ....*S. arctica*	
10.a	Leaves green beneath, not fleshy..	11
10.b	Leaves pale beneath (pale waxy layer can be scraped with the fingernail revealing green plant tissue beneath), fleshy...............*S. setchelliana*	
11.a	Plant minute, less than 5 cm high; leaves at most 1.5 cm long, roundish, not toothed at the margin; female catkins short, bearing 4 to 15 pistils.............. ....*S. rotundifolia*	
11.b	Plant 10 cm high or more; leaves oval, more than 1.5 cm long, margin finely toothed..*S. myrtillifolia*	
12.a	Catkins appearing before the leaves open, borne directly on the stem or on short few-leafed branchlets...	13
12.b	Catkins appearing at the same time as the leaves open or later, borne on developed leafy branchlets..	18
13.a	Ovary hairless..	14
13.b	Ovary hairy	15
14.a	Stipules well-developed, leaflike, persisting several years, young leaves green, when present: *Pontania*-induced densely woolly bean-shaped leaf galls 5-15 mm long... *S. richardsonii*	
14.b	Previous year's stipules not present on the plant, new stipules rounded, young leaves reddish: when present, *Pontania*-induced bald bean-shaped leaf gall ....*S. pseudomonticola*	
15.a	Leaves densely white woolly beneath, shiny bright green above.............. ....*S. alaxensis*	
15.b	Leaves not woolly beneath..	16

16.a	Leaves hairless beneath, shiny green above; stipules linear, persisting several years on the stems..*S. pulchra*	
16.b	Leaves hairy beneath, previous year's stipules not persisting on the plant. 17	

17.a	Low, dense shrub in wet subalpine thickets, stipules and buds oily, leaves with long white hairs beneath..*S. barrattiana*	
17.b	Tall shrub or tree in forest zone, stipules and buds not oily, canopy leaves with reddish hair appearing as a reddish hue; stipules not persisting several years...*S. scouleriana*	

18.a	Ovary hairless.. 19	
18.b	Ovary hairy.. 23	

19.a	Leaves narrow, 6 or more times as long as wide, leaf margin finely toothed. Sandbars on major rivers...*S. interior*	
19.b	Leaves less than 5 times as long as wide, leaf edge toothed or not toothed. Various habitats 20	

20.a	Leaves lance-shaped; 5 stamens in each male floret; leaf petiole glandular; large shrub or small tree. Trunk bark blackish, deeply furrowed. Riverbanks and wetlands ....*S. lasiandra*	
20.b	Leaves not lance-shaped; 1 or 2 stamens in each male floret; no glands on the petiole. Bark variable. Various habitats 21	

21.a	Leaves green beneath 22	
21.b	Leaves pale beneath (pale waxy layer can be scraped with fingernail, showing green plant tissue) ....*S. barclayi*	

22.a	Small shrub less than 1 m tall; flexible stems trailing in the vegetation, leaves hairless; minute stipules 1-2 mm; style 0.3-0.5 mm.................. ....*S. myrtillifolia*	
22.b	Erect shrub 0.5-4 m; stipules 1-5 mm; styles 0.3-0.5 mm.................. ....*S. pseudomyrsinites*	

23.a	Leaf underside densely white woolly, uncommon plant.....*S. candida*	
23.b	Leaf underside not densely woolly... 24	

24.a	Reddish hairs scattered on both sides of the leaves especially on young leaves..*S. athabascensis*	
24.b	No reddish hairs on the leaves.. 25	

25.a	Stipes 2-5 mm; catkins loose, often a few remaining on the shrub through the winter; leaves upper side shiny, with veins impressed............................. ....*S. bebbiana*	
25.b	Stipe much shorter; catkins dense; veins not so conspicuous on the upper side of the leaves... 26	

26.a	Leaves silky beneath. 27	
26.b	Leaves hairy beneath, but not silky... 28	

27.a	Leaves narrow, 5-7 times as long as broad, margin toothed, with a small gland on each tip; 2 stamens in each male floret......	***S. arbusculoides***
27.b	Leaves elliptic, less than 3 times as long as broad, leaves appear satiny beneath (like the fur of a seal); 1 stamen in each male floret...................	***S. sitchensis***
28.a	Petioles 3-15 mm long, yellowish. Subalpine thicket, lake shores..............	***S. glauca***
28.b	Petioles short, 1-3mm long, reddish. Coastal meadows and river sandbars	***S. niphoclada***

Vegetative key to willows of interior Alaska.
Adapted from Viereck and Little (1972).

1.a	Dwarf shrub, mostly under 30 cm..	2
1.b	Shrub taller than 30 cm or tree..	10
2.a	Leaves toothed at least at their base...	3
2.b	Leaf margins not distinctly toothed...	4
3.a	Leaf margin toothed at the base only, tip of leaves broad, leaves pale beneath (pale waxy layer can be scraped with the fingernail revealing the green plant tissue beneath)...	***S. fuscescens***
3.b	Whole leaf margins finely toothed, tip of the leaves pointed, underside green..	***S. chamissonis***
4.a	Leaf veins with deeply impressed network on the upper side, long red petiole..	***S. reticulata***
4.b	Leaf vein not forming a deeply impressed network, petiole short, of various colorations..	5
5.a	Leaves thick, fleshy like those of a "jade plant," on barren glacier-fed river sandbars and glacial outwash plain............................	***S. setchelliana***
5.b	Leaves thinner, flexible, various habitats...	6
6.a	Leaves pale beneath (pale waxy layer can be scraped with the fingernail revealing the green plant tissue beneath)..	7
6.b	Leaves green beneath..	8
7.a	At least some of the leaves with long silk hairs beneath forming a beard at the tip of the leaves, branches stout, sparsely hairy...............	***S. arctica***
7.b	Leaves hairless, stems trailing, hairless........................	***S. arctophila***

8.a	Reddish-brown skeletonized leaves remain on the branches, leaf margin with small hairs................*S. phlebophylla*
8.b	No skeletonized leaves remain on the branches, leaf margin hairless.........9
9.a	Shrub densely matted, persisting dry reddish-brown leaves at the base of the plant, leaves 0.4-1 cm long, with 3 prominent veins beneath.................*S. rotundifolia*
9.b	Shrub loosely matted with long trailing stems, no persisting dried leaves on the plant, leaves up to 2.5 cm long................*S. polaris*
10.a	Leaves densely woolly or felt-like underneath................11
10.b	Leaves hairy or not underneath but not felty or woolly................12
11.a	Stipules 2-3 mm long, leaves 4-7 times as long as broad, branchlets densely woolly, silvery in appearance. Uncommon in bogs of the upper Yukon and upper tributaries................*S. candida*
11.b	Stipules 4-20 mm long, leaves 2-4 times as long as wide, branchlets densely yellowish woolly or hairless with bluish bloom. Some of the leaf petioles often inflated in plants with woolly stems..........*S. alaxensis*
12.a	Leaves pale beneath (pale waxy layer can be scraped with the fingernail revealing the green plant tissue beneath)................17
12.b	Leaves green beneath................13
13.a	Leaves lance-shaped, broad at the base with an elongated tip. Tree with rough blackish bark or shrub with long shoots................*S. lasiandra*
13.b	Leaves of a different shape. Growth form variable................14
14.a	No stipules or stipules minute................*S. myrtillifolia*
14.b	Stipules present................15
15.a	Low shrub usually less than 1 m tall, stout mostly unbranched stems, dense foliage, buds and stipules oily, stems and leaves densely grey hairy................*S. barrattiana*
15.b	Branching shrub usually taller than 1 m, foliage not so dense, stipules not oily, stems and leaves not densely gray hairy................16
16.a	Leaves toothed at the margin................*S. pseudomyrsinites*
16.b	Leaves toothless at the margin, petiole reddish, often short reddish hairs on the main vein................*S. hastata*
17.a	Stipules minute or lacking................18

17.b	Stipules present (stipules fall off early in *S. bebbiana*)..................................**22**	
18.a	Leaves at least 6 times as long as wide, sandbars on major rivers..*S. interior*	
18.b	Leaves broader, various habitats..**19**	
19.a	Leaves densely covered underneath with short hairs, all oriented in the same direction, appearing satiny (like the fur of a seal). Only south of interior Alaska, riverbanks..*S. sitchensis*	
19.b	Underside of the leaves not satiny. Broader distribution............................**20**	
20.a	Sparse reddish hairs on both sides of the leaves. Uncommon, in bogs..*S. athabascensis*	
20.b	Only white hairs on leaves..**21**	
21.a	Shrub mostly less than 1 m tall, leaves narrow, leaf hair sparse, straight oriented toward the tip, petioles short...........................*S. niphoclada*	
21.b	Shrub or tree, leaves broad, leaf hairs curled, hairs mostly oriented in various directions...*S. bebbiana*	
22a	Stipules persisting several years on the stem..**23**	
22.b	Stipules not persisting several years on the stem......................................**24**	
23.a	Stipules linear, leaves elongated, diamond shaped, often dried reddish leaves remaining on the shrub, *Pontania*-induced bald round leaf gall 3-10 mm long..*S. pulchra*	
23.b	Stipules large, leaflike, remaining several years giving a coarse appearance to the shrub, *Pontania*-induced densely woolly bean-shaped leaf galls 5-15 mm long ....*S. richardsonii*	
24.a	Leaves lance-shaped, broad at the base with an elongated tip. Tree with rough blackish bark or shrub with long shoots................*S. lasiandra*	
24.b	Leaves of a different shape. Growth form variable......................................**25**	
25.a	Mature leaves hairless or hairy only on the main vein................................**26**	
25.b	Mature leaves hairy..**27**	
26.a	Mature leaves hairless, stem tips stout, finely (densely) hairy stipules, leaf like, asymmetrical, often numerous large blackish rosette galls, stem variable. South of the Alaska Range...................................*S. barclayi*	
26.b	Mature leaves upper side with small red hairs on the main vein, (dissecting microscope may be needed to see the small hairs), petiole reddish, stem tips	

thin, dark, shiny, leaves ending in a pointed tip, stipules mostly rounded. Throughout Alaskan interior..............................*S. pseudomonticola*

27.a Upper surface wrinkled by creases caused by impressed veins, textured like leather, leaf hairs long, curled..*S. bebbiana*
27.b Upper surface plane, veins not impressed, leaf hairs not curled..................**28**

28.a Upper leaf surface shiny, hairs on leaf underside short, flat and reddish, buds hairless or with few hairs...**29**
28.b Hairs long and silky, buds densely hairy.............................*S. glauca*

29.a Leaves long, margin serrated, hairs densely hairy, reddish hairs underneath, branch tips thin..*S. arbusculoides*
29.b Leaves hairs not so dense, branch tips stout...............*S. scouleriana*

Index*

ABIES		27	AGROPYRON		90	alkali-grass,
amabilis		27	alaskanum	91, 104		Alaska 75
lasiocarpa	27,	38	arcticum		91	Anderson 75
ACER		345	angustiglume		91	arctic 75
douglasii		345	caninum		92	creeping 74
glabrum			latiglume	91, 104		Haupt 74
douglasii	345, 382		pauciflorum		91	Hulten 76
macrophyllum		345	repens	90, 104		large 74
ACERACEAE		345	richardsonii		92	northern 74
ACHILLEA		484	sericeum	91, 104		Pacific 76
borealis	485, 508		smithii	90, 104		small 76
lanulosa	485, 508		spicatum	92, 104		smooth 75
millefolium		485	subsecundum	92, 104		three-flowered 75
multiflora		485	tenerum		91	ALLIUM 153
ptarmica		485	trachycaulum	91, 104		schoenoprasum 153
sibirica	485, 508		violaceum			sibiricum 153, 176
aconite,			latiglume		91	victoralis
delphinium-leaved		242	yukonense	91, 104		platyphyllum
Kamchatka		242	AGROSTEMMA			154, 176
ACONITUM		241	githago		220	ALLOCARYA
delphinifolium	242, 272		AGROSTIDEAE		50	cognata 407
chamissonianum			AGROSTIS		60	cusickii 407
		242	aequivalvis		60	hirta 407
paradoxum		242	alaskana	62,	98	ALNUS 187, 427
kamtschaticum		242	breviflora		62	alnobetula 188
maximum	242, 272		alba		61	crispa 188, 220
nivatum		242	borealis	62,	98	fruticosa 188, 220
ACONOGONUM			exarata	61,	98	sinuata 188, 220
phytolaccaefolium		196	purpurescens		62	incana 188, 220
ACTAEA		239	geminata		62	oregona 188, 220
arguta	240, 272		hiemalis		62	rubra 188
eburnea		240	idahoensis		62	sitchensis 188
rubra		240	melaleuca		62	tenuifolia 188
arguta		240	maritima		61	ALOPECURUS 55
adder's mouth, little		165	palustris	61,	98	aequalis 56, 98
adder's-tongue		7	scabra	62,	98	alpinus 55, 98
family		6	aristata		62	stejnegeri 55
white		166	geminata		62	geniculatus 56, 98
ADIANTUM		10	stolonifera	61,	98	glaucus 55
pedatum	11,	30	tenuis		61	occidentalis 55
aleuticum		11	thurberiana	61,	98	pratensis 55, 98
ADOXA		433	alder, Alaska		188	stejnegeri 55
moschatellina	433, 444		green		188	ALSINE
ADOXACEAE		433	mountain		188	calycantha 211
AGOSERIS		462	Oregon		188	crassifolia 211
aurantiaca	463, 504		red		188	crispa 210
scorzoneraefolia			alfalfa,		314	humifusa 211
	463, 504		yellow-flowered		314	laeta 210
gracilens		462	alfilaria		342	longifolia 210

* Underlined species' names are synonyms.

INDEX

ALSINE (cont.)
  longipes 210
  media 209
ALSINOPSIS
  laricifolia 214
  obtusifolia 214
alyssum, American 267
ALYSSUM 266
  americanum 267
AMELANCHIER 310
  alnifolia 310, 334
  florida 311, 334
AMMIACEAE 354
AMMODENIA
  peploides 212
AMSINCKIA 406
  lycopsoides 406, 438
  menziesii 406, 438
ANAPHALIS 483
  margaritacea 483, 508
ANDROMEDA 370
  calyculata 370
  polifolia 370, 388
ANDROSACE 376
  alaskana 377, 390
  carinata 376
  chamaejasme
    andersonii 376
    lehmanniana 376, 390
  ochotensis 377, 390
  septentrionalis 376, 390
ANEMONE 235
  deltoidea 236
  drummondii 236
  globosa 236
  multiceps 236, 272
  multifida 236, 270
  narcissiflora 235
    alaskana 236, 270
    interior 236
    sibirica 236
    villosissima 236
  parviflora 236, 270
  patens
    multifida 237, 272
    nuttalliana 237
    wolfgangiana 237
  richardsonii 235, 270
anemone,
  Alaska blue 236
  cut-leaved 236
  Drummond 236
  narcissus-flowered 235

anemone, (cont.)
  northern 236
  yellow 235
ANGELICA 358
  genuflexa 359, 386
  lucida 358, 386
angelica,
  bent-leaved 359
  sea coast 358
ANGIOSPERMAE 41
ANTENNARIA 476
  alborosea 481, 508
  alaskana 479, 508
  angustata 478, 508
  breitungii 481
  compacta 479, 508
  densifolia 479
  dioica 480, 508
  ekmaniana 479, 508
  elegans 481, 508
  howellii 478, 506
  incarnata 482, 508
  isolepis 483, 508
  laingii 482, 508
  leuchippi 482, 508
  megacephala 480
  monocephala 478, 506
  neoalaskana 479
  nitida 482, 508
  oxyphylla 481, 508
  pallida 480, 508
  pedunculata 480
  philonipha 478, 506
  pulcherrima 478, 506
    angustisquama 478
  pygmaea 482
  rosea 481, 508
  shumaginensis 482
  stolonifera 480, 508
  subcanescens 480
  subviscosa 483, 508
ANTHEMIDEAE 467
ANTHEMIS 485
  cotula 486
  tinctoria 486
ANTHOXANTHUM 54
  odoratum 54, 98
ANTICLEA
  elegans 153
ANTIPHYLLA
  oppositifolia 291
APARGIDIUM 448
  boreale 448, 504

APHRAGMUS 248
  eschscholtzianus 248
APHYLLON
  fasciculatum 428
  uniflorum 428
APOCYNUM 398
  androsaemifolium 398, 436
APOCYNACEAE 397
AQUILEGIA 240
  brevistyla 240, 272
  columbiana 240
  formosa 240, 272
ARABIDOPSIS
  mollis 265
  richardsonii 267
ARABIS 263
  ambigua 263
  arnicola 263, 276
  divaricarpa 264, 276
  drummondii 264, 276
  glabra 265, 276
  hirsuta 264
    eschscholtziana 264, 276
    pycnocarpa 264
  holboelli 264, 276
    retrofracta 264
  hookeri 265, 276
  lyrata
    kamchatica 263, 276
  rupestris 264
ARACEAE 142
ARAGALLUS
  bryophilus 321
  varians 323
  viscidulus 321
  viscidus 321
ARALIA 354
  nudicaulis 354
ARALIACEAE 354
ARCEUTHOBIUM 189
  tsugense 190, 220
ARCTAGROSTIS 63
  angustifolia 63
  arundinacea 63
  latifolia 63, 100
    angustifolia 63
    arundinacea 63
    poaeoides 63
ARCTOPHILA
  fulva 70

INDEX

ARCTORANTHIS		ARNICA (cont.)			
cooleyae	234	diversifolia	495	ash, elder-leaved	
ARCTOSTAPHYLOS	371	illiamnae	494	mountain	309
alpina	371, 388	kodiakense	495	aspen,	
rubra	371	latifolia	495, 510	American	184
rubra	371	lessingii	496, 512	quaking	184
uva-ursi	371, 388	norbergii	496	ASPERUGO	407
ARCTOUS		louiseana		procumbens	408
alpina	371	frigida	494, 510	asphodel,	
erythrocarpa	371	mendenhallii	494	false	152
ARENARIA	211	mollis	496	northern	152
arctica	214, 224	nutans	494	scotch	152
biflora	214	parryi	496	ASPIDIUM	14
capillaris	213, 224	santi-laurentii	494	fragans	15
dawsonensis	213	unalaskensis	496, 512	oreopteris	16
dicranoides	213, 224	ARRHENATHERUM	64	spinulosum	
elegans	213, 224	elatius	64, 100	dilatatum	16
humifusa	213, 224	arrow-grass		ASPLENIUM	16
laricifolia	214, 224	family	48	cyclosorum	17
lateriflora	213, 224	marsh	49	trichomanes	16, 34
macrocarpa	215, 224	seaside	49	viride	17, 34
obtusifolia	214	ARTEMISIA	487	ASTER	470, 473
peploides	212	alaskana	490, 510	alpinus	471, 506
latifolia	212	aleutica	490	vierhapperi	471
major	212, 224	arctica	489, 510	ciliolatus	472
physodes	212, 224	beringensis	489	foliaceus	471
rossii	213	comata	489	junceus	471, 506
rubella	214, 224	borealis	488, 510	junciformis	471
stricta	213, 224	canadensis	488, 510	laevis	472
ARGENTINA		dracunculoides	488	lindleyanus	472
anserina	301	dracunculus	488, 510	modestus	472, 506
argentea	301	frigida	489, 510	peregrinus	473
occidentalis	301	globularia	489, 510	richardsonii	471
subarctica	301	glomerata	489, 510	sibiricus	471, 506
ARMERIA	381	gnaphaloides	490	subspicatus	471, 506
maritima	381, 390	kruhsiana	490, 510	unalaskensis	
purpurea	381	laciniata	490	major	472
sibirica	381	macrobotrys	490, 510	yukonensis	471, 506
vulgaris arctica	381	senjavinensis	489, 510	aster, Alpine	471
ARNICA	494	tilesii	491	aster family	466
alpina	494	elatior	491, 510	aster,	
angustifolia	494	gormanii	491	great northern	472
attenuata	494, 510	unalaskensis	491	Lindley	472
linearis	494	trifurcata		rush-like	471
vestita	494	heterophylla	490, 510	Siberian	471
amplexicaulis	495, 512	tyrellii	490	smooth	472
angustifolia	494	unalaskensis	491, 510	Yukon	471
attenuata	494	aleutica	491	ASTERACEAE	447, 466
betonicaefolia	495	arum family	142	ASTEREAE	467
brevifolia	494	arum, water	143	ASTRAGALUS	316
chamissonis	495, 512	ARUNCUS	295	aboriginorum	318, 336
foliosa	495	acuminatus	295	muriei	318
cordifolia	495, 512	sylvester	295	agrestis	319
		vulgaris	295, 330	alpinus	319, 338

ASTRAGALUS (cont.)		Awlwort	246	BETULA	186
amblyodon	317	azalea,		alaskana	187
americanus	317, 336	Alpine	367	glandulosa	186, 220
eucosmus	318, 336	trailing	367	sibirica	186
falciferous	316			X nana exilis	187
harringtonii	318, 336			X resinifera	
hypoglottis	319	BARBAREA	251	(B. eastwoodae)	
linearis	318	americana	252		187, 220
littoralis	317	orthoceras	252, 274	kenaica	187, 220
macounii	318, 336	stricta	252	X nana exilis	
nutzotinensis	316, 336	BALSAMINACEAE	343	(B. hornei)	
polaris	317, 336	baneberry,			187, 220
tarletonis	319	red	240	X resinifera	187
tenellus	317, 336	white	240	nana exilis	186, 220
vicifolius	319, 338	barley,		X resinifera	
umbellatus	317, 336	bobtail	92	(B. beeniana)	187
williamsii	319, 336	cultivated	93	neoalaskana	187
yukonis	318, 336	meadow	92	papyrifera kenaica	187
ATELOPHRAGMA		squirrel-tail	93	papyrifera	
aboriginum	318	bayberry family	185	occidentalis	187, 220
collieri	318	beaked-rush, white	107	resinifera	187, 220
elegans	318	bearberry,		BETULACEAE	186
williamsii	319	Alpine	371	BIDENS	484
ATHYRIUM	17	beard, goat's	295	cernua	484, 508
alpestre	17	beard-grass, annual	56	BILDERDYKIA	
americanum	17, 34	beard-tongue,		convolvulus	195
americanum	17	diffuse	413	bindweed, black	195
cyclosorum	17	Gorman	413	birch,	
filix-femina	17	BECKMANNIA	69	Alaska	187
cyclosorum	17, 34	erucaeformis	69	dwarf Alpine	186
sitchense	17	syzigachne	69, 100	family	186
ATRIPLEX	201	bedstraw,		glandular scrub	186
alaskensis	202, 222	Brandegee	431	Kenai	187
drymarioides	201	northern	430	western paper	187
gmelini	202, 222	small	431	bistort,	
hortensis	202	sweet-scented	430	Alpine	195
patula	201	beech-fern	15	mountain meadow	195
AVENA	65	bellflower family	434	BISTORTA	
fatua	65, 100	BELLIS	469	lilacina	195
sativa	66	perennis	469	vivipara	195
striata	70	bent,		bitter-cress,	
AVENEAE	51	colonial	61	Pennsylvania	254
avens,		dense-flowered	61	purplish	253
caltha-leaved	307	Rhode Island	61	Regel	254
Drummond		bent-grass,		Richardson	254
mountain	306	Idaho	62	seaside	253
eight-petaled		northern	60	small-leaved	254
mountain	306	red	62	umbel-flowered	254
entire-leaved		berry,		BLECHNUM	9
mountain	306	baked-apple	296	spicant	9, 30
glacier	308	Lingen	372	blinks	206
large-leaved	307	Nagoon	297	BLITUM	
low	308			capitatum	200
Ross	307				

blackberry		BOSCHNIAKIA	427	BROMUS (cont.)	
dwarf red	296	glabra	427	inermis	87, 102
bladder-pod, Arctic	255	rossica	427, 442	marginatus	87, 102
bladderwort,		BOTRYCHIUM	7	mollis	88, 102
common	426	boreale	7, 30	pacificus	88, 102
family	426	obtusilobium	8	pumpellianus	87, 102
flat-leaved	427	lanceolatum	8, 30	arcticus	87
lesser	427	lunaria 7,	8, 30	villosissimus	88
blue-bead	155	multifidum		racemosus	88
blue bells of Scotland	435	robustum	8	richardsonii	88
blueberry,		silaifolium	8, 30	secalinus	88, 89
Alaska	373	virginianum	8	sitchensis	87, 102
bog	372	bottle-brush	19	aleutensis	87
dwarf	373	BOYKINIA	283	subulatus	71
early	373	richardsonii	284, 326	tectorum	89, 102
family	372	bracken, western	11	buckbean	397
swamp	373	bramble,		buckhorn	429
thin-leaved	373	Alaska	297	buckwheat,	
bluegrass,		five-leaved	296	cultivated	199
alpine	83	BRASENIA	227	family	191
annual	78, 102	schreberi	227, 270	bugleweed, northern	408
arctic	81	BRASSICA	250	bug-seed	202
big	84	arvensis	251, 274	bulrush,	
bog	81	campestris	251, 274	great	113
Canada	78	juncea	251, 274	pacific	113
Canby	84	napus	251	small-fruited	113
dune	79	BRASSICACEAE	245	bunchberry	361
Eyerdam	80	BRAYA	267	bunch-flower family	151
hispid	84	henryae	267	BUPLEURUM	357
Kentucky	80	humilis	267, 276	americanum	357, 384
Komarov	83	pilosa	268	bur-marigold,	
lanate	79	purpurescens	267	nodding	484
large-glumed	80		268, 276	burnet,	
loose-flowered	79	brome,		Menzies great	309
Merrill	82	fringed	88	official great	308
narrow-flowered	83	smooth	87	Sitka great	309
Norberg	80	brome-grass,		bur-reed	
rough	81	Alaska	87	family	42
Sandberg	84	Aleutian	87	narrow-leaved	43
timberline	83	arctic	87	northern	43
Turner	79	large mountain	87	simple-stemmed	43
wood	82	pacific	88	small	43
bluejoint	59	brooklime	416	BURSA	
bog-orchid,		broom-rape family	427	bursa-pastoris	255
Bering	162	BROMUS	86	bush, copper	366
Choriso	161	aleutensis	87, 102	butter and eggs	413
northern	162	arcticus	87	buttercup,	
slender	162	brizaeformis	88	Arctic	233
white	162	ciliatus	88, 102	Bongard	230
BORAGINACEAE	402	commutatus	89	bristly	230
borage family	402	hordaceus	88	Chamisso	234

buttercup, (cont.)			CALAMAGROSTIS (cont.)		CAPSELLA (cont.)		
Cooley		234	vaseyi	58	rubella		255, 274
creeping		229	yukonensis	58	CARDAMINE		253
Eastwood		231	CALLA	142	angulata		253, 274
Eschscholtz		231	palustris	143, 174	blaisdellii		254
Gray		232	CALLITRICHACEAE	343	bellidifolia		248
Kamchatka		234	CALLITRICHE	344			253, 274
Lapland		235	autumnalis	344	digitata		254
Macoun		230	bolanderi	344	microphylla		254, 274
northern		231	palustris	344	pennsylvanica		254
Pallas		235	verna	344, 382	pratensis		254, 274
pygmy		232	CALTHA	238	purpurea		253, 274
snow		232	biflora	238, 272	regeliana		254
straight-beaked		231	leptosepala	238, 272	richardsonii		254, 274
sulphur-colored		232	natans	239, 272	umbellata		254, 274
swamp		229	palustris	238	CAREX		114
tall		229	arctica	238	aenea		126, 170
verticillate-leaved		232	asarifolia	238, 272	albo-nigra		133
western		230	CALYPSO	166	angarae		132
butterwort,			bulbosa	166, 178	anthoxanthea		121, 170
common		426	CAMELINA	255	aquatilis		131, 172
hairy		426	sativa	256	atherodes		142
			CAMPANULA	434	athrostachya		125, 170
			aurita	435, 444	atrata		133, 172
cabbage,			dasyantha	435, 444	atrosquama		133
deer		397	heterodoxa	436	atratiformis		134
skunk		142	lasiocarpa	435, 444	atrofusca		139, 174
CABOMBACEAE		227	latisepala	436	aurea		129, 172
CACALIA		496	petiolata	436	bicolor		129, 172
auriculata	497, 512		rotundifolia	435, 444	bigelowii		130, 172
CAKILE		250	alaskana	436	bipartata		126
edentula	250, 274		dubia	436	bonanzensis		127
californica		250	scouleri	435	brunnescens		128, 170
lacustris		250	uniflora	435, 444	buxbaumii		132, 172
CALAMAGROSTIS		57	CAMPANULACEAE	434	canescens		127, 170
alaskana		59	campion,		capillaris	139, 140, 174	
aleutica		59	Menzies	217	nana		140
canadensis	59, 98		moss	217	capitata		120, 168
langsdorfii		59	pink	217	chordorrhiza		123, 170
scabra		59	Williams	217	circinata		121, 170
deschampsioides			cancer-root,		concinna		136, 172
	58, 98		clustered	428	concolor		130
holmii		58	one-flowered	428	crawfordii		125, 170
hyperborea		59	candytuft, garden	247	cryptocarpa		132
inexpansa	59, 98		canary-grass, reed	53	deflexa		136, 172
lapponica		59	CAPNOIDES		diandra		124, 170
langsdorfii		59	aureum	245	disperma		128, 170
neglecta		59	sempervirens	245	eburnea		137
borealis		59	CAPRIFOLIACEAE	431	echinata		129
nutkaensis	59, 98		CAPSELLA	255	eleocharis		123
purpurescens	58, 98		bursa-pastoris		enanderi		134
arctica		58		255, 274	filifolia		122

INDEX

CAREX (cont.)		CAREX (cont.)		CASSIOPE	
flava	140	neurochlaena	126	lycopodioides	370, 388
garberi		nigricans	122, 170	mertensiana	369, 388
bifaria	130, 172	norvegica	127	stelleriana	369, 388
glacialis	137, 172	inferalpina	132, 172	tetragona	369, 388
glareosa	127, 170	obtusata	122, 170	CASTILLEJA	418, 421
gmelini	133, 172	oederi viridula	140	annua	419
gynocrates	120, 170	pachystachya	125	chrymactis	421
hassei	130	pauciflora	123, 170	hyetophila	421, 440
heleonastes	126	paupercula	138	hyperborea	420, 440
hepburnii	120	peckii	136	miniata	421, 440
hindsii	130, 172	phaeocephala	125	muelleri	420
incurva	123	phyllomanica	129, 170	pallida	419
jacobi-peteri	120, 168	physocarpa	141, 174	auricoma	419
karaginensis	134	pluriflora	137, 174	caudata	419, 440
kelloggii	130, 172	podocarpa	135, 172	elegans	420
kokrinensis	131	praegracilis	124	mexiae	419
krausei	140	praticola	125, 170	parviflora	419, 440
lachenalii	126, 170	pribylovensis	126, 170	raupii	420, 440
laeviculmis	129, 172	pyrenaica	122, 170	unalaschcenis	421, 440
laevirostris	141	micropoda	122	transnivalis	421
lagopina	126	ramenskii	132, 172	villosissima	420, 440
lapponica	127	rariflora	137, 172	yukonis	420
laxa	138	rhyncophysa	141, 174	catchfly,	
leersii	129	rossii	136, 172	night-flowering	217
leiophylla	133	rostrata	140, 141, 174	catchweed	408
leptalea	121, 170	rotundata	141, 174	catnip	410
limosa	138, 174	rupestris	122	cat-tail, common	42
livida	138, 174	saltuensis	139	cat-tail family	42
loliacea	128, 170	scirpoidea	121, 170	cedar,	
lugens	130, 172	sitchensis	131, 172	giant	26
lyngebyei		spaniocarpa	136	western red	26
cryptocarpa	132, 172	spectabilis	135, 172	yellow	26
mackenziei	127, 170	stellulata	128, 170	CERASTIUM	207
macloviana		stenochleana	121	aleuticum	208
pachystachya		stenophylla		arcticum	209
	125, 170	eleocharis	123	arvense	208, 224
macrocephala		stipata	124, 170	beeringianum	208, 224
anthericoides	124	stygia	137	caespitosum	208, 224
macrochaeta	134, 172	stylosa	132, 172	fischerianum	208, 224
magellanica	138, 174	subspathacea	131, 172	glomeratum	208, 224
maritima	123, 170	supina		maximum	207, 222
melozitensis	141	spaniocarpa	136, 172	viscosum	208
membranacea	141, 174	tenuiflora	128, 170	vulgatum	208
membranopacta	141	utriculata	141	CHAMAEDAPHNE	370
mertensii	134, 172	vaginata	139, 174	angustifolium	350
microglochin	123, 170	viridula	140, 174	calyculata	370, 388
micropoda	122	williamsii	140	CHAMAENERION	
misandra	139, 174	carrot family	354	latifolium	350
montanensis	135, 172	CARYOPHYLLACEAE		spicatum	350
muricata	129		206	Chamaerhodos,	
nardina	120, 168	CASSANDRA		American	306
nesophila	135, 172	calyculata	370		

532　INDEX

CHAMAERHODOS 306
 erecta nuttallii 306
 nuttallii 306, 334
chamomile,
 Arctic 487
 scentless 487
CHAMOMILLA
 suaveolens 487
charlock,
 jointed 251
cheat, 88
 smooth-flowered
 soft 88
CHENOPODIUM 199
 album 200, 222
 berlandieri
  zschackei 200
 capitatum 200, 222
 gigantospermum 200
 glaucum
  salinum 200, 222
 hybridum 200
 leptophyllum 200
CHENOPODIACEAE 199
CHERLERIA
 dicranoides 213
cherry, Rocky
 Mountain wild 294
chess, 88
 downy 89
 hairy 89
 soft 88
CHAMAECYPARIS 26
 nootkatensis 26, 38
CHEIRINIA
 cheiranthoides 266
chicory family 447
chickweed,
 Aleutian 208
 Arctic 209
 Beering 208
 blue 414
 common 209
 field 208
 Fischer 208
 great 207
 larger mouse-ear 208
 low 211
 mouse-ear 208
 water 206
CHIMAPHILA 362
 umbellata
  occidentalis 362, 386

chives,
 garden 153
 wild 153
CHLORIDEAE 51
CHONDROPHYLLA
 americana 394
CHRYSANTHEMUM 486
 arcticum 486, 508
  polaris 486
 integrifolium 486, 508
 leucanthemum 486
CHRYSOPLENIUM 281
 beringianum 282
 wrightii 282, 326
 tetrandrum 282, 326
CICHORIACEAE 447
CICUTA 360
 douglasii 360, 386
 mackenzieana 361, 386
 maculata 360, 386
CINNA 57
 latifolia 57, 98
cinquefoil,
 Arctic 304
 Chamisso 305
 coast 302
 cut-leaved 303
 diverse-leaved 303
 glandular 301
 Hooker 305
 marsh 301
 one-flowered 304
 Pennsylvania 302
 pretty 304
 purple 301
 red-stemmed 303
 rough 304
 shrubby 301
 snow 305
 tall 301
 two-flowered 303
 Vahl 304
 villous 304
 wooly 302
CIRCAEA 349
 alpina 349, 382
CIRSIUM 502
 arvense 502
 edule 502
 foliosum 502, 512
 kamtschaticum 503, 512
 lanceolatum 502
 vulgare 502

CLADOTHAMNUS 366
 pyrolaeflorus 366, 388
CLAYTONIA 203
 acutifolia 204, 222
 graminifolia 204
 alsinoides 205
 arctica 204, 222
 asarifolia 205
 chamissoi 205, 222
 flagellaris 206, 222
 parviflora 205
 parvifolia 206
 perfoliata 205, 222
 sarmentosa 205, 222
 scammaniana 205, 222
 sibirica 205, 222
 tuberosa 203, 204, 222
cleavers 430
cliff-brake, slender 10
CLINTONIA 155
 uniflora 155, 176
cloudberry 296
clover,
 alsike 313
 burr 314
 coast 313
 cow 313
 Dutch 313
 hop 314
 lupine 312
 red 312
 small-headed 313
 white 313
 white-tipped 313
club, devil's 354
club-moss,
 Alaska 22
 Alpine 22
 bog 21
 family 20
 fir 20
 stiff 22
club-rush,
 few-flowered 112
 red 113
 tufted 112
CNIDIUM 358
 ajanense 358
COCHLEARIA 248
 officinalis 248, 274
  arctica 248
  oblongifolia 248
cockle, corn 220

INDEX

COELOGLOSSUM	160	cotton-grass, (cont.)		CRYPTANTHE	407
bracteatum	160	slender	111	torreyana	407, 438
viride	160, 178	tall	112	CRYPTOGRAMMA	10
bracteatum	160	thin-leaved	111	acrostichoides	10, 30
COELOPLEURUM		white	110	sitchense	10
gmelini	358	cottonwood, black	185	crispa	10
COLLINSIA	414	COTULA	492	stelleri	10, 30
parviflora	414, 438	coronopifolia	492	cucumber-root	156
COLLOMIA	400	crab-apple, western	310	currant,	
linearis	400, 436	cranberry,		American red	293
COLPODIUM	70	mountain	372	blue	292
fulvum	70, 100	swamp	374	fetid	293
effusum	71	crane's-bill,		maple-leaved	292
wrightii	71	Bicknell	342	northern black	292
columbine,		Carolina	342	trailing black	293
small-flowered	240	COMANDRA	190	CYANOGLOSSUM	402
western	240	livida	190	boreale	403
COPTIDIUM		pallida	190	CYNAREAE	468
lapponicum	235	comandra,		CYPERACEAE	107
COPTIS	239	northern	190	cypress, Alaska	26
asplenifolia	239, 272	pale	190	CYPRIPEDIUM	159
trifoliata	239, 272	COMARUM		guttatum	159, 178
CORALLORRHIZA	166	palustre	301	montanum	159
corallorrhiza	166	comfrey,		passerinum	159, 178
innata	166	northern wild	403	CYSTOPTERIS	12
mertensiana	167, 178	COMPOSITAE	447	fragilis	13, 32
trifida	166, 178	CONIOSELINUM	357, 358	montana	13, 32
coral-root,		benthami	357, 384	cystopteris, mountain	13
early	166	cnidifolium	358, 384	CYTHEREA	
Mertens	167	dawsonii	358	bulbosa	166
CORISPERMUM	202	gmelini	357		
hyssopifolium	202, 222	CONVALLARIACEAE	154		
CORNACEAE	361	CRASSULACEAE	279	DACTYLIS	69
CORNUS	361	CRATAEGUS	311	glomerata	69, 100
canadensis	361, 362, 386	douglasii	311, 334	dagger-fern	14
stolonifera	362, 386	CREPIS	464	daisy,	
baileyi	362	capillaris	465	Arctic	486
suecica	361, 362, 386	elegans	465, 504	European	469
cornel, Lapland	362	nana	465, 504	ox-eye	486
CORYDALIS	244	cress,		dandelion	448
aurea	245, 272	Alpine	253	DANTHONIA	65
pauciflora	245, 272	water-	252	intermedia	65
sempervirens	245, 272	winter	252	spicata	65, 100
corydalis,		crocus, wild	237	darnel	90
few-flowered	245	crowberry	344	DASIPHORA	
golden	245	family	344	fruticosa	301
pink	245	crowfoot,		dead nettle, white	411
cotton-grass,		celery-leaved	233	deerberry	155
Alpine	110	family	227	deer-fern	9
Arctic	111	seaside	234	delight, single	363
close-sheathed	111	smooth-leaved	231	DELPHINIUM	240
russet	110	CRUNOCALLIS		alatum	241
sheathed	111	chamissonis	205	blaisdellii	241

DELPHINIUM (cont.)		dock, (cont.)		DRABA (cont.)		
brachycentrum	241,272	golden	192	pilosa		259
brownii	241	great western	193	pseudopilosa		260
glaucum	241, 272	Siberian	194	ruaxes		259
hookeri	241	western	193	stenoloba	260, 274	
menziesii	241	DODECATHEON	375	unalaschensis		262
nutans	241	frigidum	375, 390	ventosa ruaxes		259
ruthae	241	integrifolium	376	draba, Aleutian		257
scopulorum		macrocarpum	375, 390	DRACOCEPHALUM		410
glaucum	241	alaskanum	375	parviflorum	410, 438	
splendens	241	pauciflorum	375	dragon head		410
DESCHAMPSIA	67	superbum	375	DROSERA		268
alaskana	68	viviparum	376, 390	anglica	269, 276	
atropurpurea	67, 100	dogbane family	397	longifolia		269
latifolia	68	dogbane, spreading	398	rotundifolia		
paramushirensis	68	dog-fennel	486		268, 269, 276	
patentissima	68	dogwood family	361	DROSERACEAE		268
beringensis	68, 100	dogwood, red osier	362	DRYAS		306
bottnica	68	DONDIA		drummondii	306, 334	
caespitosa	68, 100	maritima	203	integrifolia	306, 334	
glauca	68	DOUGLASIA	380	sylvatica		307
orientalis	68	arctica	380	octopetala	306, 334	
calycina	67	gormanii	380, 390	DRYMOCALLIS		
curtifolia	68	DRABA	256	arguta		301
danthonioides	67	aleutica	257, 274	DRYOPTERIS		14
elongata	67, 100	alpina	260, 274	aquilonaris		15
flexuosa	68	aurea	262, 276	austriaca	16, 32	
holciformis	68, 100	borealis	262, 276	dilatata		16
DESCURAINIA	249	caesia	258	dryopteris		15
pinnata		chamissonis	258, 274	fragrans	15, 32	
filipes	250	cinerea	261, 276	linnaeana	15, 32	
richardsonii	250, 274	crassifolia	258	oreopteris	16, 32	
sophia	249, 274	densifolia	258, 274	phegopteris	15, 32	
sophioides	249, 274	eschscholtzii	260	robertiana		15
DEYEUXIA		exaltata	259	spinulosa		16
purpurescens	58	fladnizensis	259	duckweed family		143
DIANTHUS	218	glabella	261, 276	duckweed,		
repens	218	hirta	261	ivy-leaved		143
DIAPENSIA	374	hyperborea	262, 276	lesser		143
lapponica		inserta	258	DUPONTIA		69
obovata	374, 390	kamtschatica	261	fischeri	69, 100	
diapensia family	374	lactea	260, 274	psilosantha		70
DIAPENSIACEAE	374	lanceolata	261, 276	psilosantha		70
DICOTYLEDONEAE	181	longipes	261, 274			
DIGITALIS	418	lutea	257			
purpurea	418	macouniana	260	ears, cat's		448
ditch-grass,	47	macounii	260	ECHINOPANAX		
western	47	macrocarpa	260	horridum		354
dock,		maxima	262, 276	eel-grass		48
Arctic	193	nemorosa	257, 274	family		48
beach	194	nivalis	258, 274	ELAEAGNACEAE		348
bitter	192	denudata	258, 274	ELAEAGNUS		348
curled	193	kamtschatica	261	argentea		348
garden	193	oligosperma	258	commutata	348, 382	

INDEX

ELEOCHARIS	108	EQUISETUM (cont.)		ERIOPHORUM (cont.)	
acicularis	108, 168	scirpoides	19, 34	russeolum	110
kamtschatica	108, 168	sylvaticum	19, 34	scheuchzeri	110
mamillata	109	variegatum	19, 20, 34		111, 168
nitida	108	alaskanum	19	spissum	111
palustris	109, 168	anceps	19	vaginatum	111, 168
tenuis	108	ERETRICHIUM	405	spissum	111
uniglumis	109, 168	aretioides	405, 438	viridi-carinatum	111
elephants, little	425	chamissonis	406	ERMANIA	265
elder, red-berried	433	splendens	405	borealis	265
ELYMUS	93	eretrichium, showy	405	parryoides	265
aleuticus	94	ERICACEAE	365	ERODIUM	342
arenarius		ERIGERON	472, 473	cicutarium	342
mollis	94	acris	476	ERYSIMUM	265
borealis	95	asteroides	476, 506	angustatum	266
canadensis	94, 104	debilis	476	cheiranthoides	
glaucus	94, 104	elatus	476		266, 276
hirsutus	95, 104	alaskanus	474	inconspicuum	266, 276
howellii	94	caespitosus	473, 506	officinale	249
innovatus	93, 104	compositus	475, 506	pallasii	266, 276
macounii	94, 104	discoideus	475	ERXLEBANIA	
mollis	94, 104	glabratus	475	minor	364
villosissmus	94	eriocephalus	474	EUCARICES	
virescens	94, 104	glabellus		DISTIGMATICAE	117
EMPETRACEAE	344	pubescens	473, 506	EUCARICES	
EMPETRUM	344	grandiflorus	474, 506	TRISTIGMATICAE	117
nigrum	344, 382	humilis	475, 506	EUPHRASIA	422
EPILOBIUM	349	hyperboreus	474, 506	disjuncta	422
adenocaulon	351, 384	lonchophyllus	475, 506	mollis	422, 440
anagallidifolium	351	muirii	474	subarctica	422, 440
angustifolium	350, 382	peregrinus	473, 506	EUTREMA	248
behringianum	351, 384	callianthemus	473	edwardsii	248, 274
bongardii	352	philadelphicus	475	evening-primrose	
davuricum	350, 384	purpuratus	475, 506	family	348
glandulosum	351, 384	unalaskensis	475	everlasting, pearly	483
hornemannii	352, 384	uniflorus			
lactiflorum	352	eriocephalus			
latifolium	350, 382		474, 506	FABACEAE	311
leptocarpum	351, 384	yukonensis	474, 506	FAGOPYRUM	
macounii	351	ERIOPHORUM	109	esculentum	199
luteum	350, 384	alpinum	110, 168	FATSIA	
palustre	351, 384	angustifolium	111	horrida	354
sertulatum	352		112, 168	FAURIA	397
treleaseanum	350	brachyantherum		crista-galli	397, 436
EQUISETACEAE	18		111, 168	felwort, marsh	394
EQUISETUM	18	callitrix	111	fern family	· 8
alaskanum	19	chamissonis	110	fescue,	
arvense	18, 34		111, 168	alpine	85
fluviatile	19	albidum	110	bearded	85
hiemale	20	leucothrix	110	hard	86
californicum	20, 34	gracile	111, 168	meadow	85
limosum	19, 34	medium	110, 111	red	85
palustre	19, 34	opacum	111	rough	86
pratense	18, 34	polystachyon	112	western six-weeks	86

FESTUCA	84	foxtail,(cont.)		GENTIANA (cont.)	
altaica	86, 102	marsh	56	arctophila	397
arenaria	85	meadow	55	auriculata	396, 436
arundinacea	85	mountain	55	barbata	395, 436
aucta	85	short-awned	56	calycosa	395
barbata	85	FRAGARIA	299	covillei	395
brachyphylla	85	bracteata	299, 330	detonsa	396
	86, 102	chiloensis	299, 330	douglasiana	395, 436
duriuscula	86	glauca	300, 330	frigida	395
elatior	85, 102	platypetala	300	glauca	395, 436
arundinacea	85	yukonensis	300	gormani	395
glabrata	85	fragile-fern	13	macounii	395
kitaibeliana	85	fringe cup	283	platypetala	395, 436
lanuginosa	85	FRITILLARIA	154	procera	395
megalura	86	camtchatcensis		propinqua	396
megastachya	85		154, 176		397, 436
mutica	85	FUMARIACEAE	244	prostrata	394, 436
ovina		fumewort family	244	raupii	395
duriuscula	86			romanzovii	395
richardsonii	85			serrata	396
rubra	85, 102	gale, sweet	185	tenella	396, 436
subulata	85, 102	GALEOPSIS	412	GENTIANACEAE	393
subvillosa	85	bifida	412, 438	GEOCAULON	190
FESTUCEAE	51	tetrahit	412	lividum	190, 220
fiddle-neck,	406	GALIUM	430	GERANIACEAE	341
Menzies	406	aparine	430, 442	GERANIUM	341
field-sedge,		boreale	430, 442	bicknellii	342, 382
clustered	124	brandegei	431	carolinianum	342
figwort family	412	kamtschaticum	430, 442	erianthum	341, 382
FILIX		trifidum	431, 442	robertianum	342
fragilis	13	columbianum	431	sanguineum	342, 382
montana	13	triflorum	430, 442	geranium,	
fir,		GAULTHERIA	370	family	341
alpine	27	miqueliana	371	northern	341
lovely	27	shallon	371, 388	GEUM	307
silver	27	gentian,		calthifolium	307, 334
fireweed,	350	Aleutian	396	X Geum rossii	308
dwarf	350	Arctic	397	glaciale	308, 334
flag	157	auricled	396	macrophyllum	307, 334
flax,		broad-petaled	395	pentapetalum	308, 334
false	256	family	393	rossii	307, 334
family	343	four-parted	396	GILIA	400
Lewis wild	343	glaucous	395	capitata	400
flower,		moss	394	ginseng family	354
cuckoo	254	northern	396	glasswort, slender	203
star	381	slender	396	GLAUX	380
wax	363	smaller fringed	395	maritima	308, 390
foamflower,		swamp	395	GLECOMA	410
trifoliate	284	whitish	395	hederacea	410
unifoliate	284	GENTIANA	394	GLEHNIA	359
forget-me-not	405	acuta	396, 436	leiocarpa	360
foxglove	418	plebeja	396	littoralis	
foxtail,		aleutica	396, 436	leiocarpa	360
glaucous	55	albida	395, 436		

GLYCERIA		71	GYNOPHORARIA		HEDYSARUM	323
borealis	72,	100	falcata	316	alpinum	
grandis	73,	100	GYMNOSPERMAE	24	americanum	
leptostachya	72,	100				323, 338
maxima grandis		73			grandiflorum	323
pauciflora	72,	100	HABENARIA	160, 161	auriculatum	323
pulchella		73	bracteata	160	boreale	323
striata stricta		72	chorisiana	161	mackenzii	323, 338
GNAPHALIUM		483	dilitata	162	truncatum	323
uliginosum	483,	508	hyperborea	162	HELIANTHEAE	467
golden rod, northern		469	obtusata	161	HELIANTHUS	484
goldthread,			orbiculata	160	annuus	484
fern-leaved		239	saccata	162	hellebore,	
trifoliate		239	unalaschensis	163	American white	152
GOODYERA			viridis interjecta	160	European white	152
decipiens		165	HACKELIA	403	hemlock,	
repens		165	leptophylla	403	mountain	28
gooseberry family		291	hair-grass,		spotted water	360
gooseberry,			annual	67	western	28, 190
northern		292	Bering	68	hemlock-parsley,	
swamp		292	California	68	Dawson	358
goosefoot family		199	mountain	67	western	357
goosefoot,			slender	67	HERACLEUM	358
maple-leaved		200	tufted	68	lanatum	358, 384
narrow-leaved		200	wavy	68	herb, cow	219
oak-leaved		200	HALERPESTIS		herb Robert	342
Zschacke		200	cymbalaria	234	HESPEROPEUCE	
goose-tongue		428	HALORAGIDACEAE	352	mertensiana	28
GORMANIA			HAPLOPAPPUS	469	HEUCHERA	283
oregona		279	macleanii	469	glabra	283, 326
grape-fern,			harebell,	435	heuchera, Alpine	283
lance-leaved		8	Arctic	435	HIERACIUM	465
leathery		8	mountain	435	albiflorum	466, 504
northern		7	Scouler	435	canadense	466
Virginia		8	HARRIMANELLA		gracile	466, 504
grass family		49	stelleriana	369	alaskanum	466
grass,			hawkweed,		yukonense	466
blue-eyed		158	slender	466	triste	466, 504
orchard		69	white	466	tristeforme	466
rabbit-foot		56	wooly	466	HIEROCHLOE	52
rattlesnake		88	hawthorn, black	311	alpina	52, 96
scurvy		248	hay-sedge, Fernald	126	odorata	53, 96
velvet		63	heal-all	411	pauciflora	53, 96
grass-of-Parnassus,			heath family	365	highbush cranberry,	
fringed		281	heather,		few-flowered	433
Kotzebue		281	Alaska	369	HIPPURIS	352
Montana		281	Aleutian	369	montana	353, 384
northern		281	purple	368	tetraphylla	353, 384
GROSSULARIA			red	368	vulgaris	353, 384
oxycanthoides		292	yellow	369	HOLCUS	63
GROSSULARIACEAE		291	hedysarum,		lanatus	63, 100
ground cedar		21	American	323	holly-fern	14
groundsel, common		501			holy-grass	53
ground-pine		21				

INDEX

holy-grass,		IRIS	157	JUNIPERUS	25	
Alpine	52	arctica	157	communis		
Arctic	53	setosa	157, 178	montana	25, 36	
HOMALOBUS		interior	157	sibirica	25	
amblyodon	317	platyrincha	157	horizontalis	25, 38	
tenellus	317	ISOETES	23	nana	25	
HONCKENYA		braunii	23, 36	prostrata	25	
peploides	212	maritima	23	sibirica	25	
honeysuckle family	431	echinospora				
honeysuckle, glaucous	432	truncata	23			
hopclover, low	312	ISOETACEAE	23	KALMIA	368	
HORDEAE	52	ivy, ground	410	microphylla	368	
HORDEUM	92			occidentalis	368	
brachyantherum				polifolia	368, 388	
	92, 93, 104	jewel-weed family	343	kinnikinnick	371	
boreale	92	JUNCACEAE	143	Kneshenaka	297	
caespitosum	92, 104	JUNCOIDES	149	knotweed,		
jubatum	93, 104	arcticum	150	Alaska	197	
nodosum	92	arcuatum	150	bushy	198	
vulgare	93	campestris	151	common	198	
horehound,	410	hyperborium	150	Fowler	197	
western water	408	parviflorum	149	proliferous	198	
horsetail family	18	spicatum	150	various-leaved	198	
horsetail,		JUNCUS	144	kobresia, Bellard	114	
common	18	albescens	147	KOBRESIA	113	
little	19	alpinus	147	bellardii	114	
marsh	19	nodulosus	148, 176	myosuroides	114, 168	
meadow	18	arcticus	145, 174	simpliciuscula	114	
swamp	19	ater	146	Koeler-grass,		
thicket	18	balticus	145, 146	Cairnes	65	
wood	19	sitchensis	146, 174	Yukon	64	
huckleberry, red	373	biglumis	147, 174	KOELERIA	64	
HYDROPHYLLACEAE		bufonius	146, 174	cairnesiana	65	
	400	castaneus	148, 176	yukonensis	64	
HYPOCHAERIS	448	drummondii	145, 174	KOENIGIA	191	
radicata	448	effusus	145, 174	islandica	191, 220	
HYPOPITYS	365	ensifolius	146, 174	Koenigia	191	
latisquama	365, 388	falcatus		KRUHSEA	156	
		sitchensis	148, 176	streptopoides	156, 176	
		filiformis	145, 174	Kruhsea	156	
		leocochlamys	148			
IBERIS		macer	146, 174			
amara	247	mertensianus	147, 174	Labrador tea,	366	
IBIDIUM		nodosus	148, 176	narrow-leaved	366	
romanzoffianum	163	nodulosis	148	LACTUCA	463	
IMPATIENS	343	oreganus	147	pulchella	464	
noli-tangere	343, 382	richardsonianus	147	spicata	464, 504	
occidentalis	343	stygius		tartarica	464	
Indian pipe family	364	triglumis		virosa	464	
INULEAE	467	americanus	146, 174			
IRIDACEAE	157	triglumis	147, 174	ladies' slipper,		
iris family	157	juniper,		mountain	159	
iris, wild	157	creeping	25	northern	159	
		low	25	pink	159	

INDEX 539

ladies' tresses,	**LEPTASEA**	**LINNAEA** 431
hooded 163	alaskana 290	borealis 431
lady-fern, 17	funstonii 290	americana 432, 442
Alpine 17	**LEPTAXIS**	longiflora 432
LAGOTIS 417	menziesii 283	**LINUM** 343
glauca 418, 440	**LESQUERELLA** 255	lewisii 343
stelleri 418	arctica 255, 274	perenne
lamb's quarters 200	scammanae 255	lewisii 343, 382
LAMIACEAE 408	lettuce,	liquorice,
LAMIUM 411	large-flowered blue	northern wild 430
album 411	464	**LISTERA** 164
LAPPULA 403	prickly 464	borealis 164, 178
echinata 403	tall blue 464	caurina 164, 178
myosotis 403, 438	licorice-fern 11	convallarioides
occidentalis 403	**LIGUSTICUM** 359	164, 178
redowski 403, 438	hultenii 359, 386	cordata 164, 178
LAPSANA 447	macounii 359	nephrophylla 165
communis 448	mutellinoides	nephrophylla 165
larch, American 27	alpinum 359, 386	little club-moss
LARIX 27	scoticum 359	family 22
alaskensis 27	**LILAEOPSIS** 356	**LLOYDIA** 154
americana 27	occidentalis 356	serotina 154, 176
laricina 27, 38	**LILIACEAE** 153	**LOBELIA** 436
larkspur,	lily family 153	kalmii 436, 444
glaucus 241	lily,	**LOISELEURIA** 367
northern dwarf 241	Alp 154	procumbens 367, 388
LATHYRUS 325	black 154	**LOLIUM** 89
japonicus 325	yellow pond 242	multiflorum 89, 104
maritimus 325, 338	lily-of-the-valley	perenne 89
palustris	family 154	temulentum 90
pilosus 325, 338	**LIMNIA**	**LOMARIA**
venosus 325	parviflora 205	spicant 9
laurel, swamp 368	sarmentosa 205	**LOMATOGONIUM** 394
leather-leaf 370	sibirica 205	rotatum 394, 436
LEDUM 365	**LIMNORCHIS** 161	**LONICERA** 432
decumbens 366, 388	behringiana 162	glaucescens 432
groenlandicum 366, 388	chorisiana 161, 178	involucrata 432, 444
pacificum 366	convallariaefolia	loosestrife, tufted 380
palustris	162, 178	**LORANTHACEAE** 189
decumbens 366	dilatatoides 162	**LUETKEA** 295
LEMNA 143	dilitata 162	pectinata 295, 330
minor 143, 174	angustifolia 163	lungwort,
trisulca 143, 174	chlorantha 163	Asiatic 404
LEMNACEAE 143	leucostachys	Eastwood 404
LENTIBULARIACEAE	163, 178	sea 404
426	hyperborea 162, 178	tall 404
LEPARGYREA	leptoceratitis 163	lupine,
canadensis 348	stricta 162, 178	Arctic 315
LEPIDIUM 247	**LIMOSELLA** 414	large-leaved 315
densiflorum 247, 274	aquatica 415, 440	Nootka 315
sativum 247	**LINACEAE** 343	prairie 315
virginicum 247	**LINARIA** 413	silky 316
LEPTARRHENA 284	vulgaris 413	**LUPINUS** 315
pyrolifolia 284, 326		arcticus 315, 336

LUPINUS (cont.)			LYCOPODIUM (cont.)		marigold,		
kiskensis		315	porophilum	20	mountain		238
lepidus	315,	336	sabinaefolium		yellow marsh		238
nootkatensis	315,	336	sitchense	22	MARRUBIUM		
polyphyllus	315,	336	selago	20, 36	vulgare		410
sericeus	316,	336	adpressum	21	marsh-fleabane		498
LUZULA		148	sitchense	22, 36	marsh marigold,		
arcuata	150,	176	LYCOPUS	408	broad-leaved		238
campestris		151	lucidus	408, 438	floating		239
carolinae		149	uniflorus	408, 438	MARUTA		
confusa		150	LYSIAS	160	cotula		486
hyperborea	150,	176	orbiculata	160, 178	MATTEUCIA		
japonica		149	LYSICHITUM	142	struthiopteris		10
multiflora	151,	176	americanum	142, 174	MATRICARIA		487
comosa		151	LYSIELLA	161	ambigua	487,	508
frigida		151	obtusata	161, 178	grandiflora		487
kobayasii		151	LYSIMACHIA	380	inodora		487
nivalis	150,	176	thyrsiflora	380, 390	suaveolens	487,	508
latifolia		150			mayweed		486
parviflora	149,	176			meadow-rue,		
divaricata		150	MACOUNASTRUM	191	Arctic		237
pilosa		149	islandicum	191	few-flowered		237
rufescens	149,	176	madder family	430	western		238
saltuensis		149	MADIA	484	MEDICAGO		314
spicata	150,	176	glomerata	484, 508	falcata	314,	336
wahlenbergii	149,	176	madwort, German	408	hispida	314,	336
LYCHNIS		218	MAIANTHEMUM	155	lupulina	314,	336
affinis		219	bifolium		sativa	314,	336
apetala	218,	224	kamtschaticum	155	MELANDRIUM		
dawsonii		219	dilitatum	155, 176	apetalum		218
furcata		219	maid, mist	401	dawsonii		219
macrosperma		218	maiden-hair fern	11	furcatum		219
soczavianum		219	MAIRANIA		macrospermum		218
taylorae		219	alpina	371	soczavianum		219
triflora dawsonii		219	MALAXIS	165	taylorae		219
lychnis,			monophylla	166, 178	MELANTHACEAE		151
Arctic		219	paludosa	165, 178	melic, false		70
Dawson		219	MALUS	310	MELICA		71
large-seeded		218	diversifolia	310	subulata		71
nodding		218	fusca	310, 334	MELILOTUS		313
Taylor		219	manna-grass,		alba	314,	336
LYCOPODIACEAE		20	American	73	officinalis	314,	336
LYCOPODIUM		20	Davy	72	MENTHA		409
alpinum	22,	36	fowl	72	arvensis	409,	438
annotinum	22,	36	northern	72	canadensis		409
pungens		22	weak	72	piperita		409
clavatum	21,	36	maple,		spicata		409
monostachyon		21	broad-leaved	345	MENYANTHES		397
complanatum			Douglas	345	crista-galli		397
	21, 22,	36	maple family	345	trifoliata	397,	436
dendroideum		21	mare's-tail,		MENZIESIA		367
inundatum	21,	36	common	353	ferruginea	367,	388
obscurum	21,	36	mountain	353	menziesia, rusty		367
dendroideum		22					

MERCKIA		212	mitrewort,		MYOSOTIS		404
physodes		212	Alpine	282	arvensis		405
MERTENSIA		404	stoloniferous	282	alpestris		
asiatica		404	MOEHRINGIA		asiatica	405, 438
eastwoodae		404	lateriflora	213	palustris		405
maritima	404, 438	MONESES		362	MYRICA		185
paniculata	404, 438	uniflora	363, 386	gale	185, 220
MICRANTHES			monkey-flower,		tomentosa		186
flabellifolia		288	Lewis	414	MYRICACEAE		185
galacifolia		288	yellow	414	MYRIOPHYLLUM	353
yukonensis		287	MONOCOTYLEDONEAE		alterniflorum	353
MICROSTERIS		400		41	spicatum	353, 384
gracilis		400	MONOLEPIS	201			
MICROSTYLIS			nuttalliana	201, 222			
monophyllos		166	monolepis, Nuttall	201	NABALUS		
milk vetch,			MONOTROPA	364	hastatus		464
Alpine		319	uniflora	364, 388	NAIOCRENE		
Arctic		317	MONOTROPACEAE	364	flagellaris		206
hairy Arctic		317	MONTIA	206	parvifolia		206
Harrington		318	chamissonis	205	NAUMBURGIA		
Indian		318	flagellaris	206	thyrsiflora		380
loose-flowered	317	fontana	206	neckweed		417
Macoun		318	hallii	206	needle and thread	54
polar		317	lamprosperma	206, 222	needle-grass,		
pretty		318	parviflora	205	columbia		54
purple		319	parvifolia	206	NEPETA		
sickle-pod		316	sarmentosa	205	cataria		410
vetch-leaved		319	sibirica	205	hederacea		410
Williams		319	moonwort	7	NEPHROPHYLLIDIUM	
Yukon		318	moschatel	433	crista-galli		397
milkwort, sea		380	family	433	NESLIA		256
MIMULUS		414	mountain ash,		paniculata	256, 274
guttatus	414, 438	European	310	NESODRABA		
langsdorfii		414	Sitka	309	grandis		262
lewisii	414, 438	western	310	nettle family		189
mint family		408	mountain heather,		nettle,		
mint, wild		409	blue	368	Emerson hedge	411
MINUARTIA			club-moss	370	hedge		411
arctica		214	four-angled	369	hemp		412
biflora		214	Mertens	369	Lyall		189
elegans		213	mud-disk	492	slender		189
laricifolia		214	mudweed	415	niggerheads		111
macrocarpa		215	MUSCARIA		nightshade, enchanter's	
obtusifolia		214	silenefolia	287			349
rubella		214	mustard family	245	ninebark, Pacific	294
stricta		213	mustard,		nipplewort		448
mistletoe,			ball	256	nonsuch		314
dwarf hemlock	190	hedge	249	NORTA		
mistletoe family	189	Indian	251	altissima		249
MITELLA		282	tansy	249	NUPHAR		242
nuda	282, 326	tower	265	polysepalum	242, 272
pentandra	282, 326	tumble	249	variegatum		242
			wild	251	NYMPHAEA		242
			wormseed	266	polysepala		242

NYMPHAEA (cont.)		ORTHOCARPUS	421	OXYTROPIS (cont.)	
tetragona		hispidus	421	viscidula	321, 338
leibergi	243, 272	OSMORRHIZA	355		
NYMPHAEACEAE	242	chilense	356, 384		
		obtusa	356, 384	paintbrush, lesser	421
		purpurea	356, 384	PANICULARIA	
oak-fern,	15	OSMUNDA		americana	73
scented	15	spicant	9	borealis	72
oat,		ostrich-fern	10	nervata stricta	72
cultivated	66	OXYCOCCUS	374	pauciflora	72
wild	65	intermedia	374	PAPAVER	243
oat-grass,		microcarpus	374, 390	alaskanum	244
downy	66	oxycoccus	374	alboroseum	243
poverty	65	OXYGRAPHIS		macounii	244, 272
tall	64	glacialis	234	mcconellii	244
timber	65	OXYRIA	194	nudicaule	244, 272
OENANTHE	360	digyna	194, 220	radicatum	244, 272
sarmentosa	360, 386	oxytrope,		alaskanum	244
oleaster family	348	blackish	321	walpolei	243, 272
ONAGRACEAE	348	deflexed-podded	320	PAPAVERACEAE	243
onion-grass, Alaska	71	foliose	320	PARNASSIA	280
ONOCLEA		Kokrines Mountains		fimbriata	281, 326
struthiopteris	10		321	kotzebuei	281, 326
OPHIOGLOSSACEAE	6	Maydell	322	montanensis	281
OPHIOGLOSSUM	7	Mertens	320	palustris	281, 326
alaskanum	7	northern yellow	322	neogaea	281
vulgatum	7	Roald	322	parviflora	281
OPHRYS		Scamman	322	PARRYA	268
borealis	164	showy	323	nudicaulis	268, 276
caurina	164	variable	323	parsley-fern	10
convallarioides	164	vicid	321	parsley, water	360
cordata	164	OXYTROPIS	319	parsnip,	
nephrophylla	165	arctica	322	common garden	357
OPLOPANAX	354	borealis	321	cow	358
horridus	354, 384	campestris	322	pasque-flower	237
orache,		deflexa	320, 338	PASTINACA	357
garden	202	erecta	322	sativa	357
spear	201	foliolosa	320	pea family	311
orchid family	158	gracilis	322, 338	pea,	
orchid,		kokrinensis	321	beach	325
large round-leaved	160	leucantha	321, 338	veiny	325
long-bracted	160	maydelliana	322, 338	wild	325
small northern bog	161	mertensia	320, 338	pearlwort,	
ORCHIDACEAE	158	nigrescens	321, 338	Arctic	215
ORCHIS	159	bryophila	321, 322	beach	215
aristata	159, 178	pygmaea	321, 322	fleshy	216
rotundifolia	160, 178	pygmaea	321, 322	snow	215
orchis,		retrorsa	320	western	215
rose-purple	159	richardsonii	323	PECTIANTHIA	
round-leaved	160	roaldi	322	pentandra	282
OROBANCHACEAE	427	scammaniana	322, 338	PEDICULARIS	422
OROBANCHE	427	splendens	323, 338	arctica	424
fasciculata	428, 442	varians	323, 338	capitata	423, 440
uniflora	428, 442	viscida	321, 338	chamissonis	425, 442

INDEX

PEDICULARIS (cont.)		PHACELIA	401	PINGUICULA	426
euphrasioides	424	franklinii	401, 436	villosa	426, 442
flammaea	423, 440	mollis	402, 436	vulgaris	426, 442
groenlandica	425	phacelia,		pink family	206
labradorica	424, 442	Franklin	401	pink,	
lanata	423, 440	silky	402	moss	217
langsdorfii	424, 440	PHALARIDEAE	50	northern	218
lapponica	424, 442	PHALARIS	53	sea	381
oederi	423, 440	arundinacea	53, 98	PINUS	26
ornithorhyncha		canariensis	53	contorta	26, 38
	424, 442	PHEGOTERIS		latifolia	26
parviflora	424, 442	dryopteris	15	murrayana	26
pedicellata	424	phegopteris	15	murrayana	26
pennellii	425, 442	PHIPPSIA	57	pipe, Indian	364
sudetica	423, 440	algida	57, 98	PIPERIA	163
verticillata	425, 442	PHLEUM	56	unalaschensis	163, 178
penny cress,		alpinum	56, 98	piperia, Alaska	163
Arctic	247	americanum	56	pipsissewa	362
field	247	pratense	57, 98	PLAGIOBOTRYS	406
PENTSTEMON	413	PHLOX	399	cognatus	407, 438
diffusus	413, 438	hoodii	399	cusickii	407
gormanii	413, 438	sibirica	399, 436	hirtus	407
procerus	413, 438	phlox family	398	orientalis	406
pepper, water		196	phlox,		PLANTAGINACEAE	428
pepper-grass,		moss	399	PLANTAGO	428
common	247	Siberian	399	aristata	429
cultivated	247	PHYLLODOCE	368	asiatica	429
wild	247	aleutica	368, 369, 388	canescens	429, 442
peppermint	409	coerulea	368, 388	eriopoda	429
PERAMIUM	165	empetriformis	368, 388	lanceolata	429
decipiens	165, 178	glanduliflora	369, 388	macrocarpa	429, 442
menziesii	165	PHYLLOSPADIX	48	major	429, 442
repens		scouleri	48	maritima	
ophioides	165, 178	PHYSOCARPUS	294	juncoides	428, 442
PERSICARIA		capitatus	294	septata	429
amphibia	196	PICEA	28	plantain family	428
maculosa	197	canadensis	28	plantain,	
persicaria,		glauca	28, 38	common	429
dock-leaved	197	mariana	28, 38	large-bracted	429
Pennsylvania	197	sitchensis	29, 38	saline	429
tomentose	196	PICRIS	448	seashore	429
water	196	hieracoides		PLATANTHERA	160, 161
PETASITES	492	kamtschatica		behringiana	162
frigidus	493, 510		448, 504	chorisiana	161
hyperboreus	493, 510	PINACEAE	24	dilitata	162
palmatus	493, 510	pine family	24	hyperborea	162
sagittatus	493, 510	pine,		obtusata	161
vitifolius	493	lodgepole	26	orbiculata	160
PHACA		prince's	362	stricta	162
americana	317	scrub	26	tipuloides	162
littoralis	317	tamarack	26	unalaschensis	163
polaris	317	pinesap	365	PLEUROGYNE	
				rotata	394

INDEX

PLEUROPOGON		POA (cont.)		POLYGONUM (cont.)	
sabinii	95	turneri	79, 102	tomentosa	196
PLUMBAGINACEAE	381	williamsii	81	viviparum	195, 222
plumbago family	381	wrightii	71	POLYPODIACEAE	8
PNEUMARIA		POACEAE	49	POLYPODIUM	11
maritima	404	PODAGROSTIS		falcatum	11
POA	76	aequivalvis	60	glycyrrhiza	11
abbreviata	82	thurberiana	61	vulgare	
acutiglumis	83	PODISTERA		occidentalis	11, 30
alpigena	80	macounii	359	POLYPOGON	56
alpina	83, 102	POLEMONIACEAE	398	monspeliensis	56, 98
ampla	84	POLEMONIUM	398	POLYSTICHUM	13
annua	78, 102	acutiflorum	398, 436	alaskense	14
arctica	81, 102	boreale	399, 436	aleuticum	14
longiculmis	81	macranthemum		andersonii	14
williamsii	81		398, 399	braunii	14, 32
brachyanthera	82	richardsonii	399	lonchites	14, 32
canbyi	84	fasciculatum	399	munitum	14, 32
cenisia	81	lanatum	399	pondweed family	43
compressa	78, 102	lindleyi	399	pondweed,	
confinis	79	occidentale	398	clasping-leaved	46
crocata	82	pulcherrimum	399, 436	fennel-leaved	47
eminens	79, 102	lindleyi	399	filiform	46
eyerdamii	80	richardsonii	399	floating	44
glacialis	82	rotatum	399	Fries	45
glauca	83, 102	POLYGONACEAE	191	horned	47
glumaris	79	POLYGONUM	194	interior	46
gracillima	84	achoreum	198, 222	northern	45
hispidula	84, 102	alaskanum	196, 222	Nuttall	45
aleutica	84	alpinum		Porsild	45
irrigata	81, 102	alaskanum	196	sheathed	46
komarovii	83	amphibium		small	45
lanata	79, 102	laevimarginatum		various-leaved	44
laxiflora	79		196, 222	white-stemmed	46
leptocoma	81, 102	bistorta		poplar, balsam	185
elatior	79	plumosum	195, 222	poppy family	243
scabrinervis	81	buxiforme	198, 222	poppy,	
macrocalyx	80, 102	caurianum	197, 222	Arctic	244
merrilliana	82, 102	coccineum	196	Iceland	244
nemoralis	82, 102	convolvulus	195, 222	Macoun	244
norbergii	80	fowleri	197, 222	McConnell	244
palustris	82, 102	heterophyllum	198, 222	Walpole	243
paucispicula	81	hydropiper	196	POPULUS	184
pratensis	80, 102	hydropiperoides	197	balsamifera	185
alpigena	80	lapathifolium		candicans	185
angustifolia	80	nodosum	197	tacamahacca	185, 220
rupicola	83, 102	natans	196	tremuloides	
sandbergii	84	neglectum	198, 222		184, 185, 220
secunda	84	nodosum	197, 222	tricocarpa	185, 220
stenantha	83, 102	pennsylvanicum	197	poque	427
vivipara	83	persicaria	197, 222	PORTULACEAE	203
triflora	82	prolificum	198	POTAMOGETON	44
trinii	79	ramosissimum	198	alpinus	
trivialis	81, 102	scabrum	196	tenuifolius	45, 96

INDEX

POTAMOGETON (cont.)		POTENTILLA (cont.)		PUCCINELLIA	73
epihydrus	45	uniflora	304, 332	alaskana	75, 100
nuttallii	45	vahliana	304, 332	andersoni	75, 100
filiformis	46, 96	virgulata	302, 332	borealis	74, 100
friesii	45, 96	villosa	304, 332	glabra	75
gramineus	44, 96	yukonensis	301	grandis	74, 100
heterophyllus	44	prairie-rocket,		hauptiana	74, 100
interior	46	small-flowered	266	hulteni	76, 100
natans	44, 96	PRENANTHES	464	kamtschatica	
pectinatus	47, 96	alata	464, 504	sublaevis	76
perfoliatus	46, 96	lessingii	464	nutkaensis	76, 102
gracilis	46	PRIMOCAREX	114	paupercula	75, 100
richardsonii	46	primrose family	374	phryganodes	74, 100
porsildiorum	45, 96	primrose,		pumila	76, 100
praelongus	46	Chukch	379	triflora	75
pusillus	45, 96	Greenland	378	PULSATILLA	
richardsonii	46	Lake Mistassini	379	ludoviciana	237
vaginatus	46	northern	378	multiceps	236
POTAMOGETONACEAE		Siberian	379	purse, shepherd's	255
	43	small-leaved	379	purslane family	203
POTENTILLA	300	snow	379	purslane, sea	202
alaskana	303	wedge-leaved	377	PYROLA	363
anserina	301, 332	PRIMULA	377	asarifolia	363
sericea	301	borealis	378, 390	incarnata	364, 386
arguta	301, 332	ajanensis	378	borealis	363
biflora	303, 332	cuneifolia	377	chlorantha	363, 386
blaschkeana	303	saxifragaefolia		gormanii	363
chamissonis	305		378, 390	grandiflora	363, 386
diversifolia	303, 332	egalikensis	378	minor	364, 386
egedii		eximia	379	occidentalis	363
groenlandica	301	incana	378	secunda	364, 388
elegans	304	macounii	379	obtusata	364
emarginata	304, 332	mistassinica	379	uliginosa	363
fruticosa	301, 332	nivalis	379	uniflora	363
glaucophylla	303	parvifolia	379	PYROLACEAE	362
gracilis	303, 332	sibirica	379, 390	PYRUS	
hippiana	302	stricta	378	diversifolia	310
hookeriana	305, 332	tschuktschorum		rivularis	310
monspeliensis	304, 332		379, 390		
multifida	303, 332	PRIMULACEAE	374		
nana	304	PRUNELLA	410	quackgrass	90
nivea	305, 332	vulgaris		quillwort, Braun's	23
norvegica		aleutica	411	quillwort family	23
monspeliensis	304	lanceolata	411, 438		
nuttallii	303	PRUNUS	294		
pacifica	301, 332	melanocarpa	294	RADICULA	
palustris	301, 332	PTERETIS		palustris	252
pectinata	302, 332	nodulosa	10	radish, garden	251
pennsylvanica	302, 332	PTERIDIUM	11	RAMISCHIA	
glabrata	302	aquilinum		secunda	364
strigosa	302	lanuginosum	11, 30	RANUNCULACEAE	227
pulchella	302	pubescens	11	RANUNCULUS	228
rubricaulis	303	PTERIDOPHYTA	6	acris	229, 270
				frigidus	229

INDEX

RANUNCULUS (cont.)		RANUNCULUS (cont.)		RIBES (cont.)	
abortivus	231, 270	tricophyllus	234	bracteosum	292, 328
affinis	231	turneri	230	echinatum	292
aquatilis		unalaschensis	233	glandulosum	293
capillaceus	234, 270	verecundus	231	howellii	292
eradicatus	234	verticillatus	232, 270	hudsonianum	292, 328
hispidulus	234	yukonensis	233	lacustre	292, 328
bongardii	230, 270	rape	251	laxiflorum	293, 328
tenellus	230	RAPHANUS	251	oxycanthoides	292, 328
chamissonis	234, 270	raphanistrum	251	prostratum	293
cooleyae	234, 270	sativus	251	triste	293, 328
confervoides	234	raspberry,		ribgrass	429
cymbalaria	234, 270	American red	297	rice, Indian	154
douglasii	230	western black	297	riverweed	350
eastwoodianus	231, 270	rattlebox	425	rock-cress,	
eschscholtzii	231, 270	rattlesnake plantain,		Arctic	263
flammula	233, 270	lesser	165	Drummond	264
filiformis	233	Menzies	165	hairy	264
ovalis	233	RAZOUMOFSKYA		Holboell	264
gelidus	232	douglasii		Hooker	265
glacialis		tsugensis	190	Kamchatka	263
chamissonis	234	tsugensis	190	northern	267
gmelini	233	redtop,	61	spreading-pod	264
terrestris	233, 270	Alaska	62	rocket,	
yukonensis	233	spike	61	sea	250
grayanus	234	Thurber	61	yellow	252
grayi	232	reed-grass,		ROMANZOFFIA	401
hyperboreus	233, 270	Holm	58	sitchensis	401, 436
kamchaticus	234	Lapland	59	unalaschcensis	
lapponicus	235, 270	narrow	59		401, 436
macounii	230, 270	northern	59	glabriuscula	401
oreganus	230	Pacific	59	root,	
nivalis	232, 270	purple	58	musk	433
occidentalis	230, 270	slender	57	rattlesnake	464
brevistylis	230	RHINANTHUS	425	RORIPPA	252
insularis	230	borealis	425	barbareaefolia	252, 274
nelsoni	230	crista-galli	425	clavata	252
turneri	230	minor groenlandicus		nasturtium-aquaticum	
orthorhynchus			425, 442		252, 274
alachensis	231, 270	RHODIOLA		palustris	252, 274
pallasii	235, 270	alaskana	280	williamsii	252
pedatifidus	231, 270	integrifolia	279	ROSA	298
pennsylvanicus		rosea	279	acicularis	298, 330
	230, 270	RHODODENDRON	366	aleutensis	299
purshii	233	kamtschaticum	366, 388	nutkana	299, 330
pygmaeus	232, 270	glandulosum	367	woodsii	299, 330
repens	229, 270	lapponicum	367, 388	ROSACEAE	293
reptans	233	parvifolium	367	rose family	293
septentrionalis	229	rhododendron,		rose,	
pacifica	229	Kamchatka	366	Nootka	299
sceleratus		rhubarb, wild	196	prickly	298
multifidus	233, 270	RIBES	291	woods	299
sulphureus	232, 270	acerifolium	292	yellow	301

rose bay, Lapland	367	rush (cont.)		SAMBUCUS	432	
rosemary, bog	370	chestnut	148	racemosa		
roseroot	279	dagger-leaved	146	pubens	433, 444	
rosewort	279	Drummond	145	sandalwood family	190	
Rowan tree	310	knotted	148	sand spurry,		
RUBIACEAE	430	Mertens	147	Canadian	216	
RUBUS	296	moor	146	purple	216	
acaulis	297	mountain	146	sandwort,		
alaskensis	297, 330	Oregon	147	Alpine	214	
arcticus	297, 330	Richardson	147	Arctic	214	
chamaemorus	296	sickleleaved	148	beautiful	213	
idaeus		slender	146	blunt-leaved	213	
canadensis	297	thread	145	long-podded	215	
sachalinensis	297	three-flowered	147	low	213	
leucodermis	297	toad	146	matted	213	
nutkanus	298	two-flowered	147	rock	213	
parviflorus	298, 330	rutabaga	251	Ross	213	
pedatus	296, 330	rye, cultivated	89	sea-beach	212	
pubescens	296	rye-grass,		two-flowered	214	
spectabilis	298, 330	Aleutian	94	SANGUISORBA	308	
stellatus	297, 330	beach	94	latifolia	309	
strigosus	297, 330	Canada	94	menziesii	309, 334	
subarcticus	297	downy	93	microcephala	308	
RUMEX	191	English	89	officinalis	308, 334	
acetosa	192, 220	Italian	89	sitchensis	309, 334	
alpestris	192	Macoun	94	SANICULA	355	
acetosella	192, 220	northern	95	marylandica	355	
angiocarpus	192	Pacific	94	SANTALACEAE	190	
arcticus	193, 220	western	94	SAPONARIA	219	
perlatus	193	RYNCHOSPORA	107	vaccaria	219	
crispus	193, 220	alba	107, 168	sarsaparilla, wild	354	
domesticus	193, 220			SAUSSUREA	501	
fenestratus	193, 220			americana	501, 512	
puberulus	193	sagewort, prairie	489	angustifolia	501, 512	
graminifolius	192	SAGINA	215	nuda	501, 512	
maritimus	192, 220	crassicaulis	216, 224	subsinuata	501	
fueginus	192	intermedia	215, 224	vicida	501	
occidentalis	193	linnaei	215, 224	yukonensis	502, 512	
obtusifolius		litoralis	215	SAVASTANA		
agrestis	192, 220	nivalis	215	alpina	52	
persicarioides	192	occidentalis	215	odorata	53	
sibiricus	194	saginoides	215	pauciflora	53	
transitorius	194, 220	salal	371	SAXIFRAGA	285	
running pine	21	SALICACEAE	184	aestivalis	288	
RUPPIA	47	SALICORNIA	203	adscendens		
canadensis	47	europea	203	oregonensis	286, 326	
lacustris	47	herbacea	203, 222	aleutica	289, 328	
spiralis	47, 96	pacifica	203, 222	bongardii	289	
rush family	143	SALIX	185	bracteata	286, 326	
rush,		salmonberry	298	bronchialis		
Arctic	145	saltweed,		cherlerioides	290	
Baltic	145	Alaska	202	funstonii	290, 328	
bog	145	Gmelin	202	brunoniana	289	

INDEX

SAXIFRAGA (cont.)
  caespitosa
    sileneflora 287, 326
  cernua 286, 326
  comosa 289
  davurica
    grandipetala 287, 326
  eschscholtzii 290, 328
  ferruginea 289, 328
    macounii 289
  flagellaris 290, 328
  foliolosa 289, 328
  hieracifolia 287, 326
    rufopilosa 287
  hirculus 290, 328
  integrifolia 289
  lyallii 288, 328
  mertensiana 291, 328
  nelsoniana 288
  nivalis 287, 326
  nudicaulis 291, 328
  oppositifolia 291, 328
  punctata 288
    insularis 288
    nelsoniana 288, 328
    pacifica 288
  radiata 286, 326
  reflexa 287, 328
  rivularis 286, 326
  rufidula 287
  serpyllifolia 289, 328
    purpurea 289
  spicata 288, 328
  tolmiei 290, 328
  tricuspidata 290, 328
  unalaschensis 288, 328
SAXIFRAGACEAE 280
saxifrage family 280
saxifrage,
  Alaska 289
  Aleutian 289
  Alpine 287
  Alpine brook 286
  bracted 286
  brook 288
  ciliate 290
  flagellate 290
  foliose 289
  hawkweed-leaved 287
  Hooker 289
  leather-leaf 284
  naked-stemmed 291
  nodding 286

saxifrage, (cont.)
  purple mountain 291
  red-stemmed 288
  Richardson 284
  rusty 287
  spiked 288
  spotted 290
  three-toothed 290
  thyme-leaved 289
  Tolmie 290
  tufted 287
  Unalaska 288
  wedge-leaved 286
  wood 291
  yellow marsh 290
  Yukon 287
SCHEUCHZERIA 49
  palustris 49
  americana 49
SCHEUCHZERIACEAE 48
SCHIZACHNE 70
  purpurescens 70, 100
SCIRPUS 112
  acicularis 108
  americanus 112, 168
  caespitosus
    austriacus 112
    callosus 112, 168
  kamtschaticus 108
  lacustris 113
  mamillatus 109
  microcarpus 113, 168
  nitidus 108
  pacificus 113, 168
  palustris 109
  paucifloris 112
  rufus 113, 168
  uniglumis 109
  validus 113, 168
SCORZONELLA
  borealis 448
scouring-rush, 20
  Alaska 19
  northern 19
SCROPHULARIACEAE 412, 427
SCUTELLARIA 409
  epilobifolia 410
  galericulata 410, 438
sea-blite, low 203
sea lovage, Hulten 359
SECALE
  cereale 89
sedge family 107

sedge,
  Alaska long-awned 134
  Anderson 120
  Arctic hare's-foot 126
  awl-fruited 124
  awned 142
  beaked 140
  Bering Sea 135
  Bigelow 130
  black 134
  black and white-
    scaled 133
  blackish 122
  black-scaled 133
  bristle-leaved 137
  bristle-stalked 121
  bog 138
  brownish 128
  Buxbaum 132
  capitate 120
  Carcross 133
  coiled 121
  coastal stellate 129
  Crawford 125
  creeping 123
  curved 123
  dark-brown 139
  Enander 134
  few-flowered 123
  fragile 141
  Garber 130
  glacier 137
  gold-fruited 129
  Gmelin 133
  green 140
  Hepburn 120
  involute-leaved 123
  Karaginsk Island 134
  Krause 140
  hair-like 139
  Hinds 130
  Hoppner 131
  Hudson Bay 126
  Kellogg 130
  Kokrines mountain 131
  Lapland 127
  large-headed 124
  lesser panicled 124
  little prickly 128
  livid 138
  loose-flowered
    Alpine 137
  low northern 136
  Lyngbye 132

INDEX

sedge, (cont.)
   meadow 125
   Mertens 134
   Montana 135
   mountain hare 125
   northern 136
   northern bog 120
   northern clustered 126
   northern single-spike 121
   Norway 127
   Peck 136
   Pribylof 126
   pyrenean 122
   Ramenski 132
   rock 122
   Ross 136
   round-fruited 141
   sheathed 139
   shore 138
   short-leaved 139
   short-stalk 135
   showy 135
   silvery 127
   slender-beaked 125
   smooth-stemmed 129
   sparse-flowered 128
   Sitka 131
   soft-leaved 128
   thick-headed 125
   thread-leaved 122
   two-color 129
   variegated 132
   water 131
   weak 138
   weak Arctic 136
   weak clustered 127
   Williams 140
   yellow 140
   Yukon 127
SEDUM 279
   oregonum 279, 326
   roseum 279, 326
      frigidum 280
      integrifolium 279, 280
      stenopetalum 280, 326
sedum,
   narrow-petaled 280
   Oregon 279
SELAGINELLA 22
   schmidtii 23
   selaginoides 23, 36
   sibirica 23, 36

selaginella,
   low 23
   northern 23
SELAGINELLACEAE 22
SENECIO 497
   alaskanus 498
   atropurpureus 498
   dentatus 498
   frigidus 498, 512
   tomentosus 498
   congestus
      palustris 498, 512
   conterminus 499, 512
   cymbalarioides 499
      borealis 499
   denali 498
   fuscatus 498, 512
   hyperborealis 498, 512
   indecorus 499, 512
   kjellmanii 498
   lindstroemii 498
   lugens 500, 512
   palmatus 500, 512
   pauciflorus 499
   pauperculus 499, 512
   pseudo-arnica 500, 512
   resedifolius 499, 512
   sheldonensis 500
   triangularis 500, 512
   vulgaris 501
   yukonensis 498, 512
SENECIONEAE 467
service-berry,
   northwestern 310
   Pacific 311
shamrock 312
SHEPHERDIA 348
   canadensis 348, 382
shield-fern,
   Aleutian 14
   Anderson's 14
   fragrant 15
   prickly 14
SIBBALDIA 305
   procumbens 305, 332
SIEVERSIA
   calthifolia 307
   glacialis 308
   macrantha 308
   pentapetala 308
   rossii 307
SILENE 217
   acaulis 217, 224

SILENE (cont.)
   menziesii 217, 224
   noctiflora 217
   repens 217, 224
   williamsii 217, 224
silverberry 348
silverweed,
   common 301
   Pacific 301
SINAPSIS
   arvensis 251
SISYMBRIUM 249
   altissimum 249, 274
   officinale 249, 274
   sophia 249
SISYRINCHIUM 157
   littorale 158, 178
SIUM 361
   cicutaefolium 361
   suave 361, 386
skullcap, marsh 410
slough-grass 69
SMELOWSKIA 262
   calycina
      integrifolia 263, 276
SMILACINA 155
   racemosa 156, 176
   stellata 156, 176
snakeroot, black 355
snowberry 432
soapberry 348
SOLIDAGO 469
   decumbens
      oreophila 470, 506
   elongata 470, 504
   lepida 470, 506
   multiradiata 469, 504
      arctica 470
Solomon's seal,
   star-flowered 156
SONCHUS 463
   arvensis 463
   asper 463, 504
   oleraceus 463, 504
soopolallie 348
SOPHIA
   sophia 249
   sophioides 249
SORBUS 309
   alaskana 310
   andersonii 310
   aucuparia 310
   sambucifolia 309, 334
   scopulina 310, 334
   sitchensis 309, 334

sorrel,			SPIRAEA	294	STELLARIA (cont.)	
grass-leaved		192	beauverdiana	295, 330	laxmanni	210
green		192	menziesii	294, 330	longifolia	210, 224
mountain		194	stevenii	295	longipes	210, 224
sheep		192	spiraea,		media	209, 224
sow-thistle,			Beauverd	295	monantha	210
common		463	Menzies	294	ruscifolia	
field		463	SPIRANTHES	163	aleutica	210, 224
spiny		463	romanzoffiana	163, 178	sitchana	211, 224
SPARGANIACEAE		42	spleenwort,		bongardiana	211
SPARGANIUM		42	green	17	STENOTUS	
affine		43	maidenhair	16	borealis	469
angustifolium	43,	96	spring beauty,		stickseed,	
hyperboreum	43,	96	Alaska	205	European	403
minimum	43,	96	Arctic	204	Redowski	403
multipedunculatum		43	Bering Sea	204	STIPA	54
simplex		43	long-branched	206	columbiana	54, 98
spear-grass,			Scamman	205	comata	54, 98
glaucous		83	Siberian	205	stone-crop family	279
large-flowered		79	small-flowered	205	strawberry,	
low		82	small-leaved	206	beach	299
spearmint		409	tuberous	204	bracted	299
spearwort, creeping		233	spruce,		Yukon	300
speedwell,			black	28	STREPTOPUS	156
Alpine		417	Sitka	29	amplexifolius	156, 176
corn		417	white	28	roseus	
Germander		416	spurry	216	curvipes	157, 176
large-flowered		415	STACHYS	411	streptopoides	156
low		416	emersonii	411	STRUTHIOPTERIS	9
Persian		417	palustris		filicastrum	10, 30
skullcap		416	pilosa	411, 438	spicant	9
Steller		416	starwort,		SUAEDA	202
thyme-leaved		416	Alaska	209	maritima	203, 222
SPERGULA		216	crisp	210	SUBULARIA	246
arvensis	216,	224	fleshy	211	aquatica	246, 274
SPERGULARIA		216	larch-leaved	214	sundew family	268
canadensis	216,	224	long-leaved	210	sundew,	
rubra		216	long-stalked	210	long-leaved	269
SPERMATOPHYTA		23	ruscus-leaved	210	round-leaved	268
SPHENOPHOLIS		64	shining	210	sunflower, common	484
intermedia	64,	100	Sitka	211	surf-grass, Scouler	48
spikenard,			STATICE		SVIDA	
false		156	armeria	381	instolonea	362
wild		156	STELLARIA	209	sweet-cicely,	
spike-rush,			alaskana	209, 224	blunt-fruited	356
creeping		109	borealis	211	Chile	356
Kamchatka		108	calycantha	211, 224	Sitka	356
needle		108	ciliatosepala	210	sweet clover,	
pale		109	crassifolia	211, 224	white	314
one-bracted		109	crispa	210, 224	yellow	314
slender		108	dicranoides	213	sweet coltsfoot,	
spinach, strawberry		200	humifusa	211, 224	Arctic	493
			laeta	210, 224	arrow-leaf	493

sweet-grass	53	TARAXACUM (cont.)		thistle,		
sweet pea, wild	323	leptoglossum	457	bull	502	
SWERTIA	393	leptopholis	457	Canada	502	
perennis	393, 436	microceras	458	common	502	
SYMPHORICARPOS	432	mitratum	458	edible	502	
albus	432	multesimum	458	Kamchatka	503	
rivularis	432, 444	murolepium	457	THLASPI	247	
SYNTHYRIS	415	ochraceum	458	arcticum	247	
borealis	415, 440	oncophorum	458	arvense	247, 274	
		ovinum	458	THORORHODION		
		paralium	459	kamtschaticum	366	
tails, kitten	415	patagiatum	459	thorough-wort,		
tamarack	27	pellianum	459	American	357	
TANACETUM	491	phalolepis	459	three-square	112	
bipinnatum	492, 510	phymatocarpum		THUJA	25	
huronense	492		453, 455	gigantea	26	
vulgare	492, 510	pribilofense	459	plicata	26, 38	
tansy, common	492	retroflexum	461	thumb, lady s	197	
tansy mustard,		scanicum	460	TIARELLA	284	
mountain	250	scotostigma	459, 504	trifoliata	284, 326	
northern	249	sibiricum	462	unifoliata	284, 326	
western	250	signatum	460	ticklegrass	62	
TARAXACUM	448	speirodon	460	timothy,		
alaskanum	461, 504	sublacerum	460	common	57	
andersonii	453, 504	trigonolobium	460, 504	mountain	56	
angulatum	453	undulatum	461	TINIARIA		
arietinum	454	vagans	461	convolvulus	195	
aureum	454	tarweed	484	TISSA		
caligans	454	TAXACEAE	24	canadensis	216	
callorhinorum	454	TAXUS	24	rubra	216	
carneocoloratum	462	brevifolia	24	toad-lily	205	
chamissonis	454	TELLIMA	282	TOFIELDIA	151	
chlorostephum	454	grandiflora	283, 326	coccinea	152, 176	
chromocarpum	455	THALESIA		intermedia	151, 152	
cinericolor	461	fasciculata	428	nutans	152	
collinum	462	uniflora	428	occidentalis	151, 176	
dahlstedtii	461	THALICTRUM	237	palustris	152	
decoraifolium	461	alpinum	237, 272	pusilla	152, 176	
demissum	455	hultenii	237, 272	tofieldia, western	151	
dumetorum	455	kemense	237	TOLMIEA	283	
eurylepium	455	minus		menziesii	283, 326	
eyerdamii	455, 504	kemense	237	TORRESIA		
fabbeanum	455	occidentale	238, 272	odorata	53	
festivum	456	sparsiflorum	237, 272	touch-me-not,		
flavovirens	456	THELYPTERIS		western	343	
hyperarcticum	453	dryopteris	15	TRIENTALIS	381	
hypochoeropsis	456	oreopteris	16	europea	381	
integratum	456, 504	phegopteris	15	arctica	381, 390	
kamtschaticum	453, 504	robertiana	15	TRIFOLIUM	312	
kodiakense	456	THEREFON		dubium	312	
lacerum	456, 504	richardsonii	284	fimbriatum	313	
lateritium	457, 504	thimbleberry	298	hybridum	313, 336	
latilimbatum	457			lupinaster	312, 334	

TRIFOLIUM (cont.)		UTRICULARIA	426	VERONICA (cont.)	
microcephalum	313	intermedia	427, 442	persica	417
pratense	312	macrorhiza	426, 442	scutellata	416
procumbens	312	minor	427, 442	serpyllifolia	416
repens	313, 336	vulgaris	426, 427	stelleri	416, 440
variegatum	313			glabrescens	417
TRIGLOCHIN	49			tenella	416, 440
palustris	49, 96	VACCARIA		wormskjoldii	417, 440
maritima	49, 96	segetalis	219	vetch,	
TRISETUM	66	VACCINIACEAE	372	American	324
alaskanum	66	VACCINIUM	372	cow	324
cernuum	66, 100	alaskensis	373, 390	hairy	324
sibiricum	66, 100	caespitosa	373	narrow-leaved	325
spicatum	66, 100	membranaceum		Sitka	324
subspicatum	66		373, 390	VIBURNUM	433
trisetum,		ovalifolium	373, 390	edule	433, 444
nodding	66	paludicola	373, 390	pauciflorum	433
Siberian	66	parvifolium	373, 390	VICIA	324
TRITICUM		uliginosum	372, 390	americana	324, 338
aestivum	89	vitis-idea	372, 388	angustifolia	325
TROLLIUS	239	minus	372	cracca	324
riederianus	239	VAGNERA	155	gigantea	324, 338
TSUGA	27	racemosa	156	sitchensis	324
heterophylla		stellata	156	villosa	324
	28, 38, 190	VAHLODEA		VIGNEA	115
mertensiana	28, 38	atropurpurea	67	VIOLA	345
turnip	251	valerian family	433	achyrophora	347
TURRITIS		valerian,		adunca	347, 382
glabra	265	capitate	434	biflora	346, 382
twayblade,		northern	434	epipsila	
broad-leaved	164	Sitka	434	repens	347, 382
heart-leaved	164	VALERIANA	433	glabella	346, 382
northern	164	capitata	434, 444	langsdorfii	347, 382
western	164	septentrionalis	434	nephrophylla	346
twinberry, black	432	sitchensis	434, 444	orbiculata	346
twin-flower	431	VALERIANACEAE	433	palustris	347
twisted-stalk,		VERATRUM	152	renifolia	347
clasping	156	album		brainerdii	347, 382
simple-stemmed	157	oxysepalum	152	rugulosa	346
TYPHA	42	eschscholtzii	152, 176	selkirkii	347, 382
latifolia	42, 96	viride	152	VIOLACEAE	345
TYPHACEAE	42	vernal-grass, sweet	54	violet family	345
		VERONICA	415	violet,	
		americana	416, 440	Alaska	347
UMBELLIFERAE	355	alpina		bog	426
uncinia, false	123	unalaschkensis	416	great-spurred	347
UNIFOLIUM	155	wormskjoldii	417	hook-spurred	347
dilitatum	155	arvensis	417	kidney-leaved white	
eschscholtzianum	155	buxbaumii	417		347
URTICA	189	chamaedrys	416	northern bog	346
gracilis	189, 220	grandiflora	415, 440	northern marsh	347
lyallii	189, 220	humifusa	416	stream	346
URTICACEAE	189	peregrina		tall-stemmed	346
		xalapensis	417, 440	two-flowered	346
				western round-leaved	
					346

WAHLBERGELLA		willow family	184	yarrow,	
apetala	218	willow-herb,		northern	485
wallflower,		Bering	351	Siberian	485
narrow-leaved	266	Davurian	350	yellow-cress,	
Pallas	266	glandular	351	marsh	252
WASHINGTONIA		Hornemann	352	round-podded	252
divaricata	356	northern	351	yew family	24
obtusa	356	Pimpernel	351	yew, western	24
purpurea	356	swamp	351	YOUNGIA	
water carpet,		thin-capsuled	351	elegans	465
Bering Sea	282	thin-leaved	352	nana	465
northern	282	Trelease	350	youth-on-age	283
water-crowfoot,		yellow	350		
white	234	wind-flower,			
water hemlock,		columbia	236		
Mackenzie	361	wintergreen family	362	ZANNICHELLIA	47
western	360	wintergreen,		palustris	47
water-leaf family	400	greenish-flowered	363	ZOSTERA	48
water lily family	242	large-flowered	363	marina	48, 96
water lily, white	243	lesser	364	ZOSTERACEAE	48
water-milfoil family	352	liver-leaf	363	ZYGADENUS	152
water-milfoil,		one-sided	364	chloranthus	153
loose-flowered	353	wood-fern,		elegans	153, 176
spiked	353	mountain	16	zygadenus, glaucus	153
water parsnip,		spreading	16		
hemlock	361	wood-rush,			
water pepper, mild	197	Alpine	150		
water-shield	227	hairy	149		
water-shield family	227	many-flowered	151		
water starwort family	343	northern	150		
water-starwort,		small-flowered	149		
Bolander	344	snow	150		
northern	344	spiked	150		
vernal	344	Wahlenberg	149		
weed,		WOODSIA	12		
pineapple	487	alpina	12, 14, 32		
poverty	201	glabella	12, 30		
wedgegrass, slender	64	ilvensis	12, 32		
wheat, common	89	scopulina	12		
wheat-grass,		woodsia			
Alaska	91	Alpine	12		
bearded	92	Rocky Mountain	12		
bluebunch	92	rusty	12		
slender	91	smooth	12		
western	90	wormwood,			
Yukon	91	Arctic	489		
Whitlow-grass, wood	257	Canada	488		
widgeon-grass	47	linear-leaved	488		
		northern	488		

www.ingramcontent.com/pod-product-compliance
Lightning Source LLC
Chambersburg PA
CBHW070027040426
42333CB00040B/877